THE PELLET HANDBOOK

The Pellet Handbook

The Production and Thermal Utilisation of Pellets

Ingwald Obernberger and Gerold Thek

London • Washington, DC

First published in 2010 by Earthscan

Copyright © BIOS BIOENERGIESYSTEME GmbH, 2010

The moral right of the authors has been asserted.

All rights reserved. No part of this publication may be reproduced, stored in a retrieval system, or transmitted, in any form or by any means, electronic, mechanical, photocopying, recording or otherwise, except as expressly permitted by law, without the prior, written permission of the publisher.

Disclaimer: "The statements, technical information and recommendations contained herein are believed to be accurate as of the date hereof. Since the conditions and methods of use of the products and of the information referred to herein are beyond our control, IEA Bioenergy, Task 32, Earthscan and the authors expressly disclaim any and all liability as to any results obtained or arising from any use of the products or reliance on such information. The opinions and conclusions expressed are those of the authors."

Earthscan Ltd, Dunstan House, 14a St Cross Street, London EC1N 8XA, UK
Earthscan LLC, 1616 P Street, NW, Washington, DC 20036, USA

Earthscan publishes in association with the International Institute for Environment and Development

For more information on Earthscan publications, see www.earthscan.co.uk or write to earthinfo@earthscan.co.uk

ISBN: 978-1-84407-631-4

Typeset by Gerold Thek
Cover design by Yvonne Booth
Cover photos: left and centre image © BIOS; right image © Jörg Ide, KWB Biomass Heating Systems

A catalogue record for this book is available from the British Library

Library of Congress Cataloging-in-Publication Data has been applied for

At Earthscan we strive to minimize our environmental impacts and carbon footprint through reducing waste, recycling and offsetting our CO_2 emissions, including those created through publication of this book. For more details of our environmental policy, see www.earthscan.co.uk.

This book was printed in the UK by MPG Books, an ISO 14001 accredited company. The paper used is FSC certified.

Preface

Pellets are a solid biomass fuel with consistent quality – low moisture content, high energy density and homogeneous size and shape. The problems of conventional biomass fuels as an alternative to coal, oil or gas, which are attributed mainly to their low energy density, high moisture content and heterogeneity, can be lessened or even prevented altogether by the use of pellets. Consistent fuel quality makes pellets a suitable fuel type for all areas of application, from stoves and central heating systems to large-scale plants, and with practically complete automation in all these capacity ranges. It was not until such a homogeneous biomass fuel with regard to shape and size was introduced to the market that the development of fully automatic biomass furnaces for small-scale applications with a user comfort similar to modern oil or gas heating systems was possible. That is why pellets have shown an enormous development during the last two decades, from a more or less unknown product in the early 1990s to internationally and inter-continentally traded merchandise.

Despite the rapid development of the pellet market, there is still a lack of public awareness. In countries like Austria, Germany, Sweden or Denmark, where the use of pellets is already quite common, this is not as pronounced. In countries where the pellet market is at the beginning of its development, like Canada or the UK, there is still a considerable lack of information. A lot of potential pellet users may not know about the existence and benefits pellets can offer. International exchange of knowledge is thus especially important.

This handbook addresses all the players of the pellet market – from raw material producers or suppliers, pellet producers and traders, manufacturers of pellet furnaces and pelletisation systems, installers, engineering companies, energy consultants up to the end users – as it tries to provide a comprehensive overview about pellet production, energetic utilisation, ecological and economic aspects, as well as market developments and ongoing research and development.

IEA Bioenergy Task 32, BIOS BIOENERGIESYSTEME GmbH and Landesenergieverein Steiermark (LEV) supported the production of this handbook financially, and it is the result of a collective effort of the members of IEA Bioenergy Task 32, with additional inputs from experts from IEA Bioenergy Tasks 29, 31 and 40, as well as from many external pellet experts (see List of Contributors). We herewith express our deepest thanks to all who have contributed to this handbook and we trust this handbook will contribute substantially to the international exchange of information, and to a further increase of pellet utilisation within the energy sector by appropriate distribution of information.

Gerold Thek	Ingwald Obernberger
BIOS BIOENERGIESYSTEME GmbH	Institute for Process and Particle Engineering,
Graz, Austria	Graz University of Technology
	and
	IEA Bioenergy Task 32
	"Biomass Combustion and Cofiring"
	(Austrian representative)

List of Contributors

Contributions to this handbook were provided by the following:

Members of IEA Bioenergy Task 32:

- Anders Evald, Force Technology, Denmark
- Hans Hartmann, Technology and Support Centre of Renewable Raw Materials (TFZ), Germany
- Jaap Koppejan, Procede Biomass BV, Netherlands
- William Livingston, Doosan Babcock Energy Limited, UK
- Sjaak van Loo, Procede Biomass BV, Netherlands
- Sebnem Madrali, Department of Natural Resources, Canada
- Thomas Nussbaumer, Verenum, Switzerland
- Ingwald Obernberger, Graz University of Technology, Institute for Process and Particle Engineering
- Øyvind Skreiberg, SINTEF Energy Research, Norway
- Michaël Temmerman, Walloon Agricultural Research Centre, Belgium
- Claes Tullin, SP Swedish National Testing and Research, Sweden

Members of IEA Bioenergy Task 29 (provided Section 10.12):

- Gillian Alker, TV Energy, UK
- Julije Domac, North-West Croatia Regional Energy Agency, Croatia
- Clifford Guest, Tipperary Institute, Ireland
- Kevin Healion, Tipperary Institute, Ireland
- Seamus Hoyne, Tipperary Institute, Ireland
- Reinhard Madlener, E.ON Energy Research Center, RWTH Aachen University, Germany
- Keith Richards, TV Energy, UK
- Velimir Segon, North-West Croatia Regional Energy Agency, Croatia
- Bill White, Natural Resources Canada, Canadian Forest Service, Canada

Members of IEA Bioenergy Task 31 (provided Sections 10.10.1 and 10.10.2):

- Blas Mola-Yudego, University of Joensuu, Faculty of Forest Sciences; Finnish Forest Research Institute, Finland
- Robert Prinz, Finnish Forest Research Institute, Finland
- Dominik Röser, Finnish Forest Research Institute, Finland
- Mari Selkimäki, University of Joensuu, Faculty of Forest Sciences, Finland

Members of IEA Bioenergy Task 40 (provided Sections 10.10.3 and 10.11):

- Doug Bradley, Climate Change Solutions, Canada
- Fritz Diesenreiter, Institute of Power Systems and Energy Economics, Vienna University of Technology, Austria
- André Faaij, Copernicus Institute for Sustainable Development, Utrecht University, Netherlands
- Jussi Heinimö, Lappeenranta University of Technology, Finland

- Martin Junginger, Copernicus Institute for Sustainable Development, Utrecht University, Netherlands
- Didier Marchal, Walloon Agricultural Research Centre, Belgium
- Erik Tromborg, Department of Ecology and Natural Resource Management (INA), Norwegian University of Life Sciences, Norway
- Michael Wild, European Bioenergy Services - EBES AG, Austria

External partners:
- Eija Alakangas, Technical Research Centre of Finland (VTT), Finland
- Mehrdad Arshadi, Swedish University of Agricultural Sciences (SLU), Sweden
- Göran Blommé, Fortum Hässelby Plant, Sweden
- Per Blomqvist, SP Technical Research Institute of Sweden, Sweden
- Christoffer Boman, Umeå University, Sweden
- Dan Boström, Umeå University, Sweden
- Jan Burvall, Skellefteå Kraft, Sweden
- Marcel Cremers, KEMA Nederland BV, Netherlands
- Jan-Olof Dalenbäck, Chalmers University of Technology, Sweden
- Waltraud Emhofer, BIOENERGY 2020+ GmbH, Austria
- Michael Finell, Swedish University of Agricultural Sciences (SLU), Sweden
- Lennart Gustavsson, SP Technical Research Institute of Sweden, Sweden
- Walter Haslinger, BIOENERGY 2020+ GmbH, Austria
- Bo Hektor, Svebio, Sweden
- Jonas Höglund, Swedish Association of Pellet Producers (PiR), Sweden
- Tomas Isaksson, Swedish Association of Pellet Producers (PiR), Sweden
- Torbjörn A. Lestander, Swedish University of Agricultural Sciences (SLU), Sweden
- Bengt-Erik Löfgren, Pellsam, Sweden
- Staffan Melin, Wood Pellets Association of Canada, Canada
- Anders Nordin, Umeå University, Sweden
- Marcus Öhman, Luleå University of Technology, Sweden
- Heikki Oravainen, VTT Expert Services Ltd. (VTT Group), Finland
- Susanne Paulrud, SP Technical Research Institute of Sweden, Sweden
- Henry Persson, SP Technical Research Institute of Sweden, Sweden
- Klaus Reisinger, Technology and Support Centre of Renewable Raw Materials (TFZ), Germany
- Marie Rönnbäck, SP Technical Research Institute of Sweden, Sweden
- Peter-Paul Schouwenberg, Nidera Handelscompagnie BV, Netherlands
- Gerold Thek, BIOS BIOENERGIESYSTEME GmbH, Austria
- Bas Verkerk, Control Union Canada Inc., Canada
- Emiel van Dorp, Essent, Netherlands
- Wim Willeboer, Essent, Netherlands

The support from Ms. Sonja Lukas for translations and proofreading is gratefully acknowledged.

Table of contents

1 INTRODUCTION 1
2 DEFINITIONS AND STANDARDS 5
   2.1 DEFINITIONS 5
      2.1.1 General definitions 6
      2.1.2 CEN solid biofuels terminology 7
      2.1.3 CEN fuel specifications and classes 13
         2.1.3.1 Classification of origin 14
         2.1.3.2 Fuel specification 15
      2.1.4 International convention on the harmonized commodity description and coding system (HS convention) 20
      2.1.5 International Maritime Organization (IMO) code for pellets 20
   2.2 PELLET PRODUCT STANDARDS IN EUROPE 21
   2.3 PELLET ANALYSIS STANDARDS IN EUROPE 24
   2.4 PELLET QUALITY ASSURANCE STANDARDS IN EUROPE 28
   2.5 STANDARDS FOR PELLET TRANSPORT AND STORAGE FOR RESIDENTIAL HEATING SYSTEMS 31
   2.6 CERTIFICATION SYSTEM ENPLUS 34
   2.7 ISO SOLID BIOFUELS STANDARDISATION 35
   2.8 STANDARDS FOR PELLET FURNACES IN THE RESIDENTIAL HEATING SECTOR 36
   2.9 ECODESIGN DIRECTIVE 44
   2.10 SUMMARY/CONCLUSIONS 45
3 PHYSIO-CHEMICAL CHARACTERISATION OF RAW MATERIALS AND PELLETS 47
   3.1 RELEVANT PHYSICAL CHARACTERISTICS OF RAW MATERIALS AND PELLETS 47
      3.1.1 Size distribution of raw materials 47
      3.1.2 Dimensions of pellets 48
      3.1.3 Bulk density of pellets 48
      3.1.4 Stowage factor 49
      3.1.5 Particle density of pellets 50
      3.1.6 Angle of repose and angle of drain for pellets 50
      3.1.7 Mechanical durability of pellets 51
      3.1.8 Pellets internal particle size distribution 52
   3.2 RELEVANT CHEMICAL CHARACTERISTICS OF RAW MATERIALS AND PELLETS 52
      3.2.1 Content of carbon, hydrogen, oxygen and volatiles of pellets 52
      3.2.2 Content of nitrogen, sulphur and chlorine of pellets 53
      3.2.3 Gross calorific value, net calorific value and energy density of pellets 54
      3.2.4 Moisture content of raw materials and pellets 57
      3.2.5 Ash content of raw materials and pellets 58
      3.2.6 Major ash forming elements relevant for combustion 59
      3.2.7 Content of natural binding agents of raw materials and pellets 60
      3.2.8 Possible contaminations of raw materials 61
         3.2.8.1 Mineral contamination 61
         3.2.8.2 Heavy metals 61
         3.2.8.3 Radioactive materials 62

	3.2.8.3.1	Sources for radioactivity in the environment and in biomass fuels	62
	3.2.8.3.2	Radioactivity in biomass fuels	63
	3.2.8.3.3	Radioactivity in ashes from biomass combustion	63
	3.2.8.3.4	Legal framework conditions	65

3.3 EVALUATION OF INTERDEPENDENCIES BETWEEN DIFFERENT PARAMETERS ... 66
   3.3.1 Interrelation between abrasion and particle density of pellets ... *66*
   3.3.2 Interrelation between abrasion and moisture content of pellets ... *67*
   3.3.3 Interrelation between abrasion and starch content of pellets ... *68*
   3.3.4 Influence of raw material storage time on bulk density, durability and fines of pellets as well as on energy consumption during pelletisation ... *69*
   3.3.5 Influence of the contents of sulphur, chlorine, potassium and sodium on the corrosion potential of pellets ... *69*
   3.3.6 Correlation between measured and calculated gross calorific value ... *71*

3.4 LIGNO-CELLULOSIC RAW MATERIALS FOR PELLETS ... 72
   3.4.1 Softwood and hardwood ... *72*
   3.4.2 Bark ... *75*
   3.4.3 Energy crops ... *76*

3.5 HERBACEOUS RAW MATERIALS FOR PELLETS (STRAW AND WHOLE CROPS) ... 77
3.6 ADDITIVES ... 78
   3.6.1 Organic additives ... *78*
   3.6.2 Inorganic additives ... *79*
	3.6.2.1	Fuels with low content of phosphorus	80
	3.6.2.2	Pellets mixed with peat	81
	3.6.2.3	Fuels with high content of phosphorus	81

3.7 SUMMARY/CONCLUSIONS ... 82

4 PELLET PRODUCTION AND LOGISTICS ... 85

4.1 PELLET PRODUCTION ... 85
   4.1.1 Pre-treatment of raw material ... *87*
	4.1.1.1	Size reduction	87
	4.1.1.2	Drying	89
		4.1.1.2.1 Basics of wood drying	90
		4.1.1.2.2 Natural drying	90
		4.1.1.2.3 Forced drying	91
		4.1.1.2.3.1 Tube bundle dryer	91
		4.1.1.2.3.2 Drum dryer	92
		4.1.1.2.3.3 Belt dryer	94
		4.1.1.2.3.4 Low temperature dryer	95
		4.1.1.2.3.5 Superheated steam dryer	97
	4.1.1.3	Conditioning	99

   4.1.2 Pelletisation ... *100*
   4.1.3 Post-treatment ... *102*
	4.1.3.1	Cooling	102
	4.1.3.2	Screening	103

   4.1.4 Special conditioning technologies of raw materials ... *104*
	4.1.4.1	Steam explosion pre-treatment of raw materials	104
	4.1.4.2	Torrefaction	104

4.2 LOGISTICS ... 108

4.2.1		Raw material handling and storage		108
4.2.2		Transportation and distribution of pellets		109
	4.2.2.1	Consumer bags		110
	4.2.2.2	Jumbo or big bags		111
	4.2.2.3	Trucks		112
	4.2.2.4	Bulk containers		114
	4.2.2.5	Railcars		115
	4.2.2.6	Ocean transportation		116
4.2.3		Pellet storage		118
	4.2.3.1	Small-scale pellet storage at residential end user sites		118
		4.2.3.1.1	Pellet storage room	119
			4.2.3.1.1.1 Storage room design	119
			4.2.3.1.1.2 Storage room dimensioning	119
			4.2.3.1.1.3 Comparison of storage room demand for pellets and heating oil	120
		4.2.3.1.2	Underground pellet storage tanks	121
		4.2.3.1.3	Storage tanks made of synthetic fibre	122
	4.2.3.2	Medium- and large-scale pellet storage		123
		4.2.3.2.1	Types of storages	123
			4.2.3.2.1.1 Vertical silo with tapered (hopper) bottom	123
			4.2.3.2.1.2 Vertical silo with flat bottom	124
			4.2.3.2.1.3 A-frame flat storage	124
			4.2.3.2.1.4 General purpose flat storage	125
		4.2.3.2.2	Requirements and examples of pellet storage at producer and commercial end user sites	125
4.2.4		Security of supply		127
4.3	SUMMARY/CONCLUSIONS			129

5 SAFETY CONSIDERATIONS AND HEALTH CONCERNS RELATING TO PELLETS DURING STORAGE, HANDLING AND TRANSPORTATION 133

5.1	DEFINITIONS RELATED TO SAFETY AND HEALTH ASPECTS			133
5.1.1		Safety related terms		133
5.1.2		Health related terms		134
5.2	SAFETY CONSIDERATIONS FOR PELLETS			135
5.2.1		Safe handling of pellets		135
	5.2.1.1	Fines and dust from pellets		135
	5.2.1.2	Airborne dust from pellets		136
		5.2.1.2.1	Explosibility of airborne dust	138
		5.2.1.2.2	Mitigation measures	140
		5.2.1.2.3	Flammability (burning rate) of airborne dust	142
5.2.2		Pellets expansion through moisture sorption		143
5.2.3		Self-heating and spontaneous ignition		144
	5.2.3.1	Wet solid biomass fuels		144
	5.2.3.2	Dry solid biomass fuels		146
	5.2.3.3	Self-heating – main risks and recommendations		147
5.2.4		Off-gassing		148
	5.2.4.1	Non-condensable gases		148
	5.2.4.2	Condensable gases		150
	5.2.4.3	Oxygen depletion		151
	5.2.4.4	Relevance of off-gassing for small-scale pellet storage units		151

5.2.5 Fire risks and safety measures ... 152
5.2.5.1 External ignition sources ... 153
5.2.5.2 Safety measures related to storage of pellets ... 154
5.2.5.3 Temperature and moisture control and gas detection ... 154
5.2.5.3.1 Indoor storage in heaps ... 157
5.2.5.3.2 Storage in silos ... 157
5.2.5.4 Extinguishing fire in pellet storages ... 158
5.2.5.4.1 Fire fighting in heaps in indoor storage ... 158
5.2.5.4.2 Fire fighting in silos ... 159
5.2.5.5 Anatomy of silo fires ... 163
5.3 HEALTH CONCERNS WITH HANDLING OF PELLETS ... 165
5.3.1 Exposure to airborne dust generated during handling of pellets ... 166
5.3.1.1 Entry routes and controls ... 168
5.3.1.1.1 Inhalation of dust ... 168
5.3.1.1.2 Skin contact ... 168
5.3.1.2 Effects on the human body ... 169
5.3.1.2.1 Irritation of eyes, nose and throat ... 169
5.3.1.2.2 Dermatitis ... 169
5.3.1.2.3 Effects on the respiratory system ... 169
5.3.1.2.4 Cancer ... 170
5.3.2 Exposure to off-gassing emissions and control measures ... 170
5.3.3 Exposure to oxygen depletion and control measures ... 171
5.4 MSDS FOR PELLETS – BULK AND BAGGED ... 172
5.4.1 Recommended format for MSDS ... 172
5.4.2 Recommended data set for pellet MSDS ... 173
5.4.2.1 MSDS data set for pellets in bulk ... 174
5.4.2.2 MSDS data set for pellets in bags ... 174
5.4.3 Example of MSDS – pellets in bulk ... 175
5.4.4 Example of MSDS – pellets in bags ... 175
5.5 SUMMARY/CONCLUSIONS ... 175

6 WOOD PELLET COMBUSTION TECHNOLOGIES ... 179
6.1 SMALL-SCALE SYSTEMS (NOMINAL BOILER CAPACITY < 100 kW_{TH}) ... 179
6.1.1 Classification of pellet combustion systems ... 179
6.1.1.1 Furnace type ... 179
6.1.1.1.1 Pellet stoves ... 180
6.1.1.1.2 Pellet furnaces with external burners ... 180
6.1.1.1.3 Pellet furnaces with inserted or integrated burners ... 183
6.1.1.2 Pellet feed-in system ... 183
6.1.1.2.1 Underfeed burners ... 184
6.1.1.2.2 Horizontally fed burner ... 185
6.1.1.2.3 Overfeed burner ... 186
6.1.1.3 Pellet burner design ... 187
6.1.2 Major components of pellet combustion systems ... 188
6.1.2.1 Conveyor systems ... 188
6.1.2.2 Ignition ... 192
6.1.2.3 Burn-back protection ... 192
6.1.2.4 Furnace geometry ... 193

	6.1.2.5	Combustion chamber materials	195
	6.1.2.6	Control strategies	196
	6.1.2.7	Boiler	198
	6.1.2.8	De-ashing	199
	6.1.2.9	Innovative concepts	201
	6.1.2.9.1	Pellet furnaces with flue gas condensation	201
	6.1.2.9.1.1	Basics of flue gas condensation	201
	6.1.2.9.1.2	Legal framework conditions for pellet furnaces with flue gas condensation	204
	6.1.2.9.1.3	Types of pellet furnaces with flue gas condensation	205
	6.1.2.9.1.3.1	Pellet furnace with integrated condenser	206
	6.1.2.9.1.3.2	External condensers for pellet furnaces	208
	6.1.2.9.1.3.2.1	Racoon	208
	6.1.2.9.1.3.2.2	Öko-Carbonizer	208
	6.1.2.9.1.3.2.3	BOMAT Profitherm	209
	6.1.2.9.1.3.2.4	Schräder Hydrocube	210
	6.1.2.9.2	Multi fuel concepts	211
	6.1.2.9.3	Pellet fired tiled stoves	212
	6.1.2.9.4	Pellet furnace and solar heating combination	213
6.2	MEDIUM-SCALE SYSTEMS (NOMINAL BOILER CAPACITY 100 - 1,000 kW_{TH})		216
	6.2.1	Combustion technologies applied	216
	6.2.2	Innovative concepts	216
6.3	LARGE-SCALE SYSTEMS (NOMINAL BOILER CAPACITY > 1,000 kW_{TH})		218
	6.3.1	Combustion technologies applied	218
	6.3.2	Innovative concepts	219
6.4	COMBINED HEAT AND POWER APPLICATIONS		220
	6.4.1	Small-scale systems (nominal boiler capacity < 100 kW_{th})	220
	6.4.2	Medium-scale systems (nominal boiler capacity 100 - 1,000 kW_{th})	222
	6.4.2.1	Stirling engine process	222
	6.4.2.2	ORC process	224
	6.4.2.3	Fixed bed gasification	226
	6.4.3	Large-scale systems (nominal boiler capacity > 1,000 kW_{th})	226
6.5	COMBUSTION AND CO-FIRING OF BIOMASS PELLETS IN LARGE PULVERISED COAL FIRED BOILERS		227
	6.5.1	Technical background	227
	6.5.2	The conversion of coal mills for processing sawdust pellets	229
	6.5.3	Co-firing biomass by pre-mixing with coal and co-milling	231
	6.5.4	Direct injection biomass co-firing systems	232
	6.5.4.1	Dedicated biomass burners	232
	6.5.4.2	Direct injection through a modified coal burner	233
	6.5.4.3	Direct injection to the pulverised coal pipework	233
	6.5.4.4	Gasification of the raw biomass with co-firing the syngas	236
	6.5.5	The impacts of biomass firing and co-firing on boiler performance	236
6.6	SUMMARY/CONCLUSIONS		237
7	**COST ANALYSIS OF PELLET PRODUCTION**		**241**
7.1	COST CALCULATION METHODOLOGY (VDI 2067)		241
7.2	ECONOMIC EVALUATION OF A STATE-OF-THE-ART PELLET PRODUCTION PLANT		242
	7.2.1	General framework conditions	242
	7.2.2	General investments	243

7.2.3	Drying	243
7.2.4	Grinding	245
7.2.5	Pelletisation	246
7.2.6	Cooling	247
7.2.7	Storage and peripheral equipment	248
7.2.8	Personnel	251
7.2.9	Raw material	251
7.2.10	Total pellets production costs	253
7.2.11	Pellet distribution costs	255
7.2.12	Sensitivity analysis	258

7.3 ECONOMIC COMPARISONS OF PELLET PRODUCTION PLANTS UNDER DIFFERENT FRAMEWORK CONDITIONS 268

7.4 SUMMARY/CONCLUSIONS 272

8 COST ANALYSIS OF PELLET UTILISATION IN THE RESIDENTIAL HEATING SECTOR .. 275

8.1 RETAIL PRICES FOR DIFFERENT FUELS IN THE RESIDENTIAL HEATING SECTOR 275

8.2 ECONOMIC COMPARISON OF DIFFERENT RESIDENTIAL HEATING SYSTEMS 278

8.2.1	General framework conditions	279
8.2.2	Pellet central heating system	280
8.2.3	Pellet central heating system with flue gas condensation	282
8.2.4	Oil central heating system	284
8.2.5	Oil central heating system with flue gas condensation	285
8.2.6	Natural gas heating system with flue gas condensation	286
8.2.7	Wood chips central heating system	288
8.2.8	Biomass district heating	290
8.2.9	Comparison of the different systems	291
8.2.10	Sensitivity analysis	295

8.3 EXTERNAL COSTS OF RESIDENTIAL HEATING BASED ON DIFFERENT HEATING SYSTEMS .. 299

8.4 SUMMARY/CONCLUSIONS 303

9 ENVIRONMENTAL EVALUATION WHEN USING PELLETS FOR RESIDENTIAL HEATING COMPARED TO OTHER ENERGY CARRIERS 305

9.1 INTRODUCTION 305

9.2 POLLUTANTS CONSIDERED FOR THE EVALUATION 305

9.3 FUEL/HEAT SUPPLY 306

9.4 AUXILIARY ENERGY DEMAND FOR THE OPERATION OF THE CENTRAL HEATING SYSTEM .. 308

9.5 UTILISATION OF DIFFERENT ENERGY CARRIERS IN DIFFERENT HEATING SYSTEMS FOR THE RESIDENTIAL HEATING SECTOR 309

9.5.1	Emission factors from field measurements	309
9.5.2	Emission factors from test stand measurements	311

9.6 TOTAL EMISSION FACTORS FOR THE FINAL ENERGY SUPPLY FOR ROOM HEATING 313

9.7 CONVERSION EFFICIENCIES 318

9.8 TOTAL EMISSION FACTORS OF USEFUL ENERGY SUPPLY FOR ROOM HEATING 320

9.9 BASICS OF ASH FORMATION AND ASH FRACTIONS IN BIOMASS COMBUSTION SYSTEMS .. 321

9.10 FINE PARTICULATE EMISSIONS 323

9.10.1	Definition of fine particulates	324
9.10.2	Health effects of fine particulates	325
9.10.3	Fine particulate emissions from biomass furnaces	326

	9.10.4	Fine particulate emissions from pellet furnaces in comparison to the total fine particulate emissions of Austria *328*

9.11 SOLID RESIDUES (ASH) 330
9.12 SUMMARY/CONCLUSIONS/RECOMMENDATIONS 331

10 CURRENT INTERNATIONAL MARKET OVERVIEW AND PROJECTIONS 335

10.1 AUSTRIA 335
    10.1.1 Pellet associations 335
    10.1.2 Pellet production, production capacity, import and export 336
    10.1.3 Pellet production potential 339
    10.1.4 Pellet utilisation 341
        10.1.4.1 General framework conditions 341
            10.1.4.1.1 Small-scale users 341
            10.1.4.1.2 Medium- and large-scale users 344
        10.1.4.2 Pellet consumption 345
        10.1.4.3 Pellet consumption potential 346
            10.1.4.3.1 Framework conditions needed for further market growth ... 346
            10.1.4.3.2 Small-scale applications 347
            10.1.4.3.3 Medium- and large-scale applications 349

10.2 GERMANY 350
    10.2.1 Pellet associations 350
    10.2.2 Pellet production, production capacity, import and export 350
    10.2.3 Production potential 351
    10.2.4 Pellet utilisation 351
        10.2.4.1 General framework conditions 351
            10.2.4.1.1 Small-scale users 352
            10.2.4.1.2 Medium- and large-scale users 353
        10.2.4.2 Pellet consumption 353
        10.2.4.3 Pellet consumption potential 354

10.3 ITALY 355
10.4 SWITZERLAND 357
10.5 SWEDEN 359
    10.5.1 Pellet production, production capacity, import and export 359
    10.5.2 Pellet utilisation 359
        10.5.2.1 General framework conditions 359
        10.5.2.2 Pellet consumption 361
            10.5.2.2.1 Small-scale users 361
            10.5.2.2.2 Medium- and large-scale users 363

10.6 DENMARK 363
10.7 OTHER EUROPEAN COUNTRIES 366
10.8 NORTH AMERICA 371
10.9 OTHER INTERNATIONAL MARKETS 373
10.10 INTERNATIONAL OVERVIEW OF PELLET PRODUCTION POTENTIALS 374
    10.10.1 Pellet production plants in Europe 374
        10.10.1.1 Distribution of pellet production plants and market areas 374
        10.10.1.2 The development of the Austrian market 376
        10.10.1.3 Production in Sweden and Finland 378
    10.10.2 Evaluation of alternative raw material potentials in Europe 380

10.10.3 Evaluation of the worldwide sawdust potential available for pellet production 383
   10.10.3.1 Forest biomass resources and wood use in forest industry 383
      10.10.3.1.1 An overview of forest biomass resources and mechanical wood processing 383
      10.10.3.1.2 Use of wood as raw material and energy in forest industry 386
         10.10.3.1.2.1 Forest industry's solid by-products 386
         10.10.3.1.2.2 Forest industry's liquid by-products (black liquor) 386
         10.10.3.1.2.3 By-products in energy production 387
   10.10.3.2 Evaluation of global raw material potential for wood pellets from sawdust 388
      10.10.3.2.1 Modelling the wood streams of forest industry at country level 388
      10.10.3.2.2 Sawdust excess from forest industry 389
10.11 INTERNATIONAL PELLET TRADE 391
  10.11.1 Main global trade flows *391*
  10.11.2 The history of intercontinental wood pellet trade – the case of Canada *394*
  10.11.3 Wood pellet shipping prices, shipping requirements and standards *395*
  10.11.4 Prices and logistic requirements for truck transport *398*
  10.11.5 Future trade routes *400*
  10.11.6 Opportunities and barriers for international pellet trade *403*
   10.11.6.1 Fossil fuel prices 404
   10.11.6.2 Policy support measures 405
   10.11.6.3 Feedstock availability and costs 405
   10.11.6.4 Sustainability criteria, certified production and traceable chain management 406
   10.11.6.5 Technical requirements for industrial wood pellets 407
   10.11.6.6 Logistics 409
  10.11.7 Case study of a supply chain of western Canadian (British Columbia) wood pellets to power plants in Western Europe *410*
   10.11.7.1 Production plant 410
   10.11.7.2 Transport to port 411
   10.11.7.3 Terminal north Vancouver/Prince Rupert 412
   10.11.7.4 Loading wood pellets 413
   10.11.7.5 Ocean voyage 415
   10.11.7.6 Discharging wood pellets 415
   10.11.7.7 Transhipment of wood pellets 416
   10.11.7.8 Storage at inland terminal (in VARAGT zone/Western Europe) 416
   10.11.7.9 Barging to final destination 417
   10.11.7.10 Unloading at power plant 417
10.12 SOCIO-ECONOMIC ASPECTS OF PELLET PRODUCTION AND UTILISATION 417
10.13 SUMMARY/CONCLUSIONS 421

11 CASE STUDIES FOR THE USE OF PELLETS FOR ENERGY GENERATION 427
11.1 CASE STUDY 1 – SMALL-SCALE APPLICATION: PELLET STOVE (GERMANY) 427
  11.1.1 Plant description *427*
  11.1.2 Technical data *428*
  11.1.3 Economy *429*
11.2 CASE STUDY 2 – SMALL-SCALE APPLICATION: PELLET CENTRAL HEATING (AUSTRIA) 430
  11.2.1 Plant description *430*
  11.2.2 Technical data *432*
  11.2.3 Economy *432*

11.3 CASE STUDY 3 – SMALL-SCALE APPLICATION: RETROFITTING EXISTING BOILER WITH A PELLET BURNER (SWEDEN) 433
   11.3.1 Plant description *433*
   11.3.2 Technical data *435*
   11.3.3 Economy *435*
11.4 CASE STUDY 4 – MEDIUM-SCALE APPLICATION: 200 kW SCHOOL HEATING PLANT JÄMSÄNKOSKI (FINLAND) 435
   11.4.1 Plant description *435*
   11.4.2 Technical data *438*
   11.4.3 Economy *438*
11.5 CASE STUDY 5 – MEDIUM-SCALE APPLICATION: 500 kW HEATING PLANT IN STRAUBING (GERMANY) 439
   11.5.1 Plant description *439*
   11.5.2 Technical data *440*
   11.5.3 Economy *441*
11.6 CASE STUDY 6 – MEDIUM-SCALE APPLICATION: 600 kW DISTRICT HEATING PLANT IN VINNINGA (SWEDEN) 442
   11.6.1 Plant description *442*
   11.6.2 Technical data *443*
   11.6.3 Economy *443*
11.7 CASE STUDY 7 – LARGE-SCALE APPLICATION: 2.1 MW DISTRICT HEATING PLANT KÅGE (SWEDEN) 444
   11.7.1 Plant description *444*
   11.7.2 Technical data *445*
   11.7.3 Economy *445*
11.8 CASE STUDY 8 – LARGE-SCALE APPLICATION: 4.5 MW DISTRICT HEATING PLANT HILLERØD (DENMARK) 446
   11.8.1 Plant description *446*
   11.8.2 Technical data *447*
   11.8.3 Economy *448*
11.9 CASE STUDY 9 – CHP APPLICATION: CHP PLANT HÄSSELBY (SWEDEN) 449
   11.9.1 Plant description *449*
   11.9.2 Technical data *450*
   11.9.3 Economy *451*
11.10 CASE STUDY 10 – LARGE-SCALE POWER GENERATION APPLICATION: POWER PLANT LES AWIRS (BELGIUM) 451
   11.10.1 Plant description *451*
   11.10.2 Technical data *453*
   11.10.3 Economy *454*
11.11 CASE STUDY 11 – CO-FIRING APPLICATION: AMER POWER PLANT UNITS NO. 8 AND NO. 9 IN GEERTRUIDENBERG (THE NETHERLANDS) 454
   11.11.1 Plant description *454*
   11.11.2 Technical data *456*
   11.11.3 Economy *457*
11.12 SUMMARY/CONCLUSIONS 457

12 RESEARCH AND DEVELOPMENT **459**
12.1 PELLET PRODUCTION 459
   12.1.1 Use of raw materials with lower quality *459*
      12.1.1.1 Herbaceous biomass 459

  12.1.1.2 Short rotation crops 460
  12.1.1.3 Increasing the raw material basis 461
 12.1.2 Pellet quality and production process optimisation *461*
  12.1.2.1 Influence of production process parameters 461
  12.1.2.2 Mitigation of self-heating and off-gassing 463
  12.1.2.3 Pellet production process optimisation 464
 12.1.3 Torrefaction *464*
 12.1.4 Decentralised pellet production *465*
12.2 Pellet utilisation 465
 12.2.1 Emission reduction *465*
  12.2.1.1 Fine particulate emissions 465
   12.2.1.1.1 Fine particulate formation and characterisation 465
   12.2.1.1.2 Primary measures for particulate emission reduction 467
   12.2.1.1.3 Fine particulate precipitation 467
   12.2.1.1.4 Health effects of fine particulate emissions 468
  12.2.1.2 Gaseous emissions 470
 12.2.2 New pellet furnace developments *470*
  12.2.2.1 Pellet furnaces with very low nominal boiler capacities 470
  12.2.2.2 Multi fuel concepts 471
 12.2.3 Increase of annual efficiencies *471*
 12.2.4 Micro- and small-scale CHP systems based on pellets *471*
 12.2.5 Utilisation of pellets with lower quality *472*
 12.2.6 Furnace optimisation and development based on CFD simulations *473*
 12.2.7 Pellet utilisation in gasification *475*
12.3 Support of market developments 476
12.4 Summary/conclusions 477

APPENDIX A: EXAMPLE OF MSDS – PELLETS IN BULK **479**

APPENDIX B: EXAMPLE OF MSDS – PELLETS IN BAGS **493**

REFERENCES **501**

List of figures

Figure 2.1:	CEN TC 335 within the biomass-biofuel-bioenergy field	9
Figure 2.2:	Classification of woody biomass	15
Figure 2.3:	Wood pellet classification according to Part 2 of prEN 14961	16
Figure 2.4:	Example of fuel specification according to EN 14961-1	19
Figure 2.5:	Scheme and picture of a tester for mechanical durability of pellets according to EN 15210-1	26
Figure 2.6:	Ligno-Tester LT II	26
Figure 2.7:	Scheme of the abrasion tester according to the Swedish standard	27
Figure 2.8:	Correlation between durability determinations according to EN 15210-1 and ÖNORM M 7135	27
Figure 2.9:	Supply chain covered by the prEN 15234-1	28
Figure 2.10:	Requirements for boiler efficiency derived from nominal boiler capacity according to ÖNORM EN 303-5	37
Figure 3.1:	Relative effect of shock impact on volume compared to a non-shock application in bulk density determination	49
Figure 3.2:	Visualisation of angle of repose and angle of drain	51
Figure 3.3:	Comparison of different calculation methods for the NCV	55
Figure 3.4:	Gross and net calorific values of densified biomass fuels	56
Figure 3.5:	Energy densities of pellets	57
Figure 3.6:	Specific activity of ^{137}Cs in biomass fuels	63
Figure 3.7:	Specific activities of ^{137}Cs in bottom and coarse fly ashes as well as aerosols	65
Figure 3.8:	Relation between particle density and durability	67
Figure 3.9:	Abrasion of wood pellets as a function of a varying moisture content	68
Figure 3.10:	Molar ratio of sulphur in the fuel in relation to available alkali compounds and chlorides ($M_{S/AC+Cl}$) as an indicator for high temperature chlorine corrosion potential during combustion	70
Figure 3.11:	Extended corrosion diagram	71
Figure 3.12:	Correlation between calculated and measured gross calorific value of densified biomass fuels	72
Figure 4.1:	Process line of pelletisation	86
Figure 4.2:	Hammer mill	88
Figure 4.3:	Working principle of a hammer mill	89
Figure 4.4:	Tube bundle dryer	91

Figure 4.5:	Tube bundle	92
Figure 4.6:	Drum dryer	93
Figure 4.7:	Cross section of a drum dryer with three ducts	93
Figure 4.8:	Belt dryer	94
Figure 4.9:	Working principle of a belt dryer	94
Figure 4.10:	Pre-assembled drying cell of a low temperature dryer	95
Figure 4.11:	Working principle of a low temperature dryer	96
Figure 4.12:	Working principle of a superheated steam dryer ("exergy dryer")	98
Figure 4.13:	Fluidised bed dryer with superheated steam circuit	99
Figure 4.14:	Blender for the conditioning by steam or water	100
Figure 4.15:	Designs of pellet mills	101
Figure 4.16:	Pellet mill	101
Figure 4.17:	Counter flow cooler	102
Figure 4.18:	Working principle of a counter flow cooler	103
Figure 4.19:	Energy demand for grinding of torrefied biomass in comparison with untreated biomass and bituminous coal	105
Figure 4.20:	Energy demand for grinding of torrefied biomass in correlation with torrefaction temperature	106
Figure 4.21:	Flow sheet of the BO_2-technology	107
Figure 4.22:	Typical small bags in North America and Europe	110
Figure 4.23:	Big bag as it is filled	111
Figure 4.24:	Typical jumbo or big bags	111
Figure 4.25:	Typical European tank truck with pneumatic feed	113
Figure 4.26:	Typical North American "stinger" truck (B-train)	113
Figure 4.27:	Dump truck	113
Figure 4.28:	Standard truck	114
Figure 4.29:	Semi-trailer truck with hydraulic unloading	114
Figure 4.30:	Bulk loading containers on weigh scales	115
Figure 4.31:	Typical hopper railcar in North America	115
Figure 4.32:	Typical transatlantic bulk carrier	116
Figure 4.33:	Shiploader with choke spout	117
Figure 4.34:	Clam bucket during loading	117
Figure 4.35:	Receiving hopper with dust suppression fans	118
Figure 4.36:	Cross section of a pellet storage space	119

Figure 4.37:	Pellet globe for underground pellet storage	121
Figure 4.38:	Underground pellet storage with discharge from the top	122
Figure 4.39:	Tank made of synthetic fibre for pellet storage	122
Figure 4.40:	Example of a vertical silo with tapered (hopper) bottom	123
Figure 4.41:	Example of a vertical silo with flat bottom	124
Figure 4.42:	Example of an A-frame flat storage	125
Figure 4.43:	Example of a general purpose flat storage	125
Figure 4.44:	Plane storage building at the landing stage at Öresundskraft AB in Helsingborg (Sweden)	126
Figure 5.1:	Size distribution of airborne dust	136
Figure 5.2:	Particle sedimentation time in still air	137
Figure 5.3:	Illustration of possible impact points in a bunker with steep slopes	138
Figure 5.4:	Fire triangle and the explosion pentagon indicating factors for fire and explosion to occur	139
Figure 5.5:	Effect of water application to pellets and equilibrium moisture content for wood pellets	143
Figure 5.6:	CO concentrations in the headspace of a pellet storage at different temperatures over time due to off-gassing	149
Figure 5.7:	CO_2 concentrations in the headspace of a pellet storage at different temperatures over time due to off-gassing	149
Figure 5.8:	CH_4 concentrations in the headspace of a pellet storage at different temperatures over time due to off-gassing	150
Figure 5.9:	Examples of embedded temperature monitoring systems and comparison of single cable and multi cable solutions	155
Figure 5.10:	Permeability for pellets with various aspect ratios	156
Figure 5.11:	Mobile fire fighting unit for silo fire fighting	160
Figure 5.12:	Principle sketch of distributed gas injection in silos	161
Figure 5.13:	Steel lances used for gas injection in a silo	161
Figure 5.14:	Flames on the outside of a silo caused by an opening in the silo wall	162
Figure 5.15:	Fire ball movement within a column of pellets	164
Figure 5.16:	Fire ball seen from underneath in the test silo	164
Figure 5.17:	Agglomerated pellets above the fire ball in the test silo	165
Figure 5.18:	Regional particle deposition in the human respiratory system	166
Figure 6.1:	Stove fed with pellets	180
Figure 6.2:	Pellet furnace with external burner	181
Figure 6.3:	Boiler retrofitted for the use of pellets	182

Figure 6.4:	Horizontal stoker burner principle	182
Figure 6.5:	Basic principles of wood pellet combustion systems	183
Figure 6.6:	Underfeed furnace	184
Figure 6.7:	Horizontally fed pellet furnace	185
Figure 6.8:	Overfeed pellet furnace	186
Figure 6.9:	Different types of pellet burners	187
Figure 6.10:	Rotary grate pellet burner	188
Figure 6.11:	Conveyor system with conventional screw	189
Figure 6.12:	Conveyor system with flexible screw	189
Figure 6.13:	Pneumatic pellet feeding system	190
Figure 6.14:	Combination of feeding screw and pneumatic feeding system	191
Figure 6.15:	Combination of feeding screw and agitator	191
Figure 6.16:	Rotary valve	192
Figure 6.17:	Fireproof valve	193
Figure 6.18:	Self-initiating fire extinguishing system	193
Figure 6.19:	Principle of an overfeed pellet furnace with staged air supply and optimised mixing of flue gas and secondary air	194
Figure 6.20:	Example of an optimised secondary air nozzle design by CFD simulation	195
Figure 6.21:	Correlation scheme of CO emissions and excess air coefficient λ in small-scale biomass furnaces	196
Figure 6.22:	Emission of fine particulates, CO and TOC during load change of a modern pellet furnace	198
Figure 6.23:	Fully automatic heat exchanger cleaning system	199
Figure 6.24:	Ash compaction system	200
Figure 6.25:	External ash box	200
Figure 6.26:	Dependency of efficiencies on the outlet temperature from the condenser and different moisture contents	201
Figure 6.27:	Dependency of efficiencies on the outlet temperature from the condenser and different O_2 contents of the flue gas	202
Figure 6.28:	Efficiencies of pellet furnaces with and without flue gas condensation	203
Figure 6.29:	Fine particulate emissions of pellet furnaces with and without flue gas condensation	204
Figure 6.30:	Scheme of a pellet boiler with flue gas condensation	207
Figure 6.31:	Pellet boiler with flue gas condensation	207
Figure 6.32:	Racoon	208
Figure 6.33:	Öko-Carbonizer	209

Figure 6.34:	BOMAT Profitherm	210
Figure 6.35:	Schräder Hydrocube	211
Figure 6.36:	Combined boiler for the use of pellets and firewood	212
Figure 6.37:	Pellet and solar heating system using roof integrated solar collectors	214
Figure 6.38:	Solar and pellet heating system with pellet boiler	215
Figure 6.39:	Solar and pellet heating system with a pellet stove and a small buffer store	215
Figure 6.40:	TDS Powerfire 150	217
Figure 6.41:	PYROT rotation furnace	218
Figure 6.42:	Bioswirl® burner	219
Figure 6.43:	SPM Stirlingpowermodule	221
Figure 6.44:	Principle of thermoelectric electricity generation	222
Figure 6.45:	Prototype of a thermoelectric generator designed for utilisation in a pellet furnace	222
Figure 6.46:	Stirling engine process – scheme of integration into a biomass CHP plant	223
Figure 6.47:	Pictures of a pilot plant and the 35 kW$_{el}$ Stirling engine	224
Figure 6.48:	Scheme of the ORC process as integrated into the biomass CHP plant Lienz	225
Figure 6.49:	Biomass co-firing options at large pulverised coal fired power plants	227
Figure 7.1:	Price development of sawdust from December 2003 to August 2009	252
Figure 7.2:	Pellet production costs and their composition according to the different cost factors when sawdust is used as raw material	254
Figure 7.3:	Pellet production costs and their composition according to VDI 2067 when sawdust is used as raw material	254
Figure 7.4:	Energy consumption of pellet production when sawdust is used as raw material	255
Figure 7.5:	Total and specific pellet transport costs versus transport distance	256
Figure 7.6:	Influence of investment costs on the specific pellet production costs of the base case scenario of different plant components	258
Figure 7.7:	Influence of the utilisation periods of different plant components on the specific pellet production costs	259
Figure 7.8:	Influence of maintenance costs on the specific pellet production costs of different plant components	260
Figure 7.9:	Influence of plant availability and the simultaneity factor of electric equipment on the specific pellet production costs	261
Figure 7.10:	Influence of electricity price on the specific pellet production costs	262
Figure 7.11:	Influence of specific heat costs on the specific pellet production costs	262
Figure 7.12:	Influence of annual full load operating hours on the specific pellet production costs	263
Figure 7.13:	Influence of personnel and hot water demand for conditioning on the specific pellet production costs	264

Figure 7.14:	Influence of raw material costs on the specific pellet production costs	265
Figure 7.15:	Influence of interest rate on the specific pellet production costs	265
Figure 7.16:	Influence of throughput on the specific pellet production costs	266
Figure 7.17:	Influence of personnel demand for marketing and administration on the specific pellet production costs	267
Figure 7.18:	Overview of the effects of parameter changes on the specific pellet production costs	268
Figure 7.19:	Composition of the specific pellet production costs according to different cost factors when sawdust and wood shavings are used as a raw material	271
Figure 8.1:	Average prices of different fuels based on NCV from 2006 to 2008	275
Figure 8.2:	Price development of pellets, heating oil and natural gas from June 1999 to September 2009 in Austria	276
Figure 8.3:	Price development of pellets in Germany	277
Figure 8.4:	Comparison of investment costs for different heating systems	291
Figure 8.5:	Comparison of annual fuel and heat costs	293
Figure 8.6:	Comparison of specific heat generation costs of different heating systems	294
Figure 8.7:	Comparison of annual heat generation costs broken down into costs based on capital, consumption costs, operating costs and other costs	294
Figure 8.8:	Influence of fuel or heat price on specific heat generation costs	295
Figure 8.9:	Influence of investment costs on specific heat generation costs	296
Figure 8.10:	Influence of annual efficiencies on specific heat generation costs	297
Figure 8.11:	Specific heat generation costs of central heating systems with external costs for the scenario "emission trade"	300
Figure 8.12:	Specific heat generation costs of central heating systems with external costs based on local emission prognoses	301
Figure 8.13:	Specific heat generation costs of central heating systems with external costs based on global emission prognoses	302
Figure 9.1:	Emission factors of different central heating systems based on field measurements	310
Figure 9.2:	Comparison of test stand and field measurements of Austrian pellet furnaces	311
Figure 9.3:	Development of CO emissions from Austrian pellet furnaces from 1996 to 2008	312
Figure 9.4:	Development of particulate emissions from Austrian pellet furnaces from 1996 to 2008	313
Figure 9.5:	Emission factors of final energy supply of different heating systems	315
Figure 9.6:	Emission factors of final energy supply for different heating systems as well as their composition	317
Figure 9.7:	Comparison of boiler and annual efficiencies of the systems compared	318
Figure 9.8:	Annual efficiencies and useful heat demands of pellet boilers based on field measurements	320

Figure 9.9:	Emission factors of useful energy supply for different heating systems	321
Figure 9.10:	Ash formation during biomass combustion	322
Figure 9.11:	Excesses of the fine particulate emission limit in Austria from 2005 to 2008	324
Figure 9.12:	Aerosol emissions from medium- and large-scale biomass furnaces compared to aerosol forming elements in the fuel	326
Figure 9.13:	Composition of fine particulate emissions from old and modern small-scale biomass furnaces at nominal load	328
Figure 9.14:	Fine particulate emissions in Austria according to sources	329
Figure 10.1:	Pellet production sites in Austria and their capacities	337
Figure 10.2:	Development of Austrian pellet production capacities from 1996 to 2009	337
Figure 10.3:	Development of Austrian pellet production, consumption and export from 1995 to 2009	338
Figure 10.4:	Development of pellet stoves in Austria from 2001 to 2008	342
Figure 10.5:	Development of pellet central heating systems in Austria from 1997 to 2009	343
Figure 10.6:	Development of annually installed nominal boiler capacity of pellet central heating systems in Austria from 1997 to 2008	343
Figure 10.7:	Development of medium- and large-scale wood chip furnaces in Austria from 1997 to 2008	344
Figure 10.8:	Development of pellet consumption in Austria from 1997 to 2008	345
Figure 10.9:	Gross domestic consumption of renewable fuels (without hydropower) in Austria (2007)	346
Figure 10.10:	Heating systems in Austrian homes from 1980 to 2006	348
Figure 10.11:	Annual boiler installations in Austria from 1997 to 2008	349
Figure 10.12:	Pellet production and production capacities in Germany from 1999 to 2008	350
Figure 10.13:	Development of pellet central heating systems in Germany from 1999 to 2010	352
Figure 10.14:	Pellet consumption in Germany from 1999 to 2008	353
Figure 10.15:	Annual boiler installations in Germany from 1998 to 2008	354
Figure 10.16:	Development of pellet stoves in Italy from 2002 to 2008	356
Figure 10.17:	Pellet production and use in Italy from 2001 to 2009	356
Figure 10.18:	Cumulated pellet furnace installations in Switzerland	358
Figure 10.19:	Pellet production and use in Switzerland from 2000 to 2007	358
Figure 10.20:	Pellet production, import and export in Sweden from 1997 to 2012	359
Figure 10.21:	Cumulated pellet central heating and pellet stove installations in Sweden from 1998 to 2007	361
Figure 10.22:	Development of pellet consumption in Sweden from 1995 to 2012	362

Figure 10.23:	Total use of wood pellets in detached and semi-detached houses in Sweden from 1999 to 2007 362
Figure 10.24:	Cumulative number of residential pellet boiler installations 364
Figure 10.25:	Development of pellet consumption, production capacity, production and net import in Denmark from 2001 to 2008 365
Figure 10.26:	Pellet production and utilisation in selected European countries 366
Figure 10.27:	Development of pellet consumption in North America from 1995 to 2010 372
Figure 10.28:	Development of pellet production in North America from 1995 to 2010 372
Figure 10.29:	Location of the pellet production plants in Europe (left) and market analysis using percent volume contours (right) 375
Figure 10.30:	Pellet production and production capacity in Austria for the period 1994 to 2006 and projection of production until 2016 377
Figure 10.31:	Location of sawmills and pellet production plants in Sweden and Finland 379
Figure 10.32:	Correlation between sawmill and pellet production plant capacity aggregated using 80 km radius in Sweden and Finland 379
Figure 10.33:	Estimated pellet production capacity of several European countries compared to the annual sawlog production. 380
Figure 10.34:	Theoretical forest fuel potential for the EU27 from logging residues and potential sustainable surplus of commercial growing stock (annual change rate) 381
Figure 10.35:	The consumption of logs and the production of sawn timber and plywood from 1985 to 2004 385
Figure 10.36:	Wood streams in the Finnish forest industry in 2007 387
Figure 10.37:	Illustration of the wood stream model and its main parameters 388
Figure 10.38:	The largest producers of by-products from sawmills and plywood mills 389
Figure 10.39:	Comparison of the production of solid by-products in the sawmill and plywood industry and the demand for raw material in the particle board and fibreboard industry 390
Figure 10.40:	Theoretical excess of solid by-products from the mechanical wood processing industry 390
Figure 10.41:	Overview of pellet production, consumption and trade flows for the most important pellet markets in 2007 392
Figure 10.42:	Overview of main wood pellet trade flows in and towards Europe 393
Figure 10.43:	Wood pellet spot prices CIF ARA 396
Figure 10.44:	Pellets stapled on a pallet 399
Figure 10.45:	Expectations for the main growth in wood pellet production in the coming five years by wood pellet experts. 401
Figure 10.46:	Expectations for the main growth in wood pellet demand in the coming five years by wood pellet experts 402

Figure 10.47:	Main barriers for international wood pellet trade in the coming five years as stated by wood pellet experts 403	
Figure 10.48:	Main drivers for international wood pellet trade in the coming five years as stated by wood pellet experts 404	
Figure 10.49:	Anticipated logistical challenges to be tackled for more efficient wood pellet supply chains as estimated by wood pellet experts..409	
Figure 10.50:	Logistical chain of western Canadian wood pellets to Western Europe 410	
Figure 10.51:	Forestry area in British Columbia, Canada 411	
Figure 10.52:	Discharging railcar..412	
Figure 10.53:	Cascading spout in vessel's hold 415	
Figure 10.54:	Typical barge used for pellet transport 416	
Figure 10.55:	Specific pellet consumption in different countries..423	
Figure 11.1:	Pellet stove (8 kW) in the living room..428	
Figure 11.2:	Pellet central heating system in St. Lorenzen/Mürztal, Austria 430	
Figure 11.3:	Old combination boiler with new pellet burner 433	
Figure 11.4:	Week's storage connected to the pellet burner by a screw conveyor..................................434	
Figure 11.5:	Koskenpää elementary school in Jämsänkoski town 436	
Figure 11.6:	Modified heating centre of Koskenpää elementary school 437	
Figure 11.7:	New pellet boiler of Koskenpää elementary school..437	
Figure 11.8:	Underground pellet silo...438	
Figure 11.9:	Pellet boiler at the IFH in Straubing 440	
Figure 11.10:	District heating plant in Vinninga behind the pellet silo...442	
Figure 11.11:	District heating plant in Kåge 444	
Figure 11.12:	District heating plant in Hillerød 447	
Figure 11.13:	CHP plant Hässelby 449	
Figure 11.14:	Les Awirs power plant near Liège 451	
Figure 11.15:	Amer co-firing power plant in Geertruidenberg 455	
Figure 11.16:	Biomass unloading station at Amer power plant in Geertruidenberg 455	
Figure 12.1:	Iso-surfaces of flue gas temperature [°C] in horizontal cross sections of the furnace 474	
Figure 12.2:	Iso-surfaces of CO concentration in the flue gas [ppmv] in cross sections of the furnace 475	

List of tables

Table 2.1:	EN standards for solid biofuels published and under preparation under committee TC335 from CEN	8
Table 2.2:	Classification of woody biomass according to EN 14961-1	14
Table 2.3:	Specification of wood pellets for non-industrial use	17
Table 2.4:	Specification of normative properties for pellets according to EN 14961-1	18
Table 2.5:	Specification of normative/informative properties for pellets according to EN 14961-1	19
Table 2.6:	Comparison of pellet standards	22
Table 2.7:	Requirements for boiler efficiency derived from nominal boiler capacity according to ÖNORM EN 303-5	37
Table 2.8:	Emissions limits defined by ÖNORM EN 303-5	38
Table 2.9:	CO emission limits according to EN 303-5 and in different countries	39
Table 2.10:	NO_x emission limits according to EN 303-5 and in different countries	40
Table 2.11:	OGC emission limits according to EN 303-5 and in different countries	41
Table 2.12:	Particulate matter emission limits according to EN 303-5 and in different countries	42
Table 3.1:	Concentrations of C, H, O and volatiles in different biomass materials	53
Table 3.2:	Guiding values for N, S and Cl for various biomass fuels	54
Table 3.3:	Typical ash contents of different types of biomass	59
Table 3.4:	Concentrations of major ash forming elements in biomass ashes	60
Table 3.5:	Typical concentrations of heavy metals in various types of biomass fuels	62
Table 3.6:	Overview of different woody biomass fractions with regard to their use in pelletisation	74
Table 3.7:	Parameters for the production of compressed bark	76
Table 3.8:	Typical ash, N, S and Cl contents in poplar and willow	77
Table 3.9:	Guiding values for straw and whole crops in comparison with values from prEN 14961-2 and general guiding values for the production of class A1 and A2 pellets	77
Table 3.10:	Overview of evaluation criteria for possible raw materials for pelletisation and pellet characteristics	83
Table 4.1:	Raw material demand for production of 1 t of pellets	86
Table 4.2:	Fibre saturation ranges of a few wood species	90
Table 5.1:	Results from testing dust (< 63 µm) from white pellets and bark pellets	139
Table 5.2:	Recommended precautionary measures in the presence of metal dust and related minimum ignition energy requirements	141
Table 5.3:	Burning rate of pellet dust of less than 63 µm	142

Table 5.4:	CO concentrations in small-scale pellet storage units at residential end user sites	151
Table 5.5:	Reasons for accidents or incidents with different types of dust	153
Table 5.6:	Summary of toxicological data concerning the exposure value limits recommended by various regulatory bodies	167
Table 5.7:	Examples of TWA and STEL for CO, CO_2 and CH_4 in Canada and Sweden	170
Table 6.1:	Heavy metal contents of the condensate of a pellet furnace with flue gas condensation in comparison with limiting values of the Austrian waste water emission act	205
Table 7.1:	General framework conditions for the calculation of the pellet production costs for the base case scenario	242
Table 7.2:	Calculation of full costs for general investments of a pellet production plant	243
Table 7.3:	Moisture contents before and after drying of different raw materials for pelletisation	244
Table 7.4:	Framework conditions for full cost calculation of drying in a belt dryer	244
Table 7.5:	Full cost calculation of a belt dryer	245
Table 7.6:	Framework conditions for full cost calculation of raw material grinding in a hammer mill	246
Table 7.7:	Full cost calculation of grinding in a hammer mill	246
Table 7.8:	Framework conditions for full cost calculation of a pellet mill	247
Table 7.9:	Full cost calculation of a pellet mill	247
Table 7.10:	Framework conditions for full cost calculation of cooling in a counterflow cooler	248
Table 7.11:	Full cost calculation of a counterflow cooler	248
Table 7.12:	Framework conditions for full cost calculation of raw material and pellet storage at the producer's site	249
Table 7.13:	Full cost calculation of raw material and pellet storage at the producer's site	250
Table 7.14:	Framework conditions for full cost calculation of peripheral equipment	250
Table 7.15:	Full cost calculation for peripheral equipment	250
Table 7.16:	Price range of possible raw materials for pellets	251
Table 7.17:	Overview of the composition of the total pellet production costs	253
Table 7.18:	Basic data for the calculation of transport costs per silo truck	256
Table 7.19:	Total costs of pellet distribution	257
Table 7.20:	Total costs of pellet supply	257
Table 7.21	Key parameters of the scenarios considered in comparison to the base case scenario	270
Table 8.1:	General framework conditions for full cost calculation of different heating systems	280
Table 8.2:	Basic data for the full cost calculation of a pellet central heating system	280
Table 8.3:	Full cost calculation of a pellet central heating system	281
Table 8.4:	Basic data for full cost calculation of a pellet central heating system with flue gas condensation	283

Table 8.5:	Full cost calculation of a pellet central heating system with flue gas condensation	283
Table 8.6:	Basic data for full cost calculation of an oil central heating system	284
Table 8.7:	Full cost calculation of an oil central heating system	285
Table 8.8:	Basic data for full cost calculation of an oil central heating system with flue gas condensation	285
Table 8.9:	Full cost calculation of an oil central heating system with flue gas condensation	286
Table 8.10:	Basic data for full cost calculation of a natural gas heating system with flue gas condensation	287
Table 8.11:	Full cost calculation of a natural gas central heating system with flue gas condensation	288
Table 8.12:	Basic data for full cost calculation of a wood chip central heating system	288
Table 8.13:	Full cost calculation of a wood chip central heating system	289
Table 8.14:	Basic data for the full cost calculation of biomass district heating	290
Table 8.15:	Full cost calculation of biomass district heating	291
Table 8.16:	Different scenarios and their effects on specific heat generation costs	298
Table 9.1:	Basic data for the calculation of the emission factors along the pellet supply chain	306
Table 9.2:	Energy consumption of the pellet production process steps for pelletisation of wood shavings and sawdust	307
Table 9.3:	Emission factors of the pellet supply chain	307
Table 9.4:	Emission factors of the supply of heating oil, natural gas and wood chips as well as the supply of district heat	308
Table 9.5:	Emission factors of auxiliary energy use during operation of central heating systems	308
Table 9.6:	Emission factors of different central heating systems based on field measurements	309
Table 9.7:	Emission factors of the final energy supply of different heating systems in order of supply steps	314
Table 9.8:	Typical Ca and nutrient contents of different biomass ashes	330
Table 10.1:	Cumulated number of combustion equipment in the residential sector	360
Table 10.2:	Average sales of some combustion equipment between 2003 and 2007	360
Table 10.3:	Estimates for theoretical forest fuel and short rotation coppice potential	382
Table 10.4:	World production of industrial log wood, logs, sawn timber and plywood	384
Table 10.5:	World top 15 countries in the production of logs, sawn timber and plywood in 2004	384
Table 10.6:	World top 15 countries in the production of particle board and fibreboard in 2004	385
Table 10.7:	Overview of vessel specifications	396
Table 10.8:	Maximum and minimum charter rates	397
Table 10.9:	Sample calculation for estimating the freight rates for 22,000 t pellets by bulk cargo for a shipment from Indonesia to Italy through the Suez canal	398

Table 10.10:	Example of a quality standard for pellets to be used in a large power plant	408
Table 10.11:	CO_2 balance of pellet supply from different countries and regions	408
Table 10.12:	General socio-economic aspects associated with local pellet production and utilisation	419
Table 11.1:	Technical data of the pellet stove in Straubing	428
Table 11.2:	Economic data of the pellet stove in Straubing	429
Table 11.3:	Emissions and emission limits of the pellet central heating system	431
Table 11.4:	Technical data of the pellet central heating system	432
Table 11.5:	Economic data of the pellet central heating system	432
Table 11.6:	Technical data of the retrofitted burner in Sweden	435
Table 11.7:	Technical data of the pellet heating plant at the IFH in Straubing	440
Table 11.8:	Economic data of the pellet heating plant at the IFH in Straubing	441
Table 11.9:	Technical data of the district heating plant in Vinninga	443
Table 11.10:	Technical data of the district heating plant in Kåge	445
Table 11.11:	Technical data of the district heating plant in Hillerød	448
Table 11.12:	Technical data of the CHP plant Hässelby	450
Table 11.13:	Emissions of the Les Awirs power plant fired with wood pellets	453
Table 11.14:	Technical data of the Les Awirs power plant	453
Table 11.15:	Technical issues related to different process steps of Amer power station units no. 8 and no. 9	456
Table 11.16:	Technical data of the Amer units no. 8 and no. 9	456

Abbreviations, formula symbols, chemical formulas, units, prefixes and indices

Abbreviations

A	ash content
ACGIH	American Conference of Governmental Industrial Hygienists
AED	auxiliary energy demand
Ae.d.	aerodynamic diameter
AEV	Abwasseremissionsverordnung (Austrian waste water emission act)
AIEL	Associazione Italiana Energie Agriforestali
ALARP	as low as reasonably practicable
ARA	Amsterdam, Rotterdam, Antwerp
BDI	Baltic Dry Index
BFI	Baltic Freight Index
BM-DH	biomass district heating
BP	bark pellets
CCP	critical control point
CEN	European Committee for Standardization
CEV	ceiling exposure value
cf.	confer (compare or consult)
CFD	computational fluid dynamics
CGE	computable general equilibrium
CHP	combined heat and power
CIF	cost insurance freight
CN	combined nomenclature
CRF	capital recovery factor
CTI	Comitato Termotecnico Italiano
D	diameter
d.b.	dry basis
DeNOx	NO_x removal
DEPI	Deutsches Pelletinstitut (German pellet institute)
DEPV	Deutscher Energieholz- und Pellet-Verband e.V. (German pellet association)
DH	district heat
DIN	Deutsches Institut für Normung (German Institute for Standardization)
DKK	Danish Kroner

DT	deformation temperature
DU	mechanical durability
e.g.	exempli gratia (for example)
ed.	editor
EMC	equilibrium moisture concentration
ESCO	energy service company
etc.	et cetera
excl.	Excluding
F	Fines
FAAS	flame atomic absorption spectrometry
FAO	Food and Agriculture Organization of the United Nations
FE	final energy
FGC	flue gas condensation
FOB	free on board loading port
FSC	Forest Stewardship Council
FT	flow temperature
FWC	forest wood chips
GCV	gross calorific value
GDP	gross domestic product
GHG	greenhouse gas
GmbH	Gesellschaft mit beschränkter Haftung (limited liability company)
HDPE	high density polyethylene
HHV	higher heating value
HPLC	high pressure liquid chromatography
HS	harmonised system
HT	hemisphere temperature
IARC	International Agency for Research on Cancer
ICP-MS	Inductively coupled plasma mass spectrometry
IEA	International Energy Agency
IFH	Institute for Aurally Handicapped Persons
IFO	intermediate fuel oil
IMO	International Maritime Organization
incl.	including
ISO	International Organization for Standardization
IT	initial temperature
ITEBE	Institut des bioénergies (International Association of Bioenergy Professionals)

L	length
lcm	loose cubic metre
LEL	lower explosive limit
LHV	lower heating value
max	maximum
MCA	multi criteria analysis
MDO	marine diesel oil
MEC	minimum explosible concentration
MHB	material hazardous in bulk
MIE	minimum ignition temperature
min	minimum
MSDS	material safety data sheet
NAICS	North American Industry Classification System
NCV	net calorific value
NIOSH	National Institute for Occupational Safety and Health (USA)
NMVOC	non-methane volatile organic compounds
no.	number
OGC	organic gaseous carbon
ÖKL	Österreichisches Kuratorium für Landtechnik und Landentwicklung
ORC	Organic Rankine Cycle
OSHA	Occupational Safety and Health Administration (USA)
PAH	polycyclic aromatic hydrocarbon
PC	pulverised coal
PCDD/F	polychlorinated dibenzodioxins and furans
PCJ	Pellet Club Japan
PEFC	Programme for the Endorsement of Forest Certification Schemes
PEL	permissible exposure level
PM10	particulate matter (< 10 μm)
PPE	personal protective equipment
PV	photovoltaics
PVA	Pelletsverband Austria (former Austrian pellets association)
PVC	percent volume contours
R&D	research and development
REL	recommended exposure limit
RH	relative humidity
SCBA	self-contained breathing apparatus
SHGC	specific heat generation costs

SIC	Standard Industrial Classification
SITC	Standard International Trade Classification
SNCR	selective non-catalytic reduction
SP	Technical Research Institute of Sweden
SRC	short rotation coppice
SST	shrinkage starting temperature
STEL	short term exposure limit
TLV	threshold limit value
TOC	total organic carbon
TRVB	Technische Richtlinien vorbeugender Brandschutz (Austrian guideline concerning fire protection requirements)
TSP	total suspended particulate matter
TWA	time weighted average
UE	useful energy
USC	ultra super critical
VAT	value added tax
w.b.	wet basis
$(w.b.)_p$	wet basis pellets (related to a moisture content of 10 wt.% (w.b.), unless otherwise stated)
WC	wood chips
WCO	World Customs Organization
WG	working group
WHO	World Health Organization
WP	wood pellets
WPAC	Wood Pellet Association of Canada
WTO	World Trade Organization

Country codes

AL	Albania
AT	Austria
BA	Bosnia and Herzegovina
BE	Belgium
BG	Bulgaria
BY	Belarus
CAN	Canada
CH	Switzerland

CS	Serbia and Montenegro
CZ	Czech Republic
DE	Germany
DK	Denmark
EE	Estonia
ES	Spain
EU	European Union
FI	Finland
FR	France
GR	Greece
HR	Croatia
HU	Hungary
IE	Ireland
IT	Italy
LT	Lithuania
LU	Luxembourg
LV	Latvia
MD	Moldova
MK	Macedonia
MT	Malta
NL	Netherlands
NO	Norway
PL	Poland
PT	Portugal
RO	Romania
RU	Russian Federation
SE	Sweden
SI	Slovenia
SK	Slovakia
TR	Turkey
UA	Ukraine
UK	United Kingdom
USA	United States of America
WB	Western Balkan

Formula symbols

AB	abrasion
c	concentration
C_F	fuel costs
$C_{F,a}$	annual fuel costs
d	transport distance
i	interest rate
m	mass
M	moisture content in wt.% (w.b.)
m_{in}	initial weight
m_{out}	output weight
$M_{S/AC+Cl}$	molar ratio of sulphur to available alkali compounds and chlorides
n	utilisation period
$O_{2,ref.}$	reference oxygen content
P_N	nominal boiler capacity
r^2	coefficient of correlation
t_f	full load operating hours
V	volume
v	void
$\bar{v}$	average speed
$V_{RG,spez}$	specific volume of the flue gas
X_i	content of component i
ΔGCV	relative difference of the gross calorific value
η_a	annual efficiency
λ	lambda (excess air ratio)
ρ_e	energy density
ρ_p	particle density
ρ_b	bulk density

Chemical symbols and formulas

As	arsenic
C	carbon
Ca	calcium
Cd	cadmium
Cl	chlorine
CO	carbon monoxide

CO_2	carbon dioxide
Cr	chromium
Cu	copper
C_xH_y	hydrocarbons
H	hydrogen
H_2O	water, steam
H_3BO_4	boric acid
HF	hydrofluoric acid
Hg	mercury
HNO_3	nitric acid
K	potassium
Mg	magnesium
N	nitrogen (elemental)
N_2	nitrogen (molecular)
Na	sodium
NaOH	sodium hydroxide
NO_x	nitrogen oxides
O	oxygen (elemental)
O_2	oxygen (molecular)
Pb	lead
S	sulphur
SiC	silicon carbide
SO_2	sulphur dioxide
Zn	zinc

Units

°C	degree Celsius
a	annum (year)
Bq	Becquerel
dB(A)	decibel (A-weighted)
g	gramme
h	hour
J	joule
K	kelvin
lcm	loose cubic metre
m	metre
m^3	cubic metre

min	minute
Nm³	normal cubic metre
p.a.	per annum (annual)
Pa	Pascal
ppmv	parts per million volume
rpm	revolutions per minute
s	second
scm	solid cubic metre
t	tonne (metric ton, 1,000 kg)
vol.%	volume percent
Wh	watt hour (3,600 J)
W	watt (J/s)
wt.%	weight percent

Prefixes

µ	micro (10^{-6})
m	milli (10^{-3})
c	centi (10^{-2})
d	deci (10^{-1})
k	kilo (10^{3})
M	mega (10^{6})
G	giga (10^{9})
T	tera (10^{12})
P	peta (10^{15})
E	exa (10^{18})

Indices

ar	as received
calc.	calculated
dr	dry
el	electric
ev.w.	evaporated water
meas.	measured
i	component
n	number
th	thermal

1 Introduction

The drawbacks of biomass as a fuel alternative to coal, oil or gas are attributed mainly to its low energy density, high moisture content and heterogeneity. The problems of conventional biofuels can be lessened or even prevented altogether by the use of pellets with consistent quality – low moisture content, high energy density and homogeneous size and shape. Consistent fuel quality makes pellets a suitable fuel type for all areas of application, from stoves and central heating systems up to large-scale plants, and with practically complete automation in all these capacity ranges. It was not until such a homogenous biofuel with regard to shape and size was introduced to the market that the development of fully automatic biomass furnaces for small-scale applications with a similar user comfort as modern oil or gas heating systems was possible. Wood pellet combustion technologies ranging from pellet stoves up to large-scale plants are explained in detail and case studies of all possible applications for energy generation are presented. The standardisation of pellets has made a major contribution to their success. International environmental obligations lead the large district heating and power producing companies to convert their large coal burning plants to the use of solid biofuels including hog fuel, chips, bark, agro-material, briquettes and pellets. In many cases these plants were originally set up to burn pulverised fuel and the only fuel suitable for the large infrastructure already in place has been pelletised material that can be ground to powder before being injected into the furnaces in the same way as coal.

The focus of this book lies on the production and energetic utilisation of wood pellets, starting from the raw materials via the production process, characteristics and combustion technology up to ecological and economical considerations. Whenever it is relevant, pellets made of herbaceous biomass (e.g. straw pellets) are taken into account and explicitly noted. If it is just "pellets" that are discussed, wood pellets for energetic utilisation are meant. There is a strong demand for knowledge with regard to optimised or innovative production technologies and improved logistics, distribution systems and combustion technologies. This book addresses all the players of the pellet market, ranging from raw material producers or suppliers, pellet producers and traders, manufacturers of pellet furnaces and pelletisation systems, installers, engineering companies, energy consultants up to the end users. It should contribute to a further increase of pellet utilisation within the energy sector by the appropriate distribution of information.

The production of the pellets is labour intensive and commercially challenging. Still, recent years have seen advancement in machinery and processes used for the transformation of low grade biomass to high grade solid biofuels. Pellets can be packaged and transported in a variety of forms such as consumer bags, jumbo or big bags, containers, railcars, wrapped pallets, flat bed trucks, tanker or silo trucks and ocean bulk carriers. They can also be stored in readily available flat storages as well as vertical silos. Year-round pellet supply can therefore be realised by reasonable and efficient fuel logistics.

Consistent fuel quality allows for easier transport and greater transportation distances. Increasingly large volumes of pellets are moving across the globe in bulk, which, in general, demands a comprehensive set of testing standards. In the course of this book, physio-chemical characteristics of both raw materials and pellets are evaluated in order to estimate the suitability of certain raw materials for pelletisation on the one hand (on the basis of the relevant standards and on requirements posed by the pelletisation technique) and the combustion behaviour on the other hand.

Safety and health concerns in pellet production, handling and storage are given major priority. All biomass, like any biological material, is subject to decomposition as a result of microbial activities in combination with chemical oxidation resulting in self-heating. During the decomposition, the biomass emits non-condensable gases such as CO, CO_2 and CH_4 as well as condensable gases such as hydrocarbons. These off-gassing phenomena are a cause of concern during transportation, handling and storage. Also, most solid biofuels are relatively brittle and disintegrate when exposed to attrition and impact resulting in fines. Part of the fines is very small in size and easily becomes airborne as dust. This dust is highly explosive, with wood dust being the material that causes more fires and explosions in the industry than any other material.

Regarding the acceptance of pellet heating systems, the high investment costs as compared to oil or gas heating systems were and still appear to be unfavourable. However, full cost calculation is necessary for a valid economic evaluation, which was carried out within the framework of this work for seven different residential heating systems. Possible investment subsidies that can be attained for heating systems are of relevance and hence are taken into account in the full cost calculation. The results show that pellets can be favourably compared to oil, however, gas heating systems are more economic under present framework conditions. If external costs caused by environmental impacts such as health damage, damage to flora and fauna and damage to buildings as well as climate and safety risks are taken into consideration, pellets have clear benefits.

Market conditions in different countries can vary widely. In Sweden for instance, pellets are used in medium- and large-scale systems for the most part, whereas in Austria pellets are chiefly used in small-scale furnaces. Comparisons of international framework conditions for the use of pellets are carried out and presented.

As pellets have become an internationally and intercontinentally traded good, international trade in addition to the socio-economic impacts of pellet production and utilisation need to be addressed. The strong growth of the pellet market requires an examination of the available raw material potential and consideration of alternative raw materials. In some cases, it may be more reasonable to produce high quality wood chips with defined quality criteria instead of cost and energy intensive pelletisation. Dry wood shavings have been and still are the preferred choice for pelletisation; wet sawdust from sawmills is also suitable for pelletisation and is available in great quantities, though it requires an increased number of processing steps. Other possible raw materials for pelletisation include wood chips from diverse wood processing steps, short rotation crops, log wood or bark. However, the use of these raw materials requires even more processing steps with regard to pre-processing compared to the use of wet sawdust (chipping, coarse grinding, separation of foreign matter, bark separation). For pellets containing bark and raw materials containing bark, special consideration should be given to their utilisation in small-scale furnaces due to higher ash content. The use of herbaceous biomass as raw material for pelletisation should also be carefully evaluated because of the higher ash content and high nitrogen, sulphur, chlorine and potassium contents. These elements can cause problems regarding corrosion, deposit formation and emissions.

Whether and to what extent national framework conditions concerning energy policy have an effect on pellet utilisation is analysed in detail. Sweden is mentioned as one example where high CO_2 and energy taxes boosted the consumption of biofuels. In Austria, the changeover to modern biomass heating systems as well as their installation in new buildings is supported by investment subsidies. How big an influence these subsidies have on the increasing number of pellet heating systems requires detailed investigation, especially considering the fact that the investment costs of pellet heating systems are still much above those of comparable oil heating systems.

Despite rapid pellet market development, there is still a lack of information in the public domain. In countries such as Austria, Germany or Sweden, where the use of pellets is already quite common, this is not pronounced. In countries where the pellet market is at the beginning of its development, such as Canada or the UK, there still is considerable lack of information. A lot of potential pellet users may not know about the existence and potential benefits pellets can offer. The supply of information by appropriate marketing campaigns is regarded as one of the major tasks to fulfil in the establishment of new markets and, as a result, help new markets to benefit from the experience of "old" markets. International exchange of knowhow is thus especially important and should be supported.

Ecological arguments against the use of pellets are often brought forward from the oil sector. Energy consumption for pellet production and higher emissions of some air pollutants are often stated as arguments against pellets. Especially the much argued and controversially fought discussions about fine particulate emissions have often been linked to wood furnaces since they cause more particulate emissions than oil or gas heating systems. Detailed investigation into a number of promising approaches sheds some light onto the issue. So do the current R&D trends that are concerned with the reduction of fine particulate emissions of biomass furnaces.

In general, current R&D trends are presented with regard to pellet production, further standardisation activities, and the production of pellets from herbaceous biomass and woody short rotation crops, the improvement of pellet quality, the development and promotion of small-scale pelletisation systems, and the phenomenon of self-heating and self-ignition in raw material and pellet storages. In the area of pellet utilisation there are activities concerning the further reduction of fine particulate emissions and the utilisation of pellets made of herbaceous biomass. Innovative pellet furnaces are being developed such as micro-scale furnaces with very low heating capacities for use in low energy houses, optimised combinations of the solar and pellet heating systems, flue gas condensation, multi-fuel boilers as well as micro-CHP (combined heat and power) systems. For the development of new pellet furnace technologies, computational fluid dynamics (CFD) simulation is employed to an increasing degree, whereby development risks and the number of required prototypes and test runs can clearly be reduced.

2 Definitions and standards

Pellets are a solid biofuel with consistent quality – low moisture content, high energy density and homogenous size and shape. In particular the residential pellet markets demand high quality pellets, as pellets are predominantly used in small-scale furnaces in this sector. This fact is accommodated in many countries worldwide by the existence of different standards for pellets. The different national standards and quality regulations attempt to control pellet quality in ways that, in part, differ greatly from one another. Above the national standards, work on European standards for solid biofuels has been done in recent years, which will lead to the publication of a series of European standards from 2010 onwards and consequently to a harmonisation and better comparability of pellets on an international basis. Above all, work on ISO (International Organization for Standardization) standards for solid biofuels has been in progress since 2007 and will lead to international standards within a few years. The ISO standards will finally replace all European EN standards.

The standards demonstrated in the following sections aim at the utilisation of pellets in small-scale furnaces below 100 kW_{th}, in which high standards are required to safeguard fully automated and trouble free operation. Apart from that, there are other quality classes in some countries, so-called industrial pellets, which are intended to be used in furnaces of larger than 100 kW_{th} and have lower quality requirements. With the new European standard for pellets, two additional quality classes besides the top quality class A1 are now also standardised, i.e. classes A2 and B. Class A2 might also become a relevant standard for pellets to be used in the residential heating sector as soon as pellet heating systems adapted to this class are available on the market (adaptation will be necessary due to the higher ash content). Pellets according to class B represent industrial pellets to be used in applications above 100 kW_{th}. Class B means that a standardised quality exists for the first time for this type of pellets. What makes them different from the higher quality pellets is that larger diameters, higher ash, nitrogen, sulphur and chlorine contents and lower NCVs are allowed. Industrial pellets are adapted to the requirements of large-scale users and are relatively low-cost (as compared to the high quality pellets for small-scale users). It must be explicitly stated at this point that industrial pellets should not be used in small-scale furnaces as this could lead to serious malfunctions of the systems.

Besides product standards for pellets and related analysis and quality assurance standards, there are also standards and certification systems for pellet transport and storage in the residential heating sector, both on national and European levels. Furthermore, not only is the fuel quality assured by respective standards, but furnaces using pellets as a fuel are also standardised and probably regulations for small-scale heating systems based on the European ecodesign directive will already be in force from 2011 onwards.

All these issues are discussed and described in the following sections.

2.1 Definitions

In this section, the terminology for pellets that is used in this book is described. It is partly derived from the respective standards.

2.1.1 General definitions

The term pellet stands for "a small round mass of a substance" [1]. A pellet is thus normally a small round mass, mostly made of compressed material, of a spherical or cylindrical shape. Usually, the word is used in the plural as pellets are normally not used singularly but as a bulk of material.

Various products and materials can be pelletised to be used thermally or still as a material as shown by the following list:

- Pellets made of iron ore are preliminary products in iron production.
- Animal feed pellets are produced for easier handling of animal food (e.g. fish feed pellets, horse feed pellets, etc.).
- Hops pellets are used in beer production.
- Uranium pellets are pellets made of enriched uranium that are used for producing fuel elements.
- Catalyst pellets are used as a carrier of the actual catalyst in heterogeneous catalytic chemical reactions.
- Pellets are used as a preliminary product in the pharmaceutics industry and are then pressed into tablets or filled into capsules.
- Polystyrene pellets are used as stuffing material for dolls, toy animals, beanbags, orthopaedic cushions, nursing cushions etc., as well as for packaging.
- Insulation pellets made of waste paper or old banknotes for instance are used for thermal wall insulation.
- Pellets are also used in anaerobic digestion. In this sense, pellets are 2 to 3 mm granular aggregations of anaerobic bacteria.
- Pellets made of sawdust, wood shavings, straw, hay or hemp are also used as animal bedding in stalls, cages and the like.
- Pellets for energetic utilisation can be made of wood, peat, herbaceous biomass or waste.

The list makes no claim to be complete and it is probable that pellets of different kinds and applications are used in many other areas too. This book is exclusively concerned with the last mentioned pellet form, namely pellets for energetic utilisation. The main focus is put on wood pellets. In some cases it is necessary to distinguish between white pellets, brown pellets and black pellets. White pellets are made of sawdust or planer shavings without bark, brown pellets are made of bark containing raw materials (not to be confused with bark pellets, which are produced solely from bark). Black pellets are produced from exploded wood pulp or torrefied wood. This is explicitly stated at each instance. Whenever relevant, pellets made of herbaceous biomass (e.g. straw pellets) are looked at too, whereby this is also explicitly stated at each instance. Pellets for energetic utilisation that are made of peat or waste are not considered here. All forthcoming definitions are to be regarded from this background.

The term *densified biomass fuels* denotes both pellets and briquettes made of solid biomass fuels in this book.

According to [59], *pelletisation* is the production of uniform bodies from powdery, granulous or coarse material of partly dissimilar particle size. The output is called a *pellet*.

Compressed wood is a fuel made by densification of wood particles [2]. Depending on the dimensions, it is classified as *wood pellets* or *wood briquettes*. *Compressed bark* is a fuel made by densification of bark particles. There are *bark pellets* and *bark briquettes*, depending on the dimension.

Biological additives are chemically unmodified products from primary forestry and agricultural biomass, like for instance shredded maize, maize starch or rye flour. They may be used to reduce energy consumption of the production process and to increase mechanical durability of the pellets. This definition for biological additives according to [2] is used throughout the book, unless otherwise stated.

The *moisture content M*, which is used in this book, is always based on the wet material (w.b.) and is defined by Equation 2.1.

Equation 2.1: $$M[wt.\%(w.b.)] = \frac{m_{H_2O}}{m_{H_2O} + m_{ds}} \cdot 100$$

Interim storages denote all storage spaces that are used between pellet production and end user storage [3]. *Storage containers* are storage facilities that are closed on all sides and can be set up independent of structural conditions, e.g. containers made of metal, plastics, wood or synthetic fibre [4].

2.1.2 CEN solid biofuels terminology

The European Committee for Standardization (CEN) under committee TC335 has published a number of standards and pre-standards for solid biofuels, which have been partly upgraded to full European standards (EN, cf. Table 2.1) already. When EN standards are in force the national standards have to be withdrawn or adapted to these EN-standards within a period of six months. One of the standards is prEN 14588 – "Solid biofuels – Terminology, definitions and descriptions".

This European standard contains all the terminology used in standardisation work within the scope of CEN/TC 335. According to CEN/TC 335 it applies to solid biofuels from the following sources:

- Products from agriculture and forestry.
- Vegetable waste from agriculture and forestry.
- Vegetable waste from the food processing industry.
- Waste wood, with the exception of waste wood that may contain halogenated organic compounds or heavy metals as a result of treatment with wood preservatives or coating, and this includes in particular waste wood from construction and demolition waste.
- Cork waste.
- Fibrous vegetable waste from virgin pulp production and from production of paper from pulp, if it is co-incinerated at the place of production and heat generated is recovered.

Table 2.1: EN standards for solid biofuels published and under preparation under committee TC335 from CEN

Explanations: "pr" in the front of the standard means that the standard has not yet been published (status as per February 2010); data source [7]

Number	Title	
prEN 14588	Solid biofuels -	Terminology, definitions and descriptions
EN 14774-1	Solid biofuels -	Determination of moisture content - Oven dry method - Part 1: Total moisture - Reference method
EN 14774-2	Solid biofuels -	Determination of moisture content - Oven dry method - Part 2: Total moisture - Simplified method
EN 14774-3	Solid biofuels -	Determination of moisture content - Oven dry method - Part 3: Moisture in general analysis sample
EN 14775	Solid biofuels -	Determination of ash content
prEN 14778	Solid biofuels -	Sampling
prEN 14780	Solid biofuels -	Sample preparation
EN 14918	Solid Biofuels -	Determination of calorific value
prEN 14961	Solid biofuels -	Fuel specifications and classes, multipart standard Part 1 – General requirements (published – EN 14961-1) Part 2 – Wood pellets for non-industrial use Part 3 – Wood briquettes for non-industrial use Part 4 – Wood chips for non-industrial use Part 5 – Firewood for non-industrial use Part 6 – Non-woody pellets for non-industrial use
EN 15103	Solid biofuels -	Determination of bulk density
prEN 15104	Solid biofuels -	Determination of carbon, hydrogen and nitrogen - Instrumental method
prEN 15105	Solid biofuels -	Determination of the water soluble content of chloride, sodium and potassium
EN 15148	Solid biofuels -	Determination of the content of volatile matter
EN 15149-1	Solid biofuels -	Determination of particle size distribution - Part 1: Oscillating screen method using sieve apertures of 1 mm and above
prEN 15149-2	Solid biofuels -	Determination of particle size distribution - Part 2: Horizontal screen method using sieve apertures of 3.15 mm and below
prEN 15150	Solid biofuels -	Determination of particle density
EN 15210-1	Solid biofuels -	Determination of mechanical durability of pellets and briquettes - Part 1: Pellets
prEN 15210-2	Solid biofuels -	Determination of mechanical durability of pellets and briquettes - Part 2: Briquettes
prEN 15234	Solid biofuels -	Fuel quality assurance, multipart standard Part 1 – General requirements Part 2 – Wood pellets for non-industrial use Part 3 – Wood briquettes for non-industrial use Part 4 – Wood chips for non-industrial use Part 5 – Firewood for non-industrial use Part 6 – Non-woody pellets for non-industrial use
prEN 15289	Solid Biofuels -	Determination of total content of sulphur and chlorine
prEN 15290	Solid Biofuels -	Determination of major elements
prEN 15296	Solid Biofuels -	Conversion of analytical results from one basis to another
prEN 15297	Solid Biofuels -	Determination of minor elements
prEN 15370	Solid biofuels -	Determination of ash melting behaviour

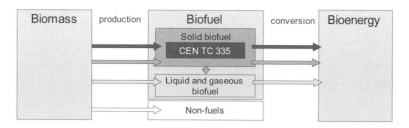

Figure 2.1: CEN TC 335 within the biomass-biofuel-bioenergy field

<u>Explanations</u>: according to EN 14961-1

The CEN/TC 335 takes into account that waste wood, including waste wood originating from construction and demolition, is included in the scope of CEN/TC 335 and in the scope of the mandate M/298 "solid biofuels" unless they contain halogenated organic compounds or heavy metals as a result of treatment with wood preservatives or coatings. For the avoidance of doubt, demolition wood is not covered within the scope of this European standard.

This section includes CEN solid biofuel terminology related to pellet raw material (woody biomass, herbaceous and fruit biomass), pellet production, fuel specification and classes as well as quality control (analysis of main properties). Consequently, this section is solely concerned with illuminating the terminology of the new European standard. The terminology used in the book is defined in Section 2.1.1 and it must be noted that it is slightly different from the one in the EN. Whenever this is the case, it is explicitly stated in the following list. Terms are listed in alphabetical order.

Additives are materials that improve the quality of the fuel (e.g. combustion properties), reduce emissions or make production more efficient. This definition is more general than the one for *biological additives* used in this book (cf. Section 2.1.1).

As received or *as received basis* is the calculation basis for material at delivery.

Ash is the residue obtained by combustion of a *fuel* (see also *total ash* and *ash fusibility*). Depending on the combustion efficiency, the ash may contain combustibles. This definition was adapted from ISO 1213-2:1992.

The *ash shrinkage starting temperature* (*SST*) is the temperature at which the first signs of shrinking of the test piece occur. The *ash deformation temperature* (*DT*) is the temperature at which first signs of rounding, due to melting, of the tip or edges of the test piece occur. The *ash hemisphere temperature* (*HT*) is the temperature at which the test piece forms approximately a hemisphere, i.e. when the height becomes equal to half the base diameter. The *ash flow temperature* (*FT*) is the temperature at which the ash is spread out over the supporting tile in a layer, the height of which is half the height of the test piece at the hemisphere temperature.

The *ash fusibility* or *ash melting behaviour* is a characteristic physical state of the *ash* obtained by heating under specific conditions. *Ash fusibility* is either determined under oxidising or reducing conditions (see also *ash shrinkage starting temperature*, *ash deformation temperature*, *ash hemisphere temperature* and *ash flow temperature*). This definition was adapted from ISO 540:1995.

Bioenergy is energy from *biomass*.

A *biofuel* is a fuel produced directly or indirectly from *biomass*.

A *biofuel blend* is a *biofuel* resulting from intentionally mixing different *biofuels*.

A *biofuel mixture* is a *biofuel* resulting from natural or unintentional mixing of different *biofuels* and/or different types of *biomass*.

Biofuel pellet is a *densified biofuel* made of *pulverised biomass* with or without *additives* usually with a cylindrical form, random length of typically 5 to 40 mm and broken ends. The raw material for biofuel pellets can be *woody biomass, herbaceous biomass, fruit biomass*, or *biomass blends* and *mixtures*. *Biofuel pellets* are usually produced in a die. The *total moisture* of *biofuel pellets* is usually less than 10 wt.% (w.b.). In this book, the term *pellets* is used synonymously (cf. Section 2.1.1).

Biomass is defined from a scientific and technical point of view as material of biological origin excluding material embedded in geological formations and/or transformed to fossil. *Biomass* is defined in legal documents in many different ways according to the scope and goal of the respective documents (e.g. directive 2001/77/EC of the European Parliament and the Council; Commission Decision (2007/589/EC) of 18 July 2007). The definition does not contradict legal definitions though (see also *herbaceous biomass, fruit biomass* and *woody biomass*).

The *bulk density* is the mass of a portion of a solid *fuel* divided by the *volume* of the container that is filled by that portion under specified conditions. This definition was adapted from ISO 1213-2:1992.

The *calorific value* or *heating value (q)* is the energy amount per unit of mass or volume released from complete combustion.

Chemical treatment is any treatment with chemicals other than air, water or heat (e.g. glue and paint).

Demolition wood is *used wood* arising from demolition of buildings or civil engineering installations. This definition was adapted from prEN 13965-1:2000.

A *densified biofuel* or *compressed biofuel* is a *solid biofuel* made by mechanically compressing *biomass* to increase its *density* and to bring the *solid biofuel* in a specific size and shape, e.g. cubes, pressed logs, *biofuel pellets* or *biofuel briquettes* (see also *biofuel briquette* and *biofuel pellet*).

Dry or *dry basis* is the calculation basis where the *solid fuel* is free from *moisture*. This definition was adapted from ISO 1213-2:1992.

The *energy density* is the ratio of net energy content to *bulk volume*. The energy density is calculated using the *net calorific value* and the *bulk density*.

Fines are defined as the aggregate of all material smaller than 3.15 mm.

Forest and plantation wood is woody biomass from forests and/or tree plantations, energy forest trees, energy plantation trees, logging residues, thinning residues, tree section and whole tree.

Fruit biomass is biomass from the parts of a plant that contain seeds (e.g. nuts, olives).

Fuel classification is the physical separation of *fuels* into defined fuel fractions. The aim of classification can be to describe the fuel, gain a particle distribution and/or to analyse single fractions separately.

A *product declaration* is a document dated and signed by the *producer/supplier* and handed out to the *retailer* or *end user*, specifying origin and source, traded form and properties of a defined *lot*.

A *fuel specification* is a document stating the requirements on the *fuel*.

The *gross calorific value* (q_{gr}) is a measured value of the specific energy of combustion for a mass unit of a *fuel* burned in oxygen in a bomb calorimeter under specified conditions. The result of combustion is assumed to consist of gaseous oxygen, nitrogen, carbon dioxide and sulphur dioxide, of liquid water (in equilibrium with its vapour) saturated with carbon dioxide under conditions of the combustion bomb reaction and of solid ash, all at the reference temperature and at constant volume. The old term is "higher heating value" (HHV). This definition was adapted from ISO1928:1995. The abbreviation alternatively used in this handbook is GCV.

Herbaceous biomass is *biomass* from plants that have a non-woody stem and that die back at the end of the growing season (see also *energy grass*). This definition was adapted from BioTech's Life Science Dictionary [5].

Impurities are materials other than the fuel itself. Examples of impurities in biofuels are stones, soil, pieces of metal, plastics, rope, ice and snow.

A *lot* is a defined quantity of *fuel* for which the quality is to be determined (see also *sub-lot*). This definition was adapted from ISO 13909:2002.

Major elements are the elements in the *fuel* that predominantly will constitute the *ash*. These include aluminium (Al), calcium (Ca), iron (Fe), magnesium (Mg), phosphorus (P), potassium (K), silicon (Si) sodium (Na) and titanium (Ti).

The *mechanical durability* is the ability of *densified fuel* units (e.g. briquettes, pellets) to remain intact, e.g. resist abrasion and shocks during handling and transport.

Minor elements are the elements in the *fuel* that are present in small concentrations only. The term trace elements is often used synonymously to minor elements. If the elements are metal, the term trace metals is also used. Concerning solid biofuels, minor elements in general include the metals arsenic (As), cadmium (Cd), cobalt (Co), chromium (Cr), copper (Cu), mercury (Hg), manganese (Mn), nickel (Ni), lead (Pb), antimony (Sb), tin (Sn), vanadium (V) and zinc (Zn).

Moisture is the water in a *fuel*.

The *net calorific value* (q_{net}) is a calculated value of the energy of combustion for a mass unit of a *fuel* burned in oxygen in a bomb calorimeter under such conditions that all the water of the reaction products remains water vapour at 0.1 MPa. The net calorific value can be determined at constant pressure or at constant volume. The net calorific value at constant pressure is requested according to EN 14961-1. The old term is "lower heating value" (LHV). The net calorific value as received ($q_{net,ar}$) is calculated by the net calorific value of dry matter ($q_{net,d}$) and the *total moisture* as received. This definition was adapted from ISO 1928:1995. The abbreviation often used alternatively for the net calorific value and in this handbook as well is NCV, its calculation can be carried out according to Equation 3.4 and Equation 3.5 in Section 3.2.3, where a comparison of results derived from these equations is also shown.

The *particle size* is the size of the *fuel* particles as determined. Different methods of determination may result in different particle sizes.

The *particle size distribution* describes the proportions of various *particle sizes* in a solid *fuel*. This definition was adapted from ISO 1213-2:1992.

A *pressing aid* is an *additive* used for enhancing the production of *densified fuels*.

The *point of delivery* is the location specified in the delivery agreement at which the proprietary rights and responsibilities concerning a *fuel lot* are transferred from one organisation or unit to another.

The *quality* is the degree to which a set of inherent characteristics fulfils requirements. This definition was adapted from ISO 9000:2005.

Quality assurance is a part of quality management, focused on providing confidence that the quality requirements will be fulfilled. This definition was adapted from ISO 9000:2005.

Quality control is a part of quality management, focused on fulfilling the quality requirements. This definition is adapted from ISO 9000:2005.

A *sample* is a quantity of material, representative of a larger quantity for which the quality is to be determined. This definition was adapted from ISO 13909:2002.

Sample preparation includes actions to obtain representative *laboratory samples* or *test portions* from the original *sample*.

Sampling is a process of drawing or constituting a *sample* (according to ISO 3534-1:1993).

Sawdust are fine particles created when sawing wood. Most of the material has a typical particle length of 1 to 5 mm.

A *solid biofuel* is a solid *fuel* produced directly or indirectly from *biomass*.

Stemwood is the part of the tree stem with the branches removed.

A *supplier* is an organisation or person that provides a product. This definition was adapted from ISO 9000:2005. A supplier may deliver to the end user directly or take responsibility for fuel deliveries from several producers as well as delivery to the end user.

The *supply chain* includes the overall process of handling and processing raw materials up to the point of delivery to the end user (cf. Figure 2.1).

The *total ash* or *ash content* is the mass of inorganic residue remaining after combustion of a *fuel* under specified conditions, typically expressed as a percentage of the mass of *dry matter* in *fuel* (see also *extraneous ash* and *natural ash*). The old term is *ash content*.

Total carbon (C) is the carbon content of moisture free *fuel*. This definition was adapted from ISO 1213-2:1992.

Total chlorine (Cl) is the chlorine content of moisture free *fuel*.

Total hydrogen (H) is the hydrogen content of moisture free fuel (dry basis). This definition is adapted from ISO 1213-2:1992.

Total moisture M_T or *moisture content* is the *moisture* in *fuel* removable under specific conditions. The reference (*dry matter/dry basis,* or total mass/*wet basis*) must be indicated to avoid confusion. For fuel specification and classes standard "M" is used for denoting moisture content as received. The old term is *moisture content*. This definition was adapted from ISO1928:1995.

Total nitrogen (*N*) is the nitrogen content of *moisture free fuel*. This definition was adapted from ISO 1213-2:1992.

Total sulphur (*S*) is the sulphur content of moisture free *fuel*. This definition was adapted from ISO 1213-2:1992.

Wet basis is the condition in which the solid fuel contains *moisture*.

Wood fuels, *wood based fuels* or *wood-derived biofuels* are all types of *biofuels* originating directly or indirectly from *woody biomass*. This definition was adapted from FAO unified bioenergy terminology (UBET) [6] (see also *fuelwood, forest fuels* and *black liquor*).

Wood processing industry by-products and residues are *woody biomass* residues originating from the wood processing as well as the pulp and paper industry. Examples are *bark*, *cork residues*, *cross-cut ends*, *edgings*, *fibreboard residues*, *fibre sludge*, *grinding dust*, *particleboard residues*, *plywood residues*, *saw dust*, *slabs* and *wood shavings*.

Wood shavings or *cutter shavings* are shavings from *woody biomass* created when planning wood.

Woody biomass is *biomass* from trees, bushes and shrubs. This definition includes forest and plantation wood, wood processing industry by-products and residues, and used wood.

2.1.3 CEN fuel specifications and classes

Fuel specification and classes standard (EN 14961) consists of the following parts, under the general title "Solid biofuel – Fuel specification and classes":

- Part 1:   General requirements (published in January 2010)
- Part 2:   Wood pellets for non-industrial use (under development)
- Part 3:   Wood briquettes for non-industrial use (under development)
- Part 4:   Wood chips for non-industrial use (under development)
- Part 5:   Firewood for non-industrial use (under development)
- Part 6:   Non-woody pellets for non-industrial use (under development*)*

Part 1 – "General requirements of prEN 14961" – includes all solid biofuels and is targeted at all user groups. Product standards (Parts 2 to 6) are developed separately for non-industrial use. However, in order to apply the product standards the standards based on and supporting EN 14961-1 must be complied with (cf. Table 2.1).

In these product standards, non-industrial use means fuel intended to be used in smaller appliances such as households and small commercial and public sector buildings.

Table 2.4 and Table 2.5 are classification tables for pellets produced from different biomass raw materials and Table 2.3 for wood pellets for non-industrial use.

2.1.3.1 Classification of origin

The classification of solid biofuels (Part 1 of EN 14961) is based on their origin and source. The fuel production chain of fuels shall be unambiguously traceable over the whole chain. The solid biofuels are divided into four sub-categories for classification in EN 14961, i.e. woody biomass (Table 2.2 and Figure 2.2), herbaceous biomass, fruit biomass and blends and mixtures.

Table 2.2: Classification of woody biomass according to EN 14961-1

Explanations: [a]...cork waste is included in bark sub-groups

1.1 Forest, plantation and other virgin wood	1.1.1 Whole trees without roots	1.1.1.1 Broadleaf
		1.1.1.2 Coniferous
		1.1.1.3 Short rotation coppice
		1.1.1.4 Bushes
		1.1.1.5 Blends and mixtures
	1.1.2 Whole trees with roots	1.1.2.1 Broadleaf
		1.1.2.2 Coniferous
		1.1.2.3 Short rotation coppice
		1.1.2.4 Bushes
		1.1.2.5 Blends and mixtures
	1.1.3 Stemwood	1.1.3.1 Broadleaf
		1.1.3.2 Coniferous
		1.1.3.3 Blends and mixtures
	1.1.4 Logging residues	1.1.4.1 Fresh/Green, broadleaf (including leaves)
		1.1.4.2 Fresh/Green, coniferous (including needles)
		1.1.4.3 Stored, broadleaf
		1.1.4.4 Stored, coniferous
		1.1.4.5 Blends and mixtures
	1.1.5 Stumps/roots	1.1.5.1 Broadleaf
		1.1.5.2 Coniferous
		1.1.5.3 Short rotation coppice
		1.1.5.4 Bushes
		1.1.5.5 Blends and mixtures
	1.1.6 Bark (from forestry operations)[a]	
	1.1.7 Segregated wood from gardens, parks, roadside maintenance, vineyards and fruit orchards	
	1.1.8 Blends and mixtures	
1.2 By-products and residues from wood processing industry	1.2.1 Chemically untreated wood residues	1.2.1.1 Without bark, broadleaf
		1.2.1.2 Without bark, coniferous
		1.2.1.3 With bark, broadleaf
		1.2.1.4 With bark, coniferous
		1.2.1.5 Bark (from industry operations)[a]
	1.2.2 Chemically treated wood residues, fibres and wood constituents	1.2.2.1 Without bark
		1.2.2.2 With bark
		1.2.2.3 Bark (from industry operations)[a]
		1.2.2.4 Fibres and wood constituents
	1.2.3 Blends and mixtures	
1.3 Used wood	1.3.1 Chemically untreated wood	1.3.1.1 Without bark
		1.3.1.2 With bark
		1.3.1.3 Bark[a]
	1.3.2 Chemically treated wood	1.3.2.1 Without bark
		1.3.2.2 With bark
		1.3.2.3 Bark[a]
	1.3.3 Blends and mixtures	
1.4 Blends and mixtures		

If appropriate, the actual species (e.g. spruce, wheat) of biomass can also be stated. Wood species can be stated according to EN 13556 "Round and sawn timber Nomenclature" for instance.

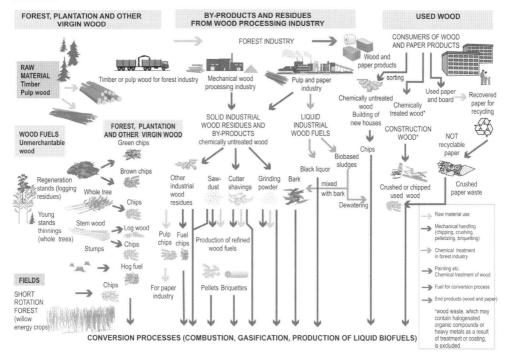

Figure 2.2: Classification of woody biomass

Explanations: data source [7]

The purpose of classification is to create the possibility to differentiate and specify raw materials based on origin with as much detail as needed. The quality classification in a table form was prepared only for major solid biofuels in trade.

2.1.3.2 Fuel specification

Properties to be specified are listed in Tables 3 to 14 of EN 14961-1 for the following forms of solid biofuels: briquettes, pellets, wood chips, hog fuel, log wood/firewood, sawdust, shavings, bark, straw bales, reed canary grass bales and miscanthus bales, energy grain, olive residues and fruit seeds. A general master table is to be used for solid biofuels not covered by Tables 3 to 14. This section concentrates on fuel specification of pellets.

In EN 14961-1 the classification of fuels is flexible, and hence the producer or the consumer may select the classification that corresponds to the produced or desired fuel quality from each property class (so-called "free classification"). An advantage of this classification is that the producer and the consumer may agree upon characteristics case-by-case. In Figure 2.3 there is an example of wood pellet classification according to Part 2 of prEN 14961.

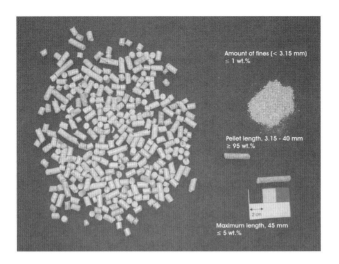

Figure 2.3: Wood pellet classification according to Part 2 of prEN 14961

<u>Explanations</u>: data source [7]

The normative characteristics shall be part of the fuel specification as per EN 14961-1. These characteristics vary for different forms of biofuels. The most significant characteristics for all solid biofuels are moisture content (M), particle size/dimensions (P or D/L) and ash content (A). For example, the average moisture content of fuels is given as a value after the symbol as in M10, which means that the average moisture content of the fuel shall be ≤ 10 wt.% (w.b.). Some characteristics, e.g. the contents of S, N and Cl, are voluntary information that can be given for all fuels that are not chemically treated (cf. Table 2.5).

In the product standards (Part 2 for wood pellets and Part 6 for non-woody pellets) all properties are normative and together they form a class, for example A1, A2 and B for wood pellets. In order to protect small-scale consumers, some heavy metals are normative for wood pellets. Property class A1 for wood pellets represents virgin woods and chemically untreated wood residues low in ash and chlorine content. Fuels with slightly higher ash content and/or chlorine content fall within grade A2. In property class B chemically treated industrial wood by-products, residues and used wood are also allowed (cf. Table 2.3). Designation symbols together with a number are used to specify property levels, e.g. M10 for a moisture content of ≤ 10 wt.% (w.b.). For designation of chemical properties chemical symbols such as S (sulphur), Cl (chlorine), N (nitrogen) are used and the value is added to the symbol, e.g. S0.20 for a sulphur content of ≤ 0.20 wt.% (d.b.).

A product declaration for the solid biofuel shall be issued by the supplier and handed to the end user or retailer (cf. Section 2.4). The fuel quality declaration shall be issued for each defined lot. The quantity of the lot shall be defined in the delivery agreement. The supplier shall date the declaration and keep the records for a minimum of one year after delivery. The fuel quality declaration shall state the quality in accordance with the appropriate part of prEN 14961. Figure 2.4 is an example of specification of wood pellets according EN 14961-1.

Table 2.3: Specification of wood pellets for non-industrial use

Explanations: based on prEN 14961-2 (final draft, March 2010); [a]...selected size of pellets to be stated; [b]...amount of pellets longer than 40 mm can be 1 wt.%, maximum length shall be 45 mm; [c]...e.g. starch, corn flour, potato flour, vegetable oil; [d]...all characteristic temperatures in oxidised conditions should be stated, i.e. SST, DT, HT and FT

Property class (analysis method)	Unit	A1	A2	B
Origin and source		1.1.3 Stemwood 1.2.1 Chemically untreated wood residues	1.1.1 Whole trees without roots 1.1.3 Stemwood 1.1.4 Logging residues 1.1.6 Bark 1.2.1 Chemically untreated wood residues	1.1 Forest, plantation and other virgin wood 1.2 By-products and residues from wood processing industry 1.3 Used wood
Diameter, D [a] and Length, L [b]	mm	D06 ± 1.0 $3.15 \leq L \leq 40$ D08 ± 1.0 $3.15 \leq L \leq 40$	D06 ± 1.0 $3.15 \leq L \leq 40$ D08 ± 1.0 $3.15 \leq L \leq 40$	D06 ± 1.0 $3.15 \leq L \leq 40$ D08 ± 1.0 $3.15 \leq L \leq 40$
Moisture, M (EN 14774-1 and -2)	wt.%$_{ar}$	M10 ≤ 10	M10 ≤ 10	M10 ≤ 10
Ash, A (EN 14775)	wt.% (d.b.)	A0.7 ≤ 0.7	A1.5 ≤ 1.5	A3.5 ≤ 3.5
Mechanical durability, DU (EN 15210-1)	wt.%$_{ar}$	DU97.5 ≥ 97.5	DU97.5 ≥ 97.5	DU96.5 ≥ 96.5
Fines at factory gate in bulk transport (at the time of loading) and in small (up to 20 kg) and large sacks (at time of packing or when delivering to end user), F (EN 15149-1)	wt.%$_{ar}$	F1.0 ≤ 1.0	F1.0 ≤ 1.0	F1.0 ≤ 1.0
Additives	wt.% (d.b.)	≤ 2 Type [c] and amount to be stated	≤ 2 Type [c] and amount to be stated	≤ 2 Type [c] and amount to be stated
Net calorific value, Q (EN 14918)	MJ/kg$_{ar}$ or kWh/kg$_{ar}$	$16.5 \leq Q \leq 19.0$ or $4.6 \leq Q \leq 5.3$	$16.3 \leq Q \leq 19.0$ or $4.5 \leq Q \leq 5.3$	$16.0 \leq Q \leq 19.0$ or $4.4 \leq Q \leq 5.3$
Bulk density, BD (EN 15103)	kg/m^3	BD600 ≥ 600	BD600 ≥ 600	BD600 ≥ 600
Nitrogen, N (prEN 15104)	wt.% (d.b.)	N0.3 ≤ 0.3	N0.5 ≤ 0.5	N1.0 ≤ 1.0
Sulphur, S (prEN 15289)	wt.% (d.b.)	S0.03 ≤ 0.03	S0.03 ≤ 0.03	S0.04 ≤ 0.04
Chlorine, Cl (prEN 15289)	wt.% (d.b.)	Cl 0.02 ≤ 0.02	Cl 0.02 ≤ 0.02	Cl 0.03 ≤ 0.03
Arsenic, As (prEN 15297)	mg/kg (d.b.)	≤ 1	≤ 1	≤ 1
Cadmium, Cd (prEN 15297)	mg/kg (d.b.)	≤ 0.5	≤ 0.5	≤ 0.5
Chromium, Cr (prEN 15297)	mg/kg (d.b.)	≤ 10	≤ 10	≤ 10
Copper, Cu (prEN 15297)	mg/kg (d.b.)	≤ 10	≤ 10	≤ 10
Lead, Pb (prEN 15297)	mg/kg (d.b.)	≤ 10	≤ 10	≤ 10
Mercury, Hg (prEN 15297)	mg/kg (d.b.)	≤ 0.1	≤ 0.1	≤ 0.1
Nickel, Ni (prEN 15297)	mg/kg (d.b.)	≤ 10	≤ 10	≤ 10
Zinc, Zn (prEN 15297)	mg/kg (d.b.)	≤ 100	≤ 100	≤ 100
Ash melting behaviour, DT [d] (prEN 15370)	°C	should be stated	should be stated	should be stated

Table 2.4: Specification of normative properties for pellets according to EN 14961-1

Explanations: [a]... amount of pellets longer than 40 (or 50 mm) can be 5 wt.%, maximum length for classes D06, D08 and D10 shall be < 45 mm; [b]...fines shall be determined by using method prEN 15149-1; [c]...the maximum amount of additive is 20 wt.% of pressing mass, type stated (e.g. starch), if amount is greater, then raw material for pellet is blend

	Master table	
	Origin: According to Table 1, 2 or 3 of EN 14961-1	Woody biomass (1), Herbaceous biomass (2), Fruit biomass (3), Blends and mixtures (4)
	Traded Form	Pellets
Normative	**Dimensions** (mm)	
	Diameter (D) and Length (L) [a]	
	D06	6 mm ± 1.0 mm and 3.15 ≤ L ≤ 40 mm
	D08	8 mm ± 1.0 mm, and 3.15 ≤ L ≤ 40 mm
	D10	10 mm ± 1.0 mm, and 3.15 ≤ L ≤ 40 mm
	D12	12 mm ± 1.0 mm, and 3.15 ≤ L ≤ 50 mm
	D25	25 mm ± 1.0 mm, and 10 ≤ L ≤ 50 mm
	Moisture, M (wt.%$_{ar}$)	
	M10	≤ 10%
	M15	≤ 15%
	Ash, A (wt.% (d.b.))	
	A0.5	≤ 0.5%
	A0.7	≤ 0.7%
	A1.0	≤ 1.0%
	A1.5	≤ 1.5%
	A2.0	≤ 2.0%
	A3.0	≤ 3.0%
	A5.0	≤ 5.0%
	A7.0	≤ 7.0%
	A10.0	≤ 10.0%
	A10.0+	> 10.0%
	Mechanical durability, DU (wt.% of pellets after testing)	
	DU97.5	≥ 97.5%
	DU96.5	≥ 96.5%
	DU95.0	≥ 95.0%
	DU95.0-	< 95.0% (minimum value to be stated)
	Amount of fines, F (wt.%, < 3.15 mm) after production when loaded or packed [b]	
	F1.0	≤ 1.0%
	F2.0	≤ 2.0%
	F3.0	≤ 3.0%
	F5.0	≤ 5.0%
	F5.0+	> 5.0% (maximum value to be stated)
	Additives (wt.% of pressing mass) [c]	
	Type and content of pressing aids, slagging inhibitors or any other additives have to be stated	
	Bulk density (BD) as received (kg/m^3)	
	BD550	≥ 550 kg/m^3
	BD600	≥ 600 kg/m^3
	BD650	≥ 650 kg/m^3
	BD700	≥ 700 kg/m^3
	BD700+	> 700 kg/m^3 (minimum value to be stated)
	Net calorific value as received, NCV (MJ/kg or kWh/kg)	
	Minimum value to be stated	

Table 2.5: Specification of normative/informative properties for pellets according to EN 14961-1

Explanations: special attention should be paid to the ash melting behaviour for some biomass fuels, for example eucalyptus, poplar, short rotation coppice, straw, miscanthus and olive stone

Normative / informative	**Sulphur, S (wt.% (d.b.))**		
	S0.02	≤ 0.02%	Normative: Chemically treated biomass (1.2.2; 1.3.2; 2.2.2; 3.2.2) or if sulphur containing additives have been used.
	S0.05	≤ 0.05%	
	S0.08	≤ 0.08%	
	S0.10	≤ 0.10%	Informative: All fuels that are not chemically treated (see the exceptions above)
	S0.20	≤ 0.20%	
	S0.20+	> 0.20% (maximum value to be stated)	
	Nitrogen, N (wt.% (d.b.))		
	N0.3	≤ 0.3%	Normative: Chemically treated biomass (1.2.2; 1.3.2; 2.2.2; 3.2.2)
	N0.5	≤ 0.5%	
	N1.0	≤ 1.0%	
	N2.0	≤ 2.0%	Informative: All fuels that are not chemically treated (see the exceptions above)
	N3.0	≤ 3.0%	
	N3.0+	> 3.0% (maximum value to be stated)	
	Chlorine, Cl (wt.% (d.b.))		
	Cl0.02	≤ 0.02%	Normative: Chemically treated biomass (1.2.2; 1.3.2; 2.2.2; 3.2.2)
	Cl0.03	≤ 0.03%	
	Cl0.07	≤ 0.07%	
	Cl0.10	≤ 0.10%	Informative: All fuels that are not chemically treated (see the exceptions above)
	Cl0.10+	> 0.10% (maximum value to be stated)	
Informative: Ash melting behaviour (°C)			Deformation temperature, DT should be stated

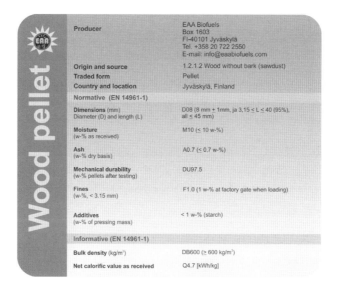

Figure 2.4: Example of fuel specification according to EN 14961-1

Explanations: data source [7]

2.1.4 International convention on the harmonized commodity description and coding system (HS convention)

The HS code is a commodity classification covering practically 98% of all commodities traded around the globe today. The World Customs Organization (WCO) maintains the international Harmonized System (HS) goods nomenclature, and administers the technical aspects of the World Trade Organization (WTO) Agreements on Customs Valuation and Rules of Origin. The current code was adopted in 1988 and is continuously being upgraded as new commodities appear on the market. The code is an important tool on which trading rules and regulations as well as vital statistics are based worldwide. The code consists of a ten digit number to cover groupings of all commodities organised in chapters, headings, subheadings and commodities and is structured as follows:

- The first two digits represent chapter
- The first four digits represent chapter and heading
- The first six digits represent chapter, heading and subheading
- The full ten digits represent chapter, heading, subheading and commodity of which the last four digits are somewhat different from one country to the next and obviously not fully harmonised at this stage. The four last digits may in some countries be different for the same commodity depending on whether the commodity is exported or imported.

The first 6 digits (4401-30-xx-xx) of the HS code are the same for wood pellets and briquettes and the last 4 digits are designated on a national level. The first two digits represent chapter 44, i.e. "Wood and articles of wood; wood charcoal". The first four digits represent chapter and heading 4401, i.e. "Fuel wood, in logs, in billets, in twigs, in faggots or in similar forms; wood in chips or particles; sawdust and wood waste and scrap, whether or not agglomerated in logs, briquettes, pellets or similar forms". The first six digits represent chapter, heading and subheading 4401-30, i.e. "Sawdust and wood waste and scrap, whether or not agglomerated in logs, briquettes, pellets or similar forms".

The reader is encouraged to look up the HS code applicable for their jurisdiction by contacting customs authorities or the department of commerce. The code is mandatory for all export and import transactions of commodities such as pellets and briquettes and should also appear on the material safety data sheet (MSDS) for the pellets (cf. Section 5.4.2).

There are some other codes such as:

- Standard Industrial Classification (SIC);
- North American Industry Classification System (NAICS), which replaces the SIC;
- Standard International Trade Classification (SITC), developed under the United Nations in 1950 and used for reports of international trade statistics.

2.1.5 International Maritime Organization (IMO) code for pellets

Shipping is perhaps the world's most international industry, serving more than 90% of global trade, and offers cost effective, clean and safe transportation. The ownership and management chain surrounding any ship can embrace many countries and ships spend their economic life moving between different jurisdictions, often far from the country of registry. The International Maritime Organization (IMO) was established in 1948 and its main task has been to develop and maintain a comprehensive regulatory framework for shipping and its remit today includes safety, environmental concerns, legal matters, technical co-operation, maritime security and the efficiency of shipping.

One of the codes that relates directly to wood pellets is the Code of Safe Practice for Solid Bulk Cargoes (the BC code). The development and inclusion of wood pellets in the BC code was requested by Canada as a result of accidents in wood pellet carrying ocean vessels in 2002. The BC code was modified in 2004 to include wood pellets as classified cargo. The present code includes description of the material characteristics that could result in hazardous conditions during ocean voyage such as oxygen depletion and off-gassing. The code also stipulates operational requirements such as entry permit, gas monitoring and fire extinguishing practices.

The BC code has been updated and was published in 2009 [8]. The modified code is already in effect on a voluntary basis and mandated starting 1 January 2011.

2.2 Pellet product standards in Europe

In recent years several national standards and quality regulations that tried to regulate the quality of densified biomass fuels have been issued, for instance in Austria, Germany, Italy or Sweden. In part, the standards and regulations differ greatly from one another.

As already mentioned, the CEN has published a number of standards and pre-standards for solid biofuels (cf. Section 2.1.2). Among others, a product standard for pellets, i.e. EN 14961-2, will be published. As soon as this product standard for pellets is in force, all the national standards have to be withdrawn or adapted to this EN standard within a period of six months. Therefore, all national standards will soon become obsolete. However, these standards were widely used in recent years and were even applied outside their countries of origin (such as ÖNORM M 7135 or DIN_{plus}). Therefore, Table 2.6 presents an overview of the forthcoming European pellets product standard in comparison to national standards for pellets in Austria, Sweden, Germany and Italy.

Pellets according to prEN 14961-2 must be made of stemwood or chemically untreated wood residues (class A1), whole trees without roots, stemwood, logging residues, bark, chemically untreated wood residues (class A2), forest, plantation and other virgin wood, by-products and residues from wood processing industry or used wood (class B). Pellets according to ÖNORM M 7135 and according to DIN_{plus} must be made solely of natural wood or bark. Pellets according to the Swedish standard SS 187120 are usually produced from logging and cutting residues, by-products from forest and timber industries, straw or paper. Pellets according to DIN 51731 are produced from wood in natural state that had only been treated mechanically. Origin and source of raw materials for pellets according to the Italian standard CTI must be classified according to EN 14961-1.

The national standards differentiate classes or groups of pellets depending on diameter and length. Table 2.6 only shows the limiting values for the smallest size classes of each standard since these are the most relevant size classes for the residential heating sector. For prEN 14961-2 all classes of pellets are shown. As concerns diameter, in markets that are dominated by residential small-scale systems (e.g. Austria, Germany or Italy) pellets of 6 mm in diameter have become more or less a convention. A lot of pellet boiler manufacturers actually call for 6 mm pellets even though the standards would allow other sizes too. Pellets used in large-scale systems (e.g. power plants) are usually of 8 mm and more in diameter. This is relevant for instance in the Netherlands or Belgium where large power plants are either fired or co-fired with pellets (cf. Chapter 10). In Sweden, pellets of 8 mm in diameter dominate the market of small-scale systems as well.

It can be seen that there are some significant differences in the standards. Concerning density for instance, sometimes the bulk density and sometimes the particle density are regulated. Other parameters such as nitrogen content, mechanical durability, fines, additives, ash melting behaviour or heavy metals are not regulated at all in some standards and the limiting values for some parameters such as the NCV are given on a different basis (e.g. dry basis versus wet basis), which makes a direct comparison of the values impossible. Here the new European standard will contribute to harmonisation and will make pellet qualities comparable on an international level.

Table 2.6: Comparison of pellet standards

Explanations: [1]...not more than 20 wt.% of pellets may have lengths of up to $7.5 \times D$; [2]...in the storage space of the producer; [3]...related to dry substance; [4]...water and ash free; [5]...solely chemically unmodified products from primary forestry and agricultural biomass; [6]...to be determined at 550°C; [7]...to be determined at 815°C; [8]...during loading according to [3]; [9]...defined as abrasion (= 100 - mechanical durability); [10]...when leaving the final point of loading for delivery to the end user, i.e. leaving the final storage point or the factory if delivering directly to the end user, the limit shall also be kept (unless there is a different agreement between the producer and their customer), even when not going directly to the end user; [11]...type and amount to be stated; [12]...all characteristic temperatures should be stated (SST, DT, HT, FT); [13]...± 1 mm; [14]...the actual diameter must be within ± 10% of the diameter stated; [15]...± 0.5 mm; [16]...for "class A without additives" no additives are allowed; [17]...total concentration of Pb, Hg, Cd and Cr ≤ 20 mg/kg as received for pellets produced from untreated raw material; [18]...to be stated; data source [2; 3; 9; 10]

Parameter	Unit	Final draft prEN 14961-2			ÖNORM M 7135	SS 187120	DIN 51731	DIN$_{plus}$	CTI
		Class A1	Class A2	Class B					
Diameter D	mm	6 or 8[13]	6 or 8[13]	6 or 8[13]	4 - 10	[18]	4 - 10	4 - 10[14]	6[15]
Length	mm	3.15 - 40	3.15 - 40	3.15 - 40	≤ 5 x D[1]	≤ 4 x D[2]	≤ 50	≤ 5xD[1]	D - 4 x D
Bulk density	kg/m³	≥ 600	≥ 600	≥ 600		≥ 600[2]			620 - 720
Particle density	kg/dm³				≥ 1.12		1 - 1.4	≥ 1.12	
Moisture content	wt.% (w.b.)	≤ 10	≤ 10	≤ 10	≤ 10	≤ 10	≤ 12	≤ 10	≤ 10
Ash content	wt.% (d.b.)	≤ 0.7[6]	≤ 1.5[6]	≤ 3.5[6]	≤ 0.5[7]	≤ 0.7[6]	≤ 1.5[7]	≤ 0.5[7]	≤ 0.7[6]
NCV	MJ/kg (w.b.)	16.5 - 19.0	16.3 - 19.0	16.0 - 19.0	≥ 18.0[3]	≥ 16.9	17.5 - 19.5[4]	≥ 18.0[3]	≥ 16.9
Sulfur content	wt.% (d.b.)	≤ 0.03	≤ 0.03	≤ 0.04	≤ 0.04	≤ 0.08	≤ 0.08	≤ 0.04	≤ 0.05
Nitrogen content	wt.% (d.b.)	≤ 0.30	≤ 0.50	≤ 1.0	≤ 0.30		≤ 0.30	≤ 0.30	≤ 0.30
Chlorine content	wt.% (d.b.)	≤ 0.02	≤ 0.02	≤ 0.03	≤ 0.02	≤ 0.03	≤ 0.03	≤ 0.02	≤ 0.03
Mechanical durability	wt.% (w.b.)	≥ 97.5	≥ 97.5	≥ 96.5	≥ 97.7[9]	≥ 99.2[2) 9)]		≥ 97.7[9]	≥ 97.5
Fines	wt.% (w.b.)	≤ 1.0[10]	≤ 1.0[10]	≤ 1.0[10]	≤ 1[8]				≤ 1.0
Additives	%	≤ 2.0[11]	≤ 2.0[11]	≤ 2.0[11]	≤ 2[5]		[11]	≤ 2.0[5]	[11) 16)]
Ash melting behaviour	°C	[12]	[12]	[12]		IT[18]			
Arsenic	mg/kg (d.b.)	≤ 1	≤ 1	≤ 1			≤ 0.8		
Cadmium	mg/kg (d.b.)	≤ 0.5	≤ 0.5	≤ 0.5			≤ 0.5		[17]
Chromium	mg/kg (d.b.)	≤ 10	≤ 10	≤ 10			≤ 8		[17]
Copper	mg/kg (d.b.)	≤ 10	≤ 10	≤ 10			≤ 5		
Mercury	mg/kg (d.b.)	≤ 0.1	≤ 0.1	≤ 0.1			≤ 0.05		[17]
Nickel	mg/kg (d.b.)	≤ 10	≤ 10	≤ 10					
Lead	mg/kg (d.b.)	≤ 10	≤ 10	≤ 10			≤ 10		[17]
Zinc	mg/kg (d.b.)	≤ 100	≤ 100	≤ 100			≤ 100		
EOX	mg/kg (d.b.)						≤ 3		

As regards limiting values for abrasion (respectively mechanical durability) in the European, Austrian and Swedish standards, it must be noted that the parameter is determined by different methods and hence the values are not comparable. The ÖNORM M 7135 stipulates the use of the Ligno-Tester LT II (cf. Figure 2.6). According to the Swedish standard SS 187120, the abrasion has to be determined as per SS 187180 [11]. This standard sets down the use of a hexagonal, rotating drum as shown in Figure 2.7 and the European standard requires the use of the ASAE tumbler (cf. Figure 2.5). A comparison of the Ligno-Tester and the ASAE tumbler showed the latter to be more reliable and reproducible, which is why this method is now part of the European standard (for details see Section 2.3).

The Italian CTI (Comitato Termotecnico Italiano) created its own pellet standards based on CEN/TS 14961:2005. Analysis methods are taken from CEN/TC 335 and pellet property tables define 4 classes (A without additives, A with additives, B and C). The Italian standards are the only standards that classify the origin according to Table 1 in CEN/TS 14961. Standardisation activities are also known from France and Japan. The International Association of Bioenergy Professionals (ITEBE) in France recently developed a quality standard that aims at safeguarding high pellet quality in France too [12; 13]. In Japan, the pellet club Japan (PCJ) seeks standardisation too [14].

Switzerland introduced its own pellet standard in 2001 [15], which was based on the German DIN 51731. However, as ÖNORM M 7135 and DIN 51731 have been accepted as appropriate standards, the Swiss standard was repealed.

In Scandinavia so-called Swan-labelling, which is the brand for the Nordic ecolabelling system, is used. This system is used for a very wide variety of products, and specific criteria covering environmental characteristics during manufacture, use and end-of-life handling are elaborated for each product type. With regard to pellets, in addition to requirements on fuel characteristics, Swan-labelling also includes requirements on manufacturing methods, transportation and storage. The aim is to define top-grade quality from an environmental perspective. This also makes the pellets easy to use and ensures that combustion does not cause adverse health or environmental effects. The requirements for pellet heating systems are described in Section 2.8. For pellets, the criteria impose requirements on raw materials, energy consumption during manufacture and fuel characteristics, as well as the manufacturer's continuous assessment of these. The requirements on fuel characteristics are very similar to those of the class A1 of the prEN 14961-2 standard. Stricter limiting values are set only for the diameter, the bulk density and the sulphur content. In addition, concerning the ash melting behaviour, both IT and HT are regulated.

The raw materials must belong to the EN 14961-1 class "Chemically untreated wood residues, wood without bark" or to class "Forest and plantation wood, stemwood". Residues from wood processing that contain adhesives or other contaminants may not be used principally. Also, chippings of municipal waste must not be used. If virgin wood is used as a raw material, at least 70% per annum of the raw material from virgin wood must come from certified forests.

The production of pellets must not consume more than 1,200 kWh of primary energy per tonne of pellets. This requirement concerns the following processes: bark separation, chipping, drying, grinding, conditioning, pelletisation, cooling and screening, as well as any intermediate stages such as electricity consumption for conveyor belts. If electricity is used, the consumption in kWh_{el} shall be multiplied by 2.5 to achieve the equivalent primary energy.

This maximum primary energy demand is quite ambitious, as the use of fresh woody raw material with a moisture content of 55 wt.% (w.b.) alone requires about 1,000 to 1,200 kWh (depending on the drying technology used) of heat for drying per tonne pellets, which is even more when related to primary energy. Including the electricity demand of the total plant multiplied by 2.5, the limit will be exceeded in any case (cf. Table 9.2). Therefore, pellet production from fresh woody biomass is not possible under Swan-labelling. The limit could probably only be kept by integrated process solutions (type biorefinery).

Fuels that are used during pellet production may produce a maximum greenhouse gas (GHG) emission of 100 kg CO_2 per tonne of pellets. This value can be achieved as long as biomass fuels are used for raw material drying. For instance, the calculated value of about 4.5 g CO_2/MJ_{NCV} for wet sawdust as a raw material (cf. Table 9.7) is equivalent to about 80 kg CO_2 per tonne of pellets.

In Belgian law, for instance, no standard related to pellet quality or utilisation has been implemented for the time being. However, a project of decree is under preparation, i.e. "King decree regulating minimal requirements and pollutant emission levels for heating devices fed by solid biofuels" and should be implemented at the end of 2010. The part of this decree dedicated to pellet quality will be based on the documents edited by CEN TC 335. The heating device part will be based on the EN 303-5, which is already included in Belgian legislation as NBN EN 303-5 [16].

Quality standards for pellets are not a very big issue in the Danish wood pellet market. In the small- and medium-scale market, pellets are to a wide extent purchased according to proprietary standards or brands such as "HP-quality", "Celsico", etc. and according to previous experiences. Such branding is popular, but proper standards such as CEN, DIN+, ÖNORM or the Nordic ecolabel are also used. However, the more the market matures the more real standards are used instead of branding. In the large-scale markets, proper standards such as CEN or DIN are commonly applied in trade and supply control of pellets. The largest market actors have company specifications for pellet quality, origin etc. that suppliers must comply with. Denmark takes part in the international standardisation process on solid biomass, and as such uses and will use the CEN documents as the main national reference for biomass fuel standards.

Concerning international distribution of standards, it has to be noted that ÖNORM M 7135 and the DIN 51731 as well as the DIN_{plus} certificate have become well established and are widely used. In their respective countries more or less all pellet producers are certified according to ÖNORM M 7135, DIN 51731 or DIN_{plus}. In addition, the mentioned standards have been established internationally and thus they generally are well known in the field of pellets [17]. The DIN_{plus} standard is especially called for internationally (e.g. by producers in many European countries, Argentina, Brazil and in Asia) [18]. This international role will, however, soon be taken over by the new European standards, in particular prEN 14961-2.

2.3 Pellet analysis standards in Europe

Apart from technical requirements on the fuel, the different standards also contain test specifications for the different parameters that characterise pellets and other biomass fuels. In this section, the methods for determining the different parameters according to EN 14961-1 are discussed. Although there are a number of standards for physical and chemical analysis of pellets in many European countries, they are not described here as they will be replaced by the forthcoming European standards. For details the reader is referred to the respective standards and to [60] where some methods according to Austrian, German and Swedish standards are described.

A European measurement standard for length and diameter is under development. In practice, the length measurement test is mainly performed manually by measuring each pellet of the test portion individually, with a calliper. The test report shall at least mention the proportion (in weight) of pellets under 40 mm and (for D06, D08 and D10 pellets) the proportion of pellets between 40 and 45 mm. Finally, a record of overlong pellets (over 45 mm long) shall be taken. Additionally, the proportion (in weight) of broken pellets or pellets having a length below their diameter is useful information.

EN 14961-1 regulates the bulk density that has to be determined according to EN 15103. A standardised container with a specified volume (e.g. 5 l for pellets) is filled with pellets and the bulk density is measured after three impacts (container falling three times from a height of 15 cm).

EN 14961-1 requires the determination of the moisture content according to EN 14774-1. A sample of at least 300 g must be used and dried at $105 \pm 2°C$.

The ash content is to be determined according to EN 14775 at 550°C. In this context it must be noted that the ash content according to ÖNORM M 7135 and DIN 51731 has to be determined at 850°C, which must be denoted too high since volatile substances such as alkaline metals already go into gaseous phase at this temperature and hence the result of the analysis is influenced. The investigations reviewed in Section 3.2.5 may be considered in this respect.

EN 14961-1 requires the determination of the net calorific value at constant pressure (cf. Section 2.1.2) from the gross calorific value according to EN 14918. Equation 3.5 has to be applied for the calculation.

European standards for the determination of sulphur, nitrogen and chlorine are prEN 15289 (S and Cl) and prEN 15104 (N).

For the determination of the mechanical durability according to EN 14961-1 the standard EN 15210-1 must be applied.

In this respect it has to be noted that for the determination of the mechanical durability, respectively the abrasion (which is the difference between 100 and the mechanical durability in percent), the Ligno-Tester (Ligno-Tester LT II of the company Borregaard Lignotech, cf. Figure 2.6) has been widely used in recent years. This was the method prescribed by the ÖNORM M 7135 and the DIN_{plus}. Moreover, another piece of equipment for the determination of the abrasion is prescribed according to the Swedish standard SS 187120 (determination as per SS 187180 [11]). This standard sets down the use of a hexagonal, rotating drum as shown in Figure 2.7.

Under the terms of EN 15210-1, the measurement is made using a pellet tester where the sample is subjected to controlled shocks by collision of pellets against each other and against the walls of a defined rotating test chamber. A scheme and a picture of such a tester are shown in Figure 2.5. The durability is calculated from the mass of the remaining sample after separation of abraded and fine broken particles. The method selected for the EN standard has been compared to the Lingo-Tester method that appeared less repeatable and reproducible [19]. Correlations between both testing equipments have been searched for, but no clear relation could be found (in particular for values keeping the threshold value of 97.5% according to prEN 14961-2, as shown in Figure 2.8). The method selected for the EN standard was found to be the more reliable method and is thus part of the European standard now. For the determination with the tester, a sample mass of 500 ± 10 g is needed that rotates in the tumbler with 50 ± 2 rpm for 10 minutes. Sieving is carried out with a 3.15 mm sieve with round holes.

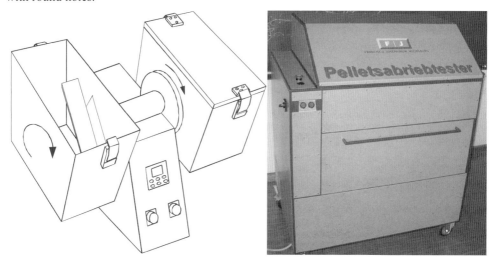

Figure 2.5: Scheme and picture of a tester for mechanical durability of pellets according to EN 15210-1

Explanations: data source: left [20], right [21]

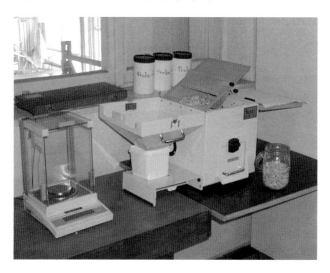

Figure 2.6: Ligno-Tester LT II

Explanations: data source [22]

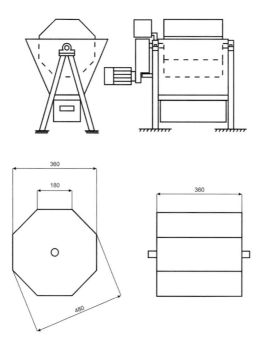

Figure 2.7: Scheme of the abrasion tester according to the Swedish standard

Explanations: values in mm; data source [11]

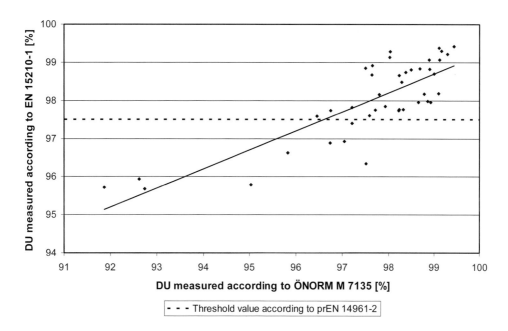

Figure 2.8: Correlation between durability determinations according to EN 15210-1 and ÖNORM M 7135

Explanations: coefficient of correlation $r^2 = 0.69$ (all values) and $r^2 = 0.27$ (values > 97.5 according to EN 15210-1); data source [19] (adapted)

The amount of fines according to EN 14961-1 shall be determined by using the hand screening operation as described in EN 15210-1.

The ash melting behaviour according to EN 14961-1 has to be determined according to prEN 15234.

The contents of As, Cd, Cr, Cu, Pb, Hg, Ni and Zn are determined according to prEN 15297.

2.4 Pellet quality assurance standards in Europe

In the past, one of the strictest quality assurance regulations was included in the Austrian ÖNORM M 7135. Conformity with this standard had to be assured by a first check (for qualification), continuous internal quality control and regular external quality control (annually without previous notification). Both first check and external quality control had to be carried out by an authorised inspection body. The results of the internal quality controls had to be documented and had be checked by the inspection body at the external reviews.

The Swedish standard SS 187120 did not comprise any such regulations whereas the German DIN 51731 standard stipulated an annual review by an authorised inspection body as a proof for standard conformity (the test mark was assigned for one year only). Unannounced external controls such as in Austria were not required though. The German certification programme DIN$_{plus}$ prescribed an external review similar to the Austrian regulation so as to ensure that the pellets distributed under this label actually possessed the quality it stands for.

The European Quality Assurance standard is also divided into different parts like the prEN 14961 standard. The objective of the quality assurance standard prEN 15234 is to serve as a tool to enable efficient trading of biofuels. Thereby:

- the end user can find a fuel that corresponds to his needs;
- the producer/supplier can produce a fuel with defined and consistent properties and describe the fuel to the customers.

The system for quality assurance may be integrated in a quality management system (e.g. ISO 9000 series) or it can be used on its own to help the supplier in documenting fuel quality and creating adequate confidence between supplier and end user.

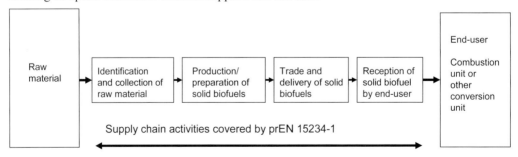

Figure 2.9: Supply chain covered by the prEN 15234-1

The quality management system according to ISO 9001 generally consists of quality planning, quality control, quality assurance and quality improvement. The forthcoming European Standard prEN 15234 – 1 covers fuel quality assurance (the part of quality management that focuses on providing confidence that the quality requirements will be fulfilled) and quality control (the part of quality management that focuses on fulfilling quality requirements).

This EN standard covers quality assurance of the supply chain and information to be used in quality control of the product, so that traceability exists and confidence is given by demonstrating that all processes along the overall supply chain of solid biofuels up to the point of the delivery to the end user are under control (cf. Figure 2.9).

Quality assurance aims to provide confidence that a steady quality is continually achieved in accordance with customer requirements.

The methodology shall allow producers and suppliers of solid biofuels to design a fuel quality assurance system that ensures that:

- traceability exists;
- requirements that influence the product quality are controlled;
- end users can have confidence in the product quality.

Some documentation is mandatory while other documentation is voluntary. Mandatory documentation of quality assurance measures includes:

- Documentation of origin (traceability of raw material);
- Steps in the process chain, critical control points (CCPs), criteria and methods to ensure appropriate control at CCPs, non-conforming products (production requirements);
- Description of transport, handling and storage;
- Product declaration/labelling (final product specification).

Critical control points are points within or between processes at which relevant properties can be most readily assessed, and the points that offer the greatest potential for quality improvement.

Methodology for production quality assurance – step-by-step:

Step 1: Document the steps in the production chain.

Step 2: Define specification(s) for the product(s).

Step 3: Analyse factors influencing product quality and company performance (this includes transportation, handling and storage).

Step 4: Identify and document CCPs for compliance with the product specification.

Step 5: Select the appropriate measures that give confidence to customers that the specification(s) is/are being realised, by:

- identifying and documenting criteria and methods to ensure appropriate control of CCPs;
- monitoring and controlling the production process and making necessary adjustments in order to comply with the quality requirements.

Step 6: Establish and document routines for separate handling of non-conforming materials and products. If any deviation from the stated specifications is noticed in the product, the deviating part shall, if possible, be removed from this specific production chain.

If the deviating part cannot be taken away, the producer shall inform the customer immediately and take the necessary corrective actions.

Appropriate methods should be applied in the production, storage and delivery of solid biofuels and care should be taken to avoid impurities and degradation in the fuel lot. Impurities can arise, for example, from stones, pieces of metal and plastic. Degradation can be caused by moisture absorption.

Factors requiring special attention include:

- Weather and climatic conditions (e.g. risk of rain and snow) during storage and the need for a cover;
- Storage conditions (e.g. ventilation, moisture absorption) and the foreseen duration of storage;
- Storage construction;
- Suitability and cleanness of all equipment;
- Effect of transportation of fuels, e.g. formation of dust;
- Professional skills of personnel.

The product declaration shall state the quality in accordance to the appropriate part of prEN 14961. The product declaration shall as a minimum include:

- Supplier (body or enterprise) including contact information;
- A reference to prEN 15234 – Fuel quality assurance (appropriate part);
- Origin and source (according EN 14961-1);
- Country/countries (locations) of origin where the biomass is harvested or first traded as biofuel;
- Traded form (e.g. pellet);
- Specification of properties:
    - Normative properties,
    - Informative properties;
- Chemical treatment yes/no;
- Signature (by operational title or responsibility), name, date and place.

The product declaration can be approved electronically. Signature and date can be approved by signing the waybill or stamping the packages in accordance with the appropriate part of prEN 14961. The quality information given in the fuel quality declaration shall be labelled on the packaging of solid biofuels.

For pellet production the important factors influencing company performance are raw material, storage of raw material, equipment, pelletising process and proficiency of the staff. The CCPs can be, for example, the following:

- Selection of the raw material (origin and source);
- Reception/storage/sampling of raw material (condition of the store, avoiding impurities, sampling methods);

- Blending of different raw materials (process control);
- Transportation (suitable conveying equipment);
- Screenings (requested particle size);
- Drying (air flow, temperature control);
- Grinding (homogenous raw material);
- Pelletising (pre-treatment/additives, equipment);
- Storage of the pellets (different quality classes, condition of the store, avoiding impurities);
- Packing of the pellets (avoid crushing of pellets);
- Delivering pellets to the retailer and/or end user (no impurities, fulfil approved quality specifications).

Measures to give confidence to customers that the pellet specifications are being realised can be, for example, the following:

- Visual inspection during the whole process chain (colour, check for odour, size, durability of pellets);
- Moisture content before pelletising and at delivery to the end user (analysis and work instructions);
- Determination of properties after production (amount of fines, dimensions, moisture content, mechanical durability, ash content);
- Production control, condition control and adjustment of the equipment;
- Measurement of certain properties after the raw material used has changed at a frequency appropriate to the process requirements;
- Equipment is repaired or changed when necessary; some parts will require changing regularly according to the nature of the production control system.

Fuel specification and classes as well as fuel quality assurance standards have been preliminary tested by several companies under the EU-funded projects BioNorm and BioNormII (pre-normative work on sampling and testing of solid biofuels for development of quality assurance systems).

2.5 Standards for pellet transport and storage for residential heating systems

To date, standards and guidelines for pellet transport and storage for residential heating systems exist only in Austria. These are the ÖNORM M 7136 "Compressed wood or compressed bark in natural state – pellets – quality assurance in the field of logistics of transport and storage", the ÖNORM M 7137 "Compressed wood in natural state – woodpellets – requirements for storage of pellets at the ultimate consumer" and the ÖKL guideline no. 66. In the near future the ENplus will also – besides the pellet quality – cover transport and storage of pellets (cf. Section 2.6).

The ÖNORM M 7136 was implemented to safeguard the quality of pellets according to ÖNORM 7135 along their way from the producer to the end user. Thus, the standard only applies to pellets that have been checked according to ÖNORM 7135.

General requirements of this standard are concerned with documentation, sole use of specific raw materials and moisture protection. According to the standard, the delivery documents have to certify that just pellets that were checked according to ÖNORM 7135 are delivered. These pellets must be stored separately from non-certified pellets, pellets of a different diameter, and other materials. Transport vehicles have to be emptied and cleaned before certified pellets may carried, if other materials were transported in the vehicle before.

Manipulation areas where pellets are handled must be equipped with a roof and be clean.

Pellets must be stored in closed warehouses with an appropriate floor (clean layer of concrete or asphalt) or in closed silos. If other materials were stored in the storage areas before, they have to be completely emptied.

Protecting pellets against moisture is particularly important. Pellets must be stored and transported in a dry way. Direct contact with snow, rain or moist walls, or condensation water, must be avoided.

Before the transport vehicle can be filled for transporting pellets to end users, the fines have to be separated, after separation their share must be less than 1%. During transport as well as during filling and discharging, the pellets have to be protected against moisture in a suitable way. The mechanical strain that the feeding system of the truck puts on the pellets may raise the amount of fines by 1% only.

Since June 2005, after a three year long transition period of the standard had passed, all transport vehicles for pellets with a useful load capacity of 8 t or more have to be equipped with an on-board weighing system.

Silo trucks must be equipped with a suction device to draw air from the storage space while the pellets are blown in, which must have a higher suction capacity than the compressor for the in-blown air in order to avoid overpressure in the storage space. For silo trucks, a filling tube of at least 30 m in length is also prescribed.

The standard also contains specifications concerning the qualification of delivery personnel. The retailer or haulier has to design a work instruction according to which the delivery personnel have to be trained. Minimal requirements as concerns the contents of this work instruction are listed in the standard.

The delivery personnel have to fill in a checklist at each end user, which itself must be part of the delivery papers. The check list has to at least take account whether the heating system was turned off, whether the storage space was closed, how many tonnes of pellets were still stored (approximation) and which tube length was used. Other remarks, such as that there was no plastic baffle plate or that there were dust accumulations in the storage space, are to be recorded in the check list too.

The standard also lays down checking methods for the different parameters. These include all areas of concern, starting from documentation and interim storage and transport up to the qualification requirements of delivery personnel.

The standard is aimed at warranting adequate handling of pellets during storage and transport, thus securing customer satisfaction by avoiding errors.

The ÖNORM M 7137 "Compressed wood in natural state – woodpellets – requirements for storage of pellets at the ultimate consumer" lays down the requirements for storage spaces at the end user site. Reliability, fire protection, static requirements and keeping the pellet quality are the issues of this standard. The standard applies to HP1 pellets according to ÖNORM M 7135 only.

Several general requirements that are relevant for all kinds of storage spaces are put down in the standard. Thus, the storage space has to be set up in a way that a tube length of no more than 30 m is required in order to keep mechanical forces on the pellets at a minimum during filling. Walls and supporting parts have to be designed in a way that allows for the static load that will be put onto them. A fuel demand of 0.6 to 0.7 m³ of pellets per kW heat load for one heating period is given as a guiding value; storage capacity should be designed for at least the fuel demand of one heating period. Water and moisture must not enter the storage space whilst it is filled. The formation of condensed water must also be inhibited. In order to avoid the intake of atmospheric moisture by the pellets, the storage space should not be aired. In addition, the installation must be dust proof. All installations in the storage space (electric, water, waste water and other installations) must be fixed according to TRVB H 118 (technical guideline for fire prevention – automatic wood furnace systems), concealed, appropriately insulated and protected against mechanical stress. Open electric installations such as lamps, plugs or light switches are not allowed for safety reasons. Fire prevention measures have to be carried out according to TRVB H 118 as well. The storage space must be accessible for service, maintenance and cleaning.

The fill-in pipe and return pipe have to be made of metallic materials and fixed in a way that keeps them from twisting. If possible, they should lead outdoors, whereby filling connections should not be longer than 10 m and changes of directions must be realised by means of arches. If fill-in pipes do not lead outdoors and filling lines pass through other rooms, they are also to be constructed as per TRVB H 118. The dimensioning of fill-in and return pipes is strictly prescribed by the standard. Blind flanges have to be used to seal the pipes closely after the filling process.

Structure-borne sound is transferred to the building by bearings, fixtures or wall bushings for discharge systems. This must be prevented by means of appropriate structural measures. With regard to the accumulation of fines as well as emptying intervals of the storage space, the manufacturer of the boiler or the discharge system has to provide appropriate information.

In storage spaces, ideally of rectangular shape, the fill-in pipe and the return pipe should be fixed at the narrow side and, if possible, at an outer wall of the room. Rising moisture in brickwork must be avoided. As concerns dust, the seals of the door should be a focus area. The door must open outwards and some pressure release structure must be in place, like for instance wooden boards that are pushed downwards into profiles mounted at the side of the door. Abrasion of the ceiling has to be avoided in order to prevent polluting of the fuel.

The storage space has to be equipped with appropriate fill-in and return pipes (at least 20 cm underneath the ceiling in the same wall, return pipe flush mounted to the inner wall, fill-in pipe reaching 30 cm into the storage space), a baffle plate made of abrasion and tearproof material (e.g. 1 mm HDPE foil), a 230 V plug for the suction fan outside the storage space, and a sloped bottom with an angle of $40 \pm 5°$ and a sleek, abrasion proof surface.

If an underground storage tank is used for pellet storage, protection against moisture and water as well as protection against electrostatic charging acquires great relevance. In order to safeguard protection against moisture and water, the tank must be built without seams and both tank and tank lid have to be made of materials that are corrosion resistant, robust against weather conditions and robust against static load during filling and use. Water has to be prevented from entering the tank and the inspection chamber at all times. Thus, the tank lid has to seal the tank watertight. The fittings on top of the lid must allow for watertight closure too. In addition, connection to the cellar must also be built in watertight manner. The manufacturer has to account for appropriate measures that protect the tank against electrostatic charging.

The discharge system of the tank has to make sure that the rest of pellets that remain in the tank amount to no more than 5% of the nominal storage capacity of the tank. The house couplings for filling the tank must be freely accessible and also closed tight in order to avoid water or moisture coming in through this path. Basic fire prevention requirements of the TRVB H 118 apply to underground storage tanks as well.

Storage tanks made of metal have to be earthed and protected against corrosion. If the storage tank is made of nonconducting materials, all conductible parts, all connexions as well as the discharge system have to be earthed. Due to possible electrostatic charging of the storage tank, a design that inhibits spark formation must be chosen, which is to be accounted for by the manufacturer. The fill-in and return pipes can lead either through the outer wall of the storage room or they can be mounted directly onto the storage tank. Again, basic fire prevention requirements as per TRVB H 118 have to be adhered too.

If only small quantities of pellets are stored, both loosely or packaged, the requirements of the standard apply correspondingly. As concerns fire prevention, the regulations may be eased by federal state law (detailed exploration of this is abstained from here).

The ÖKL guideline no. 66 [23] is a regulatory framework for the installation of wood pellet heating systems in residential buildings. Next to general arguments for wood pellet heating systems and a short definition of pellets, it contains a directory of furnace technologies that are available for the use of pellets, how fuel demand is to be determined and the way in which fuel delivery and storage are to be carried out. One m^3 of usable storage volume per kW heating load is given as a rule of thumb for the dimensioning of the storage space. For fuel delivery, a number of options are cited, ranging from delivery in 15 kg bags and delivery in big bags of up to 1,000 kg of pellets to silo trucks delivering several tonnes. Finally, some examples of actual installations are given.

2.6 Certification system ENplus

The German Pellet Institute (DEPI) will launch the certification system ENplus [24; 25], which will be based on the European pre-standard prEN 14961-2 (cf. Section 2.2). The main aim of this certification system is to secure the high quality of pellets when delivered to the end user. Therefore, not only the quality parameters of pellets but also the production process, storage and delivery to the end user will be covered by ENplus. Thus, the whole supply chain for pellets up to the point of storage at the end user is part of this certification system. Concerning fuel quality assurance, ENplus will be based on the multipart standard prEN 15234. In a first step ENplus will be implemented in Austria and Germany based on co-operation between the respective pellet associations, i.e. DEPV and proPellets Austria. Moreover, a European pellet association is about to be founded and will work on the implementation of ENplus on a European level.

For the certification of a pellet production plant, a first check from an authorised inspection body is necessary. The report from this first check is an integral part of the application for certification at the certification agency. If all requirements are fulfilled, the ENplus certificate will be issued. A re-certification is required every year.

All checks have to include the following points:

- Visual inspection of the technical facilities;
- Control of the internal documentation;
- Determination of the type and quantity of additives used by means of pellets stocked together with the delivery documents;
- Check of the competence of the quality assurance manager;
- Sampling at the last possible point before the pellets leave the production plant and subsequent analysis.

Pellet retailers and operators of intermediate pellet storages, have to apply for certification at the certification agency. However, no checks from authorised inspection bodies are necessary as long as there are no abnormalities. In case of quality problems, the certification agency can call for external checks.

ENplus also comprises a system of complete traceability along the supply chain of pellets, which makes it possible to establish who caused any quality problems. The system is based on identification numbers for each ENplus certificate holder. The numbers form a code that is displayed on the delivery documents or the bags.

For holders of ENplus certificates, it will be obligatory to participate in a monitoring system for pellets stocked. Therefore, ENplus will contribute to increase the security of supply.

The certification system ENplus will cover all property classes according to prEN 14961-2 and they will be labelled as ENplus A1, ENplus A2 and EN B. The respective quality parameters are shown in Table 2.3. Property class A1 represents the highest quality level that is particularly relevant for private end users. In property class A2, the limiting values for the ash content, the NCV, the nitrogen and chlorine content and the ash melting behaviour are less strict. This property class is mainly relevant for commercial users operating pellet boilers with higher nominal capacity. Pellets according to property class B are relevant as industrial pellets. In contrast to prEN 14961-2, the use of chemically treated wood is not allowed in property class B either.

2.7 ISO solid biofuels standardisation

In December 2007 the ISO 238 Technical Committee (ISO 238/TC) "Solid biofuels" was established and has been working on the upgrade of all CEN/EN standards concerning solid biofuels to ISO standards ever since. Within ISO 238/TC five working groups (WGs) were established:

- WG 1: Terminology
- WG 2: Fuel specifications and classes
- WG 3: Quality assurance
- WG 4: Physical and mechanical test methods

- WG 5: Chemical test methods
- WG 6: Sampling and sample preparation

The new ISO 238 standards will finally replace the EN standards related to solid biofuels (cf. Table 2.1). Currently (as of March 2010) no standards have been published from ISO 238/TC [26]. Based on an optimistic schedule the ISO 238 standards for solid biofuels could be available in 2012 [27].

2.8 Standards for pellet furnaces in the residential heating sector

For boilers used exclusively for burning solid fuels up to an actual power output of 300 kW, the EN 303-5:1999 [28] is to be applied, which is a formal European standard (EN) from CEN and is therefore also a national standard in every one of its 30 member countries. The national standards are identical with the EN apart from some stricter national deviations that are indicated in the standard. All pellet boilers need to meet minimum criteria according to EN 303-5 in order to be legally installed in the respective country. Type tests must be performed from accredited national laboratories. This standard thus includes furnaces for densified biomass fuels such as pellets or also briquettes. Such furnaces that can be fed both manually or automatically make a significant difference in the sense that limiting values for emissions and requirements for the efficiency of the boiler are not the same depending on the way in which the boilers are fed. The standard comprises terminology, technical requirements, testing and labelling regulations.

The construction requirements comprise regulations concerning fire prevention, reliability, scope and contents of documents, quality control and assurance, welding techniques, materials used as well as general safety and construction requirements.

The technical requirements for the heating system comprise regulations concerning boiler efficiency, flue gas temperature, feeding pressure, combustion time, minimal heat output, emissions and surface temperatures.

Table 2.7 and Figure 2.10 show as an example the requirements for boiler efficiency for boilers fired with solid fuels in Austria. It follows that requirements for systems with automatic fuel feeding are higher than for systems with manually fed furnaces. The required boiler efficiency rises with increasing nominal boiler capacity in both cases. The minimal heat output (minimal continuous load) may be no more than 30% of nominal heat output, whereby the requirements of Table 2.7 and Table 2.8 must be followed at minimal heat output too. In the meantime, these requirements have been surpassed by far by reality. Modern pellet furnaces have efficiencies of 90% and more (cf. Section 9.5.2). With the new ecodesign directive (cf. Section 2.9), an update of these requirements is under preparation and stricter values must be expected in the near future.

For boilers that are operated at flue gas temperatures of less than 160 K above room temperature at nominal load, information concerning the exhaust gas system design has to be provided in order to avoid possible sooting, insufficient feeding pressure and condensation.

For the feeding pressure, maximum values are derived from the nominal boiler capacity, which serve as guiding values for chimney dimensioning at the same time.

Table 2.7: Requirements for boiler efficiency derived from nominal boiler capacity according to ÖNORM EN 303-5

Explanations: data source [29]

Nominal boiler capacity	Boiler efficiency
Manual feed	
Up to 10 kW	73 %
From 10 to 200 kW	65.3 + 7.7 log P_N %
From 200 to 300 kW	83 %
Automatic feed	
Up to 10 kW	76 %
From 10 to 200 kW	68.3 + 7.7 log P_N %
From 200 to 300 kW	86 %

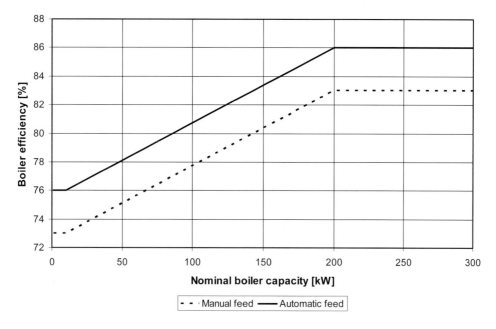

Figure 2.10: Requirements for boiler efficiency derived from nominal boiler capacity according to ÖNORM EN 303-5

Explanations: data source [29]

The emission limits for CO, NO_x, OGC and total particulate matter according to EN 303-5 and in different countries (i.e. Austria, Germany, Denmark, Switzerland, Norway, the Netherlands and Sweden) are shown in Table 2.9 to 2.12. The basis for the emission limits is the EN 303-5. However, many countries set stricter values.

As an example, Table 2.8 shows the Austrian limiting values for emissions defined by ÖNORM EN 303-5 for automatically and manually fed furnaces that are fired with solid biomass fuels. The conversion of mg/MJ to mg/Nm³ is done according to Equation 2.2. The emission limit in mg/MJ is multiplied by the NCV and then divided by the specific flue gas volume in this calculation. The specific flue gas volume depends on the chemical composition of the fuel used and is therefore specific for a certain fuel. It is related to dry flue gas and given for stoichiometric conditions. For woody biomass the specific flue gas volume is typically 4.6 Nm³ dry flue gas per kg fuel (d.b.). The moisture content of the fuel and the reference O₂ content in the flue gas must be taken into account as well in order to calculate the concentration of a certain compound in mg/Nm³ dry flue gas at the reference O₂ content.

Table 2.8: Emissions limits defined by ÖNORM EN 303-5

Explanations: [1]...relating to NCV of the fuel; [2]...applicable only in wood furnaces; [3]...at partial load of 30% nominal load the limiting value may be exceeded by 50%; [4]...conversion of mg/MJ to mg/Nm³ as per Equation 2.2, valid for dry flue gas and 10 vol.% O₂

Feeding system	Unit	CO	NO_x[2]	OGC	Dust
Manual	mg/MJ[1]	1,100	150	80	60
Automatic	mg/MJ[1]	500[3]	150	40	60
Manual[4]	mg/Nm³	2,460	330	180	135
Automatic[4]	mg/Nm³	1,120[3]	330	90	135

Equation 2.2:
$$c\left[\frac{mg}{Nm^3_{FG_{dry},O_{2,ref.}}}\right]_i = c\left[\frac{mg}{MJ}\right] \cdot \frac{NCV\left[\frac{MJ}{kg_{fuel(w.b.)}}\right]}{\frac{kg_{fuel(d.b.)}}{kg_{fuel(w.b.)}} \cdot V_{FG,spec.}\left[\frac{Nm^3_{FG_{dry}}}{kg_{fuel(d.b.)}}\right] \cdot \lambda_{O_{2,ref.}}}$$

Explanations: the specific flue gas volume depends on the kind of biomass used and can vary between 3.7 and 4.7 Nm³ dry flue gas/kg fuel (d.b.) at λ=1; for woody biomass a guiding value of 4.6 Nm³ dry flue gas/kg fuel (d.b.) may be used; $\lambda_{O2, ref.}$ = 1.91 (guiding value for O₂ content in the flue gas being 10 vol.%); data source [54]

In Germany wood pellets can be used without special permission in type-tested furnaces with a maximum thermal power output of 1,000 kW. For straw pellets the thermal power threshold for operation without special permission is 100 kW. Recently the updated German emission directive was put in force, the requirements of which are given in Table 2.9 to 2.12. The new emission limits for wood pellet furnaces are significantly lower than before and wood pellets are not treated in the same way as wood chips or firewood anymore (e.g. the emission limit for total particulate matter for wood chips and firewood furnaces is 0.1 g/Nm³ for all boilers between 4 and 1,000 kW and thus higher than for wood pellet furnaces). Until 2015 the limits for wood pellet stoves are even stricter than for wood pellet boilers (i.e. 0.4 g/Nm³ for CO, 0.03 and 0.05 g/Nm³ for total particulate matter with and without water jacket; from then onwards the limits for pellet boilers will be lowered to those of pellet stoves. For other room heating systems, higher emission limits are defined (e. g. for tiled stoves a CO limit of 2.0 and a particulate matter limit of 0.1 g/Nm³). For all boilers the emission limits have to be controlled at chimney sweep inspections now (as already required for automatically charged boilers, but the frequency was lowered from every year inspection to every two years). Pellet boilers for straw pellets or pellets from annual crops (e.g. miscanthus or grain residues) require a special type testing, which includes measurements for NO_x and PCDD/F emissions. Since 2010 they have been limited to 0.6 g/Nm³ and 0.1 ng/Nm³, respectively.

Table 2.9: CO emission limits according to EN 303-5 and in different countries

Explanations: all values in mg/Nm³; n.v....standard not valid; [a]...only class 3 considered, which requires the highest efficiencies and the lowest emissions; [b]...related to 10 vol.% O_2, dry flue gas; [c]...related to 11 vol.% O_2, dry flue gas; [d]...related to 13 vol.% O_2, dry flue gas; [e]...mg/MJ_{NCV}; [f]...manually fed; [g]...automatically fed; [h]...conversion to mg/Nm³ according to Equation 2.2; [i]...for wood pellet boilers; [j]...from 2015 onwards; [k]...from 2012 onwards; [l]...for the combustion of straw and similar crop materials; [m]...according to the Nordic Ecolabelling; [n]...for biomass fuels in general; [o]...valid for the combustion of natural wood; [p]...different emission limits for different applications and fuels, cf. [53]; [r]...no or no general limit/individual limits; [t]...according to EN 303-5; [u]...vol.% (valid for pellet stoves frequently used); data source [28; 29; 30; 31; 32; 33; 34; 53]

Nominal thermal power [kW]	EN[a]	AT[e) h]	DE	DK	CH	NL	SE
< 4	5,000[b) f] 3,000[b) g]	1,100[f] 500[g]	[r]	2,000[b) f) m] 400[b) g) m]	4,000[d]	[t]	0.04[d) u]
4 - 15	5,000[b) f] 3,000[b) g]	1,100[f] 500[g]	800[d) i] 1,000[d) l] 400[d) i) l) j]	2,000[b) f) m] 400[b) g) m]	4,000[d]	[t]	[r]
15 - 50	5,000[b) f] 3,000[b) g]	1,100[f] 500[g]	800[d) i] 1,000[d) l] 400[d) i) l) j]	2,000[b) f) m] 400[b) g) m]	4,000[d]	[t]	[r]
50 - 70	2,500[b]	1,100[f] 500[g]	800[d) i] 1,000[d) l] 400[d) i) l) j]	2,000[b) f) m] 400[b) g) m]	4,000[d]	[t]	[r]
70 - 100	2,500[b]	1,100[f] 500[g]	800[d) i] 1,000[d) l] 400[d) i) l) j]	2,000[b) f) m] 400[b) g) m]	1,000[d] 500[d) k]	[t]	[r]
100 - 120	2,500[b]	1,100[f] 500[g]	800[d) i] 400[d) i) j] 250[c) l]	1,000[b) f) m] 400[b) g) m]	1,000[d] 500[d) k]	[t]	[r]
120 - 150	2,500[b]	1,100[f] 500[g]	800[d) i] 400[d) i) j] 250[c) l]	1,000[b) f) m] 400[b) g) m] 2,500[b) n]	1,000[d] 500[d) k]	[t]	[r]
150 - 300	1,200[b]	1,100[f] 500[g]	800[d) i] 400[d) i) j] 250[c) l]	1,000[b) f) m] 400[b) g) m] 1,200[b) n]	1,000[d] 500[d) k]	[t]	[r]
300 - 500	n.v.	[p]	800[d) i] 400[d) i) j] 250[c) l]	500[b) n]	1,000[d] 500[d) k]	[r]	[r]
500 - 1,000	n.v.	[p]	500[d) i] 400[d) i) j] 250[c) l]	500[b) n]	500[d]	[r]	[r]
1,000 - 5,000	n.v.	[p]	150[c) o] 250[c) l]	625[b) n]	250[c]	[r]	[r]
5,000 - 10,000	n.v.	[p]	150[c) o] 250[c) l]	625[b) n]	250[c]	[r]	[r]
10,000 - 20,000	n.v.	[p]	150[c) o] 250[c) l]	625[b) n]	150[c]	[r]	[r]
20,000 - 50,000	n.v.	[p]	150[c) o] 250[c) l]	625[b) n]	150[c]	[r]	[r]
> 50,000	n.v.	[p]	[r]	[r]	150[c]	[r]	[r]

The emission limits that apply to the combustion of wood pellets in Denmark are the same as for other biomass fuels defined in a ministerial order. The limiting values are defined in a number of different ministerial orders and a guide from the Environmental Protection Agency. The limit values for CO, NO_x, OGC and total particulate matter have been compiled into the following tables covering all limit values for all sizes of plants.

Table 2.10: NO_x emission limits according to EN 303-5 and in different countries

Explanations: all values in mg/Nm^3; n.v....standard not valid; [a]...related to 6 vol.% O_2, dry flue gas; [b]...related to 10 vol.% O_2, dry flue gas; [c]...related to 11 vol.% O_2, dry flue gas; [d]...related to 13 vol.% O_2, dry flue gas; [e]...mg/MJ_{NCV}; [f]...related to 6 vol.% O_2, dry flue gas; [h]...conversion to mg/Nm^3 according to Equation 2.2; [i]...no limit for wood pellet boilers; [j]...no limit for the combustion of straw and similar crop material; [m]...according to the Nordic Ecolabelling; [n]...for biomass fuels in general; [p]...different emission limits for different applications and fuels, cf. [53]; [r]...no limit; [s]...no general limit/individual limits; [v]...for mass flows ≥ 2,500 g/h; [w]...for existing plants; [x]...for new plants; data source [28; 29; 30; 31; 32; 33; 34; 53]

Nominal thermal power [kW]	EN	AT [e) h)]	DE	DK	CH	NL	SE
< 4	[r)]	150	[i) j)]	340 [b) m)]	250 [d) v)]	[r)]	[r)]
4 - 15	[r)]	150	[i) j)]	340 [b) m)]	250 [d) v)]	[r)]	[r)]
15 - 50	[r)]	150	[i) j)]	340 [b) m)]	250 [d) v)]	[r)]	[r)]
50 - 70	[r)]	150	[i) j)]	340 [b) m)]	250 [d) v)]	[r)]	[r)]
70 - 100	[r)]	150	[i) j)]	340 [b) m)]	250 [d) v)]	[r)]	[r)]
100 - 120	[r)]	150	[i) p)]	340 [b) m)]	250 [d) v)]	[r)]	[r)]
120 - 150	[r)]	150	[i) p)]	340 [b) m)]	250 [d) v)]	[r)]	[r)]
150 - 300	[r)]	150	[i) p)]	340 [b) m)]	250 [d) v)]	[r)]	[r)]
300 - 500	n.v.	[p)]	[i) p)]	[r)]	250 [d) v)]	[r)]	[r)]
500 - 1,000	n.v.	[p)]	[i) p)]	[r)]	250 [d) v)]	[r)]	[r)]
1,000 - 5,000	n.v.	[p)]	[p)]	[r)]	250 [c) v)]	200 [f)]	[r)]
5,000 - 10,000	n.v.	[p)]	[p)]	300 [b) n)]	250 [c) v)]	145 [f)]	[r)]
10,000 - 20,000	n.v.	[p)]	[p)]	300 [b) n)]	150 [c)]	145 [f)]	[r)]
20,000 - 50,000	n.v.	[p)]	[p)]	300 [b) n)]	150 [c)]	145 [f)]	[r)]
50,000 - 100,000	n.v.	[p)]	[s)]	200 - 650	150 [c)]	145 [f)]	600 [a) w)] / 400 [a) x)]
100,000 - 300,000	n.v.	[p)]	[s)]	200 - 650	150 [c)]	145 [f)]	600 [a) w)] / 300 [a) x)]
300,000 - 500,000	n.v.	[p)]	[s)]	200 - 650	150 [c)]	145 [f)]	600 [a) w)] / 200 [a) x)]
> 500,000	n.v.	[p)]	[s)]	200 - 650	150 [c)]	145 [f)]	200 [a) x)]

Also, Ecolabelling Denmark, part of the Nordic co-operation on ecolabelling of products, has issued criteria for labelling of small-scale furnaces for biomass fuels, including pellets. The criteria set strict standards for emissions, which are also shown in the following tables. However, in Denmark for instance the label is of limited practical use and so far the scheme has not gained popularity among manufacturers.

Table 2.11: OGC emission limits according to EN 303-5 and in different countries

Explanations: all values in mg/Nm³; n.v....standard not valid; [a]...only class 3 considered, which requires the highest efficiencies and the lowest emissions; [b]...related to 10 vol.% O_2, dry flue gas; [c]...related to 11 vol.% O_2, dry flue gas; [e]...mg/MJ$_{NCV}$; [f]...manually fed; [g]...automatically fed; [h]...conversion to mg/Nm³ according to Equation 2.2; [i]...no limit for wood pellet boilers; [j]...no limit for the combustion of straw and similar crop material; [m]...according to the Nordic Ecolabelling; [p]...different emission limits for different applications and fuels, cf. [53]; [r]...no limit; [s]...no general limit/individual limits; [t]...according to EN 303-5; [x]...UHC (unburned hydrocarbons); [y]...C_{ges}; data source [28; 29; 30; 31; 32; 33; 34; 53]

Nominal thermal power [kW]	EN[a]	AT[e) h]	DE	DK	CH	NL	SE
< 4	150[b) f] 100[b) g]	80[f] 40[g]	i) j)	70[b) f) m] 25[b) g) m]	r)	t)	150[b) f] 100[b) g]
4 - 15	150[b) f] 100[b) g]	80[f] 40[g]	i) j)	70[b) f) m] 25[b) g) m]	r)	t)	150[b) f] 100[b) g]
15 - 50	150[b) f] 100[b) g]	80[f] 40[g]	i) j)	70[b) f) m] 25[b) g) m]	r)	t)	150[b) f] 100[b) g]
50 - 70	100[b) f] 80[b) g]	80[f] 40[g]	i) j)	70[b) f) m] 25[b) g) m]	r)	t)	100[b) f] 80[b) g]
70 - 100	100[b) f] 80[b) g]	80[f] 40[g]	i) j)	70[b) f) m] 25[b) g) m]	r)	t)	100[b) f] 80[b) g]
100 - 120	100[b) f] 80[b) g]	80[f] 40[g]	i) p)	50[b) f) m] 25[b) g) m]	r)	t)	100[b) f] 80[b) g]
120 - 150	100[b) f] 80[b) g]	80[f] 40[g]	i) p)	50[b) f) m] 25[b) g) m] 100[x]	r)	t)	100[b) f] 80[b) g]
150 - 300	100[b) f] 80[b) g]	80[f] 40[g]	i) p)	50[b) f) m] 25[b) g) m] 100[x]	r)	t)	100[b) f] 80[b) g]
300 - 500	n.v.	p)	i) p)	r)	r)	r)	r)
500 - 1,000	n.v.	p)	i) p)	r)	r)	r)	r)
1,000 - 5,000	n.v.	p)	p)	r)	r)	r)	r)
5,000 - 10,000	n.v.	p)	p)	r)	r)	r)	r)
10,000 - 20,000	n.v.	p)	p)	r)	50[c) y]	r)	r)
20,000 - 50,000	n.v.	p)	p)	r)	50[c) y]	r)	r)
> 50,000	n.v.	p)	s)	r)	50[c) y]	r)	r)

In the Netherlands, emission limits that apply for combustion of biomass pellets are the same as for other biomass fuels. The current emission limits are shown in the following tables. The mandatory requirements for emissions to air from heating installations with a heat output of up to 300 kW are given by The National Board of Housing, Building and Planning. Pellet fired boilers are classified as boilers with automatic fuel supply. These emission limit values are the same as the corresponding ones in the European standard EN 303-5. For units above 300 kW up to 20 MW, permits are required from the local government and for units above 20 MW up to 50 MW, permits are required from the county administrative board.

Table 2.12: Particulate matter emission limits according to EN 303-5 and in different countries

Explanations: all values in mg/Nm^3; n.v....standard not valid; $^{a)}$...only class 3 considered, which requires the highest efficiencies and the lowest emissions; $^{b)}$...related to 10 vol.% O$_2$, dry flue gas; $^{c)}$...related to 11 vol.% O$_2$, dry flue gas; $^{d)}$...related to 13 vol.% O$_2$, dry flue gas; $^{e)}$...mg/MJ$_{NCV}$; $^{f)}$...manually fed; $^{g)}$...automatically fed; $^{h)}$...conversion to mg/Nm^3 according to Equation 2.2; $^{i)}$...for wood pellet boilers; $^{j)}$...from 2015 onwards; $^{k)}$...from 2012 onwards; $^{l)}$...for the combustion of straw and similar crop materials; $^{m)}$...according to the Nordic Ecolabelling; $^{n)}$...for biomass fuels in general; $^{o)}$...valid for the combustion of natural wood; $^{p)}$...different emission limits for different applications and fuels, cf. [53]; $^{r)}$...no limit; $^{s)}$...no general limit/individual limits; $^{v)}$...for existing plants; $^{w)}$...for new plants; $^{x)}$...related to 6 vol.% O$_2$, dry flue gas; $^{y)}$...general advice; $^{z)}$...no general binding emission limits, but praxis are 20 to 50 mg/Nm^3 at 6 vol.% O$_2$, dry flue gas; data source [28; 29; 30; 31; 32; 33; 34; 53]

Nominal thermal power [kW]	EN$^{a)}$	AT$^{e) h)}$	DE	DK	CH	NL	SE
< 4	150$^{b)}$	60	$^{r)}$	70$^{b) f) m)}$ 40$^{b) g) m)}$	$^{r)}$	100$^{c)}$	$^{r)}$
4 - 15	150$^{b)}$	60	60$^{d) i)}$ 100$^{d) l)}$ 20$^{d) i) l) j)}$	70$^{b) f) m)}$ 40$^{b) g) m)}$	$^{r)}$	100$^{c)}$	$^{r)}$
15 - 50	150$^{b)}$	60	60$^{d) i)}$ 100$^{d) l)}$ 20$^{d) i) l) j)}$	70$^{b) f) m)}$ 40$^{b) g) m)}$	$^{r)}$	100$^{c)}$	$^{r)}$
50 - 70	150$^{b)}$	60	60$^{d) i)}$ 100$^{d) l)}$ 20$^{d) i) l) j)}$	70$^{b) f) m)}$ 40$^{b) g) m)}$	$^{r)}$	100$^{c)}$	$^{r)}$
70 - 100	150$^{b)}$	60	60$^{d) i)}$ 100$^{d) l)}$ 20$^{d) i) l) j)}$	70$^{b) f) m)}$ 40$^{b) g) m)}$	150$^{d)}$ 100$^{d) k) f)}$ 50$^{d) k) g)}$	100$^{c)}$	$^{r)}$
100 - 120	150$^{b)}$	60	60$^{d) i)}$ 20$^{d) i) j)}$ 50$^{c) l)}$	70$^{b) f) m)}$ 40$^{b) g) m)}$	150$^{d)}$ 100$^{d) k) f)}$ 50$^{d) k) g)}$	100$^{c)}$	$^{r)}$
120 - 150	150$^{b)}$	60	60$^{d) i)}$ 20$^{d) i) j)}$ 50$^{c) l)}$	70$^{b) f) m)}$ 40$^{b) g) m)}$ 150$^{b) n)}$	150$^{d)}$ 50$^{d) k)}$	100$^{c)}$	$^{r)}$
150 - 300	150$^{b)}$	60	60$^{d) i)}$ 20$^{d) i) j)}$ 50$^{c) l)}$	70$^{b) f) m)}$ 40$^{b) g) m)}$ 150$^{b) n)}$	150$^{d)}$ 50$^{d) k)}$	100$^{c)}$	$^{r)}$
300 - 500	n.v.	$^{p)}$	60$^{d) i)}$ 20$^{d) i) j)}$ 50$^{c) l)}$	300$^{b) n)}$	150$^{d)}$ 50$^{d) k)}$	100$^{c)}$	$^{r)}$
500 - 1,000	n.v.	$^{p)}$	60$^{d) i)}$ 20$^{d) i) j)}$ 50$^{c) l)}$	300$^{b) n)}$	20$^{d)}$	50$^{c)}$	100$^{d) y)}$
1,000 - 2,500	n.v.	$^{p)}$	100$^{c) o)}$ 20$^{c) l)}$	40$^{b) n)}$	20$^{c)}$	25$^{x)}$	100$^{d) y)}$
2,500 - 5,000	n.v.	$^{p)}$	50$^{c) o)}$ 20$^{c) l)}$	40$^{b) n)}$	20$^{c)}$	25$^{x)}$	100$^{d) y)}$
5,000 - 10,000	n.v.	$^{p)}$	20$^{c) l) o)}$	40$^{b) n)}$	20$^{c)}$	5$^{x)}$	100$^{d) y)}$
10,000 - 20,000	n.v.	$^{p)}$	20$^{c) l) o)}$	40$^{b) n)}$	10$^{c)}$	5$^{x)}$	$^{z)}$
20,000 - 50,000	n.v.	$^{p)}$	20$^{c) l) o)}$	40$^{b) n)}$	10$^{c)}$	5$^{x)}$	$^{z)}$
50,000 - 100,000	n.v.	$^{p)}$	$^{r)}$	30 - 50$^{b) n)}$	10$^{c)}$	5$^{x)}$	100$^{v) x)}$ 50$^{w) x)}$
100,000 - 500,000	n.v.	$^{p)}$	$^{r)}$	30 - 50$^{b) n)}$	10$^{c)}$	5$^{x)}$	100$^{v) x)}$ 30$^{w) x)}$
> 500,000	n.v.	$^{p)}$	$^{r)}$	30 - 50$^{b) n)}$	10$^{c)}$	5$^{x)}$	50$^{v) x)}$ 30$^{w) x)}$

Maximum surface temperatures for parts that are touched by hand during boiler operation may exceed room temperatures by the following maximum values:

- 35 K for metals and equivalent materials;
- 45 K for ceramics and equivalent materials;
- 60 K for plastics and equivalent materials.

In addition, regulations concerning the test itself are specified (basic test conditions, apparatus and method of analysis, test fuel, pressure test, leak test). In order to safeguard standard conformity, regulations for the performance of the test are also specified. Requirements concerning test stand setup, parameters to be tested and test durations are included. The methods how to determine the nominal boiler output, boiler efficiency and emissions are also set down. The standard also comprises specifications on the way in which test protocols and technical documentation have to be made, product labelling, the scope of delivery and the instruction manual.

Apart from the type testing of the boilers according to EN 303-5, different labels or certificates exist that set stricter requirements to the furnaces. In Denmark for instance an approval scheme is available that is optional for manufacturers and importers of pellet boilers, and includes also a documentation requirement, a listing on the official list of approved boilers, and a monitoring of all manufacturer workshops who subscribe to the scheme. The boilers tested are labelled with emission and efficiency indices. They do not reflect any quality judgement but only indicate emission and energy related facts.

In Sweden, a product can be granted permission to display the P-symbol after certification by SP Technical Research Institute of Sweden (SP). Such certification involves quality control of the product and verification that the product fulfils the requirements of standards, codes of practice for the sector concerned and other regulations, and that the continuous inspection is carried out according to all these regulations. The certification rules include the conditions for certification, technical requirements and requirements for continuous inspection and quality control of pellet appliances. The technical requirements include regulations concerning safety, efficiency and reliability as well as emission levels. Continuous quality control is performed mainly by the supplier and consists of various elements, including final inspection of the product. External control is performed by SP with the purpose of ensuring that the supplier's inspection and quality control procedures are operating properly. During these controls, product samples can be taken for subsequent performance testing.

Apart from the legal regulations in Germany there is the so-called blue angel ("Blauer Engel") label as a voluntary certification for pellet boilers. The label was issued by the Federal Environmental Agency. It can be applied for pellet boilers with nominal thermal capacities of up to 50 kW, which are exclusively fired with pellets (no multi fuel boilers), ignited and controlled fully automatically (no boilers are allowed where manual adjustments of, for example, the air supply are necessary) and that are one complete system (no retrofit pellet burners are allowed). The label prescribes stricter emission limits than the currently valid emission regulation (1.BImSchV) and higher efficiencies than the EN 303-5. The auxiliary electricity consumption is also limited. Extensive requirements are prescribed for standard settings of the furnace, the operating manual and even to the services the boiler manufacturer has to offer to installers and end users.

In Austria the so-called "Umweltzeichen", an environmental label, is available. It is a voluntary certification system for manually and automatically fed room heating systems and boilers fired with firewood, wood chips, briquettes or pellets with nominal thermal capacities of up to 400 kW. The label defines stricter efficiencies and emission limits than the respective legal requirements do. In addition, it prescribes maximum values for radiation losses and auxiliary electricity consumption. Services and information to be provided by the furnace manufacturer to installers and end users are defined in detail and requirements with regard to the content of the operating manual are made.

2.9 Ecodesign directive

The ecodesign directive 2005/32/EC of the European Parliament [35] forms the basis of a number of so-called implementing measures on the design of energy-related products. For such products, preparatory studies are performed and the European Commission can propose regulations concerning product properties, energy efficiencies and emissions of products sold on the European market based on the results of these studies. Compulsory labelling of the energy efficiency of products according to the energy labelling directive can also be required. Well-known examples for already issued regulations are, for example. refrigerators, washing machines or electric bulbs. Based on such a preparatory study for small-scale combustion installations, respective implementing measures have been under development since 2007 and should be enforced by the end of 2010. The small-scale combustion systems to be covered by this regulation are heating systems based on oil, gas and solid fuels, water heating devices, room air conditioning appliances, room heaters and hot air central heating systems [36; 37]. Consequently, pellet stoves and pellet central heating systems will also be affected by this regulation.

The aim of the ecodesign directive is the increase of the energy efficiency of energy related products and their environmental compatibility under consideration of their whole life cycle. The ecodesign directive can only regulate the placing of products on the market, not their use. Therefore, only minimum efficiencies and limiting values for emissions under test stand conditions can be prescribed. Requirements related to the operation at the end user sites such as emission measurements or requirements for the kind of fuel to be used cannot be made with the ecodesign directive. Moreover, obligations related to existing plants or the exchange of existing plants are not possible.

The implementing measures that are being discussed that must be expected for small-scale heating systems are:

- Test stand requirements concerning energy efficiency;
- Test stand requirements for emissions of important pollutants (especially particulate matter emissions are investigated in the preparatory studies);
- Labelling concerning energy efficiency;
- Control systems and special components such as heat buffer storage.

Although the details of the implementing measures have not been finally defined yet, a regulation concerning energy efficiency and emissions for small-scale heating systems on a European level will be available for the first time soon. The regulation will have impacts on national regulations and will offer a good chance for modern pellet stoves and pellet central heating systems, as they are in most cases highly efficient and show low emission levels already.

2.10 Summary/conclusions

The use of pellets poses high quality requirements on the fuel itself as well as on the furnace used with respect to failure free operation and operation with little environmental impact, in particular in small-scale systems for residential heating. As concerns pellets, this is usually secured by standards. Standards are very important to ensure quality and are imperative for homogenous fuel production with a high quality output.

There are national standards and quality regulations that try to control the quality of pellets in ways that, in part, differ greatly from one another. Apart from the national standards, work on European standards for solid biomass fuels has been done in recent years, which will lead to the publication of a series of European standards from 2010 onwards and consequently to a harmonisation and better comparability of pellets on an international basis. As soon as the European standards are issued, the national standards have to be withdrawn or adapted to these EN standards. Above all, work on ISO standards for solid biomass fuels has been in progress since 2007 and will lead to international standards in a few years. The ISO standards will finally replace all EN standards.

With all these standards and regulations in place, the end users might find themselves confronted with the question as to which kind of pellets actually warrants trouble free operation of their system. This uncertainty was partly met by boiler manufacturers in that they prescribed the use of specific high quality pellets (e.g. labelled by a certain pellet association), or else warranty would be reduced or excluded altogether. Such measures deserve the predicate of being exaggerated since two important standards are available in German speaking countries, namely the ÖNORM M 7135 and the DIN 51731, which safeguard the production of high quality pellets already. Moreover, with the introduction of the so-called DIN_{plus} certificate, a new standard was created that surely safeguards the production of pellets that allow for trouble free system operation. In the near future the leading role of the German and Austrian standards will be taken over by prEN 14961-2. For pellets to be used in the residential heating sector, the pellet class A1 according to prEN 14961-2 will be of particular relevance. The consumer is advised to pay attention to respective standards in pellet purchase or else the danger of system shortcomings or failures might arise.

Class A2 according to prEN 14961-2 might also become a relevant standard for pellets to be used in the residential heating sector as soon as pellet heating systems adapted to this class are available on the market (adaptation will be necessary due to the higher ash content). Class B represents the first standard for pellets to be used in industrial applications (the use of class B pellets in the residential heating sector should be avoided as small-scale pellet furnaces are not able to cope with many of the properties allowed for this class, e.g. lower ash melting point or higher ash content). Whether class B becomes a relevant standard for industrial pellets or not remains to be seen as many large-scale industrial consumers currently have their own pellet specifications.

Due to the fact that pellets have become an internationally and intercontinentally traded good in the meantime, the harmonized commodity description and coding system (HS convention) has to be applied to pellets that are internationally traded. The HS code is a six-digit nomenclature. Individual countries may extend an HS number to eight or ten digits for customs or export purposes. The number is mandatory for all export and import transactions of commodities and therefore also for pellets. In order to provide regulatory framework for shipping, the IMO was established in 1948. Its activities concern safety and environmental issues, legal matters, technical co-operation, maritime security and the efficiency of shipping. In the Code of Safe Practice for Solid Bulk Cargoes (the BC code) pellets were included in 2004 and it currently comprises a description of the material characteristics that could result in hazardous conditions during ocean voyage such as oxygen depletion and off-gassing and also stipulates operational requirements such as entry permit, gas monitoring and fire extinguishing practices. The BC code is currently being updated.

Beside standards and regulations for pellets, their transport, storage and trade as well as the technical requirements for furnaces using pellets are also regulated by different standards in different countries. Here too the new certification system ENplus will contribute to harmonisation at least on a European level, as ENplus will comprise not only of the pellet quality according to prEN 14961-2 but also transport and storage regulations for pellets including the end user's storage.

Similar to pellet product standards in different countries, the respective regulations concerning pellet furnaces also differ greatly from one another, in particular concerning emission limits. An international comparison of pellet related emission limits showed that a direct comparison of different emission limits of different countries is almost impossible due to different units, reference O_2 concentrations and allocations to different power ranges. Therefore, a unification at least on a European level must strongly be recommended. The regulations currently under discussion based on the European ecodesign directive (directive 2005/32/EC of the European Parliament) might clear the way in this direction in the near future. A directive for small-scale heating systems could probably be in force from 2011 onwards, which will have impacts on national regulations.

3 Physio-chemical characterisation of raw materials and pellets

3.1 Relevant physical characteristics of raw materials and pellets

In order to determine the applicability of biological raw materials for the production of pellets, evaluation criteria for possible raw materials have first to be determined.

For this purpose, specifications from prEN 14961-2 are consulted, which pose a limit on certain constituents of pellets, whereby the raw materials have to fulfil these requirements already. Raw materials can also be utilised as biological additives that must not exceed 2.0 wt.% (w.b.) according to the standard. Since those are added in such low amounts, they can show higher levels of certain parameters themselves. In any case, they have to be untreated biomass products from primary agriculture and forestry.

Further parameters for the evaluation of biological materials as raw material for pellets are predetermined by the pelletisation technique and also by the combustion technique with parameters influencing the combustion behaviour.

These particular parameters are examined in the following sections and their influence on the suitability as a raw material and on pellets is presented.

Within the framework of the EU-ALTENER project "An integrated European market for densified biomass fuels (INDEBIF)" [38] an international analysis programme was carried out in selected European countries (i.e. Austria, Sweden, Spain, Italy, Norway and the Czech Republic), where both pellets and briquettes from different producers were analysed. The analysis programme aimed to create an overview of physio-chemical characteristics of densified biomass fuels in different European countries and thus to show the variation of existing qualities as well as different standards and regulations [39; 40]. The BioNorm project "Pre-normative work on sampling and testing of solid biofuels for the development of quality assurance systems" [89] also led to numerous tests on physical and mechanical characteristics of pellets (and briquettes). The main goal of the project was the comparison of measurement methods for these properties. In the framework of this project more than 25 pellet types were tested that were selected for their representativeness with regard to the commercial market of 6 European countries (Austria, Belgium, Denmark, Finland, Germany and Spain). The Austrian bioenergy competence centre BIOENERGY 2020+ carried out analyses of 82 pellets from 20 European countries in order to provide an overview of pellet qualities in Europe [41; 42]. Selected results from these projects and activities are shown in the following sections, whenever they are relevant.

3.1.1 Size distribution of raw materials

The requirements for the particle size of raw materials depend on the diameter of the pellets, the raw material itself and finally on the pellet mill technology. In any case, the material should be as homogenous as possible. As a maximum the particle size of sawdust, i.e. 4 mm, can be stated. If a raw material has to be ground, the question of economic efficiency arises. This is dealt with in Section 7.2.4. The technical possibilities for grinding are described in

Section 4.1.1.1. So, particle size has to be kept in mind concerning necessary pre-treatment steps. However, a very big particle size is not necessarily a knock-out criterion.

3.1.2 Dimensions of pellets

The shape and particle size of a fuel usually determine the correct choice of feeding and furnace technologies as they influence the conveying and combustion behaviour of the fuel. The bigger the fuel particles are, the more robust feeding appliances have to be and the longer becomes the required time for complete combustion.

In the case of pellets, the choice and dimensioning of feeding and furnace facilities are eased by the fact that diameter and length are standardised. The pellet diameter is determined by selecting a die with the correct diameter of the die holes. The length was left to pure chance in the beginnings of wood pellet production. In the pellet mill the raw material is pressed through the die and comes out as an endlessly long string that then more or less randomly breaks into pieces depending on its stiffness. Through efforts to make pellets less abrasive and more durable at the same time, the pellets became longer and longer, which often led to blockings in the conveyor systems, especially in pneumatic conveyor facilities (in such a facility one single pellet that is too long can lead to blockings and in consequence to a system outage). For this reason, pellet producers now cut the pellets with knifes that are situated on the periphery of the die so that the length does not surpass a defined maximum.

It was not until such a homogenous biomass fuel, as regards shape and size, was introduced on the market that the development of automatic biomass small-scale furnaces, providing similar comfort as modern oil or gas heating systems, was made possible.

Most of the pellets examined in the analysis programmes are of 6 mm in diameter. A secondary quantity were pellets of 8 mm in diameter, while 10 mm pellets were examined to only a limited degree.

The length of pellets is set down to be no more than 40 mm by prEN 14961-2. The length is especially important when pneumatic feeding systems are used since one single overlong pellet can cause blockings in the feeding system. This can lead to a standstill of the whole system. In the analysis programmes it was found that all pellets analysed meet this limiting value.

3.1.3 Bulk density of pellets

The bulk density is defined by Equation 3.1. It is dependant on the particle density and the pore volume (porosity of the bulk). A rough estimation of the bulk density may be given by dividing the particle density by 2. The higher the bulk density the higher becomes their energy density and the lesser are transport and storage costs. Therefore, a high bulk density is to be aspired from the economic point of view as well as for pellet producers, retailers, intermediary distributors and after all for customers.

The European standard prEN 14961-2 sets down 600 kg (w.b.)$_p$/m³ as a minimum value. Literature values for bulk densities of pellets are between 550 and 700 kg (w.b.)$_p$/m³ [43; 44; 45; 46; 59; 60], a bulk density of 650 kg (w.b.)$_p$/m³ being assumed in most cases. This value is also stated by many pellet producers.

Equation 3.1: $\rho_b = \dfrac{m_{bulk\ good}}{V_{bulk\ good}}$

Explanations: m in kg (w.b.); V in m^3

3.1.4 Stowage factor

The stowage factor in cubic feet per tonne is used as a measure for bulk density primarily for ocean vessels. A bulk density of for example 700 kg/m^3 is converted to the stowage factor using Equation 3.2. The stowage factor is consequently the reciprocal value of the bulk density. Converted to cubic metre per tonne the value would be 1.43.

Equation 3.2: $Stowage\ factor = \dfrac{35.31\ ft^3/m^3}{700\ kg/m^3} \cdot 1{,}000\ kg/t = 50.4\ ft^3/t$

It is worth mentioning that during large bulk shipments in ocean vessels or railcars over long distance, settling (compacting) of the pellets, due to vibrations, occurs and reduces the volume by up to 3 to 5% compared to standard bulk density (cf. Section 3.1.3).

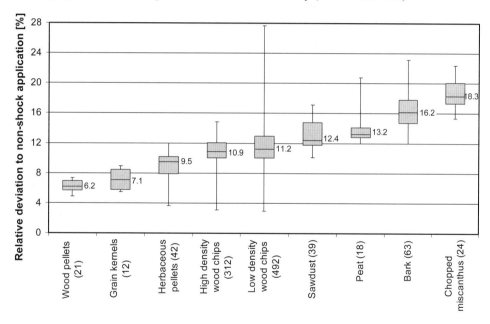

Figure 3.1: Relative effect of shock impact on volume compared to a non-shock application in bulk density determination

Explanations: numbers in brackets indicate the number of replications; given for the 50 l container, which was dropped three times before refilling, surface levelling and weighing; in the data evaluation low and high density fuels were differentiated by the boundary value of 180 kg/m^3

By exposing a bulk sample to controlled shock as described in the European Technical Specification "Solid biofuels – determination of bulk density" (EN 15103) this settling or compacting effect was measured [47] by forming the difference between the measurement with and without shock. Among the solid biofuels this compaction effect is small particularly for wood pellets. This is illustrated in Figure 3.1.

The controlled shock impact leads to a certain volume reduction, which accounts for compaction effects during the production chain (e.g. due to high pressure load in a silo or due to shock and vibration during transports). Thus, in practice the higher mass load leads to an increased load pressure and to fuel settling, which can be additionally enhanced by the vibrations during transportation. Furthermore, filling or unloading operations in practice usually involve a higher falling depth than the one chosen for the test performed here. This will also result in a respectively higher compaction due to the increased kinetic energy of the pellets falling. A procedure that applies a controlled shock to the sample was thus believed to reflect the practically prevailing bulk density in a better way than a method without shock. This is particularly true when the mass of a delivered fuel has to be estimated from the volume load of a transporting vehicle, which is common practice in many countries. For a rough estimation on how susceptible the different solid biofuels are towards shock exposure some research data are given in Figure 3.1. The data show a compaction effect between 6 and 18% for biomass fuels and between about 5 and 7% for wood pellets.

3.1.5 Particle density of pellets

The particle density is defined as the quotient of mass and volume of a pellet (cf. Equation 3.3). Generally, the particle density of pellets influences their combustion behaviour. Fuels with a higher particle density have a longer burnout time. Moreover, the bulk density rises with increasing particle density.

Equation 3.3: $$\rho_p = \frac{m_{pellet}}{V_{pellet}}$$

Explanations: m in kg (w.b.); V in m^3

Measurements of the particle densities of wood pellets within the analysis programmes and by [48] in Germany indicated a range between 1.12 and 1.30 kg/dm^3. Thus, all samples analysed have a higher particle density than prescribed by the ÖNORM M 7135 (no such limiting value exists in prEN 14961-2).

For the determination of particle density, it must be considered that the method used to measure particle density influences the result, especially the method for volume determination. Stereometric measurements were compared to liquid displacement methods, for which the hygroscopic character of pellets is supposed to be of main importance [49]. On the basis of repeatability and reproducibility, several methods were compared and the buoyancy principle was found to be the more appropriate method for the determination of the pellets' volume, if a wetting agent is added to the water prior to measurement. The method is fully described by the European standard prEN 15150.

Possible interactions of particle density with other parameters are discussed in Section 3.3.

3.1.6 Angle of repose and angle of drain for pellets

The angle of repose and angle of drain are measured as the angle between the free flowing surface of a heap of material and the horizontal plane. The angle of repose is a measure for a material poured from the top, generating a standing cone or a flat ramp on a flat surface. The angle of drain is a measure for a material draining through an orifice on the flat horizontal on which the material is located (cf. Figure 3.2). The angle of drain is usually steeper due to the forces applied tangentially as the material is congested in the lower sections of the cone of drainage. A third measure called angle of dynamics is sometimes used and refers to the free flowing surface formed in a drum during slow rotation.

The angle of repose for wood pellets depends on the aspect ratio of the individual pellet, the amount of fines and the surface friction of the pellets. Pellets traded in large bulk have an angle of repose of approximately 28 to 32 degrees. This angle is of importance when designing storage facilities for pellets. The angle of drain is approximately 33 to 37 degrees and is important when designing storage for pellets with hopper bottom drainage.

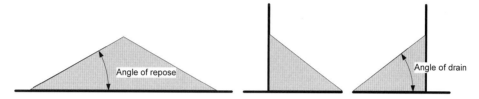

Figure 3.2: Visualisation of angle of repose and angle of drain

3.1.7 Mechanical durability of pellets

The mechanical durability is one of the most important parameters in pellet production. This is because a high amount of fines can lead to bridging in the storage facility of the customer, which can cause a stop of fuel supply. Furthermore, a lot of fines can cause a blocking of the feeding screw. This is why a low amount of fines in the storage room of the end user is of the greatest importance in view of the plant availability and thus for customer satisfaction. Furthermore, particulate matter emissions from combustion rise if there is a high amount of fines, which should be avoided from an ecological point of view.

In addition, a high amount of fines alters the bulk density and increases the losses through transport and also the dust emissions during manipulation of the fuels. Moreover, dust is also known to cause explosions during storage and handling.

The mechanical durability (DU) is defined by prEN 14588 as the "ability of densified biofuel units (e.g. briquettes, pellets) to remain intact during loading, unloading, feeding and transport". The determination of the abrasion within the analysis programmes was carried out according to ÖNORM M 7135 and converted to mechanical durability (it is the difference between 100 and the abrasion in percent). According to prEN 14961-2 the minimum value for the mechanical durability is 97.5 wt.% (w.b.) for the classes A1 and A2. It is notable that all pellets that are certified according to a standard and, above that, the majority of pellets analysed are above this value and only a few samples do not reach it. Those that are below the limit are used as industrial pellets in medium- and large-scale applications. The highest mechanical durability was found in the Norwegian pellets that come from the pilot plant described in Section 4.1.4.1, where the raw material undergoes steam explosion treatment before pelletising. With a mechanical durability of 97.6 wt.% (w.b.), one of the two straw pellets samples was slightly above the required value for wood pellets and clearly excelling some of the wood pellets. The second straw pellets sample exhibited an extremely poor mechanical durability of only about 80 wt.% (w.b.).

The results show that the majority of pellet producers are capable of producing pellets with high mechanical durability. However, serious flaws have been discovered in samples from some producers in this respect. As long as such pellets do not enter the market of small-scale systems but are used in industrial plants, this does not pose a major problem. However, utilisation of highly abrasive pellets in pellet central heating systems must be avoided by all means so as to not diminish the confidence of the end user into the product and the market as a whole.

Possible interactions of mechanical durability/abrasion with other parameters are discussed in Section 3.3.

3.1.8 Pellets internal particle size distribution

The size distribution of particles constituting the pellets is one of the specifications for users who mill the pellets before use [50]. This is usually done in power plant firing or co-firing pellets.

The measurement method of this property consists in disintegrating the pellets in water under specific conditions. Afterwards the mix is dried in a drying cabinet before being placed at room atmosphere to reach moisture equilibrium. Finally the sample is sieved using the equipment described in prEN 15149-2.

Different procedures for measuring this property have been compared on the basis of repeatability and reproducibility [51; 52]. The study reaches the conclusion that the most reliable procedure uses heated water and mixes the slurry before drying.

Published values on this property are scarce, with the values of the distributions ranging from 0.45 mm to 1.25 mm with a mean value of 0.7 mm. The Belgian utility Electrabel for instance requires 100% of the particles to be under 4 mm, 99% under 3 mm, 95% under 2 mm, 75% under 1.5 mm and 50% under 1 mm (cf. Table 10.10 in Section 10.11.6.5). Guidelines for this parameter are being discussed in WG4 CEN TC335 and have not been defined yet.

3.2 Relevant chemical characteristics of raw materials and pellets

3.2.1 Content of carbon, hydrogen, oxygen and volatiles of pellets

Table 3.1 shows average concentrations of carbon, hydrogen and oxygen as well as volatiles in different biomass materials. The applicability of the materials for pelletisation is not influenced by these elements. However, the concentrations of these elements do have an effect on the gross calorific value, hence also on the net calorific value. The volatiles influence the combustion behaviour. Carbon, hydrogen and oxygen are the main components of biomass fuels (since cellulose, hemi-cellulose and lignin consist of these elements), whereby carbon and hydrogen are the main elements responsible for the energy content due to the exothermic reaction to CO_2, respectively H_2O, during combustion. The oxygen bound in organic material covers part of the oxygen needed for the combustion; the rest that is needed for complete combustion has to be supplied by air. The concentrations of carbon and hydrogen of woody biomass are higher than those of herbaceous biomass, which gives account for the higher gross calorific value of woody biomass.

The volatiles are that part of the organic content of the fuel that is released in 7 minutes at a temperature of 900°C under exclusion of air (according to EN 15148).

The amount of volatiles in biomass fuels is high compared to coal. In woody biomass it fluctuates between 70 and 86 wt.% (d.b.); in herbaceous biomass it lies between 70 and 84 wt.% (d.b.). This high content of volatiles leads to a fast vaporisation of most of the biomass. The gases formed burn in homogenous gas phase reactions. The remaining charcoal burns relatively slowly in heterogeneous combustion reactions. This is why the volatiles have a strong impact on the thermal degradation and combustion behaviour of the biomass [53].

The determination of the concentrations of carbon and hydrogen is regulated by the prEN 15104. The oxygen content can be approximated as the difference between 100 minus the sum of carbon, hydrogen, sulphur, nitrogen and ash (in wt.% (d.b.)). The volatile content has to be determined according to EN 15148.

Table 3.1: Concentrations of C, H, O and volatiles in different biomass materials

Explanations: data source [53]

Fuel type	C	H	O	Volatiles
	wt.% (d.b.)	wt.% (d.b.)	wt.% (d.b.)	wt.% (d.b.)
Wood chips (spruce, beech, poplar, willow)	47.1 - 51.6	6.1 - 6.3	38.0 - 45.2	76.0 - 86.0
Bark (coniferous trees)	48.8 - 52.5	4.6 - 6.1	38.7 - 42.4	69.6 - 77.2
Straw (rye, wheat, triticale)	43.2 - 48.1	5.0 - 6.0	36.0 - 48.2	70.0 - 81.0
Miscanthus	46.7 - 50.7	4.4 - 6.2	41.7 - 43.5	77.6 - 84.0

3.2.2 Content of nitrogen, sulphur and chlorine of pellets

The allowed concentrations of nitrogen, sulphur and chlorine of pellets are limited by prEN 14961-2. The limiting values of the standard are based on wood as the reference input material for pelletisation, which shows significantly lower concentrations of nitrogen, sulphur and chlorine than herbaceous biomass. The elements do not have any influence on the pelletising process itself but the concentrations have to be considered when looking at potential raw materials since they follow from the natural concentrations in wood. This avoids the use of contaminated materials or materials that are not biological for the production of pellets.

Increased concentrations of these elements can be the result of a chemical contamination by, for example, insecticides, adhesives, glues, lacquer, dyestuff, wood preservatives or of admixing agricultural biomass.

Nitrogen is easily volatile and is almost completely released to the flue gas during combustion (formation of N_2 and NO_x). The formation of NO_x is a problem. It is dependent on the nitrogen content of the biomass fuel. Sulphur and chlorine are also very volatile and are mainly released into the gas phase during combustion. In gas phase reactions aerosols are then formed together with potassium and sodium (sulphates, chlorides) as well as SO_x and HCl.

There are limitations concerning these elements due to technical as well as environmental issues. The concentrations of nitrogen, sulphur and chlorine have different impacts on combustion. High levels of nitrogen, sulphur and chlorine boost the emissions of NO_x, SO_x

and HCl. Chlorine also augments the formation of polychlorinated dibenzodioxins and furans (PCDD/F). What is more, the combustion products of chlorine and sulphur have corrosive effects and are of great relevance concerning deposit formation. Guiding values for these elements for various biomass fuels are shown in Table 3.2.

Table 3.2: Guiding values for N, S and Cl for various biomass fuels

Explanations: data source [53]

Element	Unit	Wood (spruce)	Bark (spruce)	Straw (winter wheat)	Whole crops (triticale)
N	mg/kg (d.b.)	900 - 1,700	1,000 - 5,000	3,000 - 5,000	6,000 - 14,000
S	mg/kg (d.b.)	70 - 1,000	100 - 2,000	500 - 1,100	1,000 - 1,200
Cl	mg/kg (d.b.)	50 - 60	100 - 370	1,000 - 7,000	1,000 - 3,000

3.2.3 Gross calorific value, net calorific value and energy density of pellets

The definitions for the gross and net calorific value van be found in Section 2.1.2. The gross calorific value of a raw material should be as high as possible with regard to the energy density of the pellets. It is purely dependent on the material used, i.e. the chemical composition of the raw material and can therefore not be influenced. In general, the gross calorific value of woody biomass (including bark) lies around 20.0 MJ/kg (d.b.), the value for herbaceous biomass is around 18.8 MJ/kg (d.b.) [54]. The gross calorific value can be determined according to EN 14918 by using a bomb calorimeter. Furthermore, the approximate gross calorific value can be calculated with Equation 3.10 if an ultimate analysis of the fuel lies at hand.

The net calorific value depends mainly on the gross calorific value, the moisture content and the content of hydrogen in the fuel. Other parameters such as nitrogen, oxygen or ash content have a minor influence. The NCV can be calculated from the GCV, and for different fuels different equations are available in literature [55; 56; 57; 58]. For solid biomass fuels there are two important equations.

Equation 3.4 is widely used and also recommended by IEA Bioenergy, Task 32 "Biomass Combustion and Co-firing" [53; 59; 60]. It just needs the GCV, the moisture content and the hydrogen content of the fuel as input parameters. For woody biomass, such as pellets, the content of hydrogen amounts to around 6.0 wt.% (d.b.); the value for herbaceous biomass being around 5.5 wt.% (d.b.). Together with the approximation for the GCV of woody biomass as mentioned above, the NCV can be calculated as a good approximation based on the moisture content of the fuel, a parameter that can easily be determined.

Equation 3.4: $$NCV = GCV \cdot \left(1 - \frac{M}{100}\right) - 2.447 \cdot \frac{M}{100} - \frac{X_H}{200} \cdot 18.02 \cdot 2.447 \cdot \left(1 - \frac{M}{100}\right)$$

Explanations: NCV in MJ/kg (w.b.); GCV in MJ/kg (d.b.); M in wt.% (w.b.); X_H in wt.% (d.b.); data source [53]

Equation 3.5 is the required equation for the calculation of the NCV according to EN 14961-1 (cf. also Section 2.1.2). In addition to Equation 3.4, it needs the contents of oxygen and nitrogen as input parameters. This must be regarded as a disadvantage of this equation, as these parameters are usually not available and in particular the oxygen content is difficult to determine. Comparing the results of the two equations it can be seen that they deliver almost the same results. The NCV were calculated for a typical composition of pellets for a range of moisture contents between 0 and 65 wt.% (w.b.). The results as well as the relative difference between the two calculations are shown in Figure 3.3. As can be seen, the relative difference increases with increasing moisture contents. However, even at a moisture content of 65 wt.% (w.b.) the relative difference is only 0.13% and therefore negligible.

Equation 3.5: $NCV = [GCV - 0.2122 \cdot X_H - 0.0008 \cdot (X_O + X_N)] \cdot \left(1 - \dfrac{M}{100}\right) - 2.443 \cdot \dfrac{M}{100}$

Explanations: NCV in MJ/kg (w.b.); GCV in MJ/kg (d.b.); M in wt.% (w.b.); X_H, X_O and X_N in wt.% (d.b.)

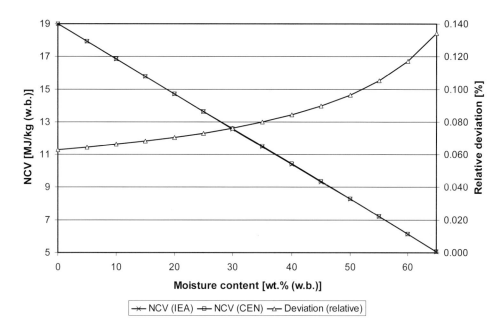

Figure 3.3: Comparison of different calculation methods for the NCV

Explanations: NCV (IEA) according to Equation 3.4; NCV (CEN) according to Equation 3.5

The energy density is the product of net calorific value and bulk density and is calculated according to Equation 3.6. The required transport and storage capacity is reduced with rising energy density, which is why a high energy density is of great relevance, especially for economic reasons.

Equation 3.6: $\rho_e = NCV \cdot \rho_b$

Explanations: ρ_e in MJ/m³; NCV in MJ/kg (w.b.); ρ_b in kg (w.b.)/m³

Gross calorific value, net calorific value (plus moisture content) and energy density have an effect on the dimensioning of the furnace and fuel storage as well as on the control system. Due to the high quality of pellets as a product, as guaranteed by standards such as prEN

14961-2, it can be assumed that these parameters show a high degree of homogeneity, which makes it possible to attune the furnace and its control system to the fuel very well.

The GCV of the analysed biomass fuels made of wood was between 19.8 and 20.7 MJ/kg (d.b.) and the GCV of fuels made of straw was between 18.6 and 19.0 MJ/kg (d.b.) (cf. Figure 3.4), which accords well with literature data (see above). The GCV of the tropical wood and eucalyptus samples lay within a similar range.

NCV, which depends on the moisture and hydrogen content of the fuel, was calculated from the GCV according to Equation 3.4 and is also shown in Figure 3.4.

The energy density of the examined wood pellets was calculated from NCV and bulk density according to Equation 3.6 and is shown for each country in Figure 3.5. The results exhibit great differences of between 8.9 and 11.5 GJ/m³. The Italian and the Spanish pellets had a lower energy density than the Swedish and Austrian pellets on average. The highest energy density was found in the Norwegian pellets, which were produced according to the technique presented in Section 4.1.4.1. The low energy density of straw pellets can be explained by the low GCV of straw, even though one of the straw pellets samples exhibited a relatively high energy density that was the result of its high bulk density.

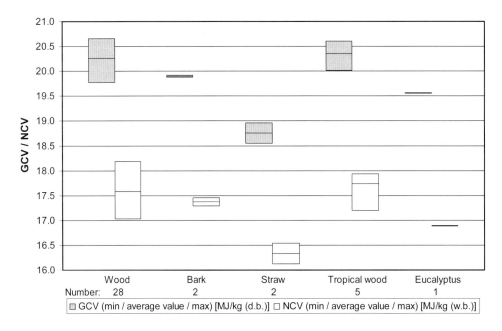

Figure 3.4: Gross and net calorific values of densified biomass fuels

A comparison between the energy densities of pellets and heating oil (around 10 kWh/l or 36 GJ/m³) shows that on average 3.5 m³ pellets as a bulk correspond to 1,000 l heating oil. The range of fluctuation exhibited by wood pellets also shows that the difference between the highest and the lowest measured value is nearly 30%, which has a strong impact on transport and storage capacities and thus also on transport and storage economy. This strikingly illustrates the importance bulk density of wood pellets has for producers, transporters, intermediaries and end users.

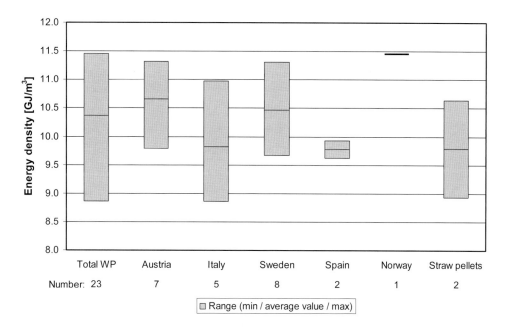

Figure 3.5: Energy densities of pellets

In comparing the energy densities of wood pellets and sawdust, that is the raw material for wood pellet production in many cases, the upgrade of wood as an energy carrier by means of pelletisation is also highlighted. Sawdust usually has a moisture content of around 50 wt.% (w.b.) and a consequential NCV of around 2.2 kWh/kg (w.b.) or 8.1 MJ/kg (w.b.). The bulk density of sawdust with this moisture content is around 240 kg (w.b.)/m³ [59], hence the energy density of sawdust is around 540 kWh/m³ or 1.9 GJ/m³. So, the energy density of wood pellets is five to six times higher than that of the raw material. Storage and fuel transport for wood pellets are thus five to six times more efficient and more economic than for sawdust.

3.2.4 Moisture content of raw materials and pellets

The moisture content needed for pelletisation (cf. definition of moisture content in Section 2.1.1, Equation 2.1) depends basically on the pelletisation technology but also on the raw materials used. If the moisture content of a raw material is too high, it needs to be dried, which mainly raises questions about the economy of the process (this problem is dealt with in Section 7.2.3). Technical possibilities for drying are examined in Section 4.1.1.2. High moisture content is therefore not a knock-out criterion for a raw material in pelletisation, but it has to be considered from an economic and technical point of view.

A guiding range of moisture content for a raw material just before entering the pellet mill lies typically between 8 and 12 wt.% (w.b.) for the pelletisation of wood. When the moisture content is below that range the frictional forces in the compression channel are so great that they render pelletisation impossible; above the value the pellets produced are not dimensionally stable. A potential conditioning of the raw material with steam or water (to achieve a more homogeneous moisture content and to improve the binding behaviour in the pellet mill; cf. Section 4.1.1.3) must be considered hereby since this can raise the moisture content by up to 2 wt.% (w.b.). The exact regulation of the moisture content is of great significance, which is confirmed by many pellet producers.

Regarding the combustion technology, the moisture content of pellets is relevant for the net calorific value, the efficiency of the furnace and the combustion temperature. Net calorific value, efficiency and combustion temperature decrease with rising moisture content.

The moisture content of wood pellets is set down to be no more than 10 wt.% (w.b.)$_p$ according to prEN 14961-2 for all classes. All analysed wood pellet samples within the analysis programmes followed this requirement.

Possible interactions of moisture content with other parameters are discussed in Section 3.3.

3.2.5 Ash content of raw materials and pellets

The ash content (fraction of minerals of the fuel in oxidised form) of raw materials does not influence the pelletisation itself (as long as the ash content is not too high, which would increase the wear and consequently reduce the lifetime of roller and die), but according to prEN 14961-2 a maximum of 0.7 wt.% (d.b.) is allowed for pellet class A1. Therefore, an ash content of more than 0.7 wt.% (d.b.) is a criterion for exclusion of a raw material for the production of class A1 wood pellets according to prEN 14961-2.

The ash content of pellets used for residential heating should be as low as possible for the comfort of the customer since low ash content means longer emptying intervals for the ash box. Assuming a pellet central heating system with a nominal boiler output of 15 kW, 1,500 annual full load operating hours and an ash content of 0.7 wt.% (d.b.), the amount of ash produced is about 35 kg/a. This should be considered in dimensioning the ash box in order to lengthen the emptying intervals and create ease of use. Some manufacturers of pellet furnaces make use of ash compaction systems as well as automatic de-ashing systems that convey the ash into an external ash container to further lengthen the emptying interval. Another essential factor that favours low ash content in the fuel is the fact that rising amounts of ash increase the danger of slag and deposit formation in the combustion chamber, which in turn can lead to operative failures. What is more, high ash contents of the fuel cause greater particulate matter emissions during combustion.

If the pellets are to be used in medium- or large-scale furnaces, such low ash contents are not absolutely necessary because bigger installations are usually built in a more robust way and are typically equipped with more sophisticated combustion and control systems.

The ash content was determined according to two different methods within the analysis programme. Firstly, loss on ignition at 550°C based on the Swedish SS 187171 standard and secondly, at 815°C based on the German DIN 51719 standard, was determined. The relative difference as relating to the determination at 550°C was calculated according to Equation 3.7.

The ash content that is determined at 815°C is generally beneath the one determined at 550°C. Relative difference in ash contents of wood pellets varies between around 15 and 32% with a median of 23%.

The lower ash content of the determination at 815°C is the result of the decomposition of carbonates as well as of partial evaporation of alkaline metals, and chlorine at high temperatures. A direct correlation between difference in ash content and potassium content could not be found though.

Equation 3.7: $\Delta_{ash} = 100 - \dfrac{X_{ash,815}}{X_{ash,550}} \cdot 100$

<u>Explanations</u>: Δ_{ash}...relative difference in ash content [%]; $X_{ash,550}$...ash content as determined at 550°C according to SS 187171; $X_{ash,815}$...ash content as determined at 815°C according to DIN 51719

Owing to these results and also on the basis of other studies that were carried out in this field [61], it is concluded that the ash content of solid biomass fuels should generally be determined at 550°C, which has been taken into account in the European standard EN 14775. It is only the ash content as determined according to SS 187171 at 550°C that is thus considered in the following investigations and interpretations.

Typical ash contents of various types of biomass are shown in Table 3.3. The ash contents determined within the analyses programmes for fuels made of bark and straw correspond quite accurately to literature values that are between 2.0 and 5.0 wt.% (d.b.) for bark and between 4.9 and 6.0 wt.% (d.b.) for straw [59; 62; 63].

The ash contents of the analysed wood samples were between 0.17 and 1.88 wt.% (d.b.) and thus partly higher than typical values for wood (cf. Table 3.3). A high level of ash content can indicate mineral contamination caused by inappropriate raw material storage and/or handling. Ash contents above the upper limit for hard wood could be due to the utilisation of other wood species (e.g. short rotation forestry) or to a certain amount of bark in the raw material.

An interesting aspect is the fact that pellets that are solely made of hardwood would not adhere to class A1 pellets according to prEN 14961-2 due to their high ash content.

Table 3.3: Typical ash contents of different types of biomass

<u>Explanations</u>: [1]...without bark; data source [59; 62; 63; 64]

Fuel type	Typical ash content wt.% (d.b.)
Softwood[1]	0.4 - 0.8
Hardwood[1]	1.0 - 1.3
Bark	2.0 - 5.0
Straw	4.9 - 6.0

It can be concluded, that the production of pellets with low ash content for utilisation in small-scale furnaces is no problem when the raw materials are carefully selected (use of softwood).

3.2.6 Major ash forming elements relevant for combustion

The elements calcium, magnesium, silicon and potassium are the main ash forming elements in wood. Primarily, the concentrations of calcium, magnesium and potassium but also sodium in the ash influence the ash melting behaviour, which is directly related to the reliability of the plant. Calcium and magnesium usually raise the ash melting point whereas potassium and sodium lower it [59]. A low ash melting point can lead to slagging and deposit formation in furnace and boiler. Silicon also influences the ash melting behaviour as low melting potassium silicates may be formed. Phosphorus is especially relevant for herbaceous fuels rich in this element. Phosphorus is semi-volatile and may also cause ash melting problems by the formation of phosphates.

Potassium, under the prerequisite of wood as the fuel and complete combustion taking place, is the main aerosol forming element (cf. Section 9.9). So, a high concentration of potassium prompts the formation of aerosols during combustion, which not only raises the emission of fine particulate matter but also increases fouling of the boiler [65]. Sodium behaves in a very similar way to potassium.

In Table 3.4 typical concentrations of silicon, phosphorus, potassium, calcium, magnesium and sodium are shown for various biomass ashes.

Table 3.4: Concentrations of major ash forming elements in biomass ashes

Explanations: data source [59; 66]

Element	Unit	Wood (spruce)	Bark (spruce)	Straw (wheat, rye)	Whole crops (wheat, triticale)
Ca	wt.% (d.b.)	26 - 38	24 - 36	4.5 - 8.0	3.0 - 7.0
K	wt.% (d.b.)	4.9 - 6.3	3.5 - 5.0	10.0 - 16.0	11.0 - 18.0
Mg	wt.% (d.b.)	2.2 - 3.6	2.4 - 5.6	1.1 - 2.7	1.2 - 2.6
Na	wt.% (d.b.)	0.3 - 0.5	0.5 - 0.7	0.2 - 1.0	0.2 - 0.5
P	wt.% (d.b.)	0.8 - 1.9	1.0 - 1.9	0.2 - 6.7	4.5 - 6.8
Si	wt.% (d.b.)	4.0 - 11.0	7.0 - 17.0	16.0 - 30.0	16.0 - 26.0

Due to the quite high potassium concentrations in straw and whole crops, problems concerning the ash melting behaviour may arise.

Magnesium, potassium and phosphorus are also of ecological relevance as they are plant nutrients and thus make the ash interesting as a fertilising and liming (calcium) agent for soils. Calcium in addition is of interest as a liming agent for soils.

3.2.7 Content of natural binding agents of raw materials and pellets

The content of binding agents, such as starch or fats, in the raw materials for pellet production is crucial for the pelletisation process itself and for the quality of the product. A high content of fats for instance decreases the energy consumption for pelletisation and the throughput of the pellet mill can be run up. This results in a decrease of operating costs. An increased amount of starch improves the binding behaviour and thus the mechanical durability of pellets. Biological additives (cf. Section 3.6) containing starch may also be used for this purpose. The utilisation of such biological additives is regulated in prEN 14961-2 (cf. Section 2.2). By adding other supplements, for instance lignosulphonate from cellulose production, similar effects can be achieved. Using such additives however modifies the physio-chemical characteristics of the biomass fuel and so stops the fuel from being natural and chemically untreated. It is therefore not allowed by prEN 14961-2 (only additives such as starch, corn flour, potato flour or vegetable oil are allowed).

Lignin is a vital constituent in view of a possible pelletisation of a raw material. It is an aromatic polymer providing wood with stiffness. If it is softened during the pelletisation process, pellets with a higher abrasion resistance can be produced. The softening point of lignin is usually not reached in common pellet mills however. Special processes, for instance steam explosion pre-treatment of the raw material (as described in Section 4.1.4), do make use of this attribute though. The softening point of lignin is around 190 to 200°C for dry wood and it sinks with increasing moisture content to around 90 to 100°C at a moisture content of 30 wt.% (w.b.). At a moisture content of 10 wt.% (w.b.) the softening temperature lies around 130°C [67]. In common pellet mills and at a moisture content of 12 wt.% (w.b.) a temperature of 80 to 90°C maximum is reached (depending on the technology), which is why the lignin softening temperature is not arrived at. Still, some manufacturers claim to have achieved temperatures of up to 130°C through special die configurations (mainly by varying the length of the press channel), which resulted in more durable pellets.

Hence, the lignin content of the raw material is significant in this respect; a conclusion that is also confirmed by [68], whereby the amount of fines of pellets decreases and the durability increases with rising lignin content. Lignin contents of various wood species are dealt with in Section 3.4.1.

3.2.8 Possible contaminations of raw materials

3.2.8.1 *Mineral contamination*

The contamination of raw materials with pebbles, soil and the like must be kept as low as possible as these contaminants influence net calorific value in a negative way and lead to an increased amount of ash both in the raw materials and the pellets. Mineral contaminants can also bring about problems in the pelletising process itself and wear off the die and rollers.

3.2.8.2 *Heavy metals*

Typical values for the concentrations of heavy metals in various biomass fuels are given in Table 3.5. Heavy metals such as cadmium and zinc are also to be limited in concentration in the ash, especially from an environmental point of view.

Herbaceous biomass generally shows much lower heavy metal concentrations than woody biomass, which can be explained by the shorter period of growth as well as the elevated pH-value of agricultural soil as compared to forest soil (less heavy metals available for the plants).

Heavy metals have a great impact on ash quality and particulate emissions from combustion. For ecological reasons, the heavy metals content of pellets has hence to be constrained. This is particularly important for small-scale furnaces because they are normally not equipped with particle precipitators and the ashes are often used as a fertiliser in private gardens. An elevated amount of heavy metals can originate from having used chemically treated raw materials such as waste wood or other disallowed foreign matter in pelletisation, but also geological constraints and the biomass species (heavy metal uptake) may cause considerable variations regarding the heavy metal concentration in biomass fuels. Moreover, due to the raw material drying process in directly heated dryers, fly ash can be accumulated in the dried raw material, which results in an elevated ash and heavy metal content of the pellets produced (cf. Section 4.1.1.2.3.2) [154]. In several European countries, regulations for biomass ash utilisation on soils exist, defining limiting values for heavy metals (cf. Section 9.11) [69; 70; 71; 72].

Table 3.5: Typical concentrations of heavy metals in various types of biomass fuels
Explanations: [1]...according to prEN 14961-2; data source [53]

Element	Unit	Limiting value[1]	Wood (spruce)	Bark (spruce)	Straw (wheat, rye)	Whole crops (wheat,
As	mg/kg (d.b.)	≤ 1	0.0 - 1.5	0.2 - 5	1.6	0.6
Cd	mg/kg (d.b.)	≤ 0.5	0.06 - 0.4	0.2 - 0.9	0.03 - 0.22	0.04 - 0.1
Cr	mg/kg (d.b.)	≤ 10	1.6 - 17	1.6 - 14	1.1 - 4.1	0.4 - 2.5
Cu	mg/kg (d.b.)	≤ 10	0.3 - 4.1	1.5 - 8.0	1.1 - 4.2	2.6 - 3.9
Pb	mg/kg (d.b.)	≤ 10	0.3 - 2.7	0.9 - 4.4	0.1 - 3.0	0.2 - 0.7
Hg	mg/kg (d.b.)	≤ 0.1	0.01 - 0.17	0.01 - 0.17	0.01	0 - 0.02
Ni	mg/kg (d.b.)	≤ 10	1.7 - 11	1.6 - 13	0.7 - 2.1	0.7 - 1.5
Zn	mg/kg (d.b.)	≤ 100	7 - 90	90 - 200	11 - 57	10 - 25

3.2.8.3 Radioactive materials

3.2.8.3.1 Sources for radioactivity in the environment and in biomass fuels

Radioactive material is found throughout nature. Natural background radiation comes from two primary sources, i.e. cosmic radiation and terrestrial sources.

Cosmic radiation primarily consists of protons (almost 90%), helium nuclei (alpha particles, almost 10%) and slightly under 1% are heavier elements and electrons from outside our solar system.

Terrestrial sources are contained naturally in the soil, rocks, water, air and vegetation. The major radionuclides of concern from terrestrial radiation are common elements with low-abundance radioactive isotopes, such as potassium and carbon, or rare but intensely radioactive elements such as uranium, thorium, radium and radon. Most of these sources have been decreasing due to radioactive decay since the formation of Earth and because there is no significant amount currently transported to Earth.

The radiation of soil is estimated to be 520 Becquerel per kg (Bq/kg) [73]. From soils natural radionuclides reach human food via water, plants and animals. The most important radionuclide in this context is ^{40}K. Consequently people contain a certain amount of natural radionuclides. The radiation from an adult person is approximately 7,000 Bq (equivalent to about 100 Bq/kg) [74] of which 4,000 Bq come from the naturally occurring ^{40}K radionuclide.

The concern for contamination of the biosphere from radionuclides (radioactive isotopes) has become an issue due to atomic bomb tests and nuclear accidents. The Caesium 137 (^{137}Cs) isotope is of particular interest since its presence in higher concentrations signifies contamination from manmade activities such as atomic bomb explosions, nuclear reactor failures and other releases from industrial processes. Large amounts of ^{137}Cs were released to the atmosphere from the Chernobyl accident in 1986 and contaminated large areas in Europe. Its radioactive half life amounts to 30.2 years. ^{137}Cs is accumulated in the upper layer of soils and therefore easily accessible for uptake in plants during their growth. Large amounts of

^{134}Cs were released during atomic bomb tests in the 1960s. Its radioactive half life amounts to 2.1 years and therefore plays a minor role today.

3.2.8.3.2 Radioactivity in biomass fuels

As already mentioned, radionuclides such as ^{137}Cs are, like heavy metals, accumulated in upper layers of soils and taken up from plants during their growth. Wood chips and bark used as fuels in different Austrian heating or CHP plants were analysed for their specific ^{137}Cs activity. The results are summarised in Figure 3.6. Moreover, the results of pellet samples analysed in Austria are also shown in this figure. It can be seen that the specific activities of Austrian wood chips are between 6 and 83 Bq/kg with a median value of 22 Bq/kg. The specific activities of the two bark samples analysed amount to 7 and 21 Bq/kg and the specific activities of pellets analysed in Austria range from 1.2 to 7.1 Bq/kg with a median value of 3.25 Bq/kg. It is worth mentioning that the limiting values for food are 370 Bq/kg for milk, dairy products and baby food and 600 Bq/kg for all other foods (cumulated radioactivity of ^{134}Cs and ^{137}Cs) [75]. This comparison clearly shows that solid biomass fuels such as wood chips, bark or pellets do not present any risk for people handling these fuels. However, it has to be considered that during combustion of biomass fuels, ^{137}Cs becomes concentrated in the ash. Similar to volatile heavy metals, ^{137}Cs is enriched in the fly ash particles. The bottom ash fraction is affected to a minor extent only [76; 77]. This issue is discussed in detail in the following sections.

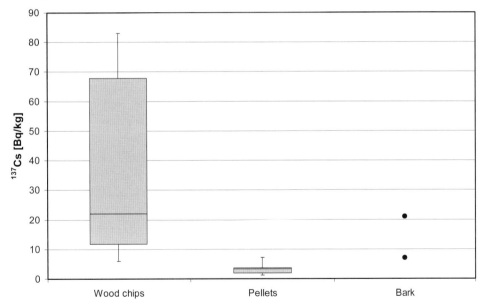

Figure 3.6: Specific activity of ^{137}Cs in biomass fuels

Explanations: wood chips: 10 samples; pellets: 10 samples; data source [76; 78]

3.2.8.3.3 Radioactivity in ashes from biomass combustion

Biomass fuels contain a certain amount of ash forming elements. During combustion, three different ash fractions are formed, i.e. bottom ash, coarse fly ash and aerosols (fine fly ash).

The bottom ash is the ash fraction remaining in the furnace after combustion. Coarse fly ash is partly precipitated on its way through the furnace and the boiler and therefore forms the so-called furnace or boiler ash. Particles leaving the boiler finally form the coarse fly ash and aerosol emissions at boiler outlet. Ash forming elements get fractionated among the different ash fractions depending on their volatility. Volatile heavy metals such as Cd or Zn for instance are enriched in fine fly ashes. Details concerning ash formation, ash fractions and fractionation of ash forming elements in biomass combustion systems as well as their precipitation in different applications are discussed in Section 9.9 and form the basis for the understanding of the behaviour of ^{137}Cs, the most important radionuclide in biomass fuels (cf. Section 3.2.8.3.1), during combustion.

^{137}Cs is, according to its vapour pressure, even more volatile than Cd and Zn. Therefore, its behaviour concerning fractionation among the different ash fractions during combustion is expected to be even more pronounced than for Pb, Cd or Zn. Consequently, ^{137}Cs is expected to be accumulated in the fly ash, especially in the aerosol fraction.

Radiation testing of ashes derived from biomass fuels are currently being done on a regular basis in some regions of the world. Results from respective measurements of the specific ^{137}Cs activities in ashes from biomass combustion plants in Austria are shown in Figure 3.7. The trend of enrichment of ^{137}Cs in the fly ash fraction has been confirmed. A very clear trend for the enrichment of ^{137}Cs with increasing specific activities from the bottom ash over the coarse fly ash to the fine fly ash (aerosols) is shown by Plant 07.

The ^{137}Cs activities in the corresponding biomass fuels were also analysed (cf. Figure 3.6). Based on these measurements, enrichment factors for ^{137}Cs from the biomass fuel to the different ash fractions were calculated (according to Equation 3.8). The enrichment factors for bottom ash amount to between 11 and 72 (average value 42), for coarse fly ash to between 57 and 682 (average value 264) and for fine fly ash 523 (only one measurement available). According to [79], enrichment factors of up to 85 were found for bottom ash and up to 113 for fly ash (based on woody biomass fuels).

Equation 3.8: $$EF = \frac{^{137}Cs\left[Bq/kg_{ash}\right]}{^{137}Cs\left[Bq/kg_{fuel}\right]}$$

From the measurement results of specific ^{137}Cs activities in pellets in Section 3.2.8.3.2 and the enrichment factors, it can be concluded that combustion of pellets with specific ^{137}Cs activities below 7.1 Bq/kg, which is the maximum value of the Austrian pellet samples analysed, will result in bottom ashes with specific ^{137}Cs activities of about 500 Bq/kg and in fly ashes with specific ^{137}Cs activities of about 4,900 Bq/kg in the worst case (based on the highest enrichment factors determined), which is still far below the limit for the definition of radioactive materials (cf. Section 3.2.8.3.4). Taking into account that average specific ^{137}Cs activities in pellets as well as average enrichment factors are considerably lower, it can be concluded that no risk concerning increased ^{137}Cs activities in ashes from pellet combustion must be expected (even not from the small amounts of emitted fine particles), based on the analyses available so far.

In general, in order to avoid any problems, it is recommended that raw material and ash samples are tested for their concentration of ^{137}Cs before new pellet plants are erected or a new source of raw material is solicited for an existing plant. However, it must also be taken into account that the problem of increased ^{137}Cs concentrations will decrease over the years, if no new and unforeseen release of radioactive elements occurs (e.g. by new reactor accidents or nuclear bomb explosions; in about ten years, the radioactive half life of ^{137}Cs resulting from the Chernobyl accident will be reached).

As a comparison to the Austrian measurement results mentioned above, the specific ^{137}Cs activities of ash samples from wood pellets produced in western Canada amounted to 50 to 100 Bq/kg [80]. According to [81], in local areas affected by the fallout from the Chernobyl accident, ^{137}Cs concentrations of up to 1,500 Bq/kg and in extreme cases 4,090 Bq/kg were found in ashes from biomass. In comparison, the radionuclides in coal ashes in the USA and Canada are reported to radiate 50–2,000 Bq/kg (on average 1,050 Bq/kg) [82], with radiation coming from uranium (^{238}U), thorium (^{232}Th) and their daughter radionuclides [83]. Ash from lignite type coal has the highest radioactivity with somewhat lower records for sub-bituminous and bituminous coal.

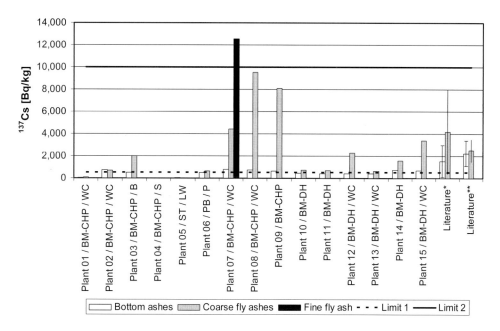

Figure 3.7: Specific activities of ^{137}Cs in bottom and coarse fly ashes as well as aerosols

Explanations: ST…stove; PB…pellet boiler; LW…log wood; P…pellets; WC…wood chips; B…bark; S…straw; limit 1…below this limit no restrictions concerning ash utilisation; limit 2…ashes below this limit can be disposed of in standard landfills; data source [76; 84]

3.2.8.3.4 Legal framework conditions

Regulation for disposal of ash containing radionuclides varies among jurisdictions. As an example, the Swedish regulations [85] stipulate that combustion plants producing more than 30 tonnes of ash per year need to determine the radiation from ^{137}Cs. Ash with a radiation exceeding 0.5 kBq/kg is considered contaminated and subject to regulations. The following rules apply to the disposal of ash in Sweden:

- Ash with a specific ^{137}Cs activity of > 0.5 and < 10 kBq/kg is allowed to be spread in the forest but not on agricultural soil. This ash is also allowed to be used for geotechnical applications (filling material).

- Ash with a specific ^{137}Cs activity of > 10 kBq/kg has to be disposed of in locations approved for such disposal.

- Ash with a specific ^{137}Cs activity of > 0.5 kBq/kg is not allowed to be disposed of in a location where the radioactivity in a closely situated water well will exceed 1.0 Bq/litre of water. For release of leach water contaminated with ^{137}Cs to a surface water recipient, the regulation stipulates that the resulting ^{137}CS concentration in the surface water recipient is not allowed to exceed 0.1 Bq/litre

Besides the regulation in Sweden there is also a recommendation from the Swedish Radiation Protection Institute. According to this regulation, ashes with specific ^{137}Cs activities above 5,000 Bq/kg should not be recycled in forests [86].

As another example, the Austrian regulation (Radiation Protection Ordinance [87]) defines substances with specific ^{137}Cs activities of more than 10,000 Bq/kg to be radioactive material. Below 500 Bq/kg there are no restrictions concerning ash utilisation. Between 500 and 10,000 Bq/kg ashes can be disposed of in standard landfills. As can be seen in Figure 3.7, all bottom and coarse fly ashes are below the limit for a substance to be considered a radioactive material. Many bottom ashes are even below the limit of 500 Bq/kg. Only one fine fly ash sample shows a specific activity of above 10,000 Bq/kg.

Review of national or local regulations is recommended as they may differ from those indicated above.

Concerning the use of ashes from pellet combustion as a fertilising and liming agent in gardens, agricultural fields and forests, it must be considered that ashes with ^{137}Cs concentrations above 10,000 Bq/kg are defined as radioactive substances (e.g. in Austria and Sweden) and must not be used on soils as fertilising and liming agents. Bottom ashes have typical specific ^{137}Cs activities below 1,000 Bq/kg and they are typically used in an amount of about 1 t/ha as a fertilising and liming agent on soils. This results in an additional specific ^{137}Cs activity of up to 100 Bq/m^2. Compared to the average ^{137}Cs activity of 20,000 Bq/m^2 from soils in Salzburg (one of the Austrian federal states), for instance, it can be seen that the use of bottom ash would increase the ^{137}Cs activity by less than 0.5%, which would be of minor relevance.

3.3 Evaluation of interdependencies between different parameters

3.3.1 Interrelation between abrasion and particle density of pellets

On the basis of numerous claims by pellet producers, a correlation between particle density and abrasion was expected. Investigations carried out by [88], where pellets were produced of the same raw materials but with different fineness and processed in a pellet mill with different compression channel lengths, did not demonstrate any interrelation between abrasion and particle density. These results lead to the assumption that other parameters, such as moisture content (cf. Section 3.3.2) or using biological additives containing starch (cf. Section 3.3.3), influence the abrasion of pellets stronger. Thus, further research needs to be done in order to clarify whether there are such interrelations.

Investigations performed by [89] revealed similar results with a better correlation efficient between durability of pellets and their particle density (i.e. 0.33, a statistically significant correlation; cf. Figure 3.8). Thus, there is an interrelation between particle density and mechanical durability of pellets.

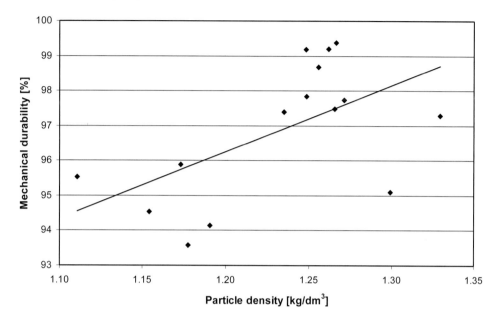

Figure 3.8: Relation between particle density and durability

Explanation: coefficient of correlation $r^2 = 0.33$ (statistically significant correlation); source [89]

3.3.2 Interrelation between abrasion and moisture content of pellets

As indicated already in Section 3.2.3 investigations by [68] showed binding properties of water in pelletisation and some correlation between moisture content and abrasion of pellets was established. The pellets investigated were produced at a pellet mill under constant framework conditions so that other parameters that could have influenced the abrasion were held constant. Investigations carried out by [90] verify the correlation. In pelletising trials with sawdust from spruce, an optimal moisture content of between 12 and 13 wt.% (w.b.), with abrasion being at a minimum, was found. Both higher and lower moisture contents cause more abrasion.

The results of [68; 90] were confirmed partly by pellet producers who agreed that moisture content has a very narrow scope for variation when producing high quality pellets. They state values of 8 to 12 wt.% (w.b.). A differing moisture content leads to worse quality and above all more abrasion.

The moisture content of pellets cannot only be influenced during pellet production but can also change during storage (which can have several effects on pellet quality). Biomass fuels are always sensitive to air humidity as they tend to either absorb or release moisture. This is also the case for wood pellets. Unfavourable storage conditions can cause moisture and weight increases. Not only the net calorific value is thus affected, but also the resistance towards mechanical stress can be reduced. This is shown in Figure 3.9. Moisture uptake was stimulated by storage in a controlled atmosphere. With increasing fuel moisture the pellet abrasion as measured in durability tests rises slowly when an atypically low moisture content of below 8 wt.% (w.b.) is given. However, the durability decreases rapidly when the moisture content level rises to 10 wt.% (w.b.) or more. Such unfavourable storage conditions should therefore be avoided.

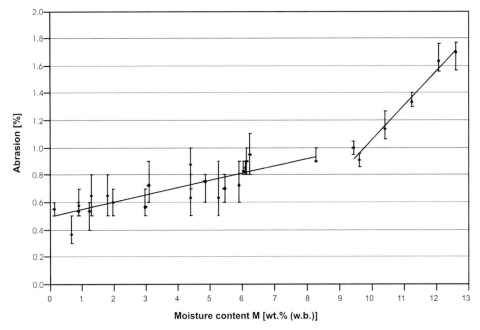

Figure 3.9: Abrasion of wood pellets as a function of a varying moisture content

Explanations: pellet samples treatment: drying oven (105°C) at variable storage times in a climate chamber; abrasion tests method according to ÖNORM M 7135; performance of 4 replications per moisture step; data source [91]

Similar results were found by [92]: above a moisture content of 10 wt.% (w.b.), abrasion rose disproportionately.

3.3.3 Interrelation between abrasion and starch content of pellets

Pelletisation trials of spruce sawdust with the addition of maize starch showed a correlation that is next to linear [90]. Abrasion declined with increased starch content. Within the analyses programmes this correlation could not be confirmed, which is, however, most probably due to the fact that the samples analysed were made by different producers. This indicates that other parameters are also of great relevance for achieving a high mechanical durability.

3.3.4 Influence of raw material storage time on bulk density, durability and fines of pellets as well as on energy consumption during pelletisation

Pelletising trials with spruce and pine showed that extended storage has positive effects on bulk density, durability and amount of fines of the pellets whereas energy consumption rises when the raw materials are stored for longer [93]. It is assumed that fatty acids and resins are degraded through oxidation processes during storage, which then increases frictional forces inside the compression channel and in turn leads to a higher bulk density, durability and energy consumption and to fewer fines. Currently, test runs are being carried out to check this assumption [94]. Similar tests were performed by [95] and pellets made of raw materials that had been stored were found to be more durable and bulk density was higher. Although a variation in resin and fatty acids concentrations was found, a direct correlation of those concentrations and the quality of pellets could not be established. This is why other differences between fresh and stored raw materials are assumed. This requires further investigation. Experiences by producers [96] also illustrate this effect. It was discovered that sawdust that had been stored for four to six weeks led to a pellet quality that could otherwise only have been achieved by using biological additives.

3.3.5 Influence of the contents of sulphur, chlorine, potassium and sodium on the corrosion potential of pellets

Looking at the corrosion potential of fuels in furnaces a key indicator was established by [53] between the molar ratio of sulphur in the fuel and available alkali compounds and chlorides ($M_{S/AC+Cl}$) according to Equation 3.9. The corrosion potential is low when the ratio exceeds a value of ten. The logic behind this is that a high concentration of sulphur in the flue gas reduces the formation of alkali chlorides and augments the formation of alkali sulphates in return. The latter are more stable and less corrosive so that the corrosion potential is lowered. The corrosion potential addresses high temperature chlorine corrosion, which is usually the most relevant corrosion process in biomass combustion plants.

Equation 3.9: $$M_{S/AC+Cl} = \frac{2 \cdot X_S}{X_{Cl,avail.} + X_{K,avail.} + X_{Na,avail.}}$$

Explanations: X_S...sulphur concentration in the fuel [mol/kg (d.b.)]; $X_{i,avail}$....concentrations of available chlorine (Cl), Potassium (K) and Sodium (Na) in the fuel [mol/kg (d.b.)]

In order to calculate the ratio, the concentrations of chlorine, potassium and sodium available in the flue gas have to be known. Those fractions that get bound in the coarse ash are not available in the flue gas. Investigations by [97] show that the availability of those elements in the flue gas differs according to the fuel used. The amount of chlorine that is available in the flue gas as compared to the chlorine content of the fuel is around 94.0% for bark, 97.0% for wood and 99.0% for straw. In the same investigations it was found that the availability of potassium is around 20% for bark and 40% for both wood and straw of the respective total amount present in the fuel. The availability of sodium also lies around 20% for bark, 40% for wood and around 50% for straw. Sodium was not analysed in the analysis programme, which is why investigations by [98] were consulted, who asserts that the content of sodium is on average 7% of the amount of potassium in biomass fuels. The usual content of sodium is so little that errors in calculation should be trivial if they occur at all due to this assumption so that it can be neglected in calculating the molar ratio of sulphur in the fuel and available alkali compounds and chlorides.

On these grounds and considering the results from the analysis of densified biomass fuels, the molar ratio of sulphur in the fuel and available alkali compounds and chlorides was calculated according to Equation 3.9. The results are illustrated in Figure 3.10.

The noticeably low values of 0.35 to 0.42 for the straw pellets as well as the elevated corrosion potential thereof was to be expected because of the high natural chlorine content of straw and the fact that straw generally causes problems concerning corrosion when used as a fuel in biomass furnaces. The two bark briquettes have a ratio of 1.3 and 2.4, the ratio of wood pellets lay between 0.59 and 2.1 and the ratio of wood briquettes between 0.54 and 1.7, whereby the median of the wood briquette ratios is notably below the median of the wood pellet ratios. That is because the briquettes put to the test probably contained some bark. Three specimen out of the wood pellets samples, the ones explicitly displayed in Figure 3.10, exhibit very high ratios, namely 7.0 to 27.8. These samples revealed high concentrations of sulphur stemming from the addition of lignosulphonate as a binding agent (according to the producers).

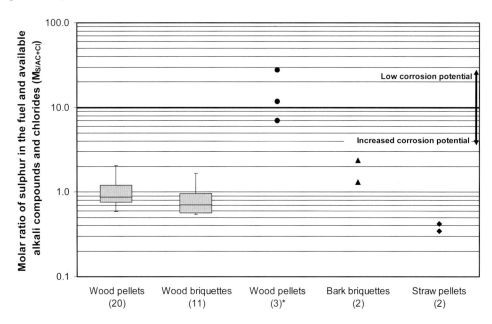

Figure 3.10: Molar ratio of sulphur in the fuel in relation to available alkali compounds and chlorides ($M_{S/AC+Cl}$) as an indicator for high temperature chlorine corrosion potential during combustion

Explanations: calculation of $M_{S/AC+Cl}$ according to Equation 3.9; *...the wood pellets investigated contained lignosulphonate as a binding agent; the numbers in brackets indicate the quantity of samples analysed

As observable in Figure 3.10, fuels made of woody biomass (wood and bark) have relatively low molar ratios of sulphur in the fuel and available alkali compounds and chlorides (apart from the three wood pellet samples displayed separately in Figure 3.10), which indicates a certain corrosion potential of these fuels. Even lower ratios, like those of the two straw pellet samples, indicate a high risk of corrosion, which coincides with the experience from plants using straw [99, 100, 101]. So, wood pellets have less corrosion potential than straw pellets.

In order for the corrosion potential illustrated in Figure 3.10 to actually cause corrosion, not only the amount of the respective elements in the fuel and their availability in the flue gas is relevant but also the thermal conditions in the furnace/boiler region. The respective interrelations are demonstrated in Figure 3.11. At high flue gas temperatures and high surface temperatures of the tubes, the risk of corrosion is high. In hot water boilers like those used in pellet furnaces, there is practically no risk of corrosion because of the low surface temperatures (typically around 100°C). Still, high flue gas temperatures can cause problems to components made of steel in uncooled combustion chambers of pellet furnaces.

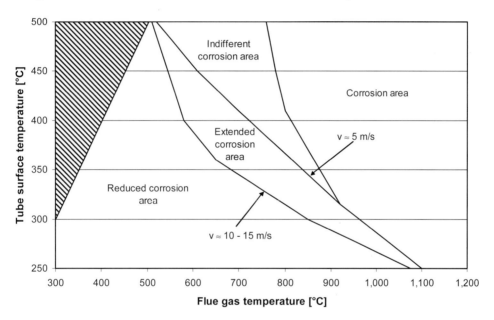

Figure 3.11: Extended corrosion diagram

Explanations: data source [102]

3.3.6 Correlation between measured and calculated gross calorific value

Equation 3.10 is an empiric formula as per [103] and is used for calculation of the gross calorific value of biomass fuels [53]. Concentrations of carbon, hydrogen, sulphur, nitrogen, oxygen and ash have to be known for this calculation. All parameters apart from oxygen were analysed in the analysis programme. The oxygen content can be approximated as the difference between 100 minus the sum of carbon, hydrogen, sulphur, nitrogen and ash (in wt.% (d.b.)).

The values calculated with Equation 3.10 were compared to measured values by means of a scatter diagram (cf. Figure 3.12). It reveals a clear correlation between the values calculated and measured, which is confirmed by the coefficient of correlation r^2 of 0.73 too. It implies that there is a highly significant correlation between measured and calculated values (probability of error < 1%) and that the gross calorific value can be adequately approximated by the calculation based on elementary analysis. It must be noted, however, that on average the measured values do lie 1.8% lower than the calculated values (calculation according to Equation 3.11).

Equation 3.10: $GCV = 0.3491 \cdot X_C + 1.1783 \cdot X_H + 0.1005 \cdot X_S - 0.0151 \cdot X_N - 0.1034 \cdot X_O - 0.0211 \cdot X_{ash}$

<u>Explanations</u>: GCV in MJ/kg (d.b.); X…concentrations of carbon (C), hydrogen (H), sulphur (S), nitrogen (N), oxygen (O) and ash; data source [103]

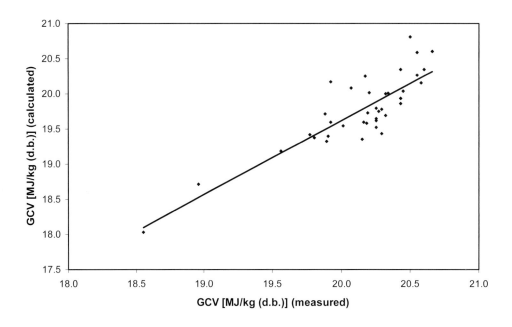

Figure 3.12: Correlation between calculated and measured gross calorific value of densified biomass fuels

<u>Explanations</u>: the gross calorific value was calculated according to Equation 3.10; $r^2 = 0.73$

Equation 3.11: $\Delta GCV = \dfrac{\sum\limits_{i=1}^{n}\left(\dfrac{GCV_{meas.,i} - GCV_{calc.,i}}{GCV_{meas.,i}}\right) \cdot 100}{n}$

<u>Explanations</u>: ΔGCV in %; $GCV_{meas.,I}$ and $GCV_{calc.,I}$ in MJ/kg (d.b.)

3.4 Ligno-cellulosic raw materials for pellets

3.4.1 Softwood and hardwood

Theoretically, any kind of woody biomass is a possible raw material for pelletisation. Differences in evaluation originate from the different characteristics and the different supply routes of the materials. Table 3.6 presents an overview of these.

Industrial wood chips are machine-ground waste and by-products from wood working and wood processing industry with the sawmill industry being the main producer. Forest wood chips come directly from the forest and are intended for thermal utilisation. Both wood shavings and sawdust result from cutting processes. Wood dust stems from mechanical surface treatment of wood.

The NCV of different wood species is no criterion for the choice of raw material. The limit for the ash content is usually held by softwood and usually exceeded by hardwood. Thus, producing pellets from hardwood would not meet the requirements of prEN 14961-2; this can be overcome by appropriate mixing of softwood and hardwood though. Still, it must be noted that hardwood is generally less suitable for pelletisation than softwood [104]. One reason for this is that hardwood has lower lignin content than softwood and hence pellets that are less durable are produced. Moreover, hardwood is more difficult to process due to its higher density, which results in increased energy consumption of the pellet mill.

The limiting values for nitrogen, sulphur and chlorine contents are complied with by softwood in most cases. It is only the sulphur content that may be too high in some cases. As concerns hardwood, beech normally exceeds the allowed nitrogen value.

Looking at industrial wood chips from sawmills, mineral contamination may be due to storage design (paved or unpaved). Industrial wood chips coming from the wood processing industry can be expected to have a low degree of mineral contamination as the industry generally uses sawn timber. Sawdust and wood shavings are normally not, or only to a small extent, contaminated by foreign matter either. Contamination of forest wood chips may occur during harvest or transport.

Owing to the particle size of industrial and forest wood chips, they have to be ground before pelletisation in any case. Sawdust and wood dust are already suitable for pelletisation in their particle size but usually they undergo grinding in a hammer mill before pelletising for reasons of homogeneity. Depending on the size, wood shavings are partly suitable and partly have to be ground. Small wood shavings from fast running machines for instance can be pelletised without prior grinding.

Drying before pelletising is necessary for industrial wood chips, sawdust as well as forest wood chips. Drying may be left out using shavings from the wood processing industry, depending on residual moisture and raw material.

Wood does not contain strong binding agents of its own, such as in the case in animal food containing starch for instance. Glide characteristics in the compression channel are not pronounced either, which causes strong frictional forces to arise very quickly inside the channels. These forces are needed though because of the low degree of self-binding.

A natural binding agent in wood is lignin. The more lignin is present in the wood, the more mechanically durable are the produced pellets. More durable pellets can thus be made of softwood such as spruce or fir, rather than of hardwood such as beech or oak. Oak can contain more lignin though, as can be seen in Table 3.6.

Regarding carbon, hydrogen, oxygen, volatile and other combustion relevant contents as well as heavy metals, see Table 3.1 and Table 3.4. Carbon, hydrogen, oxygen, and volatile contents are solely dependent on the raw materials and can thus not be influenced. Concentrations of ash forming elements and heavy metals are usually within the range of the guiding values given in Table 3.4 and Table 3.5. Higher concentrations may be a sign of having used disallowed foreign matter in pelletisation.

Table 3.6: Overview of different woody biomass fractions with regard to their use in pelletisation

Explanations: [1]...limiting value according to prEN 14961-2 for pellet class A1; [2]...spectrum of the foremost particle size according to ÖNORM 7133, extreme values being possible both beneath and above; h...as high as possible; l...as low as possible; data source [59; 64; 105; 106], own research

Parameter	Unit	Limiting/ Guideline value	Average value of the raw material	Comments
Net calorific value	kWh/kg (w.b.)	h	approx. 4.9	Hardwood (beech)
			approx. 5.2	Softwood (spruce)
Ash content	wt.% (d.b.)	0.7	1.0 - 1.3	Hard wood (beech, oak)
			0.37 - 0.77	soft wood (spruce, fir)
Nitrogen content	wt.% (d.b.)	0.3[1]	0.21 - 0.41	Hard wood (beech, oak)
			0.07 - 0.11	Soft wood (spruce, fir)
Sulphur content	wt.% (d.b.)	0.03[1]	0.02 - 0.05	Hard wood (beech, oak)
			0.01 - 0.05	Soft wood (spruce, fir)
Chlorine content	wt.% (d.b.)	0.02[1]	approx. 0.01	Hard and soft wood
Mineral contamination	wt.% (d.b.)	l	possible	IWC from BPS, depending on design of storage area, FWC caused by harvest and logging
			low	IWC from WWI
			very low	Sawdust and wood shavings
Particle size	mm	< 4	2.8 - 63[2]	IWC, FWC
			5 - 12	Wood shavings
			< 5	Sawdust
			< 0.315	Wood dust
Moisture content	wt.% (w.b.)	8 - 12	40 - 55	IWC and sawdust from BPS
			< 20	IWC and sawdust from WWI
			40 - 60	FWC at the moment of harvest
			25 - 35	FWC after usual preparation
			< 10	Wood dust
Lignin content	wt.% (d.b.)	h	21.9 - 24.0	Beech
			20.8 - 30.0	Spruce
			approx. 27.0	Fir
			26.0 - 28.6	Larch
			27.5 - 28.6	Oak

Due to the advantages of low moisture content and small particle size, shavings and wood dust from processing dried sawn timber are preferred raw materials in pellet production since grinding and drying efforts are lessened. However, almost all available quantities of dry shavings and wood dust are already being used to produce pellets. Therefore, sawdust from sawmills is at present the most important raw material for pelletisation. In the meanwhile, strong pellet market growth of recent years has already led to shortages of the "classic" raw materials. As a consequence, some pellet producers have begun to use other raw materials [107; 108; 109; 110; 111; 112; 116]. These are forest wood chips out of thinning as well as industrial wood chips but also log wood in some cases. Using these raw materials causes an increased grinding effort, which in turn leads to higher production costs. Whether pellet production from forest wood chips, industrial wood chips or log wood can be economical is discussed in Section 7.3.

3.4.2 Bark

Bark is the outermost layer of the tree stem as it accumulates during bark separation of wood. Bark separation is above all performed in the sawmill and paper industry, which is why these two are the foremost producers of bark.

Pellets containing bark or purely made of bark are assigned to pellet class A2 according to prEN 14961-2 definition. The limiting values differ compared to those for wood pellets (class A1). Table 3.7 gives an overview of concentrations found in natural bark.

The net calorific value of bark is approximately the same as for wood. The ash content is significantly higher and it can be further augmented by mineral contamination, which may be quite high as the result of the preceding process steps (harvesting, forwarding, transport and storage).

Due to the higher natural ash content of bark, the limiting value for ash in pellets according to class A2 is also higher, i.e. 1.5 wt.% (d.b.), than for wood pellets. However, the comparison of the limiting value according to prEN 14961-1 with the average values for bark shows that pure bark pellets cannot possibly be specified as A2 pellets. Only pellets made from bark containing wood fractions such as forest wood chips, industrial wood chips with bark and short rotation coppice (SRC) would comply with this pellet class. The ash content excludes bark pellets from the utilisation in small-scale installations such as residential central heating systems and stoves because the ease of use is considerably reduced when the ash box has to be emptied very frequently. The amount of ash produced by the combustion of bark pellets is 4 to 10 times higher than that of wood pellets. Bark pellets are suitable for medium- and large-scale plants in which the ash content and ease of use are not primarily important since the ash is usually automatically discharged in such plants. In small-scale plants, mixing bark with wood pellets alters the mechanical durability. Some manufacturers of pellet mills claim that by adding the proper amount of bark no additional binding agents are necessary. This is confirmed by [112] and also by the fact that bark contains a high amount of lignin (cf. Table 3.7) and the mechanical durability of pellets increases with the rise of the lignin content (cf. Section 3.2.6).

The contents of nitrogen, sulphur and chlorine are also higher in bark than in wood, which is considered in higher limiting values for class A2.

Bark has to undergo grinding before pelletising since it leaves the sawmill and paper industry in long strings and with varying particle sizes due to the processes used. It also requires drying as the moisture content of bark at sawmill can be 45 to 65 wt.% (w.b.). The outlet moisture content does not have to be as low as for wood though because the guiding value for compressed bark is 18 wt.% (w.b.). What poses more difficulties is the grinding of bark as it cannot be ground as easily as wood. Therefore, cutting mills are typically used. Because of this and because of the ash content, briquetting is preferred to pelletising if bark is to be densified.

Table 3.7: Parameters for the production of compressed bark

Explanations: [1]...limiting value according to prEN 14961-2 for pellet class A2; l...as low as possible; h...as high as possible; data source [59; 63; 64; 66]

Parameter	Unit	Limiting/ guiding value	Average value of bark	Comments
Net calorific value	kWh/kg (w.b.)	4.5	4.9 - 5.6	Bark, mixed
Ash content	wt.% (d.b.)	1.5	2 - 5	Non-contaminated bark up to 5 wt.% (d.b.), higher values caused by mineral contamination
Nitrogen content	wt.% (d.b.)	0.5[1]	0.3 - 0.45	
Sulphur content	wt.% (d.b.)	0.03[1]	0.035 - 0.055	
Chlorine content	wt.% (d.b.)	0.02[1]	0.015 - 0.02	
Mineral contamination	wt.% (d.b.)	l	relatively high	Because of preliminary processing steps
Grain size	mm	< 4	>> 4	Very inhomogeneous in the form of long strips of bark up to 0.5 m
Moisture content	wt.% (w.b.)	< 18	45 - 65	Usual moisture content of bark from sawmills
Lignin content	wt.% (d.b.)	h	16.2 - 50.0	Bark mixed

3.4.3 Energy crops

Energy crops are wood that is grown specifically for energetic utilisation. In Austria such energy crops are not yet of great relevance. Fast growing tree species that can be used for energy production are poplar, alder and willow for instance. Such wood species can normally be pelletised easily so that their applicability for pelletisation cannot be doubted. Willow and poplar have proved to be very suitable for pellet production showing good quality [113; 114; 115].

Until recently the planting of energy crops for the production of pellets was questionable from an economic point of view since the harvested wood needs to be ground and dried (cf. Chapter 4) in order to be converted into pellets (the question of economic efficiency for specific pelletising processes is dealt with in Section 7.2). The direct use of wood chips in medium- and large-scale plants seems to be more reasonable, not only from an economic but also from an emission point of view. Due to higher ash and potassium concentrations, particulate matter emissions will also increase and make dust precipitators recommendable.

Moreover, also the contents of ash, N and S would in some cases cause problems, when class A1 or A2 pellets according to prEN 14961-2 should be produced from SRC (cf. Table 3.8).

Meanwhile, the basic conditions have changed though in the sense that on the one hand prices for fossil fuels have risen dramatically (cf. Section 8.1), which rendered the pelletisation of energy crops more economical, and on the other hand market growth in the pellets sector led to a shortage of easily available raw materials such as sawdust and wood shavings, which made market players look for alternative raw materials for pellet production. A lot of research has been carried out in this respect, including some trial areas where energy crops are grown for pelletisation (cf. Section 12.1.1.2) [114; 115; 116; 117; 118; 119].

Table 3.8: Typical ash, N, S and Cl contents in poplar and willow

Explanations: [1]...for class A1 pellets according to prEN 14961-2; [2]...for class A2 pellets according to prEN 14961-2; data source [53; 64; 66]

Parameter	Unit	Limiting value[1]	Limiting value[2]	Value
Ash content	wt.% (d.b.)	0.7	1.5	0.6 - 2.3
Nitrogen content	mg/kg (d.b.)	3,000	5,000	1,000 - 9,600
Sulphur content	mg/kg (d.b.)	300	300	300 - 1,200
Chlorine content	mg/kg (d.b.)	200	200	100

3.5 Herbaceous raw materials for pellets (straw and whole crops)

Whole crops are annual energy crops of which, in contrast to straw, both stems and grains are used for energy generation. Straw is a by-product of grain harvesting, so that some experience about the pelletisation of straw is at hand already from the animal feed industry. Straw and whole crops can be summed up as herbaceous biomass. As shown in Table 3.9, the ash content of straw and whole crops is much higher than that of wood.

Table 3.9: Guiding values for straw and whole crops in comparison with values from prEN 14961-2 and general guiding values for the production of class A1 and A2 pellets

Explanations: optimum moisture content for herbaceous biomass according to [59]; data source [59; 62]

Parameter	Unit	Limiting/guiding value for class A1	Limiting/guiding value for class A2	Straw	Whole crops
Net calific value	kWh/kg (w.b.)	4.6 - 5.3	4.5 - 5.3	approx. 4.8	approx. 4.8
Ash content	wt.% (d.b.)	0.7	1.5	4 - 6	4 - 6
Nitrogen content	wt.% (d.b.)	0.3	0.5	0.3 - 0.5	0.6 - 0.9
Sulphur content	wt.% (d.b.)	0.03	0.03	0.05 - 0.11	0.1 - 0.12
Chlorine content	wt.% (d.b.)	0.02	0.02	0.25 - 0.4	0.1 - 0.3
Grain size	mm	4	4	Halm form	Halm form
Moisture content	wt.% (d.b.)	8 - 12	8 - 12	10 - 20	10 - 20

The concentrations of nitrogen, sulphur and chlorine in herbaceous biomass have to be rated more problematic than in wood with regard to the combustion technique. Other critical aspects of herbaceous biomass are its low ash melting point [117] and its elevated fly ash and aerosol emissions [65]. For those reasons the pelletisation of herbaceous biomass for use in small-scale plants is currently not recommended.

Straw and whole crops must be ground before pelletising because of their shape (stalks). During growth, the moisture content can reach more than 50 wt.% (w.b.) but it drops to below 20 wt.% (w.b.) the more the plants ripen. Harvested straw and whole crops have a moisture content between 10 and 20 wt.% (w.b.) after two or three days of drying on the field, which makes pelletisation without further drying possible.

An interesting option is to blend woody with herbaceous raw materials in order to produce pellets. Herbaceous raw materials are often available at a low price (e.g. excess straw). Pelletising herbaceous materials alone and in particular their thermal utilisation is not advisable for the reasons already mentioned (low ash melting point, high contents of elements relevant for emissions such as sulphur or chlorine, among others). Appropriate mixtures of these raw materials might lead to the reduction or even avoidance of these problems. In the laboratory, good pelletisation results were achieved by mixing willow with wheat straw or shredded wheat concerning one of the major quality criteria of prEN 14961-2, i.e. mechanical durability [115]. This has yet to be verified at industrial scale since lab results are not to be assigned directly to industrial scale without further ado. Combustion tests using such pellets have not been performed yet. [120; 121] go to similar directions in that they use different materials in pelletisation, i.e. sawdust, straw, sunflower hulls, grains and nutshells as well as different mixtures of these materials. In addition, additives such as aluminium hydroxide, kaolinite, calcium oxide and limestone were added to some extent to reduce slag formation. This also led to a good quality of the pellets. In combustion trials, however, wood pellets that were used as a reference proved to be a class on their own when looking at combustion behaviour, slag and deposit formation. All other pellets caused slagging and deposits to some degree. Further possibilities for making mixed pellets are illustrated in [122; 123].

At this point it should be realised that such approaches do not conform to prEN 14961-2. Respective standards are yet to be created and are in fact planned within Part 6 of prEN 14961. In any case the venture should be pursued since sales volumes of pellets are rising internationally, leading to shortages in the supply of the main raw materials used at present (i.e. woody biomass residues from wood processing industries).

Still, mixed pellets seem to be suitable for large-scale installations only. By mixing raw materials in the right way, advantages and disadvantages of materials can be balanced out.

3.6 Additives

3.6.1 Organic additives

According to the definition in Section 2.1.1, only unmodified biomass products from primary agriculture and forestry may be used as biological additives. They serve to improve the densification process, which in turn leads to a better energy balance and throughput or they serve to improve mechanical durability as well as the moisture resistance of the pellets.

Adding cocoa shells for instance, especially when pelletising hardwood, can notably reduce the energy demand of the process [43]. If better mechanical durability is desired, raw materials containing starch can be used, i.e. shredded maize, maize starch or rye flour [2]. Promising pelletising trials were carried out with various natural biomass materials as biological additives; lessened energy consumption, potentially higher throughputs and improved pellet quality were found [90; 115].

The reason why the use of biological additives is restricted to just natural, unmodified and unaltered products from primary agriculture and forestry is important from an ecological standpoint but most of all raising confidence in the product in the customers is of great relevance. The application of such biological additives is regulated by prEN 14961-2 (cf. Section 2.2). The use of lignosulphonate as an additive is known in Scandinavia. Trials have shown that lignosulphonate enhances the durability of pellets better than starch [124]. The problem is that lignosulphonate is a product of physio-chemical treatment of biomass and, because of that, renders the pellet inauthentic. Thus, the use of lignosulphonate is not allowed by prEN 14961-2.

The use of biological additives is essential in producing pellets with a diameter of 6 mm. Such pellets are prescribed by many manufacturers of furnaces for residential heating in order to ensure a failure free feed and uniform burnout behaviour. Bigger installations can cope with fuels of lower quality (less mechanical durability, high ash content) and larger diameter more easily than smaller installations because they are in general built to be more robust. More often than not they are equipped with automatic cleanout systems, the conveyor system is more robust and the risk of dust explosions can be managed with a system of sensors in combination with the possibility of water injections. Pellets for big furnaces are often produced with a diameter of 8 or even 10 mm so that biological additives are not in fact necessary.

Experience has shown that through storage of sawdust for pelletisation for four to six weeks, the pellets show better qualities, which can otherwise only be achieved by using biological additives (cf. Section 3.3.4). By mixing the raw materials in the right way, biological additives can be spared too. Mixing spruce and Douglasie for instance leads to pellet qualities that for spruce alone can only be realised by adding biological additives.

The use of additives in pelletisation in Denmark is very limited due to legislation that added a tax on sulphur in additives used earlier. Also, legislation prohibits the use of any non-natural biomass in pellets. Pellets containing any kind of waste would be classified as a waste product; this would generate taxation as waste and require combustion only in plants that have been approved according to the EU waste incineration directive.

In Germany, only additives made of starch, vegetable paraffin or molasses are allowed (according to the currently valid emission regulation (1.BImSchV)). The amendment of the 1.BImSchV (according to the draft from May 2009 [31]) will change this regulation and starch, vegetable stearin, molasses and cellulose fibres will be allowed to be used as additives in pellet production.

3.6.2 Inorganic additives

With regard to inorganic additives it must generally be noted that such additives increase the ash content of biomass fuels, they usually increase the operating costs of pellet production and they can also increase the emission of coarse fly ashes. These facts must be taken into account when inorganic additives are used.

3.6.2.1 Fuels with low content of phosphorus

Fuels with a low content of phosphorus are usually woody biomass fuels with phosphorus contents typically of around 0.005 wt.% (d.b.).

Problems caused by slagging occasionally occur in small- and medium-scale combustion equipment. Slagging is related to pellets with higher ash content or log wood pellets contaminated during storage or transport. The problem occurs when ash forming contents melt directly on or close to the grate. Therefore, the temperature in the area where the ash is formed, and the composition of the ash forming species, are parameters important for slagging. A study including three different burner techniques and seven pellet qualities showed that today's small-scale applications are relatively sensitive to variations in ash content and composition [125]. The amount of slag found in the burner depended on fuel and burner technique and the degree of sintering (the hardness of the slag) depended on the composition of the fuel. The results of the study indicate that the content of silica, Si, influences the initial melting temperature of the ash. In another study, the critical levels of problematic ash components in stemwood pellets regarding slagging were determined by a statistical and chemical evaluation of a database of fuels. The results showed that the problematic wood pellets had a significantly higher amount of Si in the fuel ash. The critical level of Si (given as SiO_2) was about 20 to 25 wt.% of the fuel ash, that is to say that pellets with levels in or over this range resulted in slagging problems in residential burners [154].

There are two possibilities to avoid the formation of slag. One is to reduce the temperature and the other is to alter the composition of the ash forming content. The latter can be achieved by using some sort of additive either during the pelletising process or during combustion of the fuel.

In earlier studies with conventional biomass fuels, clay minerals [126; 127; 128; 129] and lime or dolomite based additives [129; 130] were used to increase the melting temperature of the formed bottom ash. Kaolin (porcelain clay) has proved successful [129]. Combustion of wheat pellets with kaolin added to an equivalent of 20% of fuel ash content showed an increase of fusion temperature by 250°C [128]. Admixing kaolin and calcite in log wood pellet production to reduce slagging tendencies was studied [131]. The additives were directly injected into the burner, or added during the pelletising process. Slagging tendencies were eliminated at mixing degrees corresponding to 0.5 wt.% of the dry substance of the pellets. Addition of additives did not pose any technical problems. The additional costs for the additives were estimated to be around 0.3 öre/kWh (0.0003 €/kWh). The best effects were achieved when the additive was added during the pelletising process. Direct injection into the burner led to reduced slagging, but the technique has to be optimised for the purpose.

The content of critical ash forming elements differs between stem wood, bark and different forest fuel fractions, and therefore the effect and the ideal way of admixing the additives vary between these materials. The addition of inorganic additives should be even more interesting for certain bark and forest fuel fractions because earlier studies showed that some of them are prone to slagging [125; 131; 132]. The addition of 1 wt.% calcium carbonate suspension on a dry fuel basis, injected via spray nozzles during the combustion of problematic bark and forest fuels, eliminated or reduced slag [133] in a 20 kW burner. The injection of even less kaolin by spray nozzles to the same combustion also reduced slag but had a negative effect on slag formation at higher mixing degrees. The injection by nozzles proved to be a mature technique ready to implement. The costs for the additives were less than 0.6 öre/kWh (0.0006 €/kWh).

The technique of injecting calcium suspension by means of spray nozzles between pellet storage and burner is feasible for real conditions. The equipment is cheap and robust and can easily be adapted to appliances for pellets, wood chips etc., although the results show that the best effects are reached if the additive is admixed to the raw material during the pelletising process.

The reduced slagging tendencies achieved can be explained by a more favourable melting behaviour when the amount of calcium or aluminium is increased, i.e. more of the ash is found as solid phase than as liquid phase at the combustion temperatures.

Contamination by sand in bark and forest fuels resulted in severely increased slagging tendencies. A significant reduction of particle emissions, both as total particulate matter (mg/Nm^3) and as number of fine particles (< 1µm) was achieved when kaolin was added to the raw material. No such effects on particle emission were achieved when calcium was added.

The effect of kaolin on particle formation was tested in a 65 kW burner. The fuels were two ash rich agricultural fuels: wheat straw pellets and Reed Canary Grass pellets. Kaolin was mixed to the pellets in an amount of 3 and 6 wt.%. The particles formed were measured as total particulate matter and as mass and number concentrations. Particle formation and sintering tendency was reduced [134].

3.6.2.2   Pellets mixed with peat

Peat may be regarded as an inorganic additive as the ash in peat is the active substance. Peat mixed with alkali rich fuels such as straw, willow and forest residues may have both positive or negative effects on slag formation and also on particle emission. The effect depends on the ash composition of the peat. Studies indicate that the relation between silica on the one hand and calcium and magnesium on the other is crucial [133; 135; 145]. These are ongoing studies and further details will be published in the future.

3.6.2.3   Fuels with high content of phosphorus

Fuels with a high content of phosphorus are usually herbaceous biomass fuels with phosphorus contents typically of around 0.1 wt.% (d.b.).

Forest based fuels are dominated by silicates, while many agricultural fuels or by-products (cereals, oil seed, sewage, meat, bone, olives etc.) are dominated by phosphates. These fuels are basically different regarding coagulation, slagging and particle formation, and possibly also regarding corrosion. There is still a lack of knowledge and understanding about these fuels and the possibilities of making use of mixing fuels or additives to improve their combustion behaviour. Increased knowledge of additives in fuels rich in phosphorus may in the future also be used for pellet production. Studies of additives in cereal grain have shown that slagging can be reduced effectively with calcite [136; 137] and with kaolin [138]. But calcite may also increase slagging tendencies, depending on the presence of silica. Both additives decreased formation of particles in the flue gas to some extent.

3.7 Summary/conclusions

As a first step towards assessment of the physio-chemical characteristics of raw materials and pellets, evaluation criteria for raw materials as well as for pellets were established. The criteria are based on prEN 14961-2, which prescribes limitations for certain substances in pellets; thus the raw materials have to meet these requirements too. Basic requirements of combustion and pelletisation technology were also incorporated.

In Table 3.10, the evaluation criteria for possible raw materials for pelletisation as well as characteristics of pellets are summarised and their effects are stated. Mineral contamination and particle size were identified and established as relevant parameters of raw materials. The parameters content of biological additives, moisture, ash, GCV, NCV, carbon, hydrogen, oxygen, nitrogen, sulphur, chlorine and volatile contents, contents of relevant ash forming elements such as calcium, magnesium, potassium, silicon and phosphorus as well as heavy metal contents are relevant for pellets and raw materials. Dimensions, bulk and particle density, durability and energy density are parameters that are significant for pellets. Appropriate limiting and guiding values were defined for all mentioned parameters and were used as evaluation criteria for different raw materials and the pellets produced from them.

Parameters that are not relevant for small-scale pellet applications but for medium- and large-scale combustion and storage facilities are the stowage factor, the angles of repose and drain as well as the internal particle size distribution. The stowage factor is used as a measure for ocean vessel transportation and is the reciprocal value of the bulk density. The angle of repose is of importance when designing storage facilities for pellets; the angle of drain is important when designing storage for pellets with hopper bottom drainage. The internal particle size distribution of pellets is one of the specifications for users who mill the pellets before use.

Any kind of woody biomass is a possible raw material for pelletisation, however, use is often limited to softwood, with softwood pellets becoming the established norm. Producing pellets out of hardwood is possible in principal but, as a rule, hardwood pellets are not as high quality (especially with regard to durability) as softwood pellets and they are more difficult to produce because the frictional forces in the compression channels of the die are higher than in softwood pelletisation, and consequently the pellet mill is prone to blockings when hardwood is used [104]. The reduced mechanical durability of hardwood pellets is mainly caused by the lower lignin content of hardwood since durability rises with increasing lignin content. Keeping to the strict limit for ash content for class A1 pellets according to prEN 14961-2 is hardly possible when pelletising hardwood species. The right mixing of raw materials can allow for the use of hardwood assortments for pelletisation, however. Wood shavings and sawdust, which accrue at the wood processing industry and at sawmills, are the most used raw materials worldwide. Other raw materials from the wood processing industry, such as Industrial wood chips, or from forestry, such as forest wood chips, up to short rotation crops and log wood are suitable raw materials for pelletisation but they have to be pre-processed (e.g. coarse grinding, drying, fine grinding, bark separation, separation of foreign matters, as depending on the material) [111]. Since pellet production costs are raised as a result, the question of economy arises. Investigations into the economy of different raw materials are carried out in Section 7.3. Bark is excluded from the use in pellet production for small-scale furnaces because of its high ash content. Pellets made of bark would be suitable for use in medium- and large-scale furnaces but bark is at present primarily used directly as a fuel in biomass district heating plants and power plants and is thus not available for the pellet industry.

Table 3.10: Overview of evaluation criteria for possible raw materials for pelletisation and pellet characteristics

Explanations: [1]...limiting value according to prEN 14961-2; P...relevant for pellets; R...relevant for raw materials; h...as high as possible; l...as low as possible

Parameter	Unit	Limiting / guiding value	Relevance	Effect / comment
Length	mm	< 40	P	Choice of conveying and combustion technology; danger of blockings caused by overlengths; burnout time
Bulk density	kg/m^3	> 600	P	Energy density; transport and storage costs
Particle density	kg/dm^3	> 1.12	P	Burnout time; bulk density
Mechanical durability	wt.% (d.b.)	> 97.5	P	Conveying behaviour; dust emissions; transport losses
Content of natural binding agents	wt.% (d.b.)	h	R/P	Especially starch, fats and lignin of relevance; durability of pellets; throughput and economy of pelletisation; adding starch containing biological addidives of up to 2 wt.% (w.b.) is allowed according to prEN 14961-2
Moisture content	wt.% (d.b.)	8 - 12 < 10	R P	Suitability for pelletisation, durability of pellets; depends on the raw material; raw materials with higher moisture content have to be dried
Ash content	wt.% (d.b.)	0.7	R/P	Operational comfort; increased risk of slagging in the furnace
GCV	MJ/kg (d.b.)	h	R/P	Plant dimensioning and control; depends only on the raw material used; can not be influenced
NCV	MJ/kg (w.b.)	h	R/P	Plant dimensioning and control; energy density
Energy density	MJ/m^3	h	P	Plant dimensioning and control; transport and storage capacity
Carbon, hydrogen and oxygen content	wt.% (d.b.)		R/P	GCV and NCV; depends only on the raw material used; can not be influenced
Volatiles	wt.% (d.b.)		R/P	Thermal decomposition; combustion behaviour; depends only on the raw material used; can not be influenced
Nitrogen content	wt.% (d.b.)	0.3[1]	R/P	Indicator for prohibited substances; increased NO$_x$ emissions
Sulphur content	wt.% (d.b.)	0.03[1]	R/P	Indicator for prohibited substances; increased risk of corrosion and SO$_x$ emissions
Chlorine content	wt.% (d.b.)	0.02[1]	R/P	Indicator for prohibited substances; increased risk of corrosion and HCl, Cl$_2$ and PCDD/F emissions
Content of relevant ash forming elements	wt.% (d.b.)		R/P	Ca, Mg, Si, K and P of relevance; influence ash melting behaviour and therefore reliability of the plant; Ca and Mg increase, K decreases the ash melting point; K influences aerosol formation; elevated Si and P concentrations in combination with K may cause slagging
Heavy metal content	wt.% (d.b.)	l	R/P	Ash quality; ash utilisation; fine particulate emissions; indicator for prohibited substances; Zn and Cd particularly relevant for wood fuels
Mineral contamination	wt.% (d.b.)	l	R	Decrease NCV, increase ash content and wear in the pellet mill
Particle size	mm	< 4	R	Absolutely necessary for pelletisation; guiding value is approximately equivalent to the grain size of sawdust; raw materials with bigger particle size have to be ground

Straw and whole crops would be sufficiently available but due to their specific characteristics (high contents of ash, nitrogen, sulphur and chlorine, low ash melting point) they are not, or only to some extent, suitable for the pellet furnaces that are currently available on the market. In order to obtain environmentally friendly and failure free operation of pellet furnaces fed with straw pellets, further research and development (R&D) are required. One approach towards solving the problems of herbaceous biomass fuels could possibly be the mixing of herbaceous and woody biomass. Furthermore, the use of inorganic additives such as aluminium hydroxide, kaolinite, calcium oxide or limestone can positively influence ash melting behaviour.

Adding natural biological additives stemming from primary agriculture and forestry creates far reaching possibilities as concerns quality improvement of pellets and optimisation of pellet production with regard to throughput and specific energy consumption. Some substances coming from primary agriculture and forestry are successfully employed as binding agents in pelletisation already (e.g. shredded maize, maize starch and rye flour). In addition, great potential lies within materials of primary agriculture and forestry, the effects of which as biological additives have not yet been fully examined.

Utilisation of straw pellets in small-scale furnaces cannot be recommended from a present-day point of view. State-of-the-art pellet furnaces presently used in this sector are not designed to handle these fuels and hence are not suitable for their utilisation. The high ash content of straw would lead to more frequent emptying of the ash box and would thus have a negative impact on end user comfort. Moreover, the elevated contents of nitrogen, chlorine and potassium in the straw would cause emission problems as well as increased deposit formation and corrosion. Since the examined straw pellets exhibited low durability, utilisation of such pellets in conventional systems would lead to problems in their fuel conveyor systems. Straw pellets can be used in medium- or large-scale furnaces though because larger systems are usually built in a more robust way and they are typically equipped with more sophisticated combustion, control and flue gas treatment systems.

The concern for contamination of the biosphere from radionuclides (radioactive isotopes) has become an issue due to atomic bomb tests and nuclear accidents such as the Chernobyl accident in 1986. Radionuclides can be taken up from plants and remain in the ash after plants (e.g. wood) are combusted, whereby their concentration increases with decreasing particle size of ashes. Therefore, bottom ashes contain the lowest concentrations and fine fly ashes the highest. Measurements of specific ^{137}Cs activities in different biomass fuels such wood chips, bark and pellets in Austria have shown that the values are usually very low, even far below respective limits for food. Analyses of different ash fractions from biomass combustion plants have shown that the specific ^{137}Cs activities of all bottom and coarse fly ashes are below the limit at which a substance is to be considered as a radioactive material (i.e. < 10,000 Bq/kg). Many bottom ashes are even below the limit of 500 Bq/kg, where no restrictions concerning ash utilisation apply. Only fine fly ashes can exceed ^{137}Cs activities of 10,000 Bq/kg.

4 Pellet production and logistics

4.1 Pellet production

In this section a technical evaluation of pellet production from woody biomass is undertaken. It should be noted in this respect that the required process steps differ depending on the kind of raw material used.

If short rotation crops are used for pellet production, the preparation of the plantation soil, fertilising, planting, harvest and chipping as well as logistics processes before the raw material can be delivered to the pellet production site have to be taken into account. Since the raw material is delivered as wood chips, it must also be coarse ground before actual processing. If log wood is to be used for pellet production, the evaluation must start at wood harvest. The logistics of log wood transport and wood chip production have to be considered before the delivered wood chips can undergo coarse grinding. If class A1 pellets according to prEN 14961-2 are to be produced, the bark must be separated or else the strict limits of the standard cannot be kept. The accumulating bark can, for instance, serve as a fuel in a biomass furnace that generates heat for the drying process. With regard to the use of forest wood chips from thinning and industrial wood chips, the process chain would begin at coarse grinding of the wood chips. In using forest wood chips, it has to be remembered that the wood still possesses its bark and thus it cannot be used to produce class A1 pellets according to prEN 14961-2; only industrial pellets can be produced from forest wood chips. The same applies to industrial wood chips with bark. Depending on the harvest method and supply chains, separation of metals or foreign matters may be necessary for all these raw materials.

The mentioned raw materials still play a subordinate role in pelletisation. Although short rotation crops have been produced in Sweden and Italy for many years, pellets are not yet produced from short rotation crops at an industrial scale. Production of pellets from log wood has started already but it is not yet common practice. This is also true of wood chips. It is known that a small number of pellet producers use log wood or wood chips as a raw material, but they do so at a large scale. In Florida in the USA, for instance, a pellet production plant with a production capacity of 550,000 t (w.b.)$_p$/a was put into operation in May 2008, which almost exclusively uses log wood as a raw material [139; 140]. An Austrian pellet producer set up a pellet production plant with an annual capacity of 120,000 tonnes in 2007, where wood chips accruing at his own sawmill are used [141]. Due to market developments leading to an increasing demand for pellets, these raw materials are expected to gain more and more significance. Explanations as to which process steps are required by these raw materials are abstained from here. For detailed investigations into logistics chains in this field, cf. [59].

The raw materials most frequently used for pellet production are wood shavings, sawdust and wood dust. Wood shavings and wood dust are dry and thus go into fine grinding as a first process step. Sawdust is usually moist and needs drying before grinding. Pellet production consists of four or five process steps depending on whether wood shavings or sawdust are used (cf. Figure 4.1). After drying, interim storage is usually set up in order to uncouple drying from pellet production. After pellet production, storage space for the pellets is required. The process steps of pelletisation for wood shavings and sawdust, starting at drying via interim storage up to pellet storage, are explained in detail in the following sections. The quantities of the mentioned raw materials needed to produce 1 t of pellets are shown in Table 4.1.

Table 4.1: Raw material demand for production of 1 t of pellets

Explanations: relating to pellet moisture content of 10 wt.% (w.b.), bulk densities and moisture contents of raw materials as per Table 7.16

Raw material	Demand for 1 t pellets	Unit
Wood shavings, sawdust	7.5	lcm
Industrial wood chips	5.3	lcm
Forest wood chips	5.1	lcm
Log wood (spruce)	2.2	scm

In addition to raw materials discussed in the previous paragraph, wood chips are gain increasing significance [141; 142] because sawdust is not available in sufficient quantities for the further extension of pellet production capacities. Thereby, industrial wood chips with and without bark as well as forest wood chips, usually with bark, have to be differentiated. The wood chip assortments containing bark do not allow for production of class A1 pellets according to prEN 14961-2 (due to the high ash content of bark), however, the raw material acts as an alternative for industrial pellets. If wood chips are pelletised, an additional process step is required before drying, namely coarse grinding (in a hammer mill).

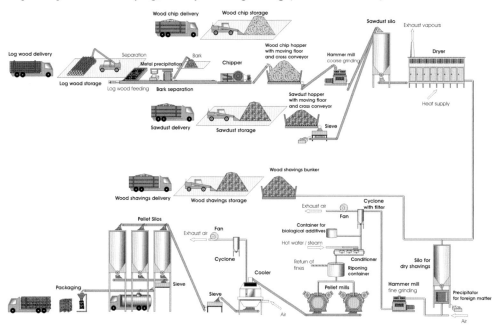

Figure 4.1: Process line of pelletisation

Explanations: data source: BIOS BIOENERGIESYSTEME GmbH

Due to the scarcity of sawdust over recent times, many pellet producers have considered producing pellets from log wood or they have even done so already [111; 112; 139; 143; 144]. If class A1 pellets according to prEN 14961-2 are to be produced, the first process step must be separation of the bark. The bark can be sensibly used as a fuel for raw material drying. The stems without bark or, in the case of industrial pellet production, the log wood with bark, must be processed in a stationary chipper to obtain wood chips. After that, the material has to be coarse ground in a hammer mill before it can follow the process line of sawdust pelletisation.

Further measures in order to extend the scope of raw materials for pellet production can be seen in Scandinavian countries, where peat is added to the woody raw materials in pelletisation [145]. The mixture leads to increased ash content, a greater tendency toward slagging and far more service effort. Concerning emissions, the gaseous emissions NO and SO_2 are augmented. It is only fine particulate emissions (< 1 μm) that can be reduced. On the whole, such mixtures are not recommendable, especially for small-scale furnaces, and so they are not dealt with in detail here.

Detailed technological evaluation of wood chips and log wood utilisation in pellet production is not carried out since sawdust and wood shavings remain the most important raw materials for pellet production. Utilisation of these raw materials is looked at from an economic standpoint though in Section 7.3.

4.1.1 Pre-treatment of raw material

4.1.1.1 Size reduction

After drying, or of course as a primary step when the material is dry already, the raw material is ground up to the required particle size.

The typical target value for the particle size of the raw materials is 4 mm when pellets of 6 mm in diameter are to be produced (6 mm is the common diameter for pellets to be used in small-scale furnaces), but deviations are possible when required by the pellet mill or the raw material itself. If larger diameter pellets are produced, the particle size of the input material may be greater too. However, not only the pellet diameter, the pellet mill or the raw material determine the required particle size but also user requirements. For example, in large power stations that were converted from the use of coal to pellets, the pellets are usually crushed in pulverisers (usually hammer mills) so that the original size fractions of the pelletisation raw material is obtained before being injected into the boiler (cf. Sections 3.1.8 and 10.11.6.5). If the particles are too large, they will not burn completely and, as a result, the stack emissions and the bottom ash will contain unburned charcoal. The smaller the particles of the input material, the better the conversion efficiency will be. However, the pellet producer would not want to grind the feedstock more than necessary because it requires more energy to produce finer fractions. In furnaces where the pellets are burned as they are, the fractional size is less important.

Grinding of raw material is generally done by hammer mills because they achieve the right fineness and homogenisation. A hammer mill is shown in Figure 4.2, while its working principle can be viewed in Figure 4.3. Hammers with a carbide metal coating are mounted onto the rotor of a hammer mill. The hammers hurl the material against the grinding bridge on the housing of the mill where the breaking up of the wood shavings takes place. The particle size of the output is determined by the screen through which the ground material has to pass (cf. Figure 4.3). The energy demand for grinding rises the smaller the particles have to be [88].

Wet feedstock is usually more difficult to grind since the material tends to blind the screens by clogging the holes, but the risk of fires and explosions is lessened by processing wet material in a hammer mill. Also, drying is faster if the particle size is smaller. Hammer milling of the feedstock before and after drying is commonly seen.

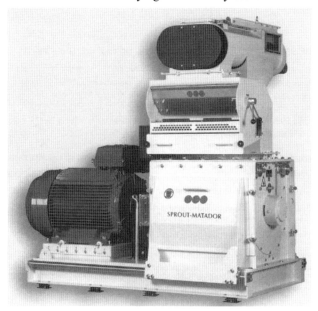

Figure 4.2:   Hammer mill

Explanations: data source [146]

If energy crops or log wood is to be pelletised, wood chips have to be produced first. This can be achieved by using mobile chippers at harvesting or by stationary chippers at the production site. If class A1 pellets according to prEN 14961-2 are to be produced the bark has to be separated before chipping.

There are drum chippers, disc chippers, screw chippers and wheel chippers [59]. Drum chippers consist of a horizontally rotating drum onto which knives are arranged in various ways. Chipped good is transported by fan or conveyor belt. Before the material is expelled it runs into standardised screens that are interchangeable and ensure a consistent output. In disc chippers, the material is conveyed to the disc in different angles and gets cut by radially mounted knives. Shovels for material transport sit on the rear of the disc. Steel combs at the outlet usually act as secondary grinding units. Screw chippers consist of an elongated, mostly conical screw with a cutting edge welded onto it. It pulls the log wood in, breaks it up and discharges it. The construction assures permanent force closure. Difficulties arise with resharpening of the cutting edge and limited input dimensions. The knives of the wheel chipper, in contrast to the drum chipper, are not put onto a drum but are on a wheel. The construction causes impulsive stress on the driving mechanics.

The disadvantages of screw chippers and wheel chippers have led to their near-extinction in the market. Big drum chippers can deal with log wood of up to 1 m in diameter; disc chippers with somewhat smaller sizes.

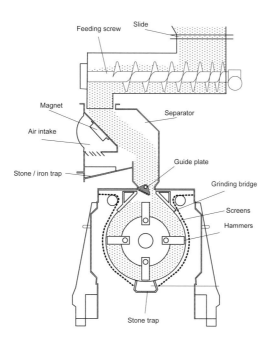

Figure 4.3: Working principle of a hammer mill

<u>Explanations</u>: data source [53]

There are two options for further treatment of wood chips produced from log wood (including forest and industrial wood chips). Either they are dried right away or they are coarse ground before drying. Drying the material directly can cause problems with inhomogeneous moisture content distribution inside the pieces (cf. Section 4.1.1.2). If the material is ground coarsely before drying it still has to be ground more finely after drying in order to make it more homogenous and disintegrate possible lumps that might have been formed during the drying process. The fine grinding is the same as for sawdust and wood shavings.

Grinding moist wood chips needs more energy than grinding dry wood chips. Moreover, throughputs decrease significantly when moist or dry wood chips (commonly broken in hammer mills) are worked with instead of sawdust. This is why a series of hammer mills or larger mills are used for grinding wood chips, which naturally increases investment costs. Despite these drawbacks coarse grinding is recommended before drying as this enables easier control.

Hammer mills are not without further complications, and may not even be suitable at all for grinding bark because moist material, and bark is usually relatively moist, can lead to blockings. So hammer mills have to be adapted and instead cutting mills are preferred since they can manage moist bark much better [147; 148].

Energy demand for grinding as well as an economic evaluation is dealt with in Section 7.2.4.

4.1.1.2 *Drying*

The process of densification in the pellet mill depends on the friction between compression channel and raw material and is amongst other things determined by the moisture content of the raw material. This is why an optimal moisture content has to be achieved according to the pelletising technology and the applied raw material (cf. Section 3.2.3). If a raw material is at hand that already has the right moisture content, drying is of course not required. This can be the case with sawdust or wood shavings that accrue at the processing of dry sawn timber or with wood dust that is generated by sanding solid wood. Therefore these raw materials are favoured in pelletisation.

If steam conditioning takes place before pelletisation, the moisture content achieved through drying should be slightly beneath the optimum moisture content for pelletisation because conditioning raises the moisture content to a certain degree (cf. Section 4.1.1.3).

A residence time of 10 to 24 h [112; 149] in intermediate storage helps balance out moisture inhomogeneities in the dry product. This is vital for wood chips because straight after drying the moisture content in the middle of the particles is higher than at the edges due to their rather large particle sizes. This can be made up for by some residence time in an interim storage facility. It is also the reason why manufacturers of dryers recommend grinding wood chips coarsely before drying. Although the effect is much less pronounced in sawdust, some manufacturers of dryers recommend interim storage after drying of sawdust too. Moreover, interim storage, which is usually done in silos, serves to uncouple drying and pelletising processes and hence it makes the whole process more flexible.

4.1.1.2.1 Basics of wood drying

Wood can contain water in different ways. Firstly, there is unbound water, also called free or capillary water. Secondly there is water that is bound inside the cell membranes. The amount of unbound water is usually greater and it can be removed easily. Bound water can only be removed with great difficulty. That is why the speed of drying decreases with sinking moisture content [150].

The state in which wood does not contain any or hardly any free water is called the fibre saturation range. It lies between 18 and 26 wt.% (w.b.) depending on the wood species [150]. An overview of fibre saturation ranges of different species is given in Table 4.2

Table 4.2: Fibre saturation ranges of a few wood species

Explanations: data source [150]

Wood species	Fibre saturation range [wt.% (w.b.)]
Beech, alder, birch, poplar	24 - 26
Spruce, fir, pine, larch	23 - 25
Oak, cherry, ash, nut, sweet chestnut	18 - 19

4.1.1.2.2 Natural drying

The simplest form of drying is of course natural drying. The material is laid off in loose heaps and turned regularly, whereby evaporation of the water is induced. This way of drying in pelletisation is only of relevance to straw and whole crops. Storage trials by [59] made clear that the optimum moisture content for pelletisation cannot be reached by natural drying of wood. It can thus only be seen as a possible step of pre-treatment that can be done before forced drying.

4.1.1.2.3 Forced drying

The utilisation of sawdust (having a moisture content of 40 to 50 wt.% (w.b.) [59]) for pelletisation is on the rise in Europe. The drying of the material down to a moisture content of 10 wt.% (w.b.) is vital and the numerous technologies that are available are presented in the following section.

4.1.1.2.3.1 *Tube bundle dryer*

For the drying of raw material, drum or tube bundle dryers are in frequent use in pelletisation. A tube bundle dryer is displayed in Figure 4.4. Tube bundle dryers are dryers that are heated indirectly, which means that there is no contact between heating medium and material to be dried. In that way the material can be dried in a gentle manner at around 90°C. Also, the low drying temperature minimises the emission of organic and odorous substances. Steam, thermal oil or hot water may be used as a heating medium. The feed temperature depends on the applied dryer and heating medium and lies between 150°C and 210°C.

Tube bundle dryers usually work with the counter flow principle, that is to say the inlet of the heating medium is on the opposite side of the inlet of the material to be dried. The heat demand adds up to roughly 1,000 kWh per tonne of evaporated water.

Figure 4.4: Tube bundle dryer

Explanations: data source [151]

The core of a dryer is the heated tube bundle that is rotating around a horizontal axis. The heating surface consists of an array of tubes that are welded in star-like rays around a central shaft. The heating medium flows through the tubes, the material to be dried is located between the tubes. At the outer end of the tube bundle there are hub and transport blades that move the material horizontally along the axis and continuously bring it into contact with the heating surface. By this recurrent trickling, i.e. contact with the heating surface, good heat transfer is achieved. A tube bundle with transport blades can be seen in Figure 4.5.

Injecting warm air or flue gas can further improve the heat transfer so that the time for drying can be reduced. Tube bundle dryers are usually equipped with an exhaust vapour suction and de-dusting system in order to follow emission limits.

Figure 4.5: Tube bundle

Explanations: data source [152]

Setting the dryer up outdoors is possible but then the dryer should be fully isolated. Tube bundle dryers can run fully automated with a measurement and control system.

Tube bundle dryers are suitable for drying wood chips, sawdust, wood shavings and more. They are the state-of-the-art in pelletisation.

Investment costs for tube bundle dryers with water evaporation rates between 2.5 and 3.5 t/h amount to between 420,000 and 550,000 € (price basis 2009), depending on dryer manufacturer and design. Maintenance costs are comparatively low, due to only a few parts being prone to wear.

4.1.1.2.3.2 Drum dryer

As already mentioned, drum dryers are also employed for the drying of raw material for pelletisation and they are state-of-the-art in this field as well. A drum dryer is displayed in Figure 4.6. In Figure 4.7 the cross section of a drum dryer with three ducts is presented.

In drum dryers, either direct or indirect heating can be applied. Dryers with direct heating pass the heating medium (flue gas or process air of an appropriate temperature) directly into the dryer. In indirectly heated dryers the drying medium (hot air in this case) is created by a heat exchanger that can be run with flue gas, process air, steam, thermal oil or hot water. The actual heating medium (flue gas, process air or hot air) is in contact with the material in both types of dryer.

The material to be dried is conveyed into the drum via rotary valves. The drum rotates with only a few revolutions per minute while the material is transported by the flow of heating gas. The drying process is supported by blades that are mounted onto the inner wall of the cylinder. They lift the material, which then drops again, which mixes the material thoroughly. At the end of the cylinder the dried material is discharged pneumatically and separated from the flow of hot gas by a cyclone.

The inlet temperature of drum dryers ranges from 300 to 600°C depending on its construction. Thereby organic emissions must be expected as volatile organic matter is released from the biomass at such high temperatures and is emitted with the exhaust vapours. In connection with nitrogen oxides and sunlight, ground ozone and harmful photo-oxidants can be formed [153]. For these reasons complex exhaust air treatment (de-dusting and afterburning) is required. Moreover, in directly heated dryers, fly ash can be accumulated in the drying material, which results in an elevated ash content of the pellets produced [154].

The heat demand of drum dryers is around 1,000 kWh per tonne of evaporated water.

Figure 4.6: Drum dryer

Explanations: data source [155]

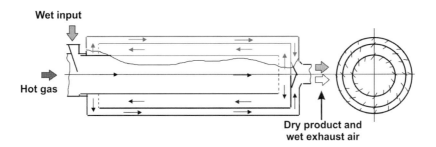

Figure 4.7: Cross section of a drum dryer with three ducts

Explanations: data source [156]

The mixing of the material is just as efficient as in tube bundle dryers, however, the heat transfer is improved due to the higher temperature gradient.

Drum dryers can be set up outdoors without any problems. Investment costs are slightly lower than those for tube bundle dryers.

4.1.1.2.3.3 Belt dryer

Belt dryers are amongst many other applications also used for drying sawdust in pelletisation. Figure 4.8 displays a belt dryer and Figure 4.9 shows its working principle. Depending on the type of dryer, the inlet temperature of the heating medium varies between 90 and 110°C and the outlet temperature between 60 and 70°C. This relatively low temperature means a gentle drying process and prevents the emission of odorous substances. Also, with the right framework conditions a de-dusting unit can be left out because the product layer on the belt acts as a filter [112]. This has to be checked on an individual basis though.

Figure 4.8: Belt dryer

Explanations: data source [157]

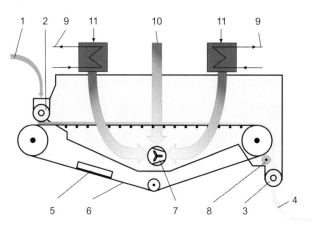

Figure 4.9: Working principle of a belt dryer

Explanations: 1…raw material; 2…feeding screw; 3…discharge screw; 4…dried material; 5…belt cleaning system; 6…belt; 7…suction fan; 8…rotating brush; energy supply by steam, thermal oil, hot water (9) or process heat (10); 11…air; data source [149]

Belt dryers are able to function with both direct heating by process air and indirect heating with a heat exchanger using steam, thermal oil or hot water. The actual heating medium is always in contact with the material. The belt is charged in an even and continuous way with the raw material by the feeding screw. An exhaust fan passes the gas for drying through the belt and the raw material. The belt is continuously cleaned by a rotating brush and, in addition, a high pressure belt cleaning system is activated at times.

The heat demand amounts to about 1,200 kWh per tonne of evaporated water, which is a little bit higher than that of tube bundle, drum or superheated steam dryers.

Belt dryers can run fully automated with an appropriate measurement and control system. The moisture content of the output is measured and the speed of the belt is regulated accordingly.

For drying sawdust, belt dryers are state-of-the-art in pelletisation. It has to be noted though that due to a lack of any mixing, the formation of lumps is possible. This can cause an inhomogeneous moisture content in the output, which can then cause problems in the pelletising process because the moisture content has to be within a certain, very narrow range as outlined in Section 3.2.3.

Belt dryers are quite big (above all long) and more expensive than tube bundle and drum dryers of the same capacity. This can, under the right basic conditions, be more than compensated for though by the belt dryer being able to work with lower temperatures, which not only reduces the expenditure of heat but also results in a greater potential to utilise waste heat. A combination of belt dryers and flue gas condensation units of biomass combustion or CHP plants for instance is an interesting option as such facilities provide heating media at rather low temperature levels. If the plant is not combined with a low temperature dryer, the low temperature heat is often redundant because it cannot be used any further.

4.1.1.2.3.4 Low temperature dryer

A new low temperature dryer has recently been developed for drying sawdust in pelletisation processes [158; 159; 160; 161]. Beside sawdust, this drying concept is also used to dry wood chips, bark or similar biomass fuels. Figure 4.10 shows a pre-assembled drying cell of such a low temperature dryer and in Figure 4.11 its working principle is shown.

Figure 4.10:   Pre-assembled drying cell of a low temperature dryer

Explanations: data source [159]

The dryer consists of two drying cells as shown in Figure 4.10. Drying is done in batch mode. Buffer storages for the input and output materials enable a continuous charge and discharge of the material. Depending on the temperature of the drying medium, two to six batches per hour are possible. The dryer works on the basis of the counterflow principle. The upper drying cell is filled with moist sawdust that is dried with warm air to a certain moisture content. As soon as the required moisture content is reached, which is detected by online moisture measurement in the drying cell, a damper under the drying cell opens and the sawdust falls into the lower drying cell. Again, drying is done by warm air and a damper under the lower drying cell opens as soon as the final moisture content is reached. The drying medium is ambient air that is heated in a heat exchanger to a certain temperature before it enters the lower drying cell. The dry warm air absorbs water from the wet sawdust and thus its temperature is reduced. In this way the moisture content of the sawdust is reduced. The humidified (relative humidity around 98%) and cooled air leaves the drying cells, passes a filter and is heated up again in a second heat exchanger before it enters the upper drying cell. Again, the air takes up moisture from the sawdust, and its temperature is again decreased. It leaves the drying cell via a metal fibre filter. On this filter a filter cake is formed, which ensures efficient dust precipitation. No further dust precipitation measures are necessary. However, some problems concerning filter blockings due to the moist air, especially if very small particles are present in the material to be dried, have been reported by the manufacturer and this filter system will soon be replaced by a newly developed system, probably based on a combination of a cyclone and a baghouse filter.

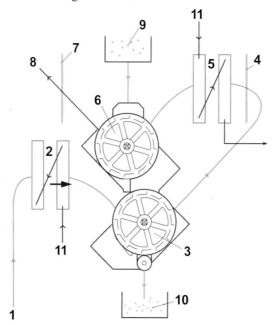

Figure 4.11: Working principle of a low temperature dryer

Explanations: 1…ambient air; 2…first heat exchanger; 3…lower drying cell; 4…intermediate filter; 5…second heat exchanger; 6…upper drying cell; 7…filter; 8…exhaust vapours; 9…input buffer storage (wet material); 10…output buffer storage (dried material); 11…heating medium; data source [161]

The air flow rate through the drying cells is controlled by maintaining a constant negative pressure in front of the suction fan. Filling and dumping the material to and from the drying cells takes around 45 seconds. During this time, the suction fan is stopped and the dryer requires no heat. As heat consumption is disrupted, a heat buffer storage or a cooler is required to get rid of the heat during this interruption of consumption. This is a clear disadvantage of this drying system.

Wheels with shovels rotate inside the drying cells so that the sawdust is exposed to the drying medium in an optimal way, ensuring efficient drying of the sawdust.

The inlet temperature of the drying medium can be as low as 50°C, but up to 100°C is possible. This very low temperature ensures a gentle drying process and the emission of odorous substances is prevented. The low temperature dryer is an indirect drying system where the actual heating medium is always air. The air is heated via a heat exchanger by different sources.

The dryer should be set up indoors. Outdoor set up is possible but requires isolation of the dryer, which might be difficult due to its complexity.

The heat demand for drying amounts to around 1,000 kWh per tonne of evaporated water (slightly less or more for lower or higher drying air temperatures), which is lower than that of belt dryers (cf. Section 4.1.1.2.3.3).

The dryer can run fully automated. The moisture content of the output material is controlled by on-line moisture measurement devices in the drying cells. Due to the efficient exposure of the sawdust to the drying air, a very homogenous output material can be achieved with regard to moisture content, which is important for the subsequent pelletisation. The required area of such dryers is slightly less than for comparable belt dryers, but they are higher and slightly more expensive. However, this could be compensated for by reduced costs for the drying heat under the right basic conditions, as lower heating medium temperatures are possible.

4.1.1.2.3.5 Superheated steam dryer

Superheated steam dryers are used to dry cellulose, mineral wool, wood chips, wood shavings, sawdust, bark, sewage sludge, fish flour and tobacco, among others. It is a proven technology for drying sawdust for pelletising [162; 163; 164; 165].

The process makes use of the drying capacity of superheated steam to remove water from a material. The example in Figure 4.12 illustrates the working principle of a so-called "exergy dryer".

Superheated steam is produced by a heat exchanger that is usually operated with saturated steam at a pressure of 8 to 15 bar. Thermal oil or hot water can also be used instead of saturated steam. The superheated steam circulates in the dryer at a pressure of two to five bars and serves as a carrier medium of the material to be dried. The dryer is fed by a rotary valve. In the dryer the material reaches temperatures of 115 to 140°C. The dried material is separated from the steam in a cyclone and discharged by a rotary valve. The steam remains in the system. Excess steam that is generated out of the material that is being dried is taken off continuously. It can be utilised in other parts of the plant (still being superheated steam) and it is of the same pressure as the steam of the system. Residence times in the dryer range from 5 to 10 s depending on the input material. For drying sawdust it is usually between 10 and 20 s [164].

Heat recovery can easily be realised by continuously taking off excess steam, which is a significant advantage of the superheated steam dryer. It can amount to up to 95% depending on the way it is used thereafter [164; 165]. The drawn off steam can be exploited in several ways, for instance in a district heating or process heating system or for the generation of electricity. What is more, the system emits neither dust nor odorous substances. However, the condensate of the reused steam has to be treated in some way (usually it is passed into the local sewer system). Finally, there is no risk of dust explosions at all as the drying takes place in a steamy atmosphere.

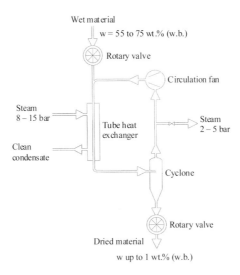

Figure 4.12: Working principle of a superheated steam dryer ("exergy dryer")

<u>Explanations</u>: data source [162]

This way of drying is quite gentle and the mixing is good.

A variety of sizes of superheated steam dryers are available. They range from small units with a water evaporation capacity of 50 kg/h to large-scale dryers with a capacity of 30,000 kg/h. The heat demand of this type of dryer is around 750 kWh per tonne of evaporated water, whereby it is possible to recover up to 95% of this heat as mentioned above [164; 165]. Another 40 to 60 kWh of electric energy per tonne of evaporated water is required for conveyor systems, rotary valves and the fan [166].

Investment costs of a dryer with a water evaporation capacity of 3,000 kg/h are about 1,700,000 € (depending on the heating medium and the material to be dried). The high costs, which are mainly caused by the fact that it is a pressurised system, are the main drawback of the superheated steam dryer. In particular, charging and discharging the dryer pose technical difficulties resulting in elevated costs. Therefore, the dryers are compatible only with large-scale systems that can use the recovered heat sensibly.

Another type of superheated steam dryers are fluidised bed dryers with superheated steam circuits (cf. Figure 4.13) [167]. To date, there are no known applications where they are used for the drying of sawdust for pelletisation. The main area of application is the drying of all sorts of sludge. The working principle, as well as pros and cons of this type of dryer, is similar to the exergy dryer, which is why the technology is not explained in detail here.

Figure 4.13: Fluidised bed dryer with superheated steam circuit

Explanations: data source [167]

4.1.1.3 Conditioning

Conditioning denotes the addition of steam or water to the prepared materials just before pelletising. Through the addition of steam or water, a liquid layer is formed on the surface of the particles. As a result, unevenness is balanced out and binding mechanisms take place during the following densification process. In order for the steam or water to penetrate the product, the moistened wood should be left in that phase for 10 to 20 min according to pellet producers' experience. The processing step also serves to further adjust the moisture content.

If conditioning is to be carried out, one has to consider that during drying a moisture content slightly underneath the optimum should be achieved as conditioning will raise it again (according to pellet producers by about 2 wt.% (w.b.)). Exact conditioning based on a control system is therefore very important for the good quality of the product.

Adding steam can also act as a means to control the right temperature needed in pelletising. This varies according to the pellet mill technology.

Another way of conditioning is the utilisation of biological additives. It too takes place just before pelletising inside a mixer, in order to achieve thorough mixing of the biological additive and raw material, which ensures it can work as it should.

One such aggregate is shown in Figure 4.14. It is designed in a way that the required residence time is also attained. If this is not the case, there should be an isolated interim storage facility between blending and pelletising to secure the right residence time.

Figure 4.14: Blender for the conditioning by steam or water

<u>Explanations</u>: data source [168]

4.1.2 Pelletisation

The next step after drying, grinding and conditioning is the actual pelletising process in the pellet mill. The pelletising technology originates in fact from the animal feed industry. It was adapted for wood to enable the production of a homogenous biomass fuel with regard to shape, particle size and moisture content. Some small-scale producers of pellets still make use of second hand mills from the animal feed industry. Such mills were designed for processing animal food though so their applicability in wood pelletising is limited.

Large-scale producers normally use ring or flat die pellet mills that are especially designed for pelletising wood; ring die mills are most common [169], as confirmed by conversations with pellets producers. Various designs of pellet mills are displayed in Figure 4.15, while Figure 4.16 shows a pellet mill.

Ring die pellet mills consist of a die ring that runs around fixed rollers. The material is fed to the rollers sideways and pressed through the bore holes of the die from the inside to the outside.

The rollers of flat die pellet mills rotate on top of a horizontal die. The material conveyed from above falls onto the platform and is pressed downwards through the die holes.

So, the main tools for pelletising are die and rollers. The raw material is fed into the pellet mill and distributed evenly. It then forms a layer of material on top of the running surface of the die. This layer gets overrun and thus densified by the rollers. By overrunning the dense material, the pressure increases persistently until the material that is in the channels already gets pushed through the channel. An infinite string comes out of the die that either breaks up into pieces randomly or gets cut into the desired length by knives.

Important parameters of pelletising are the press ratio, the quantity of bore holes and the resulting open area of holes (without considering the inlet cones).

The press ratio is the ratio of diameter of holes to length of channels. Together with the type of raw material the press ratio determines the amount of friction that is generated inside the channels, which is why it has to be adapted exactly to the raw material in order to achieve high pellet quality and throughput rate [44]. The press ratio for pelletising woody biomass (wood shavings and sawdust) is usually between 1:3 and 1:5 [115]. Variation of the press ratio is only possible by varying the length of the channels because the diameter is given by the desired diameter of the pellets. So, materials that do not have a lot of binding strength of their own call for longer compression channels. The temperature in the channels rises with increasing length so that stiffness of pellets also rises with channel length.

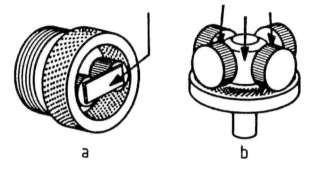

Figure 4.15: Designs of pellet mills

Explanations: data source [59]

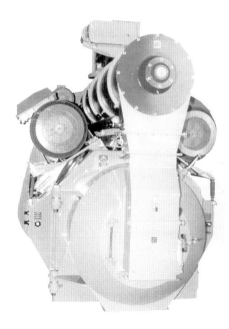

Figure 4.16: Pellet mill

Explanations: data source [168]

This is why mills that are designed for a certain raw material cannot easily be used for another material. Parameters that should, or even must, be adapted to the raw material to be pelletised are:

- thickness of the die;

- channel length (without the counter drill);
- quantity, shape and diameter of bore holes;
- quantity, diameter and width of rollers;
- shape of rollers (cylindrical or conical) of flat die mills.

The quantity of holes and the resultant open hole surface have a direct effect on the throughput together with the available driving power.

Constant feeding and homogenously ground material with constant moisture content lying between 8 and 13 wt.% (w.b.) [112; 170; 171] are prerequisites for a pelletising process without failures.

4.1.3 Post-treatment

4.1.3.1 Cooling

The last process step in pelletisation is cooling. The material gets heated up by steam or hot water conditioning before pelletising and by frictional forces in the compression channels. According to the type of pellet mill and operational parameters, the temperature of the pellets directly after the process can vary between 80 and 130°C (cf. Section 3.2.6). This is why cooling before storage is necessary. Cooling also enhances mechanical durability and it reduces the moisture content by up to 2 wt.% (w.b.).

Figure 4.17: Counter flow cooler

Explanations: data source [172]

Frequently, counter flow coolers are deployed for the process (cf. Figure 4.17) whereby dry cold air enters the cooler at the rear end and moisture laden warmer air flows through the pellets entering the cooler at the front (hence the name counter flow cooler). The working principle of a counter flow cooler is shown in Figure 4.18. Belt coolers are also in use, where the cooling air flows downwards through the pellets.

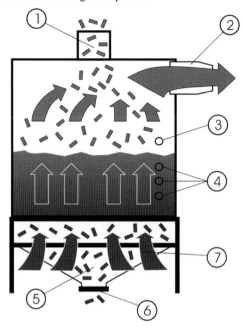

Figure 4.18: Working principle of a counter flow cooler

Explanations: 1…pellet input via rotary valve; 2… exhaust air; 3…overfilling protection sensor; 4…filling level sensors; 5…pellet outlet; 6…discharge hopper; 7…cooling air; data source adapted from [173]

4.1.3.2 Screening

At all zones of the process where dust might arise, the air is drawn off and filtered (cyclone or baghouse filter). As a rule these zones are:

- grinding;
- drying;
- after cooling;
- before packaging or loading.

Drawn off dust is returned to the production process. Screening before transport and packaging guarantees a small amount of fines in the final product.

The pellets are either filled automatically into small bags weighing 25 kg for instance (around 40 l), into big bags with a weight of around 650 kg (around 1 m³) or stored in silos or halls. The pellets are usually conveyed directly from the cooler to the storage facility via special conveyor systems (e.g. bucket conveyor, chain trough conveyor). It should be ensured that the pellets do not come into contact with water during storage. This could cause serious problems, including rendering the pellets unusable for automatic combustion plants. Pellets have a low water stability, which is why they break into pieces and so cannot be transported by automatic conveyor systems when wet.

4.1.4 Special conditioning technologies of raw materials

4.1.4.1 Steam explosion pre-treatment of raw materials

A way to get dimensionally stable and durable pellets is to treat the sawdust in a pressurised steam or steam explosion reactor [170]. Thereby the sawdust is kept under high pressure and temperature for a certain time and undergoes sudden decompression thereafter. This cracks up the wood cells and the softened lignin gets distributed evenly onto the raw material.

The moist material from the steam explosion reactor goes directly into the pellet mill. The throughput of the pellet mill is doubled by this pre-conditioning process and mechanical durability is improved by the softened lignin. The pellets produced are of a dark brown colour and they are far stiffer than common pellets, which in turn makes them less abrasive (< 0.6%) and more stable against water. Moreover, the bulk density amounts to around 630 kg/m³, which is relatively high in comparison with pellets produced in the ordinary way (cf. Section 3.1.3, Norwegian sample).

This is because the lignin gets softened by the conditioning, setting it free, whereby it lays itself onto the particle surface, achieving higher stiffness and water stability.

This novel production technology was originally applied by the Norwegian pellet producer Cambi in a trial facility, whereby the conditioning was performed in batch mode. The trial facility was shut down but the Norwegian pellet producer Norsk Pellets Vestmarka AS (owned by the company Arbaflame AS) has used it in a new production plant with a capacity of 6 t (w.b.)$_p$/h [174].

4.1.4.2 Torrefaction

Torrefaction is a thermo-chemical process for the upgrading of biomass that is run at temperatures ranging from 200°C to more than 300°C under the exclusion of oxygen and at ambient pressure. At these temperatures the biomass become almost totally dry. What is more, degradation processes take place that make the biomass lose its strength (through breaking up of the hemi-celluloses) and fibrous structure (through partial depolymerisation of the celluloses). Lignin, however, largely stays as it is and its share thus grows in the torrefied biomass. Grinding of the biomass becomes much easier, the net calorific value is increased and its hygroscopic nature swaps to hydrophobic. Moreover, its biological activity is strongly reduced. Torrefied biomass is brown to blackish brown in colour, it has a smoky smell and properties similar to coal [175; 191].

During the process, the biomass is degraded partially and diverse volatile substances change to gaseous phase (torgas), which in the end leads to a loss of mass and energy. Mass and energy yield are strongly dependant on torrefaction temperature, reaction time and type of biomass. A typical mass output is 70%; a typical energy output is 90% (both values refer to ash free dry substance) [176]. Thus, the loss of mass is much higher than the loss of energy, which leads to a greater gross calorific value of the torrefied biomass in comparison to the original biomass.

Torrefied biomass can theoretically be used for energy generation in all gasification and combustion plants. The utilisation in coal power plants and entrained-flow gasification plants is of particular interest though. In those plants the biomass has to be injected in pulverised form. If normal pellets are to be utilised for combustion or co-firing in coal power plants, substantial modifications have to be carried out, such as creating storage facilities and separate transport, milling and feeding lines (cf. Section 6.5), and these would be very expensive. Such modifications are not necessary for torrefied biomass, which can be stored on the coal yard and milled and fed together with the coal.

In contrast to conventional biomass, the energy demand for grinding drops significantly for torrefied biomass due to its brittleness (cf. Figure 4.19) and the energy demand for grinding decreases with increasing torrefaction temperature (cf. Figure 4.20). The attributes torrefied biomass exhibits in grinding are again very similar to those of coal.

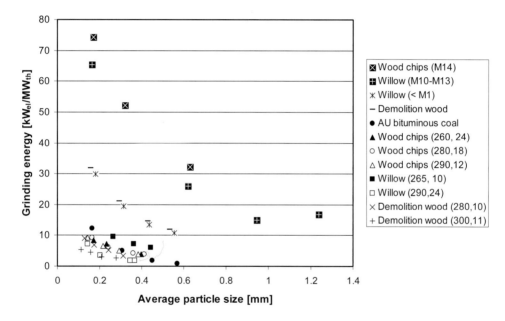

Figure 4.19: Energy demand for grinding of torrefied biomass in comparison with untreated biomass and bituminous coal

Explanations: numbers in brackets indicate the torrefaction temperature in °C and the residence time in min; data source [175]

For most applications it is preferable to combine torrefaction with a densification step, such as pelletisation. Dusting is strongly reduced, the energy density largely increased, transport, handling and storage are easier, and the existing pellet market can be served. However, the application of torrefied pellets for small-scale applications still has to be proven. The energy demand for grinding of torrefied biomass is lessened greatly because grinding is much easier, as mentioned above (depending on the type of biomass it is reduced by 50 to 85% and throughput is 2.0 to 6.5 times greater). Also, pellets made of torrefied biomass have greater energy densities (4,200 to 5,100 kWh/lcm) than conventionally produced pellets (around 3,200 kWh/lcm), which in turn has positive effects on both transport and storage efficiency. What is more, pellets made of torrefied biomass are notably more robust than conventional wood pellets, which is probably due to the elevated lignin content [175]. However, pelletisation of torrefied biomass is not straightforward. The torrefaction conditions have a large impact on the subsequent pelletisation process. In general, more extreme torrefaction conditions (e.g. higher temperature) require more binder to be added during pelletisation. Similar binders to those used in conventional wood pellet production can be applied. The general belief is that at higher torrefaction temperatures, the lignin gets affected leading to less binding capability of the torrefied material itself. Nevertheless, under proper torrefaction conditions, strong pellets can be produced without any additional binder. This requires a torrefaction process with an accurate temperature control.

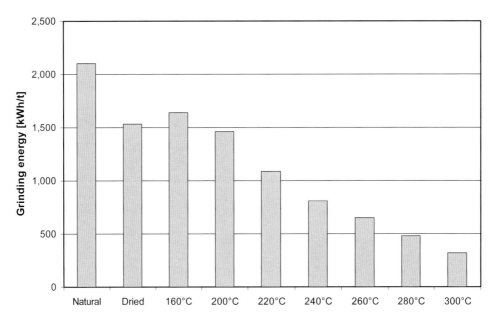

Figure 4.20: Energy demand for grinding of torrefied biomass in correlation with torrefaction temperature

<u>Explanations</u>: data source [177]

Several research groups attend to torrefaction of biomass, all with differing aims [176; 177; 178; 179; 180; 181; 182; 183; 184; 185; 186; 187; 188]. Recently, torrefaction has also gained the interest of large industry worldwide. Many companies involved in drying technology, especially in Europe and the USA, now attend to torrefaction as well, applying their drying equipment at the higher temperature level of torrefaction. Equipment being considered for torrefaction includes rotary drum dryers (e.g. Elino, Germany), belt dryers, multiple hearth furnaces (Wyssmont, USA; CMI NESA, France) and the so-called torbed reactor (Topell, Netherlands). R&D is also ongoing in Denmark by DONG Energy, DTU/Risoe and the University of Copenhagen [189]. Both direct and indirect heating concepts are applied. In the direct heating concept, flue gas is mostly used and generated from the torgas and/or a support fuel.

ECN in the Netherlands is one of the pioneers in R&D on torrefaction combined with pelletisation [175; 176; 187; 188; 190]. The start-up of a demonstration plant for pellets from torrefied biomass with a production capacity of 70,000 t/a is planned in late 2010 or 2011 [188]. Their technology, named BO$_2$-technology, aims at producing high quality torrefied pellets from a broad range of biomass feedstocks (woody biomass and agro-residues) at a high energy efficiency. A flow sheet of the BO$_2$-technology plant is shown in Figure 4.21. It consists of three main parts: drying, torrefaction and pelletising. Drying and pelletising are conventional process steps for which state-of-the-art units can be used. The torrefaction process is the great innovation. In the torrefaction reactor, i.e. a dedicated moving bed reactor, the biomass obtains the right temperature by direct contact with the hot, recycled torgas. The upstream dryer is necessary for biomass with a moisture content above 15 to 20 wt.% (w.b.). By that the heat demand of the process is diminished and combustibility of the torgas is warranted, which would otherwise be too moist. The heat stemming from torgas combustion is used for both the torrefaction process itself and the drying. If necessary, an additional fuel might be combusted together with the torgas in order to stabilise the process and deliver further energy. The output from the reactor, which is still similar to the wood chips input in shape and size, needs to be cooled and ground before it enters the pellet mill. The inlet temperature of the pellet mill is limited by the temperature constraints of the pellet mill itself and depends on the temperature increase in the pellet mill. Inlet temperatures are typically below 50°C. For size reduction, only a very mild crushing is sufficient. Similar to conventional pelletisation processes, the temperature of the pelletised material increases in the pellet mill. Therefore, subsequent cooling is also necessary in this process (not shown in Figure 4.21).

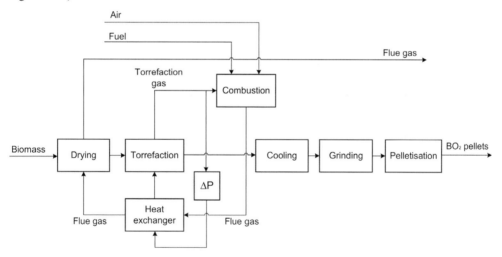

Figure 4.21: Flow sheet of the BO$_2$-technology

Explanations: data source [176] (adapted)

The efficiency of the whole process lies between 92 and 96% (energy content of the pellets related to NCV related to the energy content of the raw materials related to NCV plus energy demand of the production plant, based on moisture contents of the raw material between about 30 and 45 wt.% (w.b.), respectively) and so only slightly exceeds the efficiency of conventional wood pellet production (under the same framework conditions between 91 and 95%). If raw material with a moisture content of 55 wt.% (w.b.) is taken into account, the process efficiencies of both applications decrease to about 88%. There are two different reasons for this. First, a part of the drying heat is covered by the combustion of the torgas, which substitutes external fuel. Second, the electric energy needed for grinding and pelletising is less when the biomass has been torrefied before [176].

A second pilot plant for the production of pellets from torrefied biomass in the Netherlands is planned by the company Topell. The process is based on the so-called Torbed reactor, which can be built in a compact way because of the very intense heat and mass transfer. The technology is currently used in food processing, carbon regeneration, sludge drying and coal gasification, among others. Torrefaction is therefore an extension of the possible applications of this reactor. The input material must be ground and dried (to a moisture content of about 20 wt.% (w.b.)). The reactor is able to use raw materials with impurities such as sand, as such impurities will be separated to a substantial extent in the reactor. Biomass is inserted from above, and air, which is used as heating medium, from below. Inside the reactor there is high turbulence and consequently an efficient heat transfer, which makes fast reaction kinetics possible. Therefore, with reaction times of approximately 1 to 3 min (90 s on average), the process is significantly more rapid than other technologies based on fluidised beds or rotating screws. By adjusting the temperature and the reaction times, the NCV and the amount of volatiles left in the fuel can be adjusted to different coal types (bituminous, hard coal, etc). The torrefied biomass leaves the reactor with a temperature of between 320 and 330°C, which is the runaway temperature. In order to prevent tar formation and runaway reactions of the very porous torrefied biomass, cooling is performed directly afterwards. The material is then fed into the pellet mill. The pellets leave the pellet mill with 90 to 100°C and are subsequently cooled to about 70°C. Their typical NCV is 6.1 kWh/kg (d.b.). The process efficiency lies in a similar range to that of the ECN process described above. A first full-scale demonstration plant with an output of 60,000 tonnes is planned for construction in Duiven (near Arnhem) in the Netherlands in 2010. The price of torrefied pellets related to the NCV is claimed to be already competitive with that of conventional pellets. Torrefied pellets could completely replace coal in a power station. Tests were already carried out in cooperation with RWE in this regard. Moreover, tests in small-scale pellet boilers have also been performed already [191; 192; 193].

Another pilot plant for pelletisation of torrefied biomass is planned in Austria. It should be in operation at the end of 2010 [194; 195; 627].

4.2 Logistics

4.2.1 Raw material handling and storage

Currently, raw materials for pelletisation in many countries worldwide are mainly wood shavings and sawdust from the wood processing industry. If the pellet producer in fact belongs to the wood processing industry, the supply of raw material is part of the company's logistics and the material is commonly delivered by pipelines. Pipelines can also be used when the distance to the supplier of raw material is kept short by an appropriate choice of location. Transport costs can thus be minimised. If a pellet producer has to partly or even wholly buy their raw materials then transport becomes necessary. Lorries in drawbar or semi-trailer combination are predominantly used as the transport medium for wood shavings and sawdust, whereby the trailer has to be equipped with suitable tiltable loading platforms. For shorter distances and smaller loads, agricultural means of transport such as tractors with appropriate trailers are applicable too. One more possibility is delivery by rail but then rail connection has to be in place at the sites of both the supplier of the raw material and the pellet producer. Transport of pellets by ship is also known from Scandinavia and overseas. In [59], detailed information is given about supply chains of by-products of sawmills.

In storage one has to consider that wood shavings (with a water content of just 5 to 14 wt.% (w.b.)) are drier than sawdust (having a moisture content of up to 55 wt.% (w.b.)) at their arrival. If sawdust is going to be used as a raw material, it has to be dried and then treated in the same way as wood shavings in storage.

The moisture content of relatively dry material that is stored outdoors would rise under certain weather conditions. That is why wood shavings as well as dried sawdust have to be stored in closed facilities, i.e. silos or halls.

Moist sawdust is stored either in silos, halls, roofed areas or just open areas. Storage in open areas is not recommended though because of the risk of re-moisturising. It has to be considered that during storage many physical, chemical and biological processes might take place, which can have negative effects on the material. Fungi, spores and bacteria might unwontedly be cultivated, dry matter might degrade and the bulk as a whole might get heated up. In the worst case this can lead to self-ignition (cf. Section 5.2.3). The processes occur quite rapidly so the storage period of moist sawdust should be kept as short as possible. Thorough investigation of the storage of biomass fuels and the consequences of storage on fuels was performed by [59].

Every time pellets are handled they undergo mechanical stress that leads to abrasion and as a consequence to dust formation. So pellet handling should preferably take place in closed systems (pneumatic, feeding screws) together with separation of fines. The fines can hence be returned to the production process.

As an example, fines accrue at the return air filter during filling of storage spaces or silos. Another possibility to separate fines from the pellets is a sieving step before transport.

Depending on the distribution structure, pellet handling and hence abrasion can take place during the following activities:

- Conveying the pellets from pelletisation to storage;
- Packaging of pellets in bags or big bags;
- Filling the transport vehicle at the producer or at the intermediary (loose or packaged pellets);
- Discharge at the end user or intermediary (loose or packaged pellets);
- Conveying the pellets from the end user storage space into the furnace.

If the pellets are pneumatically conveyed, air pressure and velocity as well as the amount of air should not become too high or else the pellets undergo more abrasion, which would lead to a high amount of dust.

4.2.2 Transportation and distribution of pellets

The demand for wood pellets has been steadily increasing worldwide as wood pellets are considered carbon-neutral and renewable. Well developed transport logistics for wood pellets have been the key to their widespread use around the globe. Wood pellets are distributed in consumer-bags, big-bags, tank trucks, containers, railcars and ocean vessels depending upon end user needs and requirements. Since the introduction of pellets to the power generation industry, shipments in railcars and ocean vessels over long hauls have become the dominant transportation mode.

Pellets are either directly distributed by the producer's own transport and distribution system or they are distributed to the end customer by an intermediary. In many cases both ways are used. An important aspect of transport to the intermediary as well as to the end customer is that access of moisture has to be inhibited at all times. Re-moisturising of the pellets would cause problems for the conveying and combusting of the pellets.

For the transport of pellets from the production site to an intermediary, the same criteria are relevant as for the transport of the raw materials to the pellet producer (cf. Section 4.2.1), which is mainly carried out by lorries but also by rail and ship. In Europe for instance, the Danube offers great potential as it is a connection to important European pellet markets. Some pellet transport from Austria to Germany has already taken place along the Danube. In Germany transport by ship is expanding [196]. In the international pellet market, shipping is of great significance already [197; 198; 199; 200; 201; 202]. Pellets from South Africa and North America are delivered to Europe in ocean vessels that can carry up several 10,000 tonnes (the main port is Rotterdam but also Bremen, Wismar or Schwedt and smaller ships also deliver to recipients such as the CHP plant Hässelby in Sweden or other large power plants in Sweden, Denmark, the Netherlands, Belgium and UK). Pellets are also delivered to European power plants in the Baltic states [203]. Shipping, transport by lorry and transport by rail are all cost-saving because of the densified material in comparison to the transport of un-densified biomass fuels.

4.2.2.1 Consumer bags

Pellets in bags are normally used for pellet stoves. Typical examples of small bags used in North America and Europe are shown in Figure 4.22.

Figure 4.22: Typical small bags in North America and Europe

Explanations: left: a typical 40 lbs bag in North America; right: a typical 15 kg bag in Austria; data source [204]

The consumer bags come in many different sizes varying from 10 to 25 kg, labelled with a product quality specification, information about the manufacturer and/or distributor and safety information. The bag material is usually plastic and some bags are made of recyclable material. Consumer bags can be found in gas stations, home renovation centres, hardware stores, farm supply centres or stores selling pellet stoves, small boilers or space heaters where the customer has to come and get them. A delivery of bags by conventional truck is possible though. They are transported on pallets loaded with up to 800 kg. Pellets in bags are characterised by small amounts of dust (when treated carefully) and good moisture protection. The price of packaged pellets is generally far greater than the price of lose pellets, however.

4.2.2.2   Jumbo or big bags

Another packaging possibility are reusable big bags that usually contain 1.0 to 1.5 m^3. Examples are shown in Figure 4.24, while Figure 4.23 displays a big bag that is being filled.

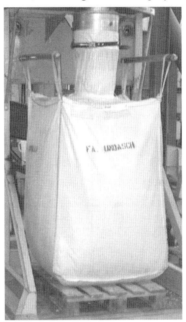

Figure 4.23:   Big bag as it is filled

Explanations: the picture was taken at a visit of the company Umdasch AG, Amstetten, Austria

Figure 4.24:   Typical jumbo or big bags

Explanations: left: with string opening; right: with lift eyes

Big bags are used extensively in the agricultural industry for transportation of animal feed or in the chemical industry for transportation of dry chemicals in bulk and more recently also for wood pellets. The bag material is usually made out of water proof fabric with re-enforced seams and strings for sealing and eyes for lifting. They can be delivered by normal trucks on pallets too. Manipulating the big bags is only possible with fork lift, front loader tractor or crane, which is far too complicated for most customers. It is thus mainly utilised by intermediary sellers or small industries. Manufacturers of pellet boilers draw their pellets partly in big bags because they can be handled with more flexibility at different sites [205].

4.2.2.3 Trucks

Supply to the end customer can be managed by the producer or an intermediary, in bags or lose. In markets that are dominated by the use of pellets in the residential heating sector, lose transport is preferred (in cases where an appropriate storage facility is in place), for instance in Austria or Germany.

For lose delivery to residential buildings, hotels, schools, greenhouses, municipal buildings, sports complexes etc. special pellet trucks, such as that shown in Figure 4.25, are in use almost exclusively. Such trucks typically carry around 15 tonnes of pellets. This has become the state-of-the-art at least for small-scale furnaces in pellet markets dominated by the residential heating sector. This is just one example of the rapid development of the pellet market and the market players' consciousness about quality, as this kind of delivery was not understood at all a few years ago. The filling of the storage space works pneumatically via a flexible tube from the truck's loading space into the storage room of the customer. In order to ensure a dust free delivery and avoid any overpressure in the storage room, air suction and filtering are imperative for exit air. Due to the pellet storage room being closed up, the charging level cannot be controlled directly during the filling process. However, the suction system will draw off pellets as soon as the storage room is full, which can be easily assessed because of the noise it causes. This is how one can find out indirectly when the storage space is filled up completely. Pellet trucks are equipped with an onboard weighing system that records the quantity of the delivered pellets, on basis of which the price is calculated. If less pellets have been ordered than the storage room can take, filling stops automatically at the given quantity. With regard to the ease of use, this kind of delivery is comparable to the delivery of heating oil and so fulfils customers' and traders' requirements. Delivery by a dipper truck with an open loading platform that dumps the pellets into a bunker should be excluded because on one hand, the open loading platform makes re-moisturising possible, and on the other hand, it would cause a lot of dust during discharge.

In North America, tank trucks carrying 40 tonnes of product are not common, but the ones in operation are called stingers and typically use augers (a screw inside a metal pipe) with an outreach of about 6.5 metre for discharge of product into a storage bin (cf. Figure 4.26). These tank trucks are typically used for delivery of pellets to greenhouses or smaller district heating installations.

Beside these special pellet trucks, other types of trucks commonly used to transport other goods can also be employed for pellet transport under certain conditions. Dump trucks, as shown in Figure 4.27, either with hydraulic unloading or as walking floor trucks, are used for lose pellet transport and can load up to 23 t. Standard trucks are employed when wood pellets are packed before delivery. The standard way of transport in such trucks is in 15 kg bags that, stapled on a pallet and wrapped by shrink foil, can load a complete standard truck but can be merged into mixed parcels as well. A truck will hold up to 23 t – depending on destination and truck type – that, again depending on packing, can be up to 40 pallets. An example of a standard truck is shown in Figure 4.28. Furthermore, semi-trailer trucks with hydraulic unloading as shown in Figure 4.29 can also be used for pellet transportation.

Figure 4.25: Typical European tank truck with pneumatic feed

Figure 4.26: Typical North American "stinger" truck (B-train)

Figure 4.27: Dump truck

Figure 4.28: Standard truck

Figure 4.29: Semi-trailer truck with hydraulic unloading
Explanations: source [206]

As wood pellet prices for end consumers are often (much) higher than wood pellets delivered in large quantities, this makes transport by truck economically possible over large distances, such as Belarus to Germany. However, transport costs can have a significant share of total costs. It could therefore be of interest to investigate increasing transport by rail, which is likely to make more sense when considering the energy balance of the entire chain and the linked greenhouse gas emissions that can possibly be avoided.

4.2.2.4 Bulk containers

With an abundance of under-utilised containers in world trade, some pellets are now transported as bulk in 20 or 40 foot containers. In both ground and ocean transportations, the containers can easily be lifted on and off flatbed trucks or flat railcars, or, into and out of cargo holds of an ocean vessel. A bulkhead is built inside the door of the container and a conveyor or spout reaching into the container is used during the filling operation. Figure 4.30 is an illustration of such a filling operation.

Figure 4.30: Bulk loading containers on weigh scales

4.2.2.5 Railcars

Railcars are used extensively in some jurisdictions for transportation of pellets from manufacturing plants to loading facilities for ocean vessels. The pellets are loaded into the railcars through hatch openings at the top of the railcar and discharged through hopper gates at the bottom, which are manually operated from the side by a screw power wrench. Typical railcars used in North America are shown in Figure 4.31.

Figure 4.31: Typical hopper railcar in North America

Railcars come in different sizes with a carrying capacity ranging from 85 to 100 tonnes in North America. The cars are either leased or rented by whoever is responsible for the transportation. To maximize efficiency, several railcars are filled and connected to trains before being picked up by a locomotive and pulled to the destination. A train set may consist of 120 railcars or sometimes even more.

Interestingly, transport by train, commonplace in North America, e.g. well known in British Columbia, is seldom or never mentioned in the European context. This probably has several reasons. In Canada, trains represent the only viable form of transport for getting wood pellets to harbours, and this is also the typical route for other wood products. In Europe, wood pellets are often destined for small-scale consumers and frequently pre-packaged in small bags (and transported on pallets). This allows for more flexibility and less logistical transactions to deliver the pellets to the end consumer or retailer.

4.2.2.6 Ocean transportation

An estimated 35% of the pellets distributed in the world in the year 2008 were transported in ocean vessels in coastal or trans-oceanic trade. A typical transatlantic bulk carrier is shown in Figure 4.32. This number is expected to increase as more pellets are used by very large power stations to replace fossil coal as a fuel for the generation of electrical power. The size of the vessels varies from 1,500 to 50,000 deadweight tonnes (dwt) sometimes fully loaded but more often partly loaded with pellets. A vessel may have between 2 and 11 cargo holds containing 700 to 7,600 tonnes each. The pellets are kept dry and protected from ocean water under hatch cover, sometimes called pontoons, with tight seals. Ventilation to the cargo holds is turned off in order to prevent moist air penetration and to minimise the risk of self-heating and decomposition of the pellets (cf. Chapter 5).

Figure 4.32: Typical transatlantic bulk carrier
Explanations: 46,000 dwt

The pellets brought to the loading port by railcars or by trucks are typically stored temporarily in a silo or flat storage close to the dock before being loaded into the ocean vessel via a conveyor system and shiploader at a speed of 100 to 2,000 tonnes per hour on a 24 hour basis during the time the vessel is berthing. A shiploader with choke spout is shown in Figure 4.33. In some cases, pellets are dumped from the railcars onto a conveyor that brings the product directly to the ocean vessel – so-called hot loading.

The point of sale for pellets in bulk is usually FOB (free on board loading port) or CIF (cargo insurance and freight included to the discharge port). A number of other terms of sale do exist under the universally used Incoterms 2000 Standard and are negotiated in order to minimise the risks for the seller or the buyer. The ocean vessels used for transportation of pellets are either on a scheduled service between ports or chartered specifically to carry the pellets from a loading port to a discharge port. Chartering part of a vessel is called slot chartering and chartering an entire vessel for multiple trips is called time chartering. The seller or the buyer may be responsible for chartering depending on the terms of sale. The discharge from an ocean vessel is typically done by cranes located on the dock in the discharge port or in some cases swivelling or gantry cranes located on the deck of the vessel. Large bucket clams holding up to 35 m^3 (cf. Figure 4.34) are connected to the crane for excavation of the cargo hold and dumping of the pellets in a hopper for further transportation by conveyor or truck to a storage facility. There are also auger systems or vacuum suction systems used for transferring the pellets from the cargo hold to a receiving hopper. Dust generated during discharge is sometimes a concern for the surrounding area and in some cases the receiving hopper is equipped with fans to create a negative pressure (dust suppression) that directs the dust into the hopper (cf. Figure 4.35).

Figure 4.33: Shiploader with choke spout

Figure 4.34: Clam bucket during loading

Explanations: source [206]

Pellets for power plants and district heating facilities are usually sold in kWh calorific value rather than per tonne. A typical calorific value is around 5 MWh/tonne (or 18 GJ/tonne), which means that for example 10 tonnes of pellets have a calorific content of 50 MWh (or 180 GJ). At the point of sale the pellets are sampled and the actual calorific value for the shipment of pellets is determined by a lab test using a bomb calorimeter instrument (see EN 14918 Standard). A supply contract usually has a formula for adjusting the actual calorific content delivered in a shipment, which in turn is used for calculating the payment for a shipment. Other adjustment formulas may be included in a bulk supply contract, such as adjustment for deviation from the actual ocean freight rate compared to a nominal freight rate set in the supply contract, adjustment for deviation of a currency conversion rate compared to a nominal conversion rate set in the supply contract, or adjustment (incentive/penalty) for the amount of fines in a shipment.

Figure 4.35: Receiving hopper with dust suppression fans

4.2.3 Pellet storage

4.2.3.1 Small-scale pellet storage at residential end user sites

Pellets are usually stored in closed storage rooms in the cellar of the end user and as close as possible to the furnace. Storage in silos is also possible but it is generally not done until greater quantities of fuel are to be stored (as for instance in larger heating or district heating systems). Smaller installations such as pellet stoves can integrate some storage space that will last for a few hours up to a few days of automatic operation. Central heating systems might also be equipped with smaller storage facilities that are integrated in the heating device that have to be placed next to the boiler and are filled up manually. This way of storage is designed so that the stored fuel will last for roughly one month.

For ease of use, storage capacity of pellet central heating systems should be big enough to store one or one and a half times the annual fuel demand, which can be achieved by an appropriate storage room in the cellar, by underground storage or by tanks made of synthetic fibre, which can be placed both inside and outside the building.

The different storage possibilities for at least the annual pellet demand are presented and discussed in the following section.

4.2.3.1.1 Pellet storage room

4.2.3.1.1.1 Storage room design

For storage rooms located in the cellar of residential houses, the respective national or regional building regulations have to be taken into account (cf. Section 2.5, where some Austrian examples are described).

Figure 4.36 shows the cross section of a properly equipped storage room with plastic baffle plate, wooden boards on the inner side of the fire door and a filling pipe that is of superior length than the return pipe for air suction whilst filling-up of the stock (so that the return pipe is still free when the storage space is completely full).

Figure 4.36: Cross section of a pellet storage space

4.2.3.1.1.2 Storage room dimensioning

In dimensioning the pellet storage room it has to be considered that elongated and narrow spaces have less dead spots (due to the required slant towards the feeding screw). Furthermore, a pellet storage room can be filled up to a certain level only depending on the position of the filling pipe and on other geometrical factors. With the right geometries, space utilisation (useable volume to total volume) can be up to 85% but it can be far less under poor conditions.

The annual fuel demand of a standard central heating system with a nominal power output of 15 kW is around 5,500 kg (w.b.)$_p$/a (1,500 annual full load operating hours, 84% annual utilisation rate, 4.9 kWh/kg (w.b.)$_p$ net calorific value). If the storage space is designed in consideration of 1 or 1.5 times the annual fuel demand and an average bulk density of 625 kg (w.b.)$_p$/m³ is taken as a basis (cf. Section 3.1.3), the required useable storage volume is 8.8 to 13.1 m³ and the total storage volume required thus is 10.3 to 15.4 m³ (with a space utilisation of 85%). If the cellar's ceiling is 2.2 m high, the base area of the storage space would have to be 4.7 to 7.0 m².

In dimensioning the storage room the ÖKL guideline 66 [23] assumes a useable pellet storage volume of 1.0 m³ per kW heating load. This would mean a useable pellet storage volume of

15 m³ in a central heating system of 15 kW. Assuming space utilisation of 85%, the required total volume of the storage space would have to be about 17.6 m³ or, assuming a ceiling 2.2 m high, the required base area would have to be 8.0 m².

According to ÖNORM M 7137, the fuel demand per heating period lies around 0.6 to 0.7 m³ per kW heating load. With 15 kW heating load it would be 9.0 to 10.5 m³ of pellets. So, with a space utilisation of 85%, the required storage volume would be 10.6 to 12.4 m³ and the base area would have to be 4.8 to 5.6 m² (ceiling height being 2.2 m).

To sum up, these guiding values for dimensioning a storage space lead to differently sized storage spaces. Especially the dimensioning given by ÖNORM creates a rather tightly dimensioned storage capacity. The dimensioning as per ÖKL guideline 66 is more generous.

An important interrelation to be thought of in this respect is that the nominal boiler capacity of the heating system has to be adapted to the actual heating load of the building. The specific heat demand of buildings can vary greatly. Buildings with poor insulation can have specific heating demands of 250 kWh/m².a and more. In modern passive houses, this can be lowered to less than 15 kWh/m².a. This is why the properly determined heating load is of primary significance for the dimensioning of storage space and not the nominal boiler capacity of the heating system, which in many cases is over dimensioned. Only if the heating system is dimensioned correctly (nominal boiler capacity according to actual heating load), the storage space capacity can be derived from the nominal boiler capacity of the system. To be on the safe side, storage capacity should account for about 1.2 times the annual fuel demand because of possible climate induced temperature shifts in the winter.

4.2.3.1.1.3   Comparison of storage room demand for pellets and heating oil

Regarding storage space, a comparison between pellet and heating oil storage space seems of interest since the end user often has to make a choice between those two or, when changing from oil to pellet heating the question arises of whether the existing oil storage space is sufficient for the exchange. The assertion that pellets (energy density of roughly 3,000 kWh/m³) need a storage volume three times greater than oil (energy density of around 10,000 kWh/m³) with regard to the energy density is right in that sense, but it is irrelevant and leads to false conclusions with regard to the necessary storage room. In fact, if the required volume of the storage room of the two heating systems is looked at, the ratio changes dramatically. Assuming an actual power output of 15 kW and 1,500 annual full load operating hours of the boiler plus the annual utilisation rate (oil furnace 90%, pellet furnace 84%), 2,500 l of heating oil or 5,500 kg pellets are needed. When 1 or 1.5 times the annual fuel demand is to be stored, the storage capacity for the pellets would be 4.7 to 7.0 m² (cf. preceding Section 4.2.3.1.1.2).

In the case of heating oil, a row of several same-sized tanks is usually used for storage so that an exact dimensioning of storage space in accordance with the actual fuel demand is not possible, as it is for pellets. So, for storing an annual fuel demand of 2,500 l, three tanks with a volume of 1,000 l each would be necessary; for storing 1.5 times the annual fuel demand, namely 3750 l, four tanks of 1,000 l would be required. The actual storage capacity would then be 1.2 and 1.6 times the annual fuel demand. Keeping to the minimum distances laid down for between walls and tanks and also in between tanks, a storage space with a base area of 4.7 m² when storing 1.2 times the annual fuel demand or 7.8 m² when storing 1.6 times the annual fuel demand becomes essential. The values can be slightly different for different manufacturers of tanks.

The required base area for a pellet storage room for 1.2 times the annual fuel demand would be 5.6 m². So the required storage area for storing pellets is only 19% larger and not 3 times as large in comparison with an equivalent heating oil storage room. Due to the fact that storage volumes for heating oil are usually not as narrowly dimensioned as described above, the storage volume of oil heating systems is mostly sufficient for the exchanging of oil for pellets.

4.2.3.1.2 Underground pellet storage tanks

Due to the growing interest in the pellet market as a whole, systems for underground storage have entered the market. Two examples of such systems are shown in Figure 4.37 and Figure 4.38. Both systems are based on pneumatic discharge of pellets, just like all other underground pellet storage systems. The system shown in Figure 4.37 has a spherical tank from which the pellets are taken from the bottom. The tank as in Figure 4.38 is a vertical cylinder, where pellets are sucked out from above. By the cylindrical shape of the storage tank the volume at hand is utilised in an optimal way. Discharge from above renders a tapered bottom unnecessary and the fuel cannot form cavities.

In case of a system failure both constructions can be accessed from above. One advantage of top discharge (cf. Figure 4.37) is that the suction tube is always on top and can thus be serviced at any time.

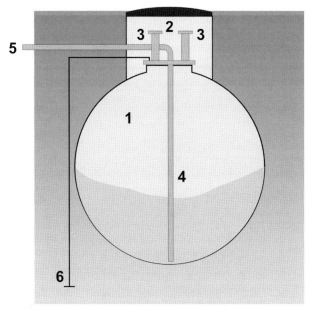

Figure 4.37: Pellet globe for underground pellet storage

Explanations: 1…pellet globe; 2…inspection chamber; 3…filling and return pipe; 4…suction lance; 5…feeding line to the integrated storage of the pellet furnace; 6…earthing; data source [207]

Advantages of underground storage systems are the facts that space is spared in the building, no safety and other installations are necessary and dust cannot enter the building when the tank is being filled. The investment costs must be mentioned as the key disadvantage of underground storage systems. An investigation into this subject and comparisons with conventional ways of pellet storage are carried out in Section 8.2.2.

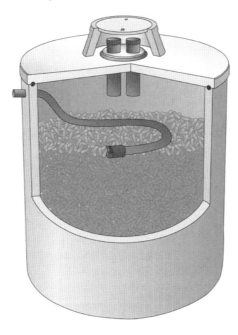

Figure 4.38: Underground pellet storage with discharge from the top

Explanations: data source [208]

4.2.3.1.3 Storage tanks made of synthetic fibre

Figure 4.39: Tank made of synthetic fibre for pellet storage

Explanations: data source [209]

Recently tanks made of synthetic fibre have been increasingly offered (cf. Figure 4.39 as an example). Depending on the design, fibre tanks can be set up both indoors and outdoors. Pellets are discharged by means of pneumatic systems or screws. Fibre tanks can be set up in humid cellars without any problems [210] considering, however, that the tank should not be in contact with damp walls. Outdoor fibre tanks have to be protected against UV light and rain, though a simple enclosure should be enough. If the tank is placed outdoors there is also the advantage of sparing space in the building's cellar. Fibre tanks are much cheaper than underground pellet tanks. An investigation in this respect is carried out in Section 8.2.2.

4.2.3.2 Medium- and large-scale pellet storage

4.2.3.2.1 Types of storages

There are a few typical types of storage used for pellets, all with their advantages and disadvantages, which are shown and explained in the following sections.

4.2.3.2.1.1 Vertical silo with tapered (hopper) bottom

Vertical silos with tapered (hopper) bottom (cf. Figure 4.40) use gravity to discharge product to a discharge tunnel underneath with a conveyor.

Figure 4.40: Example of a vertical silo with tapered (hopper) bottom

The tapered angle is somewhat greater than the "angle of drain" (cf. Section 3.1.6) in order to achieve maximum storage volume and discharge efficiency. The tapered portion is sometimes part of the metal construction of the silo and in other cases it is constructed of concrete with a metal body resting on top of the concrete hopper bottom. Sometimes an old "agriculture" silo constructed in concrete is used for storing pellets. Galvanised steel structure "agriculture" or "grain" silos may also be built for the purpose of storing pellets. A metal silo is built by rings or segments of rings stacked on top of each other with a sealant between the overlapping sheets. Dark paint should be avoided on the outside of a silo since it increases the temperature inside the silo, if exposed to the sun. Corrugated metal transfers more heat than flat metal due to the larger surface area per surface profile. Agriculture silos range in size from 50 to 10,000 m^3. The product is loaded from an overhead conveyor and falls down to the bottom, sometimes through a "bean ladder" or other mechanism in order to decrease the drop height.

Agriculture silos are often ganged together in so-called "tank farms" with a shared overhead conveyor system for loading. Remotely controlled deflectors are used to direct the product stream to a particular tank. Alternatively, a telescoping conveyor with a tripper mechanism is used to direct the product to the right bin.

4.2.3.2.1.2   Vertical silo with flat bottom

Vertical silos with flat bottoms have a circulating auger for centre feed to a discharge tunnel with conveyor. An example of such a silo is shown in Figure 4.41. This design is usually somewhat less expensive to build but is far less efficient compared to the silo with tapered bottom and, in terms of discharge rate, below the level of angle of drain (cf. Section 3.1.6). The augers require regular and frequent maintenance and make this type of silo more expensive to operate than a tapered bottom silo. However, this type of silo has a substantially lower physical profile, which sometimes is an important aspect.

Figure 4.41:   Example of a vertical silo with flat bottom

4.2.3.2.1.3   A-frame flat storage

A-frame flat storages purposely built for storing pellets in bulk (cf. Figure 4.42) are economical to erect and are used for large volume storage in the range from 15,000 to 100,000 m^3 of pellets. The pellets are loaded into the storage from a telescoping conveyor system in the ceiling and drop down to designated areas on the flat floor. In case the storage is on the same property as a power station, the pellets may be moved by front loaders to an in-feed system for a power boiler or district heating facility. If the storage is located in a port, the product may be loaded by a front loader to a hopper and on to trucks or railcars for further transportation. Handling of pellets with front loaders is common but causes a fair amount of damage to the product and generates high amounts of dust. The cabin of the front loader is usually sealed and equipped with an air conditioning system to protect the operator from exposure to fine dust. In some cases a fully automated moving scraper/re-claimer with parallel conveyors on each side of the pile may be used for retrieving the material for further transportation.

Figure 4.42: Example of an A-frame flat storage

4.2.3.2.1.4 General purpose flat storage

General purpose flat storages are used for break bulk, bulk and commodity storage. A interior view of such a storage is shown in Figure 4.43. These types of storage buildings are common in most ports and can store anywhere from 10,000 to 100,000 m^3 of product. In most cases the product has to be loaded into the storage by truck and front loaders in combination with a movable conveyor to build a storage pile. Retrieval is usually done also by front loaders and the product is dumped into a hopper for further transportation by truck or railcar.

Figure 4.43: Example of a general purpose flat storage

4.2.3.2.2 Requirements and examples of pellet storage at producer and commercial end user sites

Pellet producers and intermediaries, as well as commercial end users, should store the pellets in dry and closed up places such as silos in order to avoid the intake of moisture. Storage outdoors, including storage outdoors with a roof that has sometimes been found, is not suitable regarding the protection from re-moisturising and thus maintaining good pellet quality. While storing, any mixing with other fuels must also be prevented. Very small amounts of wood chippings for instance would cause problems for blowing the pellets into the storage room, charging the furnace and in combustion itself.

Due to seasonal fluctuations with a minimum of sales volumes at the end of the heating period in February and March and a maximum before the heating period in September [45], pellet producers need appropriate storage space. According to [43], storage capacity should be about 30% of the annual production but many producers can store less than 10% of their annual production and claim that to be sufficient [205]. This is because a lot of producers have a network of intermediaries and hence intermediate storage space at hand, which makes it possible to hold less storage capacity at the production site. Such interim storage facilities are very important and they have to be big enough to be able to balance out market fluctuations and ensure security of supply.

Commercial medium-scale consumers, such as district heating plants, would usually only store pellets for one to two weeks of operation. Larger consumers, such as Avedøreværket or Amagerværket near Copenhagen in Denmark for instance, can hold significant storage volumes, in the order of 1 to 2 months of operation. Actual storage volumes depend on pricing and other pellet market conditions. The largest storage facility in Denmark for instance is located at Avedøreværket. The covered storage space is currently being expanded to a capacity of 70,000 t (in contrast, power stations usually store coal in open air).

Figure 4.44: Plane storage building at the landing stage at Öresundskraft AB in Helsingborg (Sweden)

Explanations: data source: Öresundskraft AB

Öresundskraft AB is one of the largest individual consumers of pellets in Sweden. The company delivers each year more than 900 GWh heat to the city of Helsingborg produced from several energy sources. The heat is produced mainly from pellets and a small amounts of briquettes in a converted coal dust CHP boiler. The plant is situated at a deep harbour where Öresundskraft has a landing stage and the majority of the fuel is delivered by boat. The boats are bulk carriers and load up to 15,000 t. The number of deliveries each year varies around 30 vessels. The fuel is unloaded with a crane and transported on conveyer belts about 300 m long to a plane storage building (cf. Figure 4.44). The storage has a surface area of 6,400 m^2 and a height of 14 m. The fuel is dropped from conveyer belts along the corner nock top and is stored in piles directly on the concrete floor. The storage has room for 30,000 t. During winter 1,000 t of fuel is used each day. From the storage, the fuel is transported by wheel-loader and let into a hopper from where a conveyer belt transports the pellets to the grinder and onto the boiler.

To eliminate risks, fire detectors are installed in the storage (cf. also Chapter 5). The building is inspected four times every day for possible leakage of water that may lead to heat development in the fuel. Until now there have been no fires or fire incidents. Also, the storage is equipped with fans that ventilate the storage room. This is especially important when fresh fuel is dropped from the conveyer belts onto the floor.

4.2.4 Security of supply

The nationwide supply of pellets has been safeguarded in many countries by noteworthy pellet markets. Further market expansion requires expanding existing transport capacities. The pellet trade used to be limited to just a few specialised traders for many years. In recent years, established fuel traders, formerly concerned mainly with heating oil, have increasingly shifted their attention to the pellet trade. The use of already existing logistics of heating oil supply for efficient distribution of pellets is a logical and effective move, as pellets aim to move into the core business of the fuel trade as they often replace heating oil. This is also because the fuel trade in general already has the required storage spaces, transport capacities and rail connections as well as the right contacts to potential end customers. In Germany, for instance, the association of the German mineral oil and fuel trading industry has moved into the pellet trade [211]. The trade is seen as a reasonable supplement to the fuel and mineral oil trade and it is supported actively, for example by membership of the German pellet association (DEPV).

According to [43], production plants as well as storage and transport facilities of the mixed animal feed industry have been shut down or pooled together in recent years. The processes of this industry (drying, pelletising, storage and distribution of organic raw material) are exactly the infrastructure that is needed for the pellet industry. Making use of these capacities also makes sense due to the extensive knowledge that is at hand in the area. Some of the facilities even have rail access so that connecting to international markets would be possible. It has to be noted, however, that pellets do have to fulfil certain requirements and they also exhibit characteristics that have not been dealt with in this sector yet. Informing transporters and adhering to the standards for pelletisation are obligatory.

Despite the well developed nationwide supply structures and the activities mentioned above, security of supply became an issue in the heating period 2006/2007 for the first time in the pellet market. The reasons for this development can mainly be found in winter 2005/2006, when snowy conditions constrained the wood harvest, which led to a log wood shortage and thus less production in sawmills. As a consequence, sawmills produced less sawdust, i.e. less of the main raw material for pellet production. This raw material shortage and the resulting pellet shortage were further aggravated by a peak of pellet demand due to the harsh winter. However, although increased demand and less production at the same time led to supply bottlenecks in the winter of 2005/2006, a massive increase in the pellet price only happened in autumn 2006 (cf. Section 8.1). This delay might be due to long-term delivery contracts between producers and retailers. In addition, in spring 2006 the oil price reached a first peak, which led to a great demand for pellet heating systems. At the same time, coal fired power plants in Belgium started to co-fire large amounts of wood pellets and there were almost no pellets available on the market in 2006. This fact led to great concerns regarding a supply shortage in the coming heating season. All these facts finally led to a steep increase in pellet prices in autumn 2006 and they peaked in late 2006/early 2007. These circumstances were similar in all European countries, only the height of the price peaks was not the same. Due to the extension of pellet production capacities in many countries worldwide, re-increasing sawn timber production of sawmills and milder winters (fluctuations in demand by climatic conditions of 30% or more are possible), the situation became less dramatic. Since about 2008, the pellet market has been characterised by excess pellet supply as well as excess production capacities [212].

Similar occurrences as seen in winter 2005/2006, for example increasing demand by power plants (single coal fired power plants can have an annual pellet demand of several 100,000 t even if pellets are co-combusted to a small degree only), or, as has been the case from the beginning of the financial and economic crisis in 2008, economic fluctuations with impacts on the wood market, could rapidly cause similar supply bottlenecks. In this respect, a comparison with the oil market is of interest, where certain stocking quantities are required by law. In the pellet sector, to date there is no structured and concerted storage strategy for security of supply reasons. Large pellet producers, usually large sawmills, basically operate according to the "just-in-time" principle. Therefore, extending pellet storage is not one of their corporate purposes. It is only the wholesale sector that takes on some role in pellet stocking but its possibilities are limited by pellet storage being relatively expensive. Users of small-scale systems do not play an insignificant role in this respect as they usually hold storage capacities of one to 1.5 times the annual fuel demand. With that in hand, users are able to react in a flexible way to seasonal fluctuations and thus can create their own security of supply at least for 1 to 1.5 heating seasons. The market as a whole is relieved by that but the trend towards larger pellet heating systems in commercial applications, public buildings and apartment houses counteracts this as such systems usually cannot store the fuel demand of a whole heating period [212].

In order to avoid supply shortages of pellets in the future, different market actors think of, prepare or have already carried out appropriate measures. In Austria log wood storage was set up, which was a concerted action by pellet producers and pellet boiler manufacturers, co-ordinated by the Austrian pellet association. Storing pulp wood without bark is less expensive than pellet storage. Circulating capital is less for pulp wood due to its lower price and investment costs for storage are less compared to pellet silos because outdoor storage is possible. Also, the subsequent drying effort is lessened by natural drying of the material during storage, which reduces drying costs. In contrast, pellet storage is prone to storage losses due to the amount of fines that has to be sieved off. If supply shortages arise, these stores could be used. Another measure concerns the extension of the raw material basis towards wood chips without bark from sawmills. Many pellet producers presently invest in appropriate grinding equipment in order to be capable of pelletising this material. Another shift in the attitude of pellet producers is noted in that greater storage capacities at the production site would improve the price stability of pellets. The large price drop in spring caused by full storage spaces at production sites meant that producers have to sell pellets at a low price in order to continue production. Market observation by suitable monitoring is also seen as a significant security of supply criterion. Activities in this respect in place at proPellets Austria, where the development of supply and demand is closely watched (by continuous observation of boiler sales volumes, regular examination of pellet production prognoses and development of simulation programmes). Finally, increased activities in the pellet wholesale as well as international pellet trade via large ports such as Rotterdam with appropriate storage capacities contribute to security of supply. In addition, some pellet producers offer price, quality and security of supply warrantees to their customers [202; 212; 213; 214].

4.3 Summary/conclusions

The setup of the pellet production process is chiefly dependent on the raw materials used. If the raw materials are sufficiently dry and their particle size is small enough, the pelletisation process is narrowed to the most simple case, i.e. just pelletising itself and subsequent cooling. This may for instance be the case when dry sawdust from the wood working or wood processing industry is used. If the raw material is dry but coarse (greater particle size), it must be ground before pelletising. This is usually the case with wood shavings. If moist sawdust is used, which is mostly the case with sawdust coming out of sawmills, upstream drying is necessary. The utilisation of industrial or forest wood chips renders a grinding process step vital in order to achieve the required particle size. When dry wood chips from the wood working or wood processing industry are used, drying may not be needed. If class A1 pellets according to prEN 14961-2 are to be produced, the wood chips must be free of bark or else only industrial pellet quality can be attained. The utilisation of log wood demands further process steps apart from those required by forest or industrial wood chips, namely bark separation (if class A1 pellets according to prEN 14961-2 should be produced) as well as chipping. Moreover, most pellet producers condition the raw materials with hot water or steam just before the pelletising step in the mill. In any case, appropriate raw material, interim and pellet storage facilities that are adapted to raw material and pellet supply structures have to be in place.

Pellets are still chiefly made of wood shavings and sawdust worldwide. Sawdust usually exhibits moisture contents of between 50 and 55 wt.% (w.b.). Different drying technologies are available depending on the capacity and the framework conditions of the system as well as on the available heat source. In Austria and Germany, it is mainly belt dryers and tube bundle dryers that are employed. Other options would be dedicated low temperature dryers, drum dryers, which are frequently used in Scandinavia, and superheated steam dryers. For subsequent grinding, it is normally hammer mills that are employed. For conditioning, mixers and small interim tanks are used so as to guarantee thorough mixing and long enough residence time. For pelletising itself there are two key technologies available, namely flat and ring die technologies, whereby the ring die has become the common technology for producing wood pellets. Pellet cooling is usually performed by a counter flow cooler that, next to discharging residual moisture, also cares for the dimensional stability of the pellets. Pellets are finally stored either loosely or packaged in bags. Lose pellet storage is usually done in silos, either vertical silos with tapered (hopper) bottoms or vertical silos with flat bottoms. In addition, A-frame flat storages or general purpose flat storages can be used.

In order to create ideal framework conditions regarding logistics and energy supply, and make an economic operation of the pellet production plant possible, setting up pelletisation plants at the location of large sawmills or planning industries is recommendable. Sawmills are normally equipped with a bark separation unit. The accumulating bark can for instance serve as a fuel in a biomass furnace for heat generation or for combined heat and electricity generation. The sawdust that is used for pelletisation can be transported directly on location via pipe belt conveyor or pipelines for instance. Long transport distances and repeated raw material handling can hence be avoided, which saves costs. The flue gas of the furnace undergoes flue gas cleaning. Downstream flue gas condensation can work as a pre-heater of air for the drying process. The use of belt dryers has recently gained growing significance in this respect since belt dryers can be operated at relatively low drying temperatures. Production plants that are optimised in a logistic and energetic sense by using appropriate synergies can be created by realising such projects. Production plants can thus secure economic and ecological energy generation as well as upgrade of by-products.

Pellet producers in countries with well developed residential pellet markets (e.g. Austria or Germany) focus on the production of pellets with 6 mm in diameter that are mainly intended to be used in small-scale furnaces, thus the focus is put on high quality pellets that adhere to regulations and standards of the field. In this way, the end users can be sure that the product they purchased fulfils the highest quality requirements and their furnace can be operated on safe and environmentally friendly grounds, provided that labelled pellets are used. In countries with large-scale consumers, for instance in the Netherlands, Belgium or Sweden with large power plants co-firing pellets or large CHP plants, pellets of lower quality, so-called industrial pellets, are usually used. They often have larger diameters, namely 8 mm. Imported pellets from Canada or the USA are often used in such plants.

Distribution of pellets is usually carried out by intermediary or whole sellers, especially in the case of large-scale pellet producers who can so keep storage capacity (mostly silo storage) low. Pellets can be transported loosely via silo or tank truck or in bags. These options are commonly applied in pellet markets dominated by residential use of pellets. The use of big bags plays a minor role. In countries with large-scale consumers such as large power or CHP plants and for long-distance transports, pellets are transported again by different types of trucks or bulk containers, railcars or by ocean vessels.

Depending on availability, raw materials for pellet production are either supplied in-house through appropriate conveyor logistics or purchased through external logistics, the key transport being lorry transport. Storage of dry or dried raw materials usually takes place in closed silos or warehouses in order to prevent re-moisturising.

In many countries, for example Austria, Germany or Sweden, distributional structures have been set up nationwide. The main share of all pellets is distributed loosely in many countries. Sales of pellets in handy bags are mainly relevant for stoves, which is largely the case for example in Italy and the USA. End user supply in big bags is not common. It is only boiler manufactures who obtain parts of their pellets in big bags as these can be handled in a more flexible way at different internal test facilities. The use of silo trucks with onboard weighing systems is the state-of-the-art for loose pellet delivery. In Austria for instance it is even a requirement of different standards. Developments in recent years are remarkable in this respect since the use of vehicles equipped in this way was not even imagined just a few years ago. The use of appropriate suction facilities and filters for the exit air of storage spaces has also become state-of-the-art. The extension of transport and storage capacities that is demanded by the rapidly growing use of pellets can, and will be, obtained by further use of existing capacity in the heating oil trade, as well as the use of capacity in the area of animal feed industry.

Pellets are stored in closed systems along the whole pellet supply chain in order to keep water or moisture from coming in, which would lead to diminished quality. Depending on the framework conditions and the utilisation of pellets, they can be stored in closed warehouses, silos, storage spaces or integrated pellet reservoirs. In addition, underground storage is becoming increasingly significant because it frees up storage space in houses. Tanks made of synthetic fibre are relatively new in this field. They can be set up both indoors and outdoors. With regard to end user storage, there are certain standards and guidelines in different countries to enable safe and trouble free system operation.

5 Safety considerations and health concerns relating to pellets during storage, handling and transportation

Pellets are prone to both mechanical and biological degradation during handling and storage, and must be handled with care like all other fuels. Mechanical degradation during handling generates fines and dust. Dust from pellets can be considered a safety issue under certain circumstances since it can cause fires and explosions and is a health issue through inhalation. Biological and chemical decomposition happens gradually and generates gases, some of which are harmful to humans. This decomposition also results in the production of heat. Further, the propensity of pellets to absorb moisture can lead to self-heating, which is then accompanied by a sudden increase in temperature. These issues are dealt with in the following sections.

5.1 Definitions related to safety and health aspects

Terms and abbreviations relevant for safety and health aspects are explained in Sections 5.1.1 and 5.1.2, respectively. These terms and abbreviations are used in Sections 5.2 and 5.3.

5.1.1 Safety related terms

There are two different types of explosions and it is important to distinguish between them, namely *detonation* and *deflagration*. A *detonation* is defined as a sudden expansion of gas into a supersonic shockwave. A *deflagration* is initiated by an initial violent oxidation followed by a frontal combustion propagating outward as long as fuel and oxygen are present in sufficient quantities. Explosions related to dust from biomass and pellets are deflagrations and typically start in dust suspended in air. A deflagration in a dust cloud usually causes dust layers lodged on floors, girders and cable trays to swirl and ignite in what are referred to as secondary explosions (also a deflagration).

Auto-ignition Temperature for Dust Cloud (T_C) refers to the temperature at which dust suspended in air will ignite. Several different apparatus may be used for this test to inject dust from the top or from the side into a heated chamber where the ignition is registered.

Minimum Ignition Energy for Dust Cloud (*MIE*) refers to the least electrical discharge energy in combination with the least concentration of dust causing an ignition of the dust. A special apparatus with electrodes generating sparks penetrating a suspended dust cloud is used. This iterative process will find the minima for the energy in the spark and the dust cloud concentration. It should be mentioned that a person may exert in the range of 20 mJ of energy during summertime with relatively high humidity and as much as 60 to 80 mJ during dry winter conditions.

Maximum Explosion Pressure (P_{max}), *Pressure Rate* (dP/dt_{max}) and *Deflagration Index for Dust Cloud* (K_{St}) is tested in an enclosed chamber containing an ignitor exerting a spark with a known energy into a cloud of dust. The *Maximum Explosion Pressure* is a direct measure in the chamber and the *Pressure Rate* is a derivative of the pressure increase per unit of time. The *Deflagration Index* is a measure of the pressure per time unit normalised to m^3.

Minimum Explosible Concentration for Dust Cloud (*MEC*) is sometimes also referred to as *Lower Explosibility Limit* (*LEL*) or *Lean Flammability Limit* (*LFL*) and is a measure of the minimum concentration of airborne dust required in a space to propagate a deflagration in normal concentration of oxygen.

Limited Oxygen Concentration for Dust Cloud (*LOC*) is tested in a similar way to the *Minimum Explosible Concentration* above, with the difference that the apparatus allow for control of the oxygen concentration.

Hot Surface Ignition Temperature for Dust Layer (T_s) is a measure of the temperature at which a certain thickness of dust reaches a temperature of 50°C or more within 60 minutes of being paced above a hot plate. Different thicknesses are used to mimic the expected dust layers found in an industrial setting, while the determination of ignition temperatures for two different thicknesses is often used for interpolation or extrapolation of the respective ignition temperatures for dust layers of other thicknesses.

Auto-ignition Temperature for Dust Layer (T_L) is conducted in a tube where a basket filled with dust is located and exposed to a regulated airflow from below. The tube is heated and the ignition is deemed to have been initiated when the temperature in the dust reaches 25°C above the temperature of the tube within 5 minutes or 50°C without time restriction.

Safety Classification categorises dust in three classes according to their *Deflagration Index* value as follows:

- Class 1: > 0 to 200 bar.m/s
- Class 2: > 200 to 300 bar.m/s
- Class 3: > 300 bar.m/s

Explosion Severity (*ES*) is a measure of the relative explosibility of dust in relation to Pittsburgh seam bituminous coal dust (reference dust). If the quotient of *Maximum Explosion Pressure* x *Minimum Explosion Pressure Rate* for the dust under test divided by the same parameters for the reference dust is greater than 0.5, the dust is considered an explosion hazard. Equipment located in such an environment must meet Class II certified standard. Class II refers to the equipment classification but the dust is often referred to as Class II dust.

Background information on these testing methods can be found in [216].

5.1.2 Health related terms

Threshold limit value (*TLV*) is a registered trademark for an exposure limit developed by the American Conference of Governmental Industrial Hygienists (ACGIH), stated as a time weighted average, short exposure limit or ceiling.

Time weighted average (*TWA*) TLV according to Occupational Safety and Health Administration (OSHA)/ACGIH is the time weighted average concentration for a normal 8 hour workday or a 40 hour workweek to which nearly all workers may be repeatedly exposed, day after day, with no effect.

Short term exposure limit (*STEL*) TLV according to OSHA/ACGIH is the highest concentration to which workers can be exposed to for a short period of time without suffering from either irritation, chronic or irreversible tissue change or narcosis of sufficient degree to increase proneness, impair self-rescue or materially reduce work efficiency provided that no more than four excursions above the TWA per day are allowed within this STEL limit.

Permissible exposure limits (*PEL*) developed by OSHA to indicate maximum airborne concentration of contaminant to which an employee may be exposed over the duration specified.

Recommended exposure limits (*REL*) issued by the National Institute for Occupational Safety and Health (NIOSH) to aid to control hazards in the workplace, generally expressed as 8 or 10 hour TWAs for a 40 hour workweek and/or ceiling levels with limits ranging from instantaneous to 120 minutes.

All definitions are according to [215].

5.2 Safety considerations for pellets

5.2.1 Safe handling of pellets

5.2.1.1 Fines and dust from pellets

Mechanical degradation of pellets during transportation and storage causes fractures and breakage of the pellets, which generates fines and dust. All delivered loads of pellets have a proportion of fines. Smaller diameter pellets tend to resist degradation better than larger diameter pellets. The source and condition of the raw material as well as the densification, extrusion and other processing steps used in manufacturing have an effect on how well the pellets stand up against impact and abrasion during handling.

Fines are defined as the aggregate of all material smaller than 3.15 mm in accordance with the prEN 15149-2 standard (cf. Section 2.1.2). The prEN 14961 standard includes a definition of pellets in quality grades according to the fines content and mechanical durability. The mechanical durability of pellets is measured in accordance with the EN 15210-1 standard using a tumbler to simulate as close as possible the impact and attrition to which the material is exposed during handling and is a measure of how much material of the original volume tested ends up as material less than 3.15 mm in size at the end of the test.

The ideal fuel handling system for biomass, including pellets, should be as gentle as possible. In practice, mechanical degradation is primarily a function of the number of times the pellets are dropped, the height of each drop and the elasticity of the impact surface. For example, pellets may be exposed to as many as ten or more drops from various heights in a production plant before loading into transport. The drop height in large-scale bulk handling of pellets may range from less than a metre to 25 metres. During large-scale bulk handling, pellets are exposed to severe inelastic impacts during the initial loading into the cargo hold of an ocean vessel or during dumping of the pellets into a large silo or storage bunker. Preliminary results from defragmentation research indicates generation of 0.2 to 1.1% of fines smaller than 3.15 mm for white softwood 6 mm wood pellets as a result of dropping pellets from a height of 5.3 m to a concrete surface (inelastic impact) [216]. A more definitive impact index as a function of drop height and pellet properties is under development at the University of British Columbia.

The presence of dust in pellet handling and storage means there is always a risk of explosions, fires and adverse health effects. Precautions as outlined in this section and related standards and guidelines need to be considered carefully in pellet handling.

5.2.1.2 Airborne dust from pellets

Smaller fines easily become airborne and then settle on floors and surfaces in still air conditions or stay aloft in turbulent air conditions. This material is generally referred to as dust. Figure 5.1 illustrates a size analysis of white softwood pellets and bark pellets sampled in a pellet production and handling facility [216].

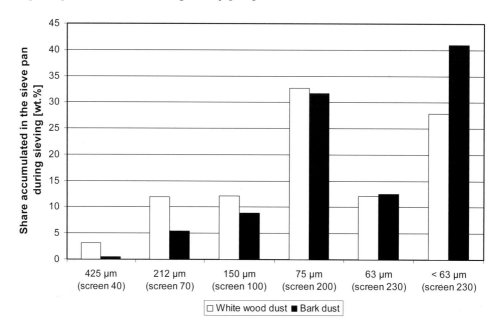

Figure 5.1: Size distribution of airborne dust

Explanations: data source [216]

In the size distribution shown in Figure 5.1, approximately 52% of the dust from bark pellets and 40% of the dust from white pellets is 63 μm and smaller. While investigations of dust size distributions continue, preliminary findings appear to indicate that bark pellets have fewer fines and dust compared to white pellets but the dust from bark pellets is smaller in size. The concentration of small dust particles relates to explosibility, as discussed in Section 5.2.1.2.1.

The settling speed for dust is proportional to the aerodynamic diameter (ae.d.), which is a function of the mass of the material in the particle, the density of the material, the shape of the particle and the density of the air. The sedimentation time has a direct impact on the concentration of particles in a given containment area as well as the time it takes to build up a sediment layer. In still air larger particles settle quicker than smaller particles, as illustrated in Figure 5.2.

The particles highlighted in Figure 5.2 have sedimentation (settling) times from a few seconds to several hours in still air. The settling time is even longer in turbulent air. In a chute or a silo where large amounts of pellets are passing through, the accumulation of lofted dust can become quite considerable and may exceed the minimum explosible concentration (MEC) (cf. Table 5.1).

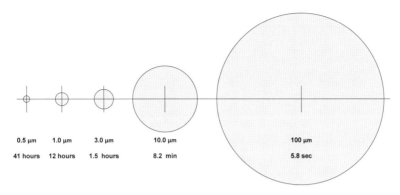

Figure 5.2: Particle sedimentation time in still air

Explanations: μm expressed as aerodynamic diameter; time to settle 5 feet (about 1.5 m) by unit density spheres; data source [217]

The dust generation caused by impact is a concern in transportation systems with many steep chutes or high drop heights where pellets are exposed to multiple inelastic impacts. The dust generated during handling adds to the inherent dust generated during previous handling and transportation of the pellets. When small particles are lofted in the air for an extended period, there is a substantial risk of explosions ignited by electrostatic discharge (cf. Section 5.2.1.2.1).

The following example illustrates the risk exposure during handling of pellets as referred to above. A bunker with steep slopes at the bottom (typical for coal bunkers in power stations burning coal) and a drop height of 25 metres as illustrated in Figure 5.3 is assumed. Using the results from the preliminary defragmentation research referenced above [216], let us assume for illustration purposes that pellets will go through a total of three inelastic impacts and fragmentation per inelastic impact against the steel is 0.7 wt.%. These impacts can result in as much as 2.1% of dust (0.7% × 3). If we add another 3% inherent fines delivered with a shipment of pellets, the total amount of fines lofted in any one period of time at the bottom of a bunker would in this case reach 5.1%. The calculation of the concentration of fines lofted in the bottom section of the bunker during the filling of pellets into the bunker would yield the following results.

For illustration purposes, the weight of the dust from 1 t of pellets can be calculated to be approximately 51 kg (1,000 kg × 5.1%).

The assumed volume in the silo filled with dust at the bottom is 142 m^3. The average dust concentration generated at 5.1% fines is approximately 51,000 g, which when divided by 142 m^3 results in a concentration of about 360 g/m^3.

About half of this dust volume may consist of 63 μm or smaller particles, which is already greater than the MEC of 70 g/m^3 for this size fraction. In summary, after dumping 1 t of pellets to the bottom of the bunker, the average concentration of accumulated dust caused by the inherent dust plus the generated dust due to the fragmentation during impacts lies well above the MEC. As additional pellets are dumped to the bottom of the bunker the concentration will further exceed this critical value. A typical coal bunker could receive 200 to 800 t of pellets. As the bunker fills up, more of the impacts will be elastic as the pellets fall on other pellets and the drop height decreases. However, even elastic impacts between pellets will defragment pellets and generate dust. The illustration above calculates the average concentration, which means that the concentration may be even higher towards the bottom as the dust settles. It should be noted that even dust with a particle size larger than 63 µm has a potential to explode if ignited.

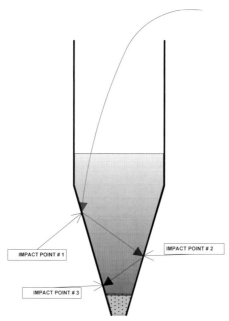

Figure 5.3: Illustration of possible impact points in a bunker with steep slopes

5.2.1.2.1 Explosibility of airborne dust

The smaller the particles are, the larger the relative surface area is for a given volume of dust, which means increased exposure to oxygen (air). The larger the oxygen exposure of combustible material is, the higher the risk for open flame combustion in the presence of an igniting source or a source of heat. Explosions occur in air-suspended dust as well as in dust deposits on hot surfaces. Explosiveness is a function of particle concentration, oxygen concentration and the energy of the ignition source or the temperature of the heat exerted on the dust. Figure 5.4 illustrates the fire triangle and the explosion pentagon often used to memorise the factors required for a fire or an explosion to occur. In brief, for a fire to occur in dust, both sufficient supply of air (oxygen) and an ignition source need to be present. In the case of explosion, in addition to the above components, the dust needs to be lofted (dispersed) and be of a sufficient concentration.

In order to emulate real conditions causing fires or explosions as accurately as possible, a number of testing standards are available. As an example, Table 5.1 summarises the results of the tests that have been conducted on dust from white pellets as well as bark pellets. Standard tests are conducted on dust clouds as well as dust layers. The result of the tests is a dust classification that is used as a guideline for the way that dust generating products should be handled and how the handling facility should be designed.

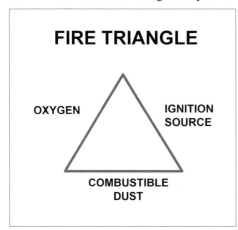 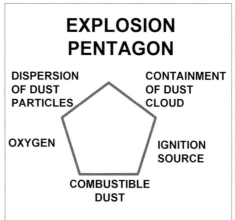

Figure 5.4: Fire triangle and the explosion pentagon indicating factors for fire and explosion to occur

When fuel handling systems are designed for large energy plants, the fuel characteristics have to be established and respective tests are usually undertaken. The deflagration index and the maximum explosion pressure are used to design systems for dust collection, explosion suppression and explosion control panels. The minimum ignition energy value is a guiding value for the way in which the electrical grounding has to be designed in order to avoid electrostatic ignition of the fuel. The autoignition and surface ignition temperatures provide guidance for the maximum temperature the airborne dust can be exposed to without igniting as a direct result of heat. The dust classification indicates which rating is required for the equipment exposed to the fuel or, in this case, the dust from pellets.

Table 5.1: Results from testing dust (< 63 μm) from white pellets and bark pellets

Test mode	Test parameter (dust < 63 μm)	Unit	White dust	Bark dust	Coal dust	Lycopodium spores	Testing standards
Dust cloud	Auto-ignition Temp (T_C) (Godbert-Greenwald)	°C	450	450	585	430	ASTM E1491
	Min Ignition Energy (MIE)	mJ	17	17	110	17	ASTM E2019
	Max Explosion Pressure (P_{max})	bar	8.1	8.4	7.3	7.4	ASTM E1226
	Min Explosion Pressure Rate (dP/dt_{max})	bar/s	537	595	426	511	ASTM E1226
	Deflagration Index (K_{St})	bar.m/s	146	162	124	139	ASTM E1226
	Min Explosible Concentration (MEC)	g/m³	70	70	65	30	ASTM E1515
	Limiting Oxygen Concentration (LOC)	%	10.5	10.5	12.5	14.0	ASTM E1515 mod
Dust layer	Hot Surface Ignition Temp (5 mm) (T_s)	°C	300	310			ASTM E2021
	Hot Surface Ignition Temp (19 mm) (T_s)	°C	260	250			ASTM E2021
	Auto-ignition Temp (T_L)	°C	225	215			USBM (Bureau of Mines) RI 5624
	Dust Class (> 0 to 200 bar.m/s)		St 1	St 1	St 1	St 1	ASTM E1226
	Dust Class (Explosion Severity (ES > 0.5)		Class II	Class II			OSHA CPL 03-00-06

The smaller the particles are, the higher the explosibility of a material. Table 5.1 indicates a slightly higher explosibility for bark than for white wood material, which contributes to a higher explosion pressure.

The MEC for dust from pellets is practically the same as for bituminous coal, such as Pittsburgh coal. Coal dust explosions can be mitigated partly by injection of incombustible mineral dust (e.g. limestone) into air intakes in order to keep the dust concentration below the critical 65% level of coal dust. This option is not practical for handling wood pellets. Coal dust has a lower electrostatic resistivity than wood dust, which means that coal dust would have a lower propensity to generate static electricity charges.

Another important observation from the test results in Table 5.1 is that the oxygen content in the ambient air needs to be higher than 10.5 vol.% for the dust to ignite. However, it is known that if the dust or pellets ignite at the oxygen concentration below 10.5 vol.%, the fire will be hard to extinguish due to the relatively high oxygen content inherent in the wood pellets, which sustains smouldering [216].

The oxygen concentration in an enclosed containment such as a storage room or a cargo space of an ocean vessel often is under the 10.5 vol.% limit within a week or two, depending on ambient temperature and pellet quality (cf. Section 5.2.4.3). Cargo spaces onboard ocean vessels are enclosed and typically sealed to minimise ingress of moisture. This also minimises self-heating and the risk of dust ignition from pellets in the containment.

5.2.1.2.2 Mitigation measures

A number of precautions need to be taken to minimise the risk of fires and explosions in large storage facilities for bulk pellets. Standard mitigation measures can be categorised as follows:

- Control of ignition;
- Inertisation;
- Explosion suppression;
- Explosion containment;
- Explosion venting.

It is beyond the scope of this handbook to cover the above listed engineering design details. However, control of ignition is closely tied to safe operation of pellets in bulk. In spaces where dust is present in a high concentration, the following strategies for combustible dust should be considered:

- Direct heating of spaces must be avoided. Indirect ambient space heating (e.g. circulating glycol systems) is a substitute.
- Smoking should be prohibited in areas where dust clouds or dust layers are present.
- Electrical motors and relay equipment should be protected in ventilated separate rooms with overpressure venting.
- Airborne dust may be carried into internal combustion engines running in dusty environments. Partly combusted dust fragments (embers) in the exhaust could ignite the dust of the surroundings and cause explosions.
- Welding and cutting in areas with dust clouds or where dust is lodged on flat surfaces are well known hazards.
- Hot surfaces under dust layers are a risk factor, as seen in Table 5.1, because they may cause ignition of the dust and acceleration of the off-gassing of dust (cf. Section 5.2.4).

The presence of metal dust may have a catalytic effect on wood, if lodged in the same place, resulting in self-heating and increased off-gassing.

- Biomass, including wood pellets, should not be stored for long periods without proper temperature monitoring, as indicated in Section 5.2.3, due to the risk of self-heating.
- Friction, impact, rubbing and mechanical sparks related to hot bearings, moving vanes and belts account for a significant part of accidents with dust.
- Electrification (electrostatic charge) of material, including dust particles, as well as surfaces is common when materials are in motion (referred to as the tribo-electric effect). This is particularly true when the surfaces or the material has a strong dielectric characteristic (low conductivity or high resistivity), such as wood pellets. The surface resistivity of pellet dusts has not yet been precisely determined but is estimated to be around 10^{12} Ω or in the same order as wood. The three steps in the creation of the electrostatic effect are charge separation, accumulation and discharge. Charge separation refers to the generation of electrostatic potential as a result of migration of electrons (a charge) from one material to the other. Accumulation refers to the increase in number of electrons (negative charge) on the surface of one of the materials and a lack of electrons on the other material, increasing the electrostatic potential. Discharge will eventually happen when the electrical field becomes strong enough to fragment the air molecules in the gap between the two materials, which results in an electrostatic discharge (spark). The key strategies to avoid creating electrostatic charge are avoiding charge separation in the first place by not using non-conductive materials in areas where the materials (pellets) are moving and to make sure conducting materials are properly grounded in order to avoid electrostatic potential from developing. Standards and guidelines for the grounding of equipment should be followed [218; 219; 245].

Table 5.2: Recommended precautionary measures in the presence of metal dust and related minimum ignition energy requirements

Minimum ignition energy (MIE)	Precautions
500 mJ	Low sensitivity to ignition. Always ground the equipment when the ignition energy is below this level.
50 mJ	Ground personnel working in direct contact with or immediate proximity of dust clouds and dust layers when or before ignition energy is at this level.
25 mJ	The majority of ignition incidents occur when ignition energy is below this level.
10 mJ	High sensitivity to ignition. Particular attention should be paid to the use of high resistivity non-conductors when ignition energy is below this level.
1 mJ	Extremely sensitive to ignition. Explosible dust clouds should be avoided whenever possible. Handling operations should be carried out so as to minimise the possibility of suspension of the dust in air. All possible steps should be taken to support the dissipation of charge and to prevent charge generation.

Table 5.2 summarises precautionary measures recommended for spaces where metal dust is present, and the related minimum ignition temperature (MIE) established by testing. In the absence of detailed facts about wood dust, similar measures are recommended [218] until further information is available for wood dust. The MIE value in Table 5.1 is valid for dust particles smaller than 63 μm (minus mesh 230) which represents about half of the airborne particles in dust from wood pellets (cf. Section 5.2.1.2). It should be noted that for larger

particles, greater MIE values are required. Electrostatic discharge from a person to a metal surface during summertime (relatively humid conditions) is typically 20 mJ and about 60 to 80 mJ during wintertime when the humidity in the surrounding air is lower.

Safety standards stipulate regular checking of grounding system. Typically, the resistance to ground should be less than 10 Ω to be on the safe side [218; 221; 222].

Total elimination of all sources of ignition is very difficult and in many cases cost prohibitive. Traditionally, risk has been assessed for each potential ignition source and then prevention measures have been implemented to make the risk as low as reasonably practicable (ALARP). As the ATEX 137 [220] is introduced in Europe the risk-zone concept is gradually implemented using the following guidelines:

- Zone 20

    A place in which an explosive atmosphere in the form of a cloud of explosive dust in the air is present continuously for long periods or frequently.

- Zone 21

    A place in which an explosive atmosphere in the form of a cloud of explosive dust in the air is likely to occur in normal operation or occasionally.

- Zone 22

    A place in which an explosive atmosphere in the form of a cloud of explosive dust in the air is not likely to occur in normal operation, but if it does occur, will persist for a short period of time.

With the relatively high frequency of accidents and incidents experienced in the pellet industry related to explosions, it is highly recommended that spaces are classified in accordance with the ATEX standard in order to heighten the awareness of potential risk of explosions.

5.2.1.2.3 Flammability (burning rate) of airborne dust

The flammability or burning rate of dust from pellets can be established in accordance with the standard specified by UN Test N.1-Class 4 Division 4.1 Substances [221] by means of measuring the distance a material has burned in 2 minutes. If the burning rate is more than 200 mm in 2 minutes, the material is considered flammable. Table 5.3 gives an example of the burning rate of dust with less than 63 μm (minus mesh 230) in size from white pellets and bark pellets [216]. It can be seen that the burning rate is clearly below 200 mm per two minutes and this dust is therefore not considered flammable.

Table 5.3: Burning rate of pellet dust of less than 63 μm

Material	Burning rate
White pellet dust as received	20 mm / 2 min
Bark pellet dust as received	22 mm / 2 min

Dust from pellets of different quality may have somewhat different burning rates. Classification of flammability is relevant for packaging requirements during transportation and, in case of the US and Canada, is regulated by the US 49 Code of Federal Regulation [222].

5.2.2 Pellets expansion through moisture sorption

Most pellets are hygroscopic, which means that they absorb water when exposed to it [223]. When fully saturated with water, compressed pellets expand about 3.5 times. This expansion precludes water as a fire-extinguishing media. Further, expansion forces may crack the containment or create an extremely hard and compact plug that requires a jack hammer to remove (cf. Section 5.2.5.4). Figure 5.5 illustrates what happens when pellets are wetted. It is not only water from extinguishment that might cause swelling and degradation of the pellets into "moist sawdust", but also condensation of humid gases formed during oxidation or by a smouldering fire inside the pellets. The condensation might occur both on the pellet surface and the wall and roof of a silo, which then affects the pellets along the wall.

Most pellets are also sensitive to humidity in the air and will absorb or expel moisture to a certain level until the equilibrium moisture concentration (EMC) for pellets is reached. The EMC varies for different qualities of pellets and is a function of relative humidity (RH) and ambient temperature. Figure 5.5 illustrates typical EMC characteristics of white pellets and bark pellets at an ambient temperature of 25°C. The fairly slow rate of sorption is due to the quality and the specific density of the pellets. However, it has been observed that pellets might also swell and "glue" together, for example in a silo, if humid air from the outside is leaking into the pellet storage.

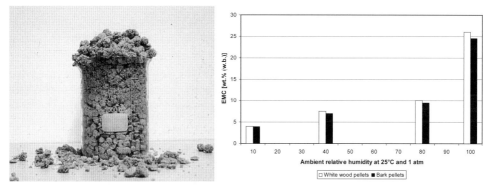

Figure 5.5: Effect of water application to pellets and equilibrium moisture content for wood pellets

Explanations: left: resulting expansion 5 minutes after water application to warm pellets; right: equilibrium moisture content for wood pellets after long term sorption in air; data source adapted from [223; 265]

The level at which the pellets start expanding depends on many factors such as particle size, pellet size, extrusion pressure, species, temperature etc. and therefore cannot be determined. Pellets with a moisture content higher than 14 to 16 wt.% (w.b.) usually attract microbes and are oxidised fast, which means that the mechanical integrity is compromised within a few days and a swelling can be seen.

Sorption should be avoided to the highest extent possible by keeping exposure to air with high relative humidity or water to a minimum. Large storages using forced ventilation should preferably have dehumidifiers to eliminate much of the moisture coming in, or they should at least feed air into the storage that is not directly exposed to outside weather conditions such as rain or mist. The sorption in bagged pellets is very low due to the protective packaging. The consequences of sorption are increased off-gassing (cf. Section 5.2.4) and self-heating potential in the pellets (cf. Section 5.2.3). Moreover, cases are known where water was brought into the silo from the top in order to extinguish a fire, which resulted in complete destruction of the silo due to the forces exerted on the walls by the expanding pellet.

5.2.3 Self-heating and spontaneous ignition

Solid biomass fuels are generally porous and susceptible to self-heating and spontaneous ignition caused by microbiological growth, chemical oxidation and moisture absorption. Due to the low moisture content of pellets, the growth of microorganisms is normally limited, but temperature build-up is often observed as a result of chemical oxidation and moisture absorption in newly produced material. A number of serious incidents of self-heating and spontaneous ignition of wood pellets in storage have occurred [224; 225]. There is also a fire risk from various external ignition sources in storage and especially in pellet handling and transport.

Self-heating in biomass is a well recognised phenomenon [226], although the chemical process involved is not well understood, particularly for pellets. The hammer milling of the raw material during manufacturing of pellets opens up the cell structure and exposes the cellulose, hemi-cellulose, lignin and the extractives (including the unsaturated fatty acids) to oxidation, which are believed to be the primary cause of off-gassing. Oxidation takes place above 5°C and generates heat, non-condensable gases (cf. Section 5.2.4.1) and a number of condensable gases (cf. Section 5.2.4.2). The higher the temperature is, the higher the rate of off-gassing becomes [241]. Oxidation of fatty acids in sawdust and other moist fuels is accelerated by microbial activity with mesophilic bacteria and fungi up to approximately 40°C and by thermophilic bacteria up to approximately 70°C. Above this temperature chemical oxidation becomes dominant, which further raises the temperature, in many cases up to an uncontrolled temperature range.

In wood pellets, the potential for microbiological activity is eliminated due to high temperature regimes applied during production processes. The drying temperature of sawdust in the pellet production process is normally above 75°C (often 100 to 200°C) and most microbes cannot endure this temperature. During pressing in the pellet mill, the temperature in the produced pellets increases to 90 to 170°C. It is known that fatty and resin acids in the produced pellets oxidise to condensable gases (aldehydes, ketones) by exothermic chemical reactions during production and storage. Restricted access to oxygen (air) suppresses self-heating, although once started, the oxygen contained in the wood (about 40 wt.% (d.b.)) will lead to a sustained level of self-heating.

Fire risks with wood pellets have been studied extensively by SP Technical Research Institute of Sweden [225; 227; 228; 229; 230; 231; 259] while gas emissions have been studied by the Swedish University of Agricultural Sciences (SLU) [240]. Work by SP is included in the Nordtest guideline for storing and handling of solid biomass fuels, NT ENVIR 010 [232]. These works form the basis of most of the text and advice given in this section.

5.2.3.1 Wet solid biomass fuels

Raw solid biomass, such as sawdust and other feedstock, used in pellet production has a moisture content exceeding 15 wt.% (w.b.), typically between 35 and 55 wt.% (w.b.), and is often stored outdoors before pelletising to secure high pellet production capacity in winter.

Storing sawdust has proved to benefit pellet quality, for example pellets made of stored pine fraction have higher bulk density (they are more compacted) and better durability properties than pellets made of fresh pine sawdust (keeping all the other process parameters the same). An industrial experiment confirmed that there is a direct correlation between stored sawdust and process parameters such as energy consumption [95], and that pellets made of stored sawdust (for 140 days) have better quality, for example higher durability than pellets made of fresh sawdust [233].

During large-scale storage of pine and spruce sawdust (dry material) the amount of fatty and resin acids was reduced after the first 12 weeks. A peak of reactivity could be observed during an initial period of time, which varies from one quality of biomass or fresh pellets to another. After this period, the smell decreases dramatically indicating a low level of reactions (oxidation). An additional 4 weeks of storage did not change the amount of fatty and resin acids, in other words, the sawdust becomes mature over 12 weeks of storage [250].

Therefore it is recommended (for the positive effect) to store lignocellulosic material such as sawdust or wood chips outdoors for a period of time before pelletising.

Wood chips, bark and other wet solid biomass fuels stored outdoors in heaps and piles are typical examples of stored fuels that exhibit self-heating. Several physical, biochemical, microbiological and chemical processes are involved in self-heating of biomaterial. The domination of one or several of these processes depends on different parameters such as temperature, moisture content, oxidation ability of the material, and so on.

The relatively high moisture content of the wet biomass fuel creates a suitable environment for microbial growth as the microorganisms feed on nutrients that are dissolved in water. The degradation of wood by fungi and bacteria results in a temperature increase of the stored fuel. Peak temperatures from microbial self-heating vary between 20 and 80°C, depending on the type of microorganism [234]. Microorganisms are divided into three groups as a function of their sensitivity to temperature, i.e. psychrophilic, mesophilic and thermophilic. Psychrophilic microorganisms have a temperature optimum of 15°C and hence are not relevant for self-heating. Mesophilic microorganisms have a temperature optimum between 20 and 40°C with very limited reproduction rate at 40°C. Thermophilic microorganisms will survive up to 70°C. Microflora existence also depends on nutrients, for example in the form of hydrolyses of carbohydrates, and increased temperature. Chemical degradation (oxidation of wood constituents) normally starts to have some influence from 40°C onwards and generally becomes the dominating process at temperatures above 50°C. As the heat generation processes proceed, heat is transported from the interior of the bulk towards the surface. The centre of the bulk dries and water is transported out of the centre condensing on the outside layers. The outcome of the self-heating process is a balance between the heat production rate and the rate of heat consumption and dissipation. Thermodynamically, the larger the size of the storage is, the greater the risk for spontaneous ignition becomes. Spontaneous ignition begins with pyrolysis in the heap in cases where the heat generated exceeds the loss of heat through conduction, convection and radiation within the heap. If sufficient oxygen (air) is present, the pyrolysis turns into open flames.

5.2.3.2 Dry solid biomass fuels

Storage of dry pellets requires a protected environment to maintain their low moisture content (normally below 10 wt.% (w.b.)) and to preserve the structure of the pellets, although there are certain types of dry fuels where a moisture content of 15 wt.% (w.b.) is acceptable. The storage conditions for dry pellets are different than those of wet fuels.

The low moisture content of the pellets limits the growth of microorganisms. Wood pellet production processes involve drying of raw sawdust at temperatures of typically 90 to 170°C (in specific cases the temperatures might be higher or lower; cf. Section 4.1.1.2.3). Pellets made from dried material have very limited microbial activity during storage. If microbial activity is observed during storage, it is likely that it was initiated by contamination. At low temperature drying processes of less than 100°C, microbial activities may not be eliminated. During pelletisation, temperatures of 100 to 170°C are reached in pellet mills because of friction. The combined effect of relatively high temperature regimes in pellet production and the low moisture content of the resultant pellets are sufficient to limit biological activities.

Self-heating of pellets does occur in large storage facilities and in some cases, in smaller piles stored at normal ambient temperatures. For wood pellets, the inclination to self-heating seems to vary among different qualities of pellets and it is most pronounced shortly after production. It has been observed that during storage of pellets, the temperature in the pile or silo can increase within a few days or even hours after production. The temperature can vary depending on the raw material and most often it is around 60 to 65°C [235; 240]. The temperature increase can sometimes be higher, for certain pellet qualities up to 90°C. At such a temperature the risk of a run-away temperature resulting in spontaneous ignition will increase, especially if the volume of the pile is large or the pellets are in a silo. Mixing pellets of different moisture content is another potential source of heat production as heat is produced in the process of balancing out the moisture in the pile.

Part of the heat generation can also be attributed to low temperature oxidation of easily oxidised components in the material, such as unsaturated fatty acids [240]. Heat is generated by oxidation of fatty acids to aldehydes and ketones. Further oxidation of these aldehydes and ketones then produces low molecular carboxylic acids [240]. These volatile organic compounds have been detected in pellet storage facilities. Fresh pine sawdust, for example, contains high amounts of unsaturated fatty acids [250].

Absorption of moisture by pellets is also an exothermic (heat generating) process that takes place in storage piles and involves two phenomena, i.e. heat produced by condensation and differential heat. Condensation heat is released when water vapour is absorbed. Differential heat of sorption is released when moisture content increases inside the pellet from its initial state up to the fibre saturation point. The heat released by condensation is much above that of differential heat when moisture is absorbed from air. It was shown that the potential of differential heat release can be predicted by near infrared spectroscopy, which is a rapid analytical tool. This method, once fully developed and established, has a potential to establish limits for a safe mixing of pellets with different moisture contents [236].

Pellets are hygroscopic and, at high air humidity, absorb water vapour from the air, especially if the pellet surface temperature is lower than that of the air [235].

The lower the initial moisture content of the pellets and/or the higher the air humidity is, the higher the risk is of heat generation via moisture absorption. Pellets with low initial moisture content (around 5 to 6 wt.% (w.b.)) are more reactive and will absorb moisture more easily than pellets with higher initial moisture content (around 8 to 10 wt.% (w.b.)). Lignocellulosic material with 5 wt.% (w.b.) moisture content at an air and material temperature of 20°C (at relative humidity of 70%) will display temperature rises of up to 100°C when the moisture content increases by 4.5 to 9.5 wt.% (w.b.) [237]. The moisture diffusion rates depend on several physical and chemical properties of pellets. Pellets with higher amount of particles (fines) will absorb moisture more easily and are thus more prone to self-heating. Parameters such as pile geometry and size, moisture in the air and the homogeneity of different pellet layers in the storage are also important for moisture absorption [236].

5.2.3.3 Self-heating – main risks and recommendations

When a fuel with a propensity to exhibit one or more of the heat generating processes is stored in a large volume, the temperature will increase within the pile, which may lead to spontaneous ignition in pellet storage facilities. The main risks resulting from the self-heating process of stored pellets are the following, in the order of occurrence:

- Release of asphyxiating (e.g. CO) and irritating gases (e.g. aldehydes and terpenes);
- Spontaneous ignition resulting in pyrolysis of bulk material and release of pyrolysis/combustion gases;
- Gas and/or dust explosion, typically as a result of approaching top compartment of a silo in fire rescue work;
- Surface fire and spread of fire, typically as a result of an explosion in a silo.

General recommendations and advice to avoid self-heating and spontaneous ignition for biomass are:

- Avoid storage and transport of large volumes if the fuel's tendency toward self-heating is unknown.
- Be conscious of the risk of self-heating and spontaneous ignition in large storage volumes.
- Avoid storing biomass with moisture contents greater than 15 wt.% (w.b.). Moisture damaged pellets from a railcar or ocean shipment should never be put into storage, instead they should be dumped in a rejection bin or directly burned.
- Avoid mixing different types of biomass fuels in the storage.
- Avoid mixing fuel batches with different moisture contents.

Specific recommendations for storage of wood pellets are:

- Avoid large amounts of fines in the fuel bulk.
- Measure and monitor the distribution of temperature and gas composition within the stored material. Frequent visual inspection is recommended.

More specific recommendations about storage of moist biomass in heaps can be found in the literature and for example maximum width and storage height will vary depending on the type of biomass [232; 234].

5.2.4 Off-gassing

Pellets decompose over time and emit non-condensable and condensable gases. Extensive research has been conducted in order to quantify the amount of gas emitted and the rate at which the gases are emitted [235; 238; 239; 241]. The quantity of emitted gases increases dramatically with increase in ambient temperature and varies among different brands of pellets, though the spectrum of gases appears to be very similar.

A comparison between newly produced (fresh) and stored pellets has shown more emissions being released from the stored pellets. In a recent study, a three week old pellet is observed to emit about 28 times more pentanal and eight times more hexanal than the reference pellets [240]. The emissions of terpenes is observed to be low in wood pellets as most of the monoterpenes are thought to leave the sawdust during high temperature drying in the pelletising process. In cases where low temperature drying is implemented, pellets may emit terpenes. However, more research is required in this area to fully understand these processes. Headspace gas analyses of pellet storages confirmed that several aldehydes and low molecular carboxylic acids are emitted from pellets.

Off-gassing also occurs during ocean transportation of pellets; emissions of aldehydes, carbon monoxide, carbon dioxide and methane are detected [247]. Emission rates of CO, CO_2 and methane at different storage temperatures have been investigated [241]. The correlation between the emissions of aldehydes and carbon monoxide is currently under investigation.

The emission of volatile organic compounds (off-gassing) often brings about a pungent smell. Some of these volatile organic compounds may have a negative impact on human health, for example irritation of the eyes and the upper airway [249; 250]. A number of fatal accidents have occurred as a result of personnel entering confined spaces where large bulks of wood pellets have been stored [241; 242; 243]. The control measures for entry into confined places are addressed in detail in Section 5.3.2. The IMO, regulating safety onboard ocean vessels, stipulates measures for carriage of wood pellets, wood chips, pulp logs, lumber and other wood products, all of which emit gases and cause oxygen depletion [244]. Similar regulations are expected to be stipulated for storage of pellets on land. Pellet manufacturers provide data on the off-gassing in the MSDS of their product (cf. Section 5.4). For information on health effects, see Section 5.3.3.

5.2.4.1 Non-condensable gases

Wood pellets as well as dust from wood pellets (white or brown) emit non-condensable gases, primarily CO, CO_2 and CH_4. Figure 5.6 to 5.8 illustrate off-gassing of CO, CO_2 and CH_4, respectively, from one type of white pellets stored in a containment without ventilation [218; 245; 246]. Measurements performed in large storage spaces and ocean vessels are consistent with these data.

The rate of off-gassing is a function of ambient temperature, raw material characteristics, drying technology and the pelletising equipment used, and varies among different brands of pellets.

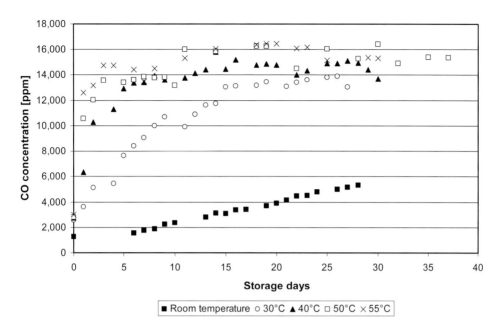

Figure 5.6: CO concentrations in the headspace of a pellet storage at different temperatures over time due to off-gassing

Explanations: pellets made from pine; data source [271; 272]

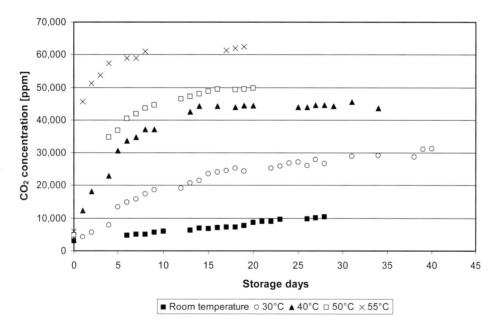

Figure 5.7: CO_2 concentrations in the headspace of a pellet storage at different temperatures over time due to off-gassing

Explanations: pellets made from pine; data source [271; 272]

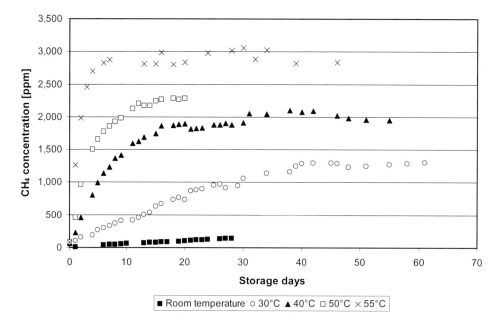

Figure 5.8: CH$_4$ concentrations in the headspace of a pellet storage at different temperatures over time due to off-gassing

Explanations: pellets made from pine; data source [271; 272]

5.2.4.2 Condensable gases

Biomass, including pellets in bulk, emits small amounts of condensable gases such as aldehydes (alcohol dehydrogenated) and ketones including hexanal and pentanal in addition to acetone and methanol [235] (total amount of all aldehydes in fresh pellets is around 1,000 to 2,900 µg/g$_{pellets}$). Pellets made of pine wood are known to emit more condensable gases than pellets made of other wood species [240; 247; 248; 249; 250]. Eleven different qualities of pellets made from a mixture of fresh pine and spruce were manufactured in an industrial-scale experiment. About 7 t of each pellet production were stored in eleven separate piles for a period of one month. The temperature in the piles made of mostly pine sawdust increased to a maximum of 55°C. During the storage period, the total amount of fatty and resin acids decreased by about 40% and aldehydes and ketones decreased by about 45% due to oxidation of fatty acids (the total amount of fatty and resin acids in pellets made of fresh pine sawdust is highest around 0.2 to 0.7% or 2,000 to 7,000 µg/g$_{pellets}$). Resin acids also get oxidised, for example most of the dehydroabietic acid oxidised to 7-oxo-dehydroabietic acid during the four weeks of storage. Pellets made of 100% spruce contain less fatty and resin acids (700 to 2,000 µg/g$_{pellets}$) and therefore emissions (off-gassing) of volatile aldehydes and ketones were limited [235]. The values can vary depending on the source of pine and spruce and on how fresh the sawdust is (during storage of sawdust, some fatty and resin acids are oxidised and their amount hence lessened).

The emissions of aldehydes and ketones from pellets during storage cause a strong smell and may have a negative impact on pellet marketing and consumption. In order to remove fatty and resin acids from sawdust before pelletising, sawdust was exposed to electron beams of different strengths in a recent investigation. The results confirm that fatty and resin acids are reduced by irradiation in laboratory-scale experiments and pellets made of the sawdust have better quality (higher density and compressive strength) than the pellets made of untreated material [251].

5.2.4.3 Oxygen depletion

Oxygen is consumed during the decomposition phenomenon. Severe depletion of oxygen in pellet storage spaces without ventilation has been observed. As discussed in detail in section 5.3.2, entering a storage area should only be permitted after having checked the oxygen as well as carbon monoxide levels. The health effect of oxygen depletion is examined in Section 5.3.3. Guiding values for allowed concentrations of carbon monoxide and oxygen can be found in Sections 5.3.2 and 5.3.3, respectively.

5.2.4.4 Relevance of off-gassing for small-scale pellet storage units

Triggered by reports on the health and safety risks associated with the emissions produced by stored wood pellets [238; 239; 240], the Austrian wood pellets industry started a field test study in 2008 to evaluate the health risks from the off-gassing of carbon monoxide from wood pellets for owners of small-scale pellet storage units at residential end user sites [252; 253].

The aim of this study was to determine the typical carbon monoxide levels occurring in small-scale pellet storage units at residential end user sites. The typical types and sizes of small-scale wood pellet storage units found in Austria have already been described in detail in Section 4.2.3.1. For this study, two types of storage facilities were investigated. The first type was a pellet storage room situated in a cellar. With a market share of almost 90%, cellar storages represent the most widespread type of pellet storage unit at residential end user sites in Austria. The second type was an underground pellet storage tank. The market share of this type of storage is only 0.5%. It is particularly prone to the build up of high concentrations of off-gases from wood pellets due to its air-tight construction. Spruce is the typical raw material for pellet production in Central Europe. Therefore, all investigated pellet storage units contained pellets made purely from spruce sawdust or pellets made from a mixture of spruce and up to 3% pine sawdust. In Table 5.4 the measured levels of carbon monoxide within the investigated pellet storage units are shown.

Table 5.4: CO concentrations in small-scale pellet storage units at residential end user sites

Explanations: [1]... recommended TWA for CO for living spaces in Canada and the USA; [2]... maximum TWA for CO for working spaces in Spain; [3]... maximum TWA for CO for working spaces in Austria and Germany

Pellet storage unit	Number of measurements	CO concentration [ppm]						
		≤ 9[1]	10 - 25[2]	26 - 30[3]	30 - 100	100 - 500	500 - 1,000	> 1,000
Pellet storage room	30	14	10	3	2	1	0	0
Underground pellet storage tank	22	6	1	0	5	4	5	1

The field test study revealed that air-tight pellet storage systems are highly prone to the build-up of carbon monoxide emissions well above international threshold limit values for human exposure. This result emphasises the importance of introducing safety measures in all cases of air-tight pellet storage systems. The measurements for pellet storage rooms situated in cellars revealed lower concentrations of carbon monoxide with 93% of all measurements below a value of 30 ppm and 47% below a value of 9 ppm. The oxygen concentration for all the tested pellet storage rooms situated in cellars was found to be unaffected from off-gassing at 20.9%. It was also shown that in a small number of cases health affecting concentrations of carbon monoxide can occur in pellet storage rooms. Considering the high diversity of constructions for this type of pellet storage unit and the small number of measurements carried out, even higher concentrations of carbon monoxide than those reported in this study might occur. Additionally, a small but given risk of leakage of CO from the pellet storage room in the cellar into an adjoining living space exists. To avoid this risk, every pellet storage room should be equipped with some sort of ventilation to the exterior of the building.

When determining the risk potentials for these types of pellet storage units, there is one more factor to consider that is related to the raw material. During this study, additional experiments dealing with the off-gassing behaviour of different wood species such as spruce and pine were conducted. The results showed that the investigated pellets made from pine sawdust emitted three to five times as much CO than the reference pellets made from spruce sawdust [252]. A change to raw material having higher fatty acid contents in wood pellets production might therefore have a significant impact on the risk potential for small-scale pellet storage units.

Furthermore, the parameter storage temperature was found to have great influence on the off-gassing behaviour of the pellets. Elevated storage temperatures (> 30°C) lead to a significant increase in the emission rate of carbon monoxide (cf. Section 5.2.3.1). Consideration should be given to the "stored pellets to storage room" volume ratio, which represents the degree of dilution of carbon monoxide within the compartment air.

The worst case scenario can be described as an air-tight, fully filled pellet storage unit in which the pellets are kept at a temperature above 30°C.

Microbial activity was found to be of no importance in the formation of off-gases from stored wood pellets in small-scale pellet storage units at residential end user sites, providing the need for dry storage conditions is abided by. Off-gassing in small-scale pellet storage units can solely be attributed to oxidative degradation of natural wood components.

A further field test study is currently carried out and will be finished in spring 2011. A particular focus of this study is to determine minimum required air exchange rates to ensure risk free access to small-scale pellet storage units. In addition, investigations on the influence that pellet dispatch temperatures have on the off-gassing behaviour of wood pellets are carried out, which may well have an influence on regulations for pellet logistics such as the ENplus certificate (cf. Section 2.6).

5.2.5 Fire risks and safety measures

Fires caused by self-heating of wood pellets and dust are not uncommon. Fires and explosions related to wood dust caused by sparks and electrostatic discharge are even more common, as illustrated in Table 5.5 [254].

Fires and explosions during the production of pellets occur primarily in the dryer or during cooling and screening. Fires and explosions also occur during handling and storage in large bulks at the pellet production plant, ocean vessel loading facilities or district heating or power plants where dust concentrations may exceed the MEC (cf. Table 5.1). Fires or explosions are not known to be a problem in the bagged pellets market. Relevant issues concerning fire risks and safety measures are discussed in the following sections.

Table 5.5: Reasons for accidents or incidents with different types of dust

Explanations: based on US data; data source [254]

Type of Dust	% of total
Wood	34
Grain	24
Synthetics	14
Metals	10
Coal / peat	10
Others	6
Paper	2
Total	100
Reason for accident or incident	**% of total**
Mechanical spark	30.0
Unknown	11.5
Static electricity	9.0
Smoulder spots	9.0
Friction	9.0
Fire	8.0
Hot surface	6.5
Self-ignition	6.0
Welding	5.0
Electrical equipment	3.5
Other	2.5
Total	100.0

5.2.5.1 External ignition sources

Except for spontaneous ignition, there are a number of possible causes of fire. Some common causes are sparks generated by metal pieces, stones, etc. that have come into contact with the bulk by accident. Other causes could be overheating of electric motors, conveyor bearings or elevator systems, friction between, for instance, conveyer belts and accumulated pellets, fines and/or dust or careless hot work. Ignition can also be instigated when material containing small pieces of smouldering material ("hot spots") is transported to a new storage location. In heat generation plants, ignition might also be the result of back-firing or sparks near the boilers. Another risk, not to be neglected, is posed by the possibility of a fire in wheel loaders, which are frequently used for taking up pellets from heaps.

The most important measures to avoid these risks are:

- Control measures for impurities when receiving the material, for example magnetic separators, sieves, etc;
- Spark detectors connected to an extinguishing system with fast acting valves at strategic locations in the transport system;
- Control schemes to check the condition of bearings (temperature measurement);

- Control schemes for cleaning in order to avoid accumulation of material in conveyors, elevators etc;
- Control schemes for hot work in the facility.

5.2.5.2   Safety measures related to storage of pellets

As mentioned in Sections 5.2.1 and 5.2.3, low temperature oxidation of pellets will result in the formation of aldehydes and low molecular carboxylic acids, CO_2, CO and methane [247]. Under certain circumstances, especially in enclosed areas with low ventilation, this might result in an acutely toxic environment with very high concentrations of the aforementioned gases.

In silos and similar storage buildings with low ventilation, a CO sensing system is recommended to monitor the atmosphere at the top of a silo complex, as well as in adjacent premises for occupational health concerns.

It is also recommended that both operating and fire rescue personnel wear a personal CO gas detector when entering these storage areas, especially if there are any signs of heat generation, sticky smell or smoke. As the 15 min threshold value for CO is 100 ppm and CO is acutely toxic at a concentration of about 1,200 ppm (0.12 vol.%), it is not sufficient to measure and rely on the oxygen concentration alone [42].

For small-scale pellet storage units that are not constructed to be air-tight, warning signs and ventilation instructions for entering the pellet storage unit are currently developed by both the German and the Austrian wood pellets associations [255] and will be subject to further investigation with the aim of developing a new standard for wood pellets storage.

5.2.5.3   Temperature and moisture control and gas detection

Pellets in storage spaces will heat up due to a combination of factors including decomposition of the pellets (cf. Section 5.2.3) and ambient conditions (such as heating of the walls from exposure to the sun). Continuous temperature control by sensors embedded in the stored product is required to minimise the risk of fire. Vertically suspended cables containing sensors at certain intervals are an example of how the temperature can be monitored and logged by a computer with alarm functions. Wireless sensors dropped into the material at various locations with a central data monitoring system are another technique that may be used. The measurement system should be able to measure temperatures of up to 100°C at a minimum. With a thermal conductivity of pellets between 0.18 and 0.24 W/mK (at a moisture content of 4 and 8 wt.% (w.b.) respectively [256]), the sensors need to be distributed at a certain distance from one to the other throughout the bulk in order to detect hot spots. The sensor configuration needs to be designed with consideration given to the thermal conductivity characteristics, the permeability and variance in permeability, ventilation fan capacity and the geometry of the fresh air inlets. When designing the ventilation system, consideration should also be given to the selected method of fire extinguishing (cf. Section 5.2.5.4). Figure 5.9 illustrates examples of embedded temperature monitoring systems.

Temperature monitoring is more complex in flat storages. As an alternative to direct temperature sensing, there are systems available for measuring carbon monoxide, certain hydrocarbons, radiated heat or smoke as precursors to overheating. The sensors may be located at the ceiling of a flat storage and thereby monitor a fairly large floor footprint.

However, these systems do not typically provide the same early warning as sensors embedded in the material. Alternatively, temperature sensors such as temperature cables may be incorporated in dividing walls, cross bars etc.

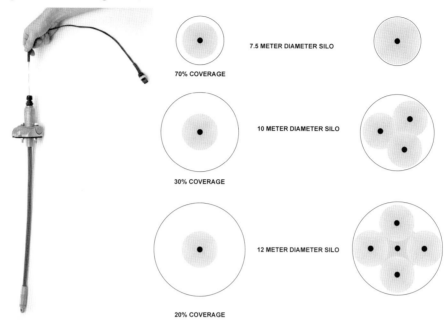

Figure 5.9: Examples of embedded temperature monitoring systems and comparison of single cable and multi cable solutions

Explanations: left: retractable cable; right: single cable versus multi cable; courtesy of OPIsystems Inc.

A modern storage facility for pellets should also contain a forced ventilation system for controlling the thermal conditions in the stored pellets. In areas with high relative air humidity, the effects of injecting outside air into the storage have to be considered in view of the EMC characteristics of pellets as illustrated in Figure 5.5. An increased humidity in the pellets will contribute to increased microbial activity. Also, in case of high ambient temperature as compared to the temperature in the pellets, the thermal content of the injected water vapour will contribute to the heating of the pellets. In other words, under certain circumstances, ventilating storage with air containing high relative humidity may in fact lead to temperature escalation in the storage rather than to temperature decrease. In order to avoid this uncontrolled condition, the ventilation system should include a dehumidifier to control the amount of moisture injected into the storage (cf. also Section 5.2.2).

Pellets in bulk contain 48 to 53% of void (empty space between pellets [247; 248]) depending on the aspect ratio of the pellets and the fines content. Permeability is a measure of the ability of air to flow through the pellets and is established by measuring the pressure drop in Pascal per metre (Pa/m) as a function of air flow rate in $m^3/s/m^2$ in a vertical containment. The air flow is affected by the viscosity of the air, which in turn is related to the temperature, gas mixture and moisture content. Figure 5.10 illustrates permeability for 6 mm pellets with various aspect ratios.

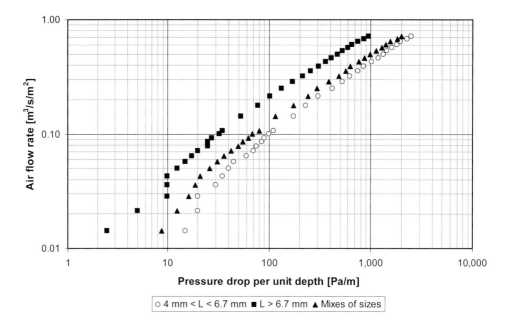

Figure 5.10: Permeability for pellets with various aspect ratios

Explanations: experiment conducted with constant temperature (20°C) and relative humidity (30%) which are the main factors affecting the viscosity of the air; data source [257]

The permeability curve [258] can be used when designing a forced ventilation system for a silo where the ventilation air is introduced from the bottom. Self-ventilation or forced ventilation from the top of a silo is less effective as this arrangement leads to parasite leakage at top sections of the storage pile and thereby decreases the efficiency of the ventilation at lower sections of the pile. Also, an effective dehumidifier can only be fitted directly to an active fan located at the bottom of a silo. Well designed forced air ventilation with dehumidifier can be combined with on/off dampers or valves and injector fixtures for fire extinguishing media (cf. Section 5.2.5.4).

As mentioned in Section 5.2.3.2, the temperature in storages for pellets in bulk can easily rise to 60°C or higher if no proper ventilation is in place. All pellet storage types need to be able to facilitate emergency discharge of pellets in case the temperature approaches the runaway temperature, which for some types of pellets is around 80°C. The emergency discharge can be done by relocating the product in another storage or in an outdoor location. This process allows pellets to be aerated, during which the pellets are cooled down and hot spots are broken up. The rule derived by experience with larger size storages in Canada is to ventilate a storage space whenever the ambient temperature is lower than the temperature inside the storage, which implies that each storage has to be equipped with forced ventilation. Large power plants using pellets as a fuel have a limit of 45°C above which shipments are rejected. As pointed out in Section 5.2.3.2, a maximum acceptable limit for the moisture content in the pellets is also often stated.

Self-heating and off-gassing are a function of several parameters such as moisture content of the pellets, reactivity related to the species (e.g. pine is generally more reactive than other species), ambient temperature, temperature transients and length of storage. Thus, it is hard to make recommendations on which storage sizes are concerned and which monitoring methods and approaches should be applied. Self-heating has been observed in storages as small as 25 tonnes and even in heaps of pellets on ground with as little as 5 tonnes of pellets. In storages equipped with forced ventilation, it is extremely important to turn the ventilation off as soon as indications for a spontaneous ignition are detected, otherwise fire development will be enhanced. Silos are often lost if the self-heating process has proceeded to ignition. Temperature monitoring is therefore essential to keep the temperature below 45°C.

5.2.5.3.1 Indoor storage in heaps

As a complement or alternative to fixed temperature measurement systems, gas detection systems could be used in combination with visual inspections. The storage volumes in indoor heaps are often very large. An ignition caused by an external ignition source is probably detected relatively quickly by attending personnel, while a spontaneous ignition may be much more difficult to detect.

The self-heating process causes the moisture to be transported to the surface, which is visible as a "white smoke" (water vapour) from the surface of the pellet pile. The water vapour might condense on the pellet surface and cause the pellets to disintegrate. These "condensation areas" are easy to detect visually. The temperature on or close to the surface can be measured with a temperature probe inserted into the pellet bulk. However, under most conditions, the naturally occurring self-heating phenomenon does not result in spontaneous ignition, making judgement on the conditions of the pile difficult. In this case, a CO analyser will provide additional information. As ventilation in the indoor storage might be considerable, the best approach is to insert a CO probe at suspected locations 0.5 to 1.0 m into the pellet bulk. Measurements of very high values of CO (> 2 to 5%) would be a strong indication of spontaneous ignition resulted from self-heating. As the CO concentrations might be very high, self-contained breathing apparatus (SCBA) should be used by the personnel.

A more general supervision of the storage can be attained with advanced fire detection systems based on gas sensors ("electronic nose"). These systems indicate when significant changes occur in the gas composition and are able to differentiate between fire gases and exhaust gases, for example from vehicles.

In order to limit the consequences of a spontaneous ignition in a large heap, it is advisable to sub-divide the heap into several cells. This could be accomplished by separate concrete walls that will prevent the pyrolysis zone from spreading over large distances inside the bulk before it is detected.

5.2.5.3.2 Storage in silos

Visual inspection of a silo by working personnel cannot be done as in the case of indoor heap storage. A silo should be equipped with both temperature probes (cf. Figure 5.9) and gas detection systems if and when it is technically and economically feasible. Also here, an advanced fire detection system based on gas sensors ("electronic nose") could be used to achieve early indications of a possible spontaneous ignition.

In the case of a suspected fire, gas analysers both for CO and O_2 should be used to measure the atmosphere in the headspace volume of the silo. CO concentrations exceeding 2 to 5%, low oxygen levels and considerable smell are strong indications of spontaneous ignition. Both the CO and O_2 concentrations would also become valuable indicators during an extinguishing operation using inert gas, as they can be used to ensure an inert atmosphere has been reached in the silo (cf. Section 5.2.5.4).

5.2.5.4 Extinguishing fire in pellet storages

Fire fighting of pellets is very different from fire fighting of most other products since water cannot be used, especially in silos. Wetted pellets swell very quickly and the resulting material becomes extremely hard, often requiring removal by a jack hammer. The fact that pellets expand to about 3.5 times of their original size (cf. Figure 5.5) when wetted, limits the choices of extinguishing media. The wet media would be suitable for surface fires where the swelling does not cause bridging or potential cracking of the containment walls. However, fires as a result of self-heating usually occur deep inside a pile or containment, and are often not detectable until the fire ball is well developed and headspace is filled with combustible gases. Methods for extinguishing fires inside silos or bunkers have been developed by the SP Technical Research Institute of Sweden [225; 230; 231; 259] over several years.

The recommended methods for extinguishing a fire in pellet storage and the actions to handle an incipient fire are different for storage in indoor located heaps and storage in silos and are summarised in Sections 5.2.5.1 and 5.2.5.2, respectively.

5.2.5.4.1 Fire fighting in heaps in indoor storage

When a fire with open flames is detected, the first measure is to suppress the flames as quickly as possible. Fire development may be quite fast and a delay in the extinguishing operation may increase the risk of total storage loss. Water usage in general should be restricted, except for the purpose of preventing dust cloud formation, which may increase the fire intensity. If possible, water sprays should be used but in case the fire is too large for the throw length of the water spray, solid water streams might be used.

If possible, the use of fire fighting foam (preferably Class A-foam) as low or medium expansion foam can improve the operation. Class A-foam is a foam specially developed to be used against class A-fires, i.e. materials causing smouldering fires. Low expansion foam is foam with an expansion ratio less than 20 (the expansion ratio is the ratio of the volume of foam to the volume of foam solution from which it was made). Medium expansion foam is foam with an expansion ratio greater than or equal to 20 but less than 200. Low expansion is used where a longer throw length is required, while medium expansion foam can be used if a close approach to the fire is possible. Several benefits of using foam are as follows:

- Foam can be applied more gently than water.
- Foam reduces the risk of dust formation.
- Foam provides a sustained cover for the pellet surface reducing heat radiation towards the surface and thereby the risk of a fast fire spread.
- Foam extinguishing requires less water, which in turn reduces water damage to the pellets.
- Foam is more effective compared to plain water, specifically in case of a fire in a wheel loader where a potential for fire to spread to oils, plastics, rubber etc. is high,

If a spontaneous ignition occurs inside the pellet bulk, the most probable locations of the smouldering fire should be identified first. The involved material has to be removed by a wheel loader to a safe place. Each bucket should be carefully inspected for the presence of smouldering material. The "safe material" should be separated from material that contains very hot or glowing matter. The latter should be spread out to allow it to cool down and smouldering material should be carefully extinguished with water spray.

During the removing operation, it is important to continuously extinguish any open fire and protect the remaining stack of pellets. The opening of the stack will provide the smouldering material with oxygen, which will result in an increase in fire intensity and possibly a "rain" of sparks. Water spray can be used but fire fighting foam is more effective. The foam layer on top of the pellet stack also limits the oxygen supply to the smouldering areas in the remaining stack and thereby reduces the possible flare up of fire.

Due to the possibility of very high concentrations of CO, SCBA should be used by all personnel, including the drivers of the wheel loaders.

5.2.5.4.2 Fire fighting in silos

The extinguishing technique for silo fires is completely different from "normal" fire fighting procedures. Extinguishing silo fire is a lengthy process that normally takes several days to complete. The technique developed by SP is based on injection of inert gas and prevention of air (oxygen) from reaching the smouldering fire zone. The inert gas should be injected close to the silo bottom to ensure inertisation of the entire silo volume, and the gas must be injected in gaseous phase. Several successfully prevented incidents in Scandinavia [260; 261; 262] provided much valuable evidence for the viability of this method. One of the key factors is to understand the anatomy of a silo fire, which is described in Section 5.2.5.5.

In order to minimise the consequences of a silo fire and ensure effective extinguishment, the following fire management aspects need to be considered:

- Suitable equipment for the gas supply must be available (gas tank for liquid gas, vaporisation unit).
- There should be possibilities to inject and distribute the gas close to the silo bottom (the silo should preferably be prepared with a fixed pipe system but in an emergency situation, penetrations of the silo wall could be made to insert lances for gas injection).
- The silo construction should be as air-tight as possible in order to reduce the infiltration of air (oxygen) into the silo compartment.
- There should be possibilities to evacuate the combustion gases at the top of the silo through a "check valve" arrangement that prevents inflow of air (oxygen) into the silo headspace.
- There should be preparations for an emergency discharge of the silo content following the inertisation of the silo.

Normally, it is not necessary to install a fixed gas tank and vaporisation unit to every silo facility. In Sweden, mobile emergency equipment is used, which is transported to the silo when a fire is suspected or detected. As a silo fire develops slowly in the early stages, and the use of temperature monitoring and gas detection systems (cf. Section 5.2.5.3) provides the possibilities for an early detection, there will normally be many hours available to bring in the mobile equipment. Figure 5.11 shows a picture of such mobile equipment, which has been successfully used at silo fires in Sweden.

Figure 5.11: Mobile fire fighting unit for silo fire fighting

Explanations: left: example of 47 m high silo fire successfully extinguished in Sweden 2007 by injection of nitrogen; right: mobile equipment used in the fire consisted of a vaporisation unit and a tank of liquefied nitrogen; data source [260]

For practical reasons, nitrogen has been used in Sweden. Nitrogen is easily available and the vaporisation unit does not need any energy supply as the vaporisation energy is taken from the surrounding air. Carbon dioxide would provide the same inertisation effect but requires a powerful external heat source for the vaporisation unit. Attempts to use carbon dioxide without a vaporisation unit have caused many unsuccessful extinguishing operations as the supply hoses/pipes, nozzles/lances and the bulk material close to the injection point tends to freeze quickly and thereby completely block further gas injection.

The gas should be injected close to the silo bottom. This ensures that the air/combustion gases are replaced by an inert atmosphere quenching the smouldering fire ball. As the fire ball has a tendency to spread downwards in the silo (cf. Section 5.2.5.5), gas injection from below ensures that the fire ball is reached by the inert gas. This technique also has the advantage of pushing the combustion gases inside the bulk material towards the headspace in the silo. Measuring the gas concentrations in the headspace (primarily CO and O_2) provides a verification of the extinguishing process as a reduced CO concentration indicates that the pyrolysis activity is controlled. Low oxygen content in the headspace is important to minimise the risk of gas or dust explosions.

In small diameter silos (5 to 6 m), one gas injection point close to the centre of the silo will normally be sufficient. At larger diameters, one inlet will not ensure an even distribution over the cross section area. Several inlets are therefore recommended as shown in Figure 5.12, and the number of required inlets will depend on the silo diameter and the gas flow rate at each inlet.

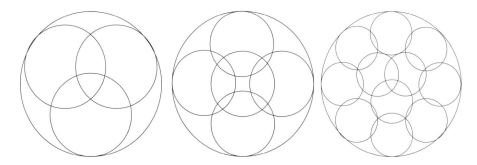

Figure 5.12: Principle sketch of distributed gas injection in silos

Explanations: the number of gas injection points will depend on the silo diameter and the gas flow rate per injection point; data source [259]

For small diameter silos, it is possible to arrange a provisional gas injection in case of an emergency by using one or several lances made of steel pipes (cf. Figure 5.13). For large diameter silos, exceeding 10 to 15 m in diameter, it will be difficult to insert the lances in the bulk material. It is therefore important to prepare the silo for gas injection during the design and construction phase. If the silo is equipped with a ventilation system, this system could be used for the gas distribution.

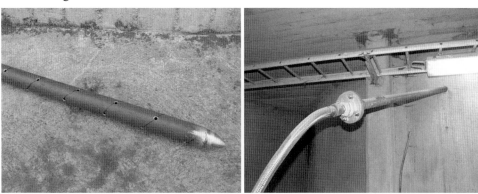

Figure 5.13: Steel lances used for gas injection in a silo

Explanations: left: a 50 mm perforated steel pipe used for gas injection in a silo fire in Sweden 2007; right: the lance was inserted through a drilled hole close to the silo bottom and connected to the vaporisation unit by a hose; data source [260]

It is important to note that water should not be used in a silo fire containing wood pellets as this will cause significant swelling of the pellets (cf. Figure 5.5). Swelling could lead pellets to stick to the silo wall and form "material bridges" high up in the silo. The extinguishing operation and subsequent unloading can easily be hindered by the swelling process. The forces from the swelling might also lead to severe damage to the silo construction. Water could also cause formation of combustible hydrogen inside the silo presenting a risk of severe explosions. The potential dangers and risks of extinguishing with water should be recognised in order to ensure the safety of the personnel involved.

From the personnel health and safety perspective, it is very important not to open up or start to unload a silo with an ongoing fire as this might cause severe gas and dust explosions. There are several examples of total losses, both of the bulk material and the entire silo construction when attempts have been made to discharge the bulk material without first controlling the pyrolysis fire. In the example shown in Figure 5.14, severe flames resulted when a concrete silo filled with wood pellets was opened up for emergency discharge. Jet flames up to 50 m long occurred due to gas explosions inside the silo.

Figure 5.14: Flames on the outside of a silo caused by an opening in the silo wall
Explanations: data source [263]

Below is a brief summary of the techniques and tactics for silo fire fighting, which are based on results and experience from the previously mentioned experiments and real silo fires. Above all, it is of utmost relevance to not open the silo during the fire fighting operation and to not use water inside a silo filled with wood pellets:

- Start with identifying the type of silo and the fire scenario and make an initial risk assessment of the situation. As there may be very high levels of carbon monoxide in the plant, there may be a need for personnel to use SCBA in certain areas. Also consider the risk of dust and gas explosions by measuring CO and O_2 concentrations in the silo top. If and when the measurements show a very high concentration of CO (> 5 vol.%, i.e. 50,000 ppm) and oxygen content exceeding 5 vol.%, personnel should not attend the silo top area unless it is absolutely necessary.

- In order to minimise the air (oxygen) entrainment, close all openings and turn off ventilation. On the silo top, there must be a small opening to allow the release of combustion gases and prevent air from coming in at the same time. A rubber cloth on an open top hatch can serve as a "check valve".

- Injection of nitrogen, close to the bottom of the silo, should be started as soon as possible. The gas should be injected in gaseous phase, and an evaporator must be used. If necessary, prepare holes close to the silo bottom and prepare lances for the injection of the gas. Holes should be kept closed until the nitrogen feed is connected. If the measurements indicate a potential risk for an explosion in the silo top, inject nitrogen into the silo headspace as well until the O_2 is below 5 vol.%. Avoid a too high flow rate to prevent dust formation.

- The injection rate at the silo bottom should be based on the silo cross sectional area. The recommended injection rate is 5 kg/m² h. The total amount of gas could be estimated based on the total gross volume of the silo (empty silo) and is likely to be in the order of 5 to 15 kg/m³.

- If possible, measure the concentration of CO and O_2 in the silo top continuously during the entire extinguishing and discharge operation. The instrument for CO must be able to measure very high concentrations, preferably at least 10 vol.% CO (i.e. 100,000 ppm) to provide relevant information.

- The discharge of the silo should not be started until there are clear signs (low levels of CO and O_2) that the fire is under control. The situation inside the silo should be continuously assessed based on the gas measurements at the silo top. An increasing concentration of CO indicates increasing "activity" inside the silo. An increasing oxygen concentration in the silo top indicates inflow of air. If the oxygen content exceeds about 5 vol.%, the unloading operation should be interrupted and the injection rate of nitrogen should be increased until the oxygen concentration is below 5 vol.% again.

- The discharge capacity might be considerably reduced compared to a normal situation and the unloading process might take many hours or even days to complete. An estimation of the time required could be made based on the normal unloading capacity multiplied by a factor two to four. Fire fighting personnel must be present at the discharge opening to extinguish any smouldering material, which means that a significant number of fire fighters and also a large number of SCBAs are required.

- The gas injection at the silo bottom should continue during the entire unloading process. During this process, the gas injection rate might be reduced when the fire is under control as long as the oxygen concentration in the silo top does not exceed 5 vol.%.

5.2.5.5 *Anatomy of silo fires*

Hot spots may develop in any part of a silo, although they are more frequently found in stratified layers of material with variance in permeability. Since the natural prevailing convection is always upward in the centre, the oxygen supply feeds the fire from underneath and the smouldering moves downward from the location of the ignition. There are cases where forensic investigation uncovered long strings of carbonised material leading down from the centre of the fire (fire ball) to a spot where oxygen (air) had entered the silo through holes at the bottom. At the same time, as the fire ball is slowly moving downward there is a wave of steam and gas (primarily carbon monoxide and other hydrocarbons) moving upwards towards the headspace of the containment driven by thermal convection. Figure 5.15 illustrates the fire ball movement within a column of pellets during experiments in a silo of 1 m in diameter and 6 m in height [225; 230]. The fire in this case started just below the centre of the 6 m column. The figures are a visualisation of the temperature measurements inside the silo and the dark colour in the point of ignition indicates combustion temperature to be around 400°C. The faint horizontal line in the upper section of the column is the surface of the pellets in the headspace. It can be clearly seen that when the gas wave breaks the surface and enters the headspace after about 20 hours, the convection pattern changes to a chimney effect resulting in a downdraft of air from the headspace along the sides of the silo wall. When the air reaches the fire ball it increases in size (30 h). Just after 30 hours the nitrogen is injected and after about 40 hours the temperature decreases, and continues to decrease gradually but slowly and so the risk of a gas explosion decreases also over time. However, biomass has a relatively high oxygen content (around 40 wt.% (d.b.)), which continues to feed the burning of the pellets unless continuous injection of inert gas is maintained.

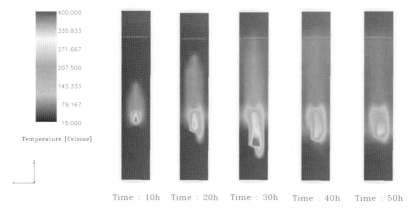

Figure 5.15: Fire ball movement within a column of pellets

Explanations: visualisation of the temperatures inside an experimental silo, 6 m high and 1 m in diameter, during fire and extinguishing tests; the smouldering fire was started in the centre of the silo and just after 30 hours, injection of inert gas was started from the bottom of the silo; data source [225; 230]

Figure 5.16: Fire ball seen from underneath in the test silo

Explanations: the photo shows the sharp limit between the fire ball involved in the pyrolysis and the surrounding pellets, which are more or less unaffected; data source [225; 230]

Looking inside the column after the fire had been extinguished revealed the nature of an encapsulated fire ball. Figure 5.16 and Figure 5.17 show the bottom of the fire ball with the char indicating a migration in the downward direction of the fire ball. Significant concentrations of carbon monoxide, carbon dioxide, unburned hydrocarbons and a reduced level of oxygen were not detected in the silo headspace until about 20 hours after ignition.

The same phenomena have been seen in real silo fire incidents where it probably took several days before spontaneous ignition was detected.

Figure 5.17 shows the pellets about 1 m above the point of ignition and the fire ball where the pellets have agglomerated due to water vapour and the gas wave that moved upwards in the bulk. As the combustion gases, including a lot of water vapour, moved upward in the silo, this caused the pellets to agglomerate above the fire ball.

Figure 5.17: Agglomerated pellets above the fire ball in the test silo

Explanations: data source [225; 230]

Ongoing research at SP, the University of British Columbia (UBC) and the industry is directed towards identifying a more sensitive trace gas generated at the runaway temperature stage or at least during the early stages of combustion.

5.3 Health concerns with handling of pellets

It is recommended that exposure to harmful compounds and materials, such as fine particulates, should be limited as they are considered to potentially affect human health. Epidemiological findings show a strong correlation between adverse health effects [264; 265; 266; 267; 268] and fine particulates (airborne) generated during the handling of pellets. The effect of combined exposure to several compounds, however, is not well researched and recommended exposure guidelines are often lacking. One example would be inhalable airborne wood particles contaminated with chemical compounds such as pentachlorophenol (a wood preservative). Another example is oxygen depletion in combination with exposure to carbon monoxide [269], which has been reported to occur in air-tight pellet storage units [238; 239]. Oxygen depletion spontaneously increases the breathing rhythm, which results in a much faster uptake of CO in the blood if CO is present in combination with oxygen depletion. The exposure to harmful compounds continues to be an important subject of ongoing research and new information becomes available regularly. As new raw materials are used for producing pellets and other products, information from testing for potentially harmful effects will continue to be published.

The pellet manufacturers provide an MSDS for their product that specifies the emissions and the toxicological information of the pellets. The stack gas emissions as a result of the power conversion are provided by the supplier of the conversion equipment. The data in the MSDS are unique for a particular product, reflecting the stability and reactivity of the product. The producers of pellets may have one MSDS for pellets in bulk and another for the bagged product. Pellets in bulk pose a far greater health risk than bagged pellets, simply due to the concentration of contaminants as a function of volume. For example, large volumes of dust are generated during handling of thousands of tonnes of pellets during the loading of an ocean vessel, while very little dust is generated when filling a heat appliance in a home from a bag. Another example is the off-gassing of pellets, which can be substantial when large volumes are stored in a silo versus the off-gassing from a bag of pellets. It should, however, be noted that even for small amounts of stored wood pellets, the gas concentrations can be substantial if the size of the storage volume is small and the storage unit is constructed to be air-tight.

5.3.1 Exposure to airborne dust generated during handling of pellets

Aside from the infectious aspect of some particles, there is also concern for health effects from exposure to particulates with an aerodynamic equivalent diameter (ae.d.) less than 100 µm (typical size distributions can be found in Figure 5.1 in Section 5.2.1.2). Depending on size, particulates can easily be deposited in various parts of our airways as we breathe (cf. Figure 5.18). Such deposits may cause illnesses such as acute reactions, chronic reactions or tumours. The most serious damage is done by particulates with less than 10 µm in ae.d. that are able to enter our bloodstream through the alveolars where the gas-to-blood exchange takes place in our lungs. The medical field uses the following classification for particulates penetration in our lungs [270]:

- Inhalable fractions: < 100 µm ae.d.

- Thoracic fractions: < 25 µm ae.d.

- Respirable fractions: < 10 µm ae.d.

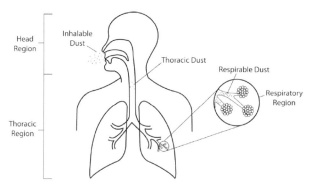

Figure 5.18: Regional particle deposition in the human respiratory system

Explanations: data source [270]

In order to put the dust contamination issue into perspective, it is estimated that an average person inhales over 10 m^3 of air in a day, which is close to 4,000 m^3 per year or 320,000 m^3 during an 80 year lifespan [268]. A typical outdoor particle concentration outside a city core is about 10 μg/m^3. The inhaled volume of particulates adds up to about 3.2 g of particulates during a lifetime, which is about three teaspoons full of particles. The air exposed surface of a human head is about 0.5 m^2, the bronchial-thoracic region about 2 m^2 and the respiratory alveoli region about 100 m^2.

The source/type of feedstock used in pellet manufacturing is the basis of the toxicological characteristics of pellets and applies primarily to the dust from wood pellets. The available toxicological data listed in Table 5.6 summarise the exposure value limits recommended by various regulatory bodies that are often referenced in the literature [264; 265; 266; 267; 268]. It should be noted that each jurisdiction may have somewhat different exposure value limits. The limits given in Table 5.6 do not make a clear distinction between whitewood and bark material.

Table 5.6: Summary of toxicological data concerning the exposure value limits recommended by various regulatory bodies

Feedstock	TWA & STEL (OHSA)	PEL (OSHA)	REL (NIOSH)	TLV (ACGIH)	Health effects
Softwood such as fir, pine, spruce and hemlock	5 mg/m^3 for 8 h at 40 h/week 10 mg/m^3 for 15 min, max. 4 times/day, each episode max. 60 min	15 mg/m^3 total dust 5 mg/m^3 respirable dust	TWA = 1 mg/m^3 for 10 h at 40 h/week	TWA = 5 mg/m^3 for 8 h at 40 h/week STEL = 10 mg/m^3 for 15 min, max. 4 times/day, each episode max. 60 min	Acute or chonic dermatitis, asthma, erythema, blistering, scaling and itching (ACGIH)
Hardwood such as alder, aspen, cottonwood, hickory, maple and poplar	1 mg/m^3 for 8 h at 40 h/week	15 mg/m^3 total dust 5 mg/m^3 respirable dust	TWA = 1 mg/m^3 for 10 h at 40 h/week	TWA = 5 mg/m^3 for 8 h at 40 h/week STEL = 10 mg/m^3 for 15 min, max. 4 times/day, each episode max. 60 min	Acute or chronic dermatitis, asthma, erythema, blistering, scaling and itching (ACGIH) Suspected tumorigenic at site of penetration (IARC)
Oak, walnut and beech	1 mg/m^3 for 8 h at 40 h/week	15 mg/m^3 total dust 5 mg/m^3 respirable dust	TWA = 1 mg/m^3 for 10 h at 40 h/week	TWA = 1 mg/m^3 for 8 h at 40 h/week	Suspected tumorigenic at site of penetration (ACGIH)
Western red cedar	-	15 mg/m^3 total dust 5 mg/m^3 respirable dust	TWA = 1 mg/m^3 for 10 h at 40 h/week TWA = 1 mg/m^3 for 10 h at 40 h/week	TWA = 5 mg/m^3 for 8 h at 40 h/week STEL = 10 mg/m^3 for 15 min, max. 4 times/day, each episode max. 60 min	Acute or chronic rhinitis, dermatitis, asthma (ACGHI)

The occupational health authority of a region regulates the permissible exposure limits. As an example, OSHA in the USA states the wood dust to be nuisance dust, but strongly recommends that employers adopt the ACGIH levels. The maximum permissible exposure for nuisance dust is 15 mg/m^3, total dust (5 mg/m^3, respirable fraction). The ACGIH lists beech, oak, birch, mahogany, teak and walnut as respiratory allergens. According to ACGIH, exposures at the STEL should not take place more than four times a day and should be separated by intervals of at least 60 minutes. Other jurisdictions may stipulate other exposure limits.

5.3.1.1   Entry routes and controls

The main entry routes for dust causing health effects are inhalation or direct skin contact. In dusty environments such as pellet production facilities, terminal loading facilities or large energy plants where pellets are handled in bulk, the aspects described in the following sections apply.

5.3.1.1.1   Inhalation of dust

Simply breathing in the dust can cause coughing and a dry throat. In concentrations above the TWA exposure value, inhalation could lead to asthma, erythema and/or allergic reactions depending on the sensitivity of the individual.

The following points should be considered:

- Adequate dust suppression and collection to keep dust level below the TWA;
- Restricted access for workers not directly involved in handling the product;
- Rotating tasks among workers to decrease the TWA over the shift or workweek;
- Continuous cleaning of areas where dust can accumulate using a vacuum;
- Proper selection, use and maintenance of respiratory protection;
- Personal protective equipment (PPE) such as a face respirator with N95 cartridges, or powered air purifying respirators with P100 filters or equivalent may be used in areas where dust concentration levels are suspected to be above the applicable TWA. Annual fitness testing and medical evaluation is required prior to the use of a respirator;
- Ban on smoking inside or close to an area where explosive dust or gases are present.

5.3.1.1.2   Skin contact

Dust settling on skin can cause dermatitis, blistering, scaling and itching. Certain hardwoods are suspected to be tumourigenic at the site of penetration according to the International Agency for Research on Cancer (IARC) [264].

The following points should be considered:

- Adequate dust suppression and collection to keep dust levels below the TWA;
- Restricted access for workers not directly involved in handling the product;
- Rotating tasks among workers to decrease the TWA over the shift or workweek;
- Continuous cleaning of areas where dust can accumulate using a vacuum;
- Cleaning of worker clothing before lunch breaks and at the end of the shifts using a vacuum to control any cross-contamination of food or personal clothing;
- No use of compressed air for cleaning of workers or worker's clothing. Showers should be provided for washing at the end of the shift;

- Full length shirts and pants or coveralls and safety glasses with side shields or goggles to protect workers from mechanical irritation and contact of dust with the skin and eyes.

It is strongly recommended that a health monitoring programme is established for workers handling and working in the vicinity of wood pellets as the extent of health effects from exposure to wood dust varies from person to person. The health monitoring programme should begin with allergy testing to determine sensitivities and worker's reactions to the material as well as the best way to control the individual's exposure. On a regular basis, workers should be re-evaluated to determine if exposure control measures are adequate for the individual.

5.3.1.2 Effects on the human body

The health effects on the human body range from nuisance to serious life threatening diseases.

5.3.1.2.1 Irritation of eyes, nose and throat

Many hardwoods and softwoods contain chemicals that can irritate the eyes, nose and throat, causing shortness of breath, dryness and soreness of the throat, sneezing, tearing and inflammation of the mucous membranes of the eye. Wood dust typically accrues in the nose, causing sneezing and a runny nose. Other observed effects include nosebleeds, impaired sense of smell and complete nasal blockage.

5.3.1.2.2 Dermatitis

Chemicals in many types of wood can cause dermatitis, a condition in which the skin can become red, itchy or dry and blisters may develop. Direct skin contact with wood dust can also cause dermatitis. With repeated exposures, a worker can become sensitised to the dust and develop allergic dermatitis. Once a worker is sensitised, exposure to even a small amount of wood dust can cause a reaction that becomes more severe with repeated exposure.

Most often allergic dermatitis is caused by exposure to tropical hardwoods. Cases of allergic dermatitis resulting from exposure to douglas fir and western red cedar have been reported. Irritant dermatitis has also been reported from exposure to western hemlock, sitka, pine and birch.

5.3.1.2.3 Effects on the respiratory system

Effects on the respiratory system include decreased lung capacity and allergic reactions in the lungs. Two types of allergic reactions can take place in the lungs:

- Hypersensitivity pneumonitis, i.e. an inflammation of the walls of the alveolars and small airways;
- Occupational asthma, i.e. a narrowing of the airways resulting in breathlessness. One of the most studied woods with respect to wood dust related asthma is western red cedar. In British Columbia some workers have reported cases of asthma attacks triggered by other substances such as ash, oak and mahogany.

Decreased lung capacity may be the result of mechanical or chemical irritation of lung tissue by the dust. This causes the airways to narrow, which reduces the volume of air taken into the lungs and results in breathlessness. Decreased lung capacity usually develops over a longer period of time (i.e. chronic exposure).

Workers who are allergic to aspirin should be aware that willow and birch contain large concentrations of salicylic acid, the predecessor of aspirin. Sensitive individuals may already react when casually exposed to these woods.

5.3.1.2.4 Cancer

Dust from certain hardwoods has been identified by IARC as a positive human carcinogen. An excess risk of nasal adeno-carcinoma has been reported mainly in those workers of the industry that are exposed to wood dusts. Several studies suggest an increased incidence in nasal cancers and Hodgkin's disease is found in workers of the sawmilling, pulp and paper and secondary wood industries. The highest risk appears to exist when workers are exposed to hardwood dust, most commonly beech and oak.

Dust from western red cedar is considered a "nuisance dust" (= containing less than 1 wt.% silicates (OSHA)) with no documented respiratory carcinogenic health effects (ACGIH). Cedar oil is a skin and respiratory irritant.

5.3.2 Exposure to off-gassing emissions and control measures

A closed containment filled with pellets needs to be ventilated properly in order to minimise the risk of serious exposure to toxic gases for personnel entering such a space. The three major gases that are emitted are carbon monoxide, carbon dioxide and methane [239; 271; 272]. These gases are non-condensable and none of them can be detected by odour. The most serious exposure is that to CO, particularly in combination with oxygen depletion. As an example, Table 5.7 lists the TWA and STEL in Canada and Sweden for these gases [264; 265; 266; 267; 268]. Similar limit values are found in most other jurisdictions. It is up to the reader to check with the occupational health authority for the exact value in the applicable region. In addition to the non-condensable gases [273], a number of condensable gases are also emitted most of which are detectable by odour and are known to cause irritation when inhaled or to penetrate both eyes and skin.

Table 5.7: Examples of TWA and STEL for CO, CO_2 and CH_4 in Canada and Sweden

Explanations: [1]...according to the Occupational Health & Safety Act; [2]...according to AFS 2005:17; there is no ceiling exposure value (CEV) for CO, or CH_4; the recommended TWA in Canada for CO for living spaces is 9 ppmv [274] (cf. Section III of MSDS wood pellets in bulk in Section 5.4); data source [275]

Component	Country	TWA (8 h/day - 40 h/week)		STEL (15 min)	
		ppmv	mg/m^3	ppmv	mg/m^3
Carbon monoxide (CO)	CAN[1]	25	29	100	115
	SE[2]	35	40	100	120
Carbon dioxide (CO_2)	CAN[1]	5,000	9,000	30,000	54,000
	SE[2]	5,000	9,000	10,000	18,000
Methane (CH_4)	CAN[1]	1,000	-	-	-
	SE[2]	-	-	-	-

In order to minimise or even prevent inhalation of non-condensable gases in areas where such gases are present, the following points should be considered:

- Spaces containing wood pellets should be ventilated before entering – ventilation must be adequate to keep CO, CO_2 and CH_4 levels at or below the TWA listed in Table 5.7.

- Access must be restricted with warning signs at entry ways.

- Personnel working in areas where off-gases are present should carry a multi meter with an alarm for both CO and oxygen (O_2). Such multi meters should be recharged and calibrated on a regular basis as per the manufacturer's instructions. Also, the sensor elements should be tested for contamination (overdose of CO or condensation of hydrocarbons). Individual meters for the two gases are not recommended.

- Self-contained breathing apparatus should be used if entry is required before the first control measure, as mentioned above, can be adequately put in place or when the gas detector indicates the presence of CO above the TWA or when the O_2 level is below 19.5 vol.% (cf. also Section 5.3.3).

- For naturally ventilated small-scale pellet storage units, warning signs and ventilation instructions for entering the pellet storage unit must be visible at the entry ways. For air-tight small-scale pellet storage units, a mandatory measurement of both CO and oxygen levels should be carried out after the ventilation of the storage unit and should guarantee risk free access (cf. Section 5.3.3).

5.3.3 Exposure to oxygen depletion and control measures

The normal oxygen concentration at sea level is 20.9 vol.% and the minimum recommended oxygen level is 19.5 vol.%. The general convention is therefore to secure a minimum oxygen content of 19.5 vol.% in a workplace. The oxygen level is easy to measure and monitor with commercially available and relatively inexpensive combined CO/O_2 meters. The use of gas meters capable of measuring both oxygen and carbon monoxide simultaneously should be a requirement. Oxygen measurement alone is insufficient and may in fact give a sense of false safety, particularly in areas where biomass or pellets are handled in bulk and where gases can be expected to mix with air. For example, an area with an oxygen reading of over 19.5 vol.% may appear to be safe and may contain up to 1.4 vol.% carbon monoxide (difference between 20.9 and 19.5). A concentration of 1.4 vol.% CO corresponds to 14,000 ppmv. This concentration of carbon monoxide is lethal within a minute or two. Exposure to a combination of low oxygen and high carbon monoxide has far more severe medical consequences than exposure to carbon monoxide at a normal oxygen concentration; as the oxygen is depleted, the body responds spontaneously by increasing the inhalation frequency, which in turn increase the intake of carbon monoxide.

Most multi gas meters on the market today are capable of measuring oxygen, carbon monoxide, hydrogen sulphide (H_2S) and detect the lower explosive limit (LEL) of combustible gases (hydrocarbons). The IMO prescribes multi gas meters onboard of ocean vessels carrying pellets in bulk [244] (also see Section 2.1.5). Furthermore, IMO regulations do not permit entry in spaces unless an oxygen level of 20.7 vol.% has been ascertained in combination with a maximum level of 100 ppmv of carbon monoxide, except if a self-sustaining breathing apparatus is used. Similar regulations are gradually being adopted on land in areas where biomass or pellets are stored or handled in bulk.

For pellets in consumer bags, the gas emissions are generally too minor to cause concern unless a number of bags are stored and sealed in a small space for a period of time. A rule of thumb is that consumer bag(s) of pellets should be stored in a space ventilated to the outside, with a volume of preferably greater than ten times the volume of the consumer bag and located in an area not accessible by small children and animals. The MSDS from the producer provides more detailed information about safe storage.

There is obviously a risk in an unventilated space of even a small amount of pellets generating a very high concentration of carbon monoxide, as can be seen from [238; 239; 240; 241; 242; 243]. The key is ventilation or a big enough room where the dilution is such that the exposure limit is not exceeded. Normally bags of pellets are stored in ventilated spaces or in areas that are not contained and therefore the risk of exposure is minimal. The MSDS provides the means to estimate the concentration and allow evaluation of the risk. Also, emissions are different from one brand of pellet to another and each producer should develop data for their own product.

5.4 MSDS for pellets – bulk and bagged

Pellet producers as per most jurisdictions are obligated to present an MSDS, sometimes simply called safety data sheet (SDS), for their product. The producer may assemble the data for the MSDS and draft the document for review by a legal counsellor. The alternative is for the producer to assemble the data and have a specialist draft the document, having it reviewed by a legal councillor before publication. An MSDS is a work in progress and subject to updates as long as new aspects related to the product or the handling and storage of the product are discovered with more experience, new or updated regulations or new scientific findings. Some jurisdictions require new information regarding the product safety to be incorporated in the MSDS within a certain time period from the date the information becomes known to the producer. Some jurisdictions require distribution of any updates of an MSDS to all clients buying or handling the product within 12 months.

The intent of the MSDS is twofold. First, the MSDS should provide sufficient information to allow safe handling, usage and storage of the product. Second, the MSDS should protect the producer from liability for incidents or accidents as a result of handling, using or storing the product in ways not intended or not compliant with recommendations in the MSDS. The best MSDS is an informative MSDS covering as many aspects as possible of the potential risks, written in a concise language, easy to read and not too long. A well developed MSDS also provides essential information about emergency and rescue operations if incidents or accidents occur. An MSDS should at all times be made easily accessible to the personnel at places where the product is handled, used or stored. An MSDS should be available on paper and electronically, written in the local language and distributed free of charge.

The characteristics of pellets from different producers are in most cases unique due to differences in the raw material and the manufacturing process. Therefore each MSDS of each product from each producer is also unique.

5.4.1 Recommended format for MSDS

The required content and format for an MSDS is often stipulated in regulations or recommendations on a national level [276]. The Workplace Hazardous Materials Information System (WHMIS) or equivalent [277; 278] often provides the guidelines for the format and content. Several internationally recognised guidelines are also available and can be used as models for almost any product [279]. However, the characteristics of the product determine the coverage in the MSDS.

An MSDS for pellets should include at least the following sections:

- Legal name of producer with complete contact information;
- Name of person(s) to contact in case of emergency;
- Name and producer's code assigned to product;
- Hazard classification of product;
- Typical chemical ingredients;
- Physical properties for easy and fast identification of the product;
- Health hazard data;
- First aid procedures;
- Fire extinguishing procedures;
- Accidental release measures;
- Safe handling and storage;
- Exposure control and personal protection (confined entry procedures);
- Stability and reactivity data;
- Exposure and toxicity data;
- Ecological information;
- Guidelines for transportation of product;
- Regulatory information;
- Notice to reader – producer liability disclaimer;
- Reference to other supporting documents available from the producer such as product specification, shipping information instructions (road, rail, marine) etc.;
- Abbreviations and nomenclature used in the MSDS.

5.4.2 Recommended data set for pellet MSDS

Wood pellets have traditionally been looked upon as a benign product and are used widely as residential fuel, animal bedding or industrial absorbent. With the increasing use of pellets as a bulk product, sometimes in very large quantities, certain risks have been experienced that prompted development of comprehensive MSDS. Wood pellets are classified as material hazardous in bulk (MHB) [244]. The major risks identified involve the potential for off-gassing, which in turn results in high concentrations of toxic gases, self-heating and generation of explosive dust. The level of risk is related to the volume of product in a given space and the ambient conditions under which the product is handled or stored. It is therefore recommended that the manufacturer issues one MSDS for pellets in small volume such as bagged pellets and another MSDS for pellets in large bulk or bagged product stored in large volume such as in a warehouse.

An MSDS is issued by the manufacturer or the supplier of the pellets and is a public document that should be available whenever there is a request for such a document. It is the responsibility of the manufacturer to properly present their product in an MSDS based on verifiable data.

5.4.2.1 MSDS data set for pellets in bulk

The following information is recommended as a minimum data set to be developed by the pellet producers and included in the MSDS of the product:

- Off-gassing emission factors for non-condensable gases such as carbon monoxide (CO), carbon dioxide (CO_2) and methane (CH_4);
- Oxygen depletion characteristics;
- Self-heating characteristics;
- Explosibility of airborne dust from the product.

For dust cloud:

- Minimum ignition temperature;
- Minimum ignition energy;
- Maximum explosion pressure;
- Specific dust constant;
- Explosion class;
- Minimum explosible concentration
- Limiting oxygen concentration.

For dust layer:

- Minimum ignition temperature;
- Flammability characteristics of fines and dust from the product.

The above data are required for proper engineering of dust suppression systems, explosion suppression systems and explosion control measures in facilities where pellets are handled in large bulk.

5.4.2.2 MSDS data set for pellets in bags

The following information is recommended as a minimum data set to be developed by the pellet producers and included in their MSDS for bagged product stored in single bags or limited numbers of small bags such as in residential use:

- Off-gassing emission factors for non-condensable gases such as carbon monoxide (CO), carbon dioxide (CO_2) and methane (CH_4);
- Oxygen depletion characteristics;
- Self-heating characteristics.

The above data are required to determine the guidelines provided in an MSDS for ventilation where bags with pellets are stored in residential spaces.

5.4.3 Example of MSDS – pellets in bulk

As an example, the Canadian MSDS for bulk pellets is shown in Appendix A. It should be noted that pellets manufactured from other species and produced using other processes such as drying may present somewhat different stability and reactivity data such as emissions factors for off-gassing, deflagration index for dust from the pellets, etc.

An MSDS is required for all deliveries of pellets to energy plants, test laboratories etc. in North America. It is also required for pellets crossing the border between the USA and Canada as well as onboard all ocean carriers worldwide. In the large European pellet markets in Austria, Germany and Sweden, neither obligatory nor voluntary use of MSDS is known.

5.4.4 Example of MSDS – pellets in bags

As an example, the Canadian MSDS for bagged pellets is shown in Appendix B. It should be noted that pellets manufactured from other species and produced using other processes such as drying may present somewhat different stability and reactivity data such as emissions factors for off-gassing, deflagration index for dust from the pellets, etc.

An MSDS is not necessarily requested in the trade of bagged products. However, pellets are classified as hazardous in bulk, which applies also to bagged pellets when transported and stored indoors on pallets, for example in stores. The MSDS from the producer of the pellets may provide more specific information. The MSDS issued in Canada for bagged pellets is an example of guidelines for storage. No use of MSDS for bagged pellets is known from Europe.

5.5 Summary/conclusions

Pellets are a compressed or densified solid biomass fuel prone to mechanical degradation, chemical decomposition and other changes such as moisture absorption during handling and storage. Mechanical degradation during handling generates fines and airborne dust, which can be considered a safety issue under certain circumstances since they can cause fires and explosions and can become a health issue when inhaled. Therefore, pellets must be handled with care, like all other fuels, in order to minimise fines and dust formation and thus to reduce both adverse health effects as well as the risk for dust explosions.

Another important safety aspect for pellets is related to the decomposition of pellet contents resulting in self-heating and self-ignition in pellet bulk storages, a known but not yet fully understood phenomenon and subject to research in many parts of the world. Unfortunately, fire incidents caused by self-ignition have been reported several times from large-scale pellet storages. The extent of self-heating is ultimately dependent on the balance between heat generated and lost in the material. The larger the storage is, the lesser is the surface area to volume ratio. Thus the heat loss via the surface area of the storage is reduced in relation to its volume and therefore the danger of self-heating and self-ignition rises with increasing storage size. Self-heating of biomass stored in heaps (piles) or silos can take place by means of biological and/or chemical oxidation. Chemical oxidation has been identified as the main decomposition mechanism leading to self-heating in storages of wood pellets, whereas biological degradation and/or chemical oxidation is the mechanism for onset of self-ignition in stored moist biomass (e.g. raw material used for pellet production).

The most important kind of biological self-heating takes place in conjunction with the respiration of aerobe bacteria and fungi in the presence of sufficient air (oxygen), temperature and moisture. Mesophilic organisms can thus produce temperatures of around 40°C maximum; above that they die off. Thermophilic organisms, by contrast, can survive temperatures of up to about 70°C maximum before they die off. Self-heating that exceeds these temperatures must hence be caused by chemical oxidation processes. The shift from biological to chemical oxidation processes is of particular relevance since self-ignition can only happen on the basis of chemical processes. Whether chemical oxidation processes can take place to such an extent that self-ignition happens not only depends on the temperature rise by biological processes but also on many other parameters such as moisture content, air flow through the heap or surface characteristics of the biomass.

Trials that were concerned especially with self-heating and self-ignition in pellet storages showed that the growth of microorganisms is normally limited by the low moisture content of pellets. However, temperature rises on the basis of chemical oxidation processes were noted especially in storages of freshly produced pellets. In some cases, these temperature rises lead to self-ignition. Self-heating appears to also depend on the raw material used for production of pellets. It is assumed that high contents of unsaturated fatty acids promote the self-heating of pellets. Pine wood for instance has a high unsaturated fatty acid content.

It is also well known that condensation heat caused by absorption of vapour in air onto lignocellulosic material such as pellets is a heat releasing (exothermic) process. Besides condensation heat, differential heat caused by the levelling of moisture content between different pellet layers is also involved in self-heating but in a much lower magnitude.

Pellets decompose over time and emit non-condensable (primarily CO, CO_2 and CH_4) and condensable gases (other toxic gases such as aldehydes and terpenes), a phenomenon called off-gassing. The danger of off-gassing is present even at lower bulk temperatures during storage of wood pellets. A number of fatal incidents in this respect have occurred when personnel entered confined spaces where wood pellets were stored in large bulk, particularly during oxygen depleted conditions. Therefore, entering a storage area that is not thoroughly ventilated should only be permitted after having checked the concentration of carbon monoxide in combination with oxygen. Deviating from current Central European standards, small-scale domestic pellets storage units should be equipped with facilities that render natural ventilation possible. Moreover, it should be obligatory to mount warning signs with ventilation instructions that have to be carried out prior to entering a pellet storage unit and should ensure risk free access.

In order to minimise the risk of fire by abnormal self-heating and spontaneous ignition, it is important to have a continuous temperature control by sensors embedded in the stored product. Vertically suspended cables containing sensors at certain intervals are one common method used in silos. It is also recommended that gas analysis equipment for detection of CO should be installed in the ceiling of the storage building or in the headspace of a silo, both from a human life and health perspective and to achieve early indications of possible spontaneous ignition. At a concentration exceeding 100 ppm, people should not enter the area without SCBA. Very high concentrations, in the order of 2 to 5 vol.% of CO in the headspace of a silo, often in combination with a strong smell, are typical indicators for an ongoing pyrolysis in the bulk. Measurements of CO in combination with the oxygen concentration are also very valuable during the extinguishing operation.

The fire fighting technique for silos differs completely from traditional fire fighting since water cannot be used and due to the anatomy of a silo fire. Wetted pellets swell very quickly to about 3.5 times of their original size and when stored within containment, for example a silo, the swelled material can form an extremely hard cake that has to be removed with a jack hammer. There is also a potential risk of bridging and cracking of the silo wall.

When spontaneous ignition occurs, it normally forms a "fire ball" of smouldering material. Since the natural prevailing convection is upwards, the fire ball is supplied with oxygen from underneath while the combustion gases, where the oxygen is to a large extent consumed, move upwards. Experiments and observations from real fires show that the fire ball therefore moves slowly downwards from the point of ignition. The combustion gases form a wave that moves very slowly upwards and experiments have shown that, depending on the silo height, it might take one or several days before the combustion gases reach the headspace and can be detected.

As it is in most cases impossible to identify exactly where the fire ball is located, silo fires should be extinguished by injection of inert gas (normally nitrogen), thus preventing air (oxygen) from reaching the smouldering fire zone. The inert gas should be injected in gaseous phase close to the silo bottom to ensure inertisation of the entire silo volume. In large diameter silos there is a need for several gas inlets in order to distribute the gas over the silo cross section area. In these cases, preparations by installing a pipe system at the silo bottom are vital for an effective fire fighting operation. Suitable equipment is required for the gas supply, for example a storage tank and a vaporisation unit. It is also important that the silo is made as air-tight as possible in order to reduce the infiltration of air into the silo compartment. The gases should be evacuated through an opening with a "check valve" arrangement at the top of the silo in order to prevent inflow of air. The discharge of the silo should not commence before gas analyses (CO and O_2) in the silo headspace indicate that the smouldering activity is effectively controlled.

General recommendations and advice to avoid self-heating and spontaneous ignition of biomass can be summarised as follows:

- Avoid storage and transport of large volumes if the fuel's tendency of self-heating is unknown.
- Be conscious of the risk of self-heating and spontaneous ignition in large storage volumes.
- Avoid mixing of different types of biomass fuels in the storage.
- Avoid mixing of fuel batches with different moisture contents.

Specific recommendations for storage of wood pellets can be summarised as follows:

- Avoid large parts of fines in the fuel bulk.
- Measure and monitor the distribution of temperature and gas composition within the stored material. Frequent visual inspection is recommended.
- Prepare silos for gas injection at the bottom of the silo in case a fire should occur.
- Pellet storage units must be equipped with size dependent, appropriate means of ventilation to control levels of carbon monoxide and carbon dioxide.

Adverse health effects must be expected from fine particulates (airborne) generated during handling and storage of pellets as well as from toxic off-gassing and oxygen depletion during storage of pellets. Epidemiological findings show a strong correlation between adverse health effects and fine particulates (airborne) generated during handling of pellets. The most serious damage is done by particulates with less than 10 µm in ae.d. that are able to enter human bloodstream, but also direct skin contact is able to cause adverse health effects. The most serious effect from off-gassing processes is the exposure to CO, particularly in combination with oxygen depletion (CO is acutely toxic at a concentration of about 1,200 ppm). Therefore, it is strongly recommended that exposure to harmful compounds, particularly from off-gassing, and to fine particulates should be limited. Recommendations for exposure limits, which might differ in different jurisdictions, should be followed, while measures recommended to minimise exposure to harmful compounds should be considered.

In most jurisdictions, pellet producers are obligated to present an MSDS for their product. An MSDS is a work in progress and subject to updates as long as new aspects related to the product or the handling and storage of the product are discovered with more experience, new or updated regulations or new scientific findings. The MSDS should provide sufficient information to allow safe handling, usage and storage of the product and should protect the producer from liability for incidents or accidents as a result of handling, using or storing the product in ways not intended or not compliant with recommendations in the MSDS. It should cover as many aspects as possible of the potential risks. The required content and format for an MSDS is often stipulated in regulations or recommendations on a national level. Several internationally recognised guidelines are also available and can be used as models for almost any product. Recommendations on format and data sets for pellet MSDS are given in this section and in Appendix A and B.

6 Wood pellet combustion technologies

The high and constant quality of pellets creates significant differences between pellet combustion technology and conventional combustion technologies. Development of automatic furnaces based on biomass fuels with a similar operational comfort to oil or gas heating systems has only been possible by the establishment of pellets as a fuel.

In comparison to wood chips, pellets are more able to flow and hence are apt for automatic operation of a furnace. The furnace itself can be adjusted in a more accurate way than in the case of wood chips because of pellets having constant moisture content and particle size. Also pellets require less storage space. Pellet heating systems have proved to be less prone to failures and more comfortable to use. These assertions are also valid when comparing pellet furnaces to firewood furnaces, indeed the differences between these two systems are more considerable. Automatic feeding of a firewood furnace over a long period of time is practically impossible, storage demand is also greater and manipulating the fuel with regard to storage filling and furnace charge is by far more work intensive.

Pellets are used in all areas ranging from small-scale furnaces with nominal capacities of up to 100 kW$_{th}$ and medium-scale furnaces with 100 to 1,000 kWh$_{th}$ to large-scale furnaces with nominal capacities of more than 1,000 kWh$_{th}$. In addition, pellets are used in CHP as well as in co-firing in fossil fuel furnaces.

In the following sections all fields of pellet applications are looked at in detail [280; 281; 282].

6.1 Small-scale systems (nominal boiler capacity < 100 kW$_{th}$)

Small-scale systems are defined as furnaces with a nominal boiler capacity of up to 100 kWh$_{th}$. The market for such pellet heating systems is experiencing continuous growth in many countries, for example Austria, Germany and Sweden (cf. Chapter 10). Such furnaces are used in the residential heating sector as either single stoves or central heating systems, as well as in micro-grids and by smaller industrial users.

The combustion technologies used in pellet furnaces must conform to the highest standards in order to guarantee failure free and easy to use operation for the end user. Combustion technologies used in pellet furnaces and special attributes for fitting are presented and discussed in detail in the following sections. Finally, examples of combustion technologies are looked at.

6.1.1 Classification of pellet combustion systems

Pellet combustion systems can be classified by the type of furnace, the type of feed-in system or according to the design. The three possibilities are looked at in this section.

6.1.1.1 Furnace type

There are basically two types of furnace, namely pellet stoves and pellet central heating systems. A pellet stove is the designation for heating systems that are placed inside the room to be heated. An example is the tiled stove common in Austria and Bavaria, but pellet stoves have also been on the market for many years. Central heating systems supply the heat for the rooms of an entire building from one central point. The heat is carried by water and released through different types of heating surfaces (radiators, floor or wall heating surfaces). Such systems can also be used in so called micro-grids that supply the heat to a series of separate buildings. Depending on the interface between boiler and burner, there are three types of central heating systems, i.e. boilers with an external, integrated or inserted burner.

6.1.1.1.1 Pellet stoves

Stoves run with pellets are equipped with an integrated storage box by which the stove self-supplies the fuel. The fuel reservoir is sufficient for a few hours to a few days of operation depending on the construction. In addition, systems have entered the market that are able to feed such stoves with pellets from a storage room located for instance in a cellar. Continuous operation can be achieved by integrated electronics. Pellet stoves can also be equipped with a water jacket acting as heat exchanger so that they can form a central heating system. The pellet stove shown as an example in Figure 6.1 contains an integrated storage space that is designed for continuous operation of up to 70 hours, the storage space being filled manually.

Figure 6.1: Stove fed with pellets

Explanations: data source [283]

6.1.1.1.2 Pellet furnaces with external burners

In pellet furnaces with external burners, pellet combustion takes place in a burner placed outside the boiler and only the flue gases enter the boiler. An example of such a furnace is shown in Figure 6.2. This approach allows separate optimisation of boiler and burner.

External pellet burners can also be used to convert already existing boiler facilities for heating oil or log wood to the use of pellets. This way of retrofitting old systems is very common in

Sweden (more than 110,000 units were installed at the end of 2007).

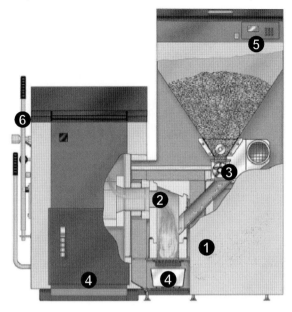

Figure 6.2: Pellet furnace with external burner

> Explanations: 1...ignition (hot air); 2...air staging by separate primary and secondary air supply; 3...rotary valve; 4...opening to ash box and heat exchanger; 5...integrated storage space; 6...handle for semi-automatic heat exchanger cleaning; data source [284]

A Swedish facility that was retrofitted in this way is shown in Figure 6.3. Pellet storage is effected in a storage space on its own, which can hold the annual fuel demand. A feeding screw feeds the pellets to a flexible tube that lets them drop into the burner. Pellets with a diameter of 8 mm are used as a fuel (which is the normal size of pellets in Sweden, even in the residential heating sector). The ash has to be removed about once a week during the heating period. The heat exchanger is cleaned by hand.

Retrofitting an existing boiler with a pellet burner presents a low-cost opportunity of changing from heating oil or log wood to pellets, without having to exchange the boiler. Looking at the concept as a whole, there are drawbacks concerning raised operational effort for cleaning the heat exchanger and emptying the ash box, as well as an increase of emissions in comparison to systems that are optimised for the use of pellets and equipped with appropriate control systems.

Special kinds of horizontal stoker burners suitable for combustion of biomass fuels have been on the market for about 30 years in Finland. The principle of a horizontal stoker burner is shown in Figure 6.4. Most of the devices are designed for an output range of 20 to 40 kW and they are used to heat single small houses and farms. Suitable fuels for the devices are wood chips, sod peat and pellets. Using a slightly different design, horizontal stoker burners have also been constructed for higher heat outputs of up to 1 MW.

The burner, which is fed by a screw conveyor, is made of cast iron with a lined refractory or a water cooled horizontal cylinder. In some burners, water cooling ensures the durability of the burner materials and improves the thermal insulation of the burner in order to reduce radiation losses. The temperature inside the burner can rise above 1,000°C when using dry fuels. The burner is mounted partially inside the furnace and partially outside, so that the whole combustion chamber of the boiler effectively takes part in radiation heat transfer.

Figure 6.3: Boiler retrofitted for the use of pellets

Explanations: 1…pellet burner; 2…pellet supply; 3…existing boiler; the picture was taken on a field trip in Sweden in January 2001

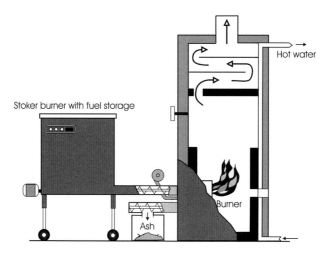

Figure 6.4: Horizontal stoker burner principle

The basic idea of horizontal stoker burners is that the fuel is fed precisely according to the heat demand. Quite a small amount of fuel is burning at any one time. Combustion air is injected via one or several nozzles. This ensures very efficient and clean combustion. The turn-down ratio of this kind of equipment is 0 to 100% (when there is no heat demand, the burner goes to an idle burning mode, where only a very small amount of pellets is burned just to keep the fire going on; in this mode no heat is taken from the boiler). This means that no separate heat buffer storage is needed. The burner is controlled by a thermostat in the boiler water using an on–off method in smaller burners and more sophisticated control methods in large burners.

This kind of stoker burner was originally designed for burning wood chips. However, wood pellets and peat pellets are even more suitable fuels for these burners since they result in very low emissions and high burning efficiency.

6.1.1.1.3   Pellet furnaces with inserted or integrated burners

Most of the central heating boilers in Central Europe, in particular Austria and Germany, are either boilers with an integrated burner or boilers with an inserted burner. Figure 6.8 displays a boiler with integrated burner. In such systems, boiler and burner form one compact unit that enables a holistic optimisation with regard to the fuel. An example of a boiler with an inserted burner is shown in Figure 6.6. The pellet burner is a self-contained unit that is inserted into the boiler. These systems too are optimised for the use of pellets and adjusted accordingly. They are described in detail in the following sections.

6.1.1.2   Pellet feed-in system

Depending on the way how the pellets are fed into the furnace, three basic principles of wood pellet combustion systems can be distinguished: underfeed burners, horizontally fed burners and overfeed burners (cf. Figure 6.5).

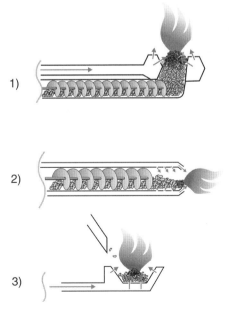

Figure 6.5:   Basic principles of wood pellet combustion systems

Explanations: 1…underfeed burner; 2…horizontally fed burner; 3…overfeed burner; data source [53]

The three types of burners are presented and illustrated in the following sections.

6.1.1.2.1 Underfeed burners

In underfeed furnaces (also called "retort furnaces" or "underfeed stoker"), a so-called stoker screw feeds the fuel horizontally into the bottom area of the retort from where the fuel is pushed upwards. Primary air is fed into the combustion chamber sideways through the retort and flows upwards and the flame burns in an upwards direction too. The ash gets discharged at the edge of the retort and falls into an ash box placed underneath. The impact on the bed of embers is low due to the slow insertion of fuel from below and no swirling of dust takes place, as would be the case in overfeed furnaces and partly in horizontally fed furnaces. However, after-smouldering of fuel and even burn-back into the storage space can happen at shutdown of the furnace because the bed of embers and fuel feed-in are always in contact. This requires appropriate measures to be taken (cf. Section 6.1.2.3).

An illustration of an underfeed burner is shown in Figure 6.6. The system shown is equipped with an integrated pellet reservoir. The fuel is conveyed by a feeding screw from the pellet reservoir via a fireproof valve in the dropshaft and to the burner by a stoker screw. Ignition takes place automatically via a hot air fan. Combustion air is injected by a fan. Distribution into primary and secondary air is preset by appropriate dimensioning of the primary and secondary air supply channels. Primary air is supplied from underneath through the retort (via openings). Secondary air is injected into the secondary combustion zone by nozzles arranged in a circle. Flue gas flow is directed upwards through the secondary combustion zone to the top as shown in Figure 6.6. The flue gas is then redirected so that it flows down to the bottom where it gets redirected once more in order to flow upwards through the smoke tubes of the boiler and through the suction fan into the chimney.

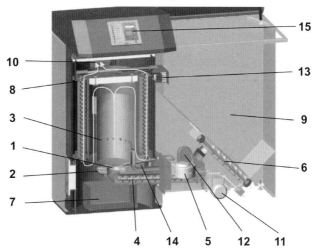

Figure 6.6: Underfeed furnace

Explanations: 1…retort; 2…primary air supply; 3…secondary air supply; 4…stoker screw; 5…combustion air fan; 6…feeding screw; 7…ash box; 8…heat exchanger with spiral scrapers; 9…pellet reservoir; 10…flue gas path; 11…main drive for fuel feeding system; 12…fireproof valve; 13…drive for automatic cleaning system; 14…automatic ignition; 15…display and control system (micro processor); data source [209]

The spiral scrapers of the heat exchanger tubes are set into motion regularly by an electric motor, hence the smoke tubes are automatically freed of possible deposits. The motor for the heat exchanger cleaning system additionally drives a grate located at the top of the ash box, which moves up and down so that the ash collected in the ash box becomes compacted, thus extending the interval between emptyings of the ash box. The control system of the plant is based on micro processors.

6.1.1.2.2 Horizontally fed burner

In contrast to underfeed furnaces, the fuel for horizontally fed furnaces is conveyed sideways only, but with the help of a stoker screw. Primary air is supplied from both underneath and above the bed of embers. In contrast to underfeed furnaces, the flame burns horizontally. The ash gets discharged at the edge of the retort and falls into an ash box placed underneath. The impact on the bed of embers is stronger than in underfeed furnaces but not as strong as in overfeed furnaces due to the sideways insertion. After-smouldering and burn-back are also possible in this construction owing to the connection between the bed of embers and fuel feed-in.

Figure 6.7 presents an example of a horizontally fed pellet furnace, though horizontal feed-in systems are not very widespread. The facility contains an integrated pellet reservoir. The fuel is conveyed by a feeding screw from the integrated pellet reservoir and through a dropshaft into the stoker screw, which in turn feeds the burner. Ignition takes place automatically via a hot air fan in the burner. Primary and secondary air supply are separated and realised by openings in the stainless steel burner. Combustion air is provided by a combustion air fan.

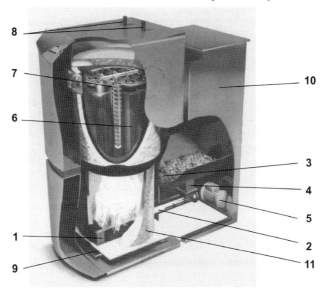

Figure 6.7:   Horizontally fed pellet furnace

Explanations: 1...stainless steel burner; 2...stoker screw; 3...conveyor screw; 4...dropshaft; 5...drive for the screws; 6...heat exchanger; 7...spiral scrapers with driving mechanics; 8...feed and return of the heating circuit; 9...ash box; 10...integrated pellet reservoir; 11...combustion chamber made of fireclay; data source [285]

The spiral scrapers placed inside the heat exchanger tubes are, depending on the construction, either set into motion by a handle from the outside that is operated manually or by a motor. They free the fire tubes of possible deposits.

The ash that is accumulated is collected in an ash box that has to be emptied regularly. The burner is made of stainless steel; the combustion chamber is built of firebricks. The furnace is regulated by a micro processor based control system.

6.1.1.2.3 Overfeed burner

In overfeed furnaces, the fuel is fed into a dropshaft by a feeding screw whereby the pellets fall onto the bed of embers on the grate. Primary air is fed from underneath the grate and flows upwards through the bed of embers. The flame burns upwards in overfeed furnaces, just as it does in underfeed furnaces. The ash falls through the grate into an ash box underneath. This type of furnace allows exact feed of fuel according to the current heat demand. Thus, only the amount of pellets needed for the actual power demand get through to the bed of embers. The dropping pellets may cause elevated particulate matter emissions and emissions of incompletely combusted particles from the bed of embers, however. Due to the spatial separation of the bed of embers and the feeding system, smouldering after shutdown and burn-back are prevented. Figure 6.8 displays such an overfeed pellet furnace.

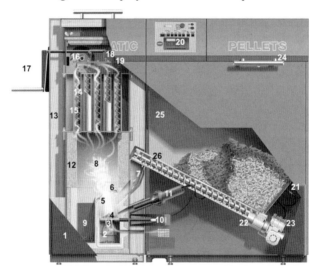

Figure 6.8: Overfeed pellet furnace

Explanations: 1…ash door; 2…grate cleaning plate; 3…primary air; 4…grate; 5…secondary air; 6…ring of nozzles; 7…dropshaft; 8…expansion zone; 9…ash box; 10…drive for grate cleaning system; 11…ignition fan; 12…ceramic insulation; 13…insulation; 14… spiral scrapers; 15…heat exchanger; 16…suction fan; 17…handle to operate cleaning system; 18…flue gas sensor; 19…lambda sensor; 20…control panel; 21…sensor for filling level; 22…motor; 23…gearbox; 24…opening; 25… integrated pellet reservoir; 26…feeding screw; data source [286]

Ignition of the pellets takes place automatically by a hot air fan. The combustion chamber of the above furnace is built of fireclay-like concrete with a high SiC content and gets cooled only to a small extent by secondary air channels. The pellets start glowing whilst they fall through the dropshaft and arrive in this state at the loose bed of embers on the grate. The air is able to flow through the bed of embers in an ideal way, leading to complete combustion. The partitioning of air into primary and secondary air is preset by the design of the channels. Primary air is fed into the burner through openings underneath the grate. Secondary air is fed in by a ring of nozzles, which creates a rotary flow. At first the secondary air acts as a cooling medium by flowing around the reaction zone, protecting it from slagging. Having been warmed up, it reaches the flame. The secondary air and combustion gases become mixed and move into the secondary combustion zone as a homogenous gas-air mixture. An appropriate residence time in the secondary combustion zone secures complete combustion. Flue gas is conveyed into the chimney by a suction fan. Ash particles that get entrained from the fuel bed can be separated by inertial forces and are collected in the ash box that has to be emptied regularly. The heat exchanger is cleaned by spiral scrapers that loosen up possible deposits inside the tubes. They are operated via a handle from outside and loosened residues fall into the ash box. An optional automatic operation of the heat exchanger cleaning system is available. An automatic grate cleaning system is built in so that primary air can get through the grate into the primary combustion zone without any problems. For this purpose, the grate tilts over once a day during an operational break and the air passages become freed of deposits by a cleaning plate. The whole installation is regulated by a micro processor based control system.

6.1.1.3 Pellet burner design

There are two major burner designs for pellets, namely retort furnaces and grate furnaces [282]. Retort furnaces are always designed as underfed burners. Grate furnaces are designed as horizontally or overfed burners. Depending on the design, grate furnaces can be subdivided into fixed grate, hinged grate and step grate furnaces. Figure 6.9 shows different types of burners that are used in Austrian pellet furnaces.

Figure 6.9: Different types of pellet burners

Explanations: upper left...retort furnace; upper right...fixed grate; bottom left...step grate; bottom right...hinged grate; data source [282]

In addition, there are other special construction pellet burners, for example rotary grate (cf. Figure 6.10) or "carousel" furnaces [287].

Figure 6.10: Rotary grate pellet burner

Explanations: data source [288]

6.1.2 Major components of pellet combustion systems

6.1.2.1 Conveyor systems

Diverse systems have been developed to meet the demands of conveying pellets from the storage space to the furnace under different framework conditions. In principle, two fundamentally different systems can be used: pneumatic feeding systems and feeding screws.

There are two types of feeding screw: conventional and flexible feeding screws. A system using the former is displayed in Figure 6.11. Taking the pellets from the storage space takes place via a horizontal screw channel that passes through the entire length of the storage space at the bottom. The difference in height between the ground level of storage space and burner is overcome by a feeding screw with cardan joint, whereby the direction is changed. The main drawback of this system is its inflexibility. The storage space should be rectangular (the longer and narrower the better) and the furnace room should be at the narrow end of the storage space. Feeding screws with a cardan joint can overcome minor alterations of the horizontal axis but as soon as the lengthwise alignment of furnace room and storage space cannot be achieved or the distance in between is too long, this conveyor system is not applicable. The main advantages of the conventional feeding screw are the fact that it is proven, it is robust and it makes little noise.

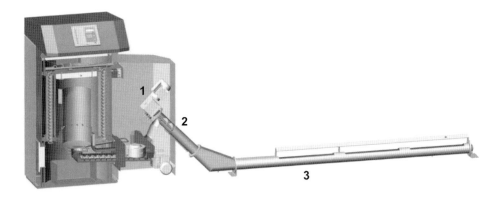

Figure 6.11: Conveyor system with conventional screw

Explanations: 1…drive for the feeding screw; 2…feeding screw; 3…screw channel; data source [209]

Figure 6.12 shows a pellet discharge system on the basis of a flexible feeding screw. Discharging of the storage space is carried out by a screw that lies at the bottom of the storage space, passing through its entire length, just as in the example above, but here the screw can move horizontally to a certain extent. This is to avoid any bridging in the storage space. With the flexible screw, the height difference between storage space and burner can be overcome without further fittings. The main benefit of flexible screws is their ability to curve, which is why installing the storage space and furnace room along one axis is not imperative. In order to keep wear low, tight curves should be avoided. Just like conventional feeding screws, flexible screws operate without much noise.

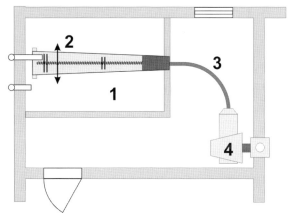

Figure 6.12: Conveyor system with flexible screw

Explanations: 1…storage space; 2…moving screw at the bottom of the storage space; 3…flexible feeding screw; 4…pellet furnace; data source [289]

Conventional and flexible screws may also be combined. Discharge of the storage space is then usually performed by a conveyor screw that is placed along the length of the storage space, while the height difference between storage space and burner, as well as possible changes of direction, are overcome by the flexible screw.

Figure 6.13 shows an example of a pneumatic feeding system of pellets from the storage room to the furnace. In this case, pellets can be taken at three positions inside the storage room, one of which is chosen automatically. A closed air circuit avoids arising of dust. Alternatively, discharge can be carried out by suction lances. If a pneumatic system is not equipped with a closed air circuit, exit air has to be filtered and the filter has to be cleaned regularly. The main advantages of pneumatic systems are flexibility with regard to the arrangement of lines and the possibility of overcoming great distances. Thus furnace room and storage facility do not necessarily have to be close to one another as in systems with feeding screws and they do not have to be aligned like in conventional screw conveyor systems. When the storage facility is underground (cf. Section 4.2.3.1.2), pneumatic feeding systems are in use exclusively because a flexible arrangement of lines and the handling of greater distances are required. Disadvantages of pneumatic feeding systems are increased noise and dust formation. Moreover, oversized pellets may lead to blockages in the tube (a single pellet that is too long can lead to a blockage). For this reason, many pellet producers have started to position knives onto the die in order to cut the pellets into the desired length (cf. Section 3.1.2). The energy demand of pneumatic systems is roughly as great as it is for feeding screws [290].

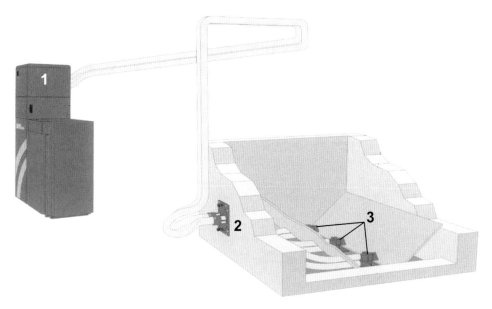

Figure 6.13: Pneumatic pellet feeding system

Explanations: 1…automatic feeding system with suction fan and control system; 2…automatic switch unit; 3…extraction units; data source [291]

All pneumatic feeding systems convey the pellets into an integrated pellet reservoir first that can contain the pellet demand of about a day. From there the pellets are transported to the burner by feeding screws. Due to the fact that pneumatic systems are quite noisy, a control system that prevents discharge from the storage room at night is recommended and generally applied.

A combined system of feeding screw for storage room discharge and pneumatic conveying from the storage room to the integrated pellet reservoir is shown in Figure 6.14. The demonstrated pneumatic conveyor system works according to the single tube principle in which the pellets are separated from the carrier air by a cyclone. The energy demand of such combinations is greater than it is for the single systems alone [292].

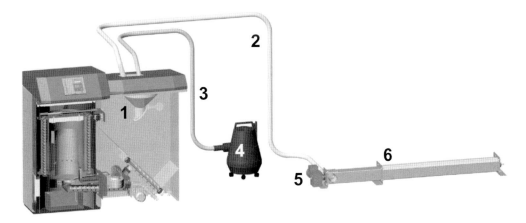

Figure 6.14: Combination of feeding screw and pneumatic feeding system

Explanations: 1…cyclone; 2…transport tube; 3…suction tube; 4…suction fan; 5…drive for feeding screw; 6…screw channel; data source [209]

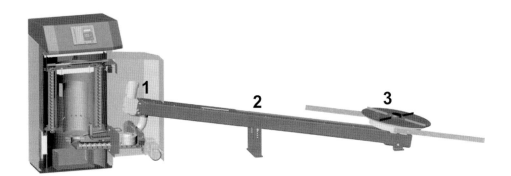

Figure 6.15: Combination of feeding screw and agitator

Explanations: 1…drive for feeding system; 2…screw channel; 3…agitator; data source [209]

A combination of feeding screw for conveying the pellets from the storage space to the furnace and storage space discharge by agitator is shown in Figure 6.15. The feeding screw is placed at a certain angle in the storage room, which renders unnecessary the use of a feeding screw equipped with cardan joint to overcome the height difference between storage room and burner. However, dead space inside the storage space increases as a result. Storage space discharge by agitator (also used in wood chip furnaces) is robust against fines and alternating fuel qualities. In contrast to all other discharge systems, a square storage room is best for this system. The agitator is placed in the middle and flush mounted into the slanted bottom together with a screw channel.

6.1.2.2 Ignition

Modern pellet furnaces are equipped with automatic ignition that normally works by an electrically driven hot air fan. An innovation in this area is the resistor ignition where the ignition element is made of silicon carbide, which during normal operation is stressed with around 900°C (it is stable up to 1200°C). Air flows around the ignition element, becomes heated and ignites the pellets. An electric power demand of 275 W [293] makes this way of ignition more economical than conventional hot air fans, which take up as much as 1,600 W during ignition.

6.1.2.3 Burn-back protection

The feeding system of furnaces has to be constructed in a way that avoids burning back to the storage facility or pellet reservoir (cf. Section 2.8). Efficient ways to avoid burn-back are achieved by furnace manufacturers through the use of rotary valves, fireproof valves or by means of self-initiating fire extinguishers, as well as by combining different measures.

An appropriate rotary valve is shown in Figure 6.16. It reliably disconnects the furnace room and storage facility.

Figure 6.16:   Rotary valve

Explanations: data source [294]

Airtight fireproof valves (cf. Figure 6.17) are usually placed inside the dropshaft between the feeding screw and stoker screw. This warrants tightness between storage room and furnace room. During operation, the valve is generally open for just a short period of time when the heating system is being fed. A buffer for a small amount of pellets is placed underneath the dropshaft, which is equipped with a capacitive level control. When a maximum filling level is reached, the valve closes automatically so that a certain distance to the valve is being kept at all times and nor can it be blocked by a high filling level. The valve not only closes at possible short circuits but also when there is any kind of failure that the system recognises.

A self-initiating fire extinguishing system is shown in Figure 6.18. There is a thermostat in the dropshaft that controls the temperature. As soon as the temperature exceeds the allowed value at the thermostat, the fire extinguisher unit that runs without external energy supply is activated.

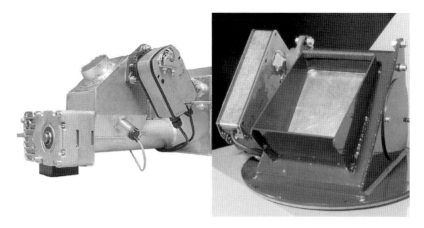

Figure 6.17: Fireproof valve

Explanations: left...overall view; right...detail (flap); data source [288]

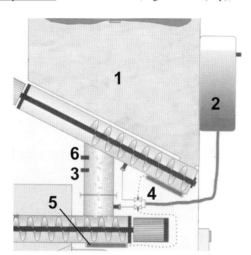

Figure 6.18: Self-initiating fire extinguishing system

Explanations: 1...integrated pellet reservoir; 2...water container; 3...thermocouple; 4...burn-back protection in the pellet reservoir; 5...burn-back protection in the stoker screw; 6...level switch; data source [295]

6.1.2.4 Furnace geometry

In order to obtain complete combustion and minimised emissions, realising a staged air supply is of great importance. Such a staged air supply is achieved separating the combustion chamber into a primary and a secondary combustion zone, both with separate air supplies. This inhibits the re-mixture of primary with secondary air and makes it possible to operate the primary combustion zone as a gasification zone with an understoichiometric air ratio, which is of great importance for the reduction of NO_x emissions since the formation of N_2 runs preferably in understoichiometric conditions. In the secondary combustion zone, complete oxidation of the flue gases takes place, whereby the mixing of the flue gases with the secondary combustion air is of great importance. It is achieved by suitable combustion chamber and nozzle design. Moreover, a long residence time of the hot flue gases in the combustion chamber, and thus a sufficiently large combustion chamber volume, is necessary for complete combustion of the flue gas. In Figure 6.19 a modern overfeed pellet furnace with staged air supply is presented.

Good utilisation of the combustion chamber (even flow distribution) as well as uniform temperature distribution (avoidance of local temperature peaks) should be achieved by optimised combustion chamber and nozzle geometries. The optimisation of combustion chamber and nozzle geometries can be accomplished by making use of CFD (computational fluid dynamics) simulations. The result of such an optimisation is shown as an example in Figure 6.20. In the base case, a very uneven flue gas velocity distribution over the cross section of the combustion chamber can be seen with a maximum in the centre of the cutting plane. This indicates bad mixing over the cross section and streak formation triggered by it. The consequences are increased excess air and CO emissions. In addition, local temperature peaks are to be expected. The optimised design displays a far more homogenous flue gas velocity distribution over the cross section of the flue gas channel as well as good mixing due to swirling flow. The consequences are less CO emissions with reduced excess air at the same time and thus higher efficiency. Local velocity and temperature peaks can be avoided, which in turn leads to reduced stress on the materials, reduced deposit formation and increased availability. Such optimisations are practically impossible without CFD simulation. The technique enables targeted and optimised improvement and hence speeds up the development process.

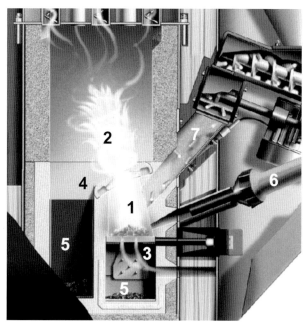

Figure 6.19: Principle of an overfeed pellet furnace with staged air supply and optimised mixing of flue gas and secondary air

Explanations: 1...primary combustion zone; 2...secondary combustion zone; 3...primary air supply (under the grate); 4...secondary air nozzles; 5...ash box; 6...ignition fan; 7...dropshaft (fuel supply); data source [286]

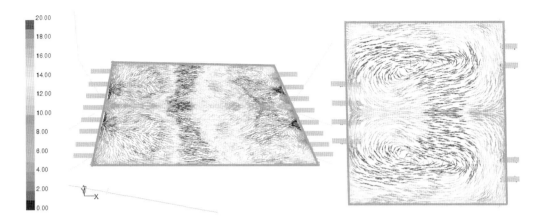

Figure 6.20: Example of an optimised secondary air nozzle design by CFD simulation

Explanations: left...basic nozzle design; right...improved nozzle design; vectors of the flue gas velocity in [m/s] in the horizontal cross-section right above the secondary air nozzles; data source [296]

Proven applications of CFD modelling are simulations in car, aerospace and biomedical industries, electronics cooling, chemical engineering, turbo machinery, combustion, heat and electricity generation. In the field of combustion, CFD simulation is increasingly used for optimisation of gas burners and pulverised coal furnaces. Making use of CFD simulation in medium-scale (100 to 1,000 kWh_{th}) and large-scale (> 1,000 kWh_{th}) biomass furnaces is relatively new but has been proven to be effective already. CFD modelling has most recently moved into the field of small-scale biomass furnaces where it is applied in the development of new product series in particular (cf. Section 12.2.6) [296].

6.1.2.5 Combustion chamber materials

With regard to operational comfort from the end user's point of view but also to failure free operation, high demands are put onto pellet furnaces. Apart from fully automatic operation, the choice of the materials used for the combustion chamber is decisive because the lifetime of a furnace can be influenced directly by utilisation of suitable and robust materials. The most used materials in combustion chambers are stainless steel, fireclay or silicon carbide.

Stainless steel is a relatively cheap material with low heat storage capacity, which makes quick start-up and shutdown processes possible. Stainless steel, however, does not possess strong resistance against corrosion and deposit formation. It is especially problematic in un-cooled areas. Combustion chambers made of fireclay are more expensive and have a high heat storage capacity. Fireclay is fairly durable with regard to deposit formation (especially at high temperatures). Silicon carbide is a very apt material for combustion chambers since it does not react with the ash and thus is very corrosion stable and not prone to deposit formations. Heat storage capacity is low, which allows for rapid start-up and shutdown just like stainless steel. Silicon carbide is a high cost material, however, which is why it is employed in combustion chambers to a limited extent only. Combinations of the designated materials are also possible.

6.1.2.6 Control strategies

There are four control circuits. These are load, combustion, temperature and pressure controls. Load control normally works with the feed temperature as the set point and it is regulated by fuel and primary air supply. Combustion control can be realised by either the O_2 content or the CO content or both contents in the flue gas as set points, with secondary air feed as the regulatory factor (lambda control, CO control or CO/lambda control). Furnace temperature is measured by thermocouples and can be controlled by flue gas recirculation, water cooled furnace walls or secondary air supply. The negative pressure of pellet furnaces is usually controlled by a revolutions per minute (rpm) controlled suction fan. Fully automatic operation with maximum efficiency and the least of emissions can be realised by a combined load and combustion control.

The minimum of CO emissions from biomass furnaces in general and from pellet furnaces is not only dependent on the excess air ratio but also on the moisture content of the fuel and the load condition of the furnace. An example of such a CO/λ characteristic is shown in Figure 6.21. CO emissions climb when the oxygen content and thus excess air (lambda/λ) is low as a consequence of local deficiencies of air in the combustion zone. When there is more excess air the combustion temperature decreases, which slows down the combustion reactions, which in turn raises CO emissions because the needed residence time for complete combustion is no longer ensured. The CO/λ characteristic is also relevant considering that CO is a leading parameter for realising complete combustion because low CO emissions imply low TOC emissions (sum of hydrocarbon emissions, also: C_xH_y). Low TOC emissions are also relevant for the reduction of fine particulate emissions (cf. Section 12.2.1.1).

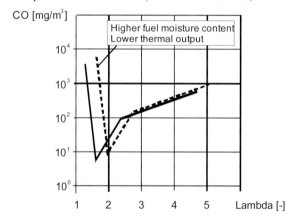

Figure 6.21: Correlation scheme of CO emissions and excess air coefficient λ in small-scale biomass furnaces

Explanations: data source [63]

The CO/λ-characteristic depends on the type of furnace and must thus be adjusted individually for each furnace. The aim is to run the furnace at an ideal level of minimal excess air ratio and CO emissions.

Most pellet furnaces available on the market use lambda control, which uses a lambda sensor for measuring the oxygen content of the flue gas. Controlling the oxygen content makes it possible to optimise the efficiency of the furnace because the efficiency rises with decreasing

O_2 content [63]. Moreover, the oxygen content in the flue gas influences the CO emissions. In lambda controls, a fixed value for the O_2 content is set. It works well as long as the moisture content of the fuel and the load of the furnace do not change. Due to the fact that CO emissions also depend on the other parameters mentioned, a change of furnace load or moisture content can lead to a considerable rise of CO emissions.

Regulating combustion, respectively burnout, by secondary air supply can also be realised solely by controlling the CO content of the flue gas. Sensors for CO can be used to measure the CO content of the flue gas, though they are relatively expensive. Measuring the CO content is not common practice for combustion control in small-scale furnaces but measuring the content of unburned components in the flue gas (TOC) is. This content provides qualitative information about the CO content of the flue gas and it can be measured with comparatively inexpensive sensors. However, when the moisture content of the fuel or the load condition of the furnace are altered, presetting a fixed value for the CO content can in the worst case lead to a massive rise in excess air and thus to a dramatic decline of efficiency.

As has been indicated already, the minimum of CO emissions in biomass furnaces in general is not only dependent on the excess air but also on the moisture content of the fuel and the load condition of the furnace, which is why a combined CO/λ control is an optimal control strategy. Thereby, excess air is varied until a CO minimum is achieved. As soon as the CO level changes, by altered moisture content or load condition for instance, the procedure repeats itself. By this permanent optimisation of the setpoint, excess air can be adjusted to any alterations of the fuel's characteristics as well as to the required power output, and at the same time excess air is kept to a required minimum. This also enhances the efficiency of the furnace.

A change in moisture content of the fuel is not to be expected in pellet furnaces due to the high quality standards for pellets. Thus, lambda control alone would be sufficient. The setpoint for the excess air could be adjusted to the fuel in such a way that CO emissions are maintained at a minimum once and then the CO content of the flue gas could not change significantly anymore because the moisture content of pellets would not change. A combined CO/λ control is advisable though because pellet furnaces are usually operated in a modularised way, i.e. also at partial load, and thus a different setpoint for the excess air ratio should be taken for partial load according to the CO/λ-characteristic (according to EN 303-5 emission limits are still to be adhered to at 30% nominal power output; cf. Section 2.8).

Adiabatic combustion chamber temperature rises with declining oxygen content [63]. This is usually controlled by cooling the combustion chamber walls or by the amount of secondary air. Control by secondary air is not the most favourable way though because the efficiency of the furnace declines with a rising amount of secondary air (the oxygen content of the flue gas is thus increased, hence the efficiency of the furnace drops) and it also cross-influences the combustion control mechanism. Flue gas recirculation is not common in small-scale furnaces.

In modern systems, all control systems that have been mentioned are based on micro processors.

Automatic control systems are of great relevance for assuring low CO, TOC and particulate matter emissions at both nominal and partial load as well as during start-up and shutdown, and during load changes. This is shown in Figure 6.22, in which particle, CO and TOC emissions, whilst changing the load from 15% partial load to nominal load, are presented. All concentrations rise during the change. Being controlled automatically, the parameters virtually reach the former level within a few minutes after the load change [65].

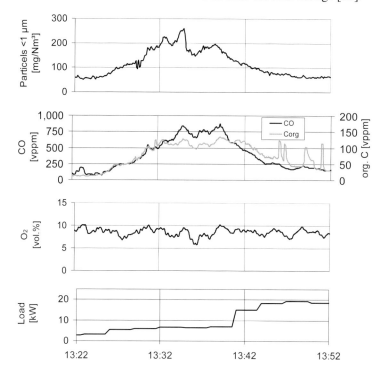

Figure 6.22: Emission of fine particulates, CO and TOC during load change of a modern pellet furnace

Explanations: change of load from 15% part load to nominal load; data source [65]

The aim of the control system is to keep emission peaks as low as possible and load changes as short as possible. Another important task of the control system is warranting low CO and TOC emissions at all load conditions. These parameters in turn influence fine particulate matter emissions because incomplete combustion leads to soot and hydrocarbon formation. Assuring complete combustion is thus an important primary measure for the reduction of particulate emissions (cf. Section 12.2.1.1).

6.1.2.7 Boiler

Usually, modern pellet boilers are fire tube boilers. A lot of pellet furnaces are equipped with an automatic boiler cleaning system. For this purpose, spiral scrapers that are driven by an electric motor are arranged inside the fire tubes of the heat exchanger and set into motion at regular intervals. An example of such a fully automatic heat exchanger cleaning system is shown in Figure 6.23. The moving spirals loosen possible deposits in the tubes, which are then collected in the ash box. Semi-automatic boiler cleaning systems are also applied. They also work with spirals inside the tubes of the heat exchanger but the spirals are operated manually with a handle from the outside. Some heat exchangers of pellet furnaces offered on the market have to be cleaned by hand.

Regular cleaning of the heat exchanger is vital to ensure that the efficiency of the furnace is steadily high. In addition, regular cleaning reduces particulate matter emissions. Fully automatic cleaning systems allow regular cleaning intervals, which is why they are to be clearly favoured. Semi-automatic systems are a low price alternative but they pose high demands onto the end user with regard to operation. The latest pellet furnaces show a clear trend towards fully automatic heat exchanger cleaning systems. Correct functioning of the automatic heat exchanger cleaning system can be controlled by checking the flue gas temperature, which should not change at certain load conditions during the heating period.

Figure 6.23: Fully automatic heat exchanger cleaning system

Explanations: 1…driving mechanics for the automatic cleaning system; 2…spiral scrapers; data source [297]

6.1.2.8 De-ashing

Removing ash is often said to be a main drawback with regard to the ease of use in pellet furnaces, contrasting with the absence of ash in gas and oil furnaces. This is why pellet furnace manufacturers attach great importance to de-ashing.

A low ash content of the fuel as set down by prEN 14961-2 is a basic prerequisite for long emptying intervals. The ash left on the grate or the retort after combustion of the pellets, the fly ash that is precipitated in the combustion chamber and ash resulting from the heat exchanger cleaning process are collected in the ash box. Depending on the dimensioning and without further measures, ash boxes that are integrated into the heating device require emptying about once a month during the heating period. In order to lengthen the emptying intervals, ash compaction systems are sometimes applied. One such ash compaction system is shown in Figure 6.24. Compaction is achieved by an up-and-down moving grate that is set into motion together with the fully automatic heat exchanger cleaning system. This allows for lengthened emptying intervals.

Even longer emptying intervals can be achieved by fully automatic de-ashing systems that convey the ash to an external, preferably mobile, ash box with a feeding screw. Such systems are currently being used in pellet furnaces to some extent (cf. Figure 6.25) and generally require emptying only once after each heating period.

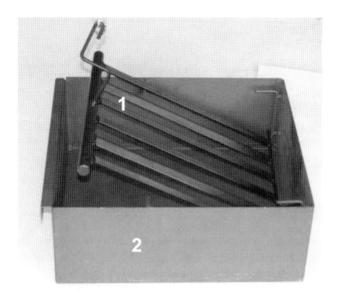

Figure 6.24: Ash compaction system

Explanations: 1...grate for ash compaction; 2...ash box; data source [209]

Figure 6.25: External ash box

Explanations: date source [288]

Under certain conditions, the ash from pellet furnaces can be used as a fertiliser in gardens (cf. Section 9.11).

6.1.2.9 Innovative concepts

6.1.2.9.1 Pellet furnaces with flue gas condensation

6.1.2.9.1.1 Basics of flue gas condensation

The most effective method for heat recovery of biomass furnaces is flue gas condensation. It is already state-of-the-art in medium-scale and large-scale biomass furnaces but in the field of small-scale furnaces it is a rather recent development. It was not until 2004 that the first pellet furnace with flue gas condensation was introduced to the market (cf. Section 6.1.2.9.1.3.1).

In flue gas condensation, the flue gas emitted from the boiler is led through a condensing heat exchanger. The return of the heating circuit cools the flue gas below dewpoint. A fraction of the flue gas' water vapour is thus condensed, which permits energetic utilisation of not just the sensible heat but also a part of the latent heat of the flue gas. The cooler the flue gas becomes the more the total efficiency of the furnace increases.

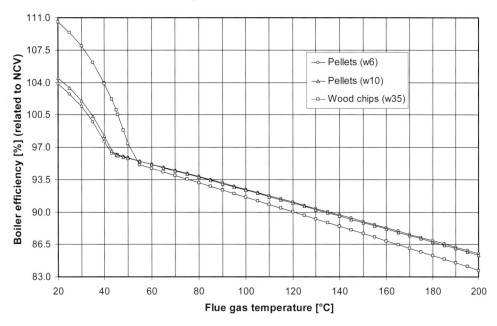

Figure 6.26: Dependency of efficiencies on the outlet temperature from the condenser and different moisture contents

Explanations: the calculation is based on pellets (M6 and M10) and wood chips (M35) under the following conditions: nominal boiler capacity 15 kWh$_{th}$; NCV = 20.2 MJ/kg (d.b.); ambient temperature 0°C; O$_2$ content of dry flue gas 10.0 vol.%; H content of fuel: 5.7 wt.% (d.b.); efficiency = sum of thermal output of boiler and condenser/energy input by the fuel based on NCV and air

Figure 6.26 shows the principle of flue gas condensation. The interrelation between the outlet temperature of the flue gas leaving the condenser and the efficiency of wood chip and pellet furnaces is shown. In conventional pellet boilers, energy losses are mainly due to the moist flue gas (and to some extent caused by radiation). The flue gas temperature of conventional pellet boilers at nominal load is usually between 120 and 160°C. Therefore, boiler efficiency (sum of thermal energy output of the boiler/energy content of the fuel based on NCV and air) is usually around 90%. In pellet boilers with flue gas condensation, the flue gas is cooled in the condenser by a suitably cold return below the dewpoint. Above the dewpoint the rise of the efficiency with falling temperature is almost linear. As soon as the temperature falls beneath the dewpoint, the rise of the efficiency with falling temperature is disproportional, because not only the sensible heat can be recovered but also the latent heat. In this way, efficiencies of more than 100% based on NCV can be achieved.

In practice, the realisable outlet temperature of the flue gas leaving the condenser and hence energy recovery by the condenser unit relies upon the cooling capacity of the heating circuit. Low temperature heating systems such as floor or wall heating systems are especially suitable for this purpose due to the low return temperature of the heating circuit. The dewpoint of the flue gas of pellet furnaces lies between 40 and 50°C. This is why return temperatures of below 30°C are needed for an efficient flue gas condensation. Temperatures such as this are more or less accomplished in modern low temperature heating systems.

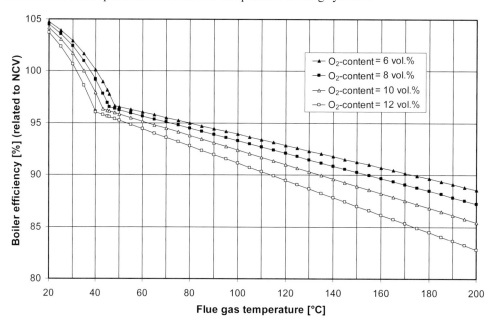

Figure 6.27: Dependency of efficiencies on the outlet temperature from the condenser and different O_2 contents of the flue gas

Explanations: the calculation is based on pellets under the following conditions: nominal boiler capacity 15 kWh$_{th}$; GCV = 20.2 MJ/kg (d.b.); ambient temperature 0°C; moisture content of fuel: 8.0 wt.% (w.b.); H content of fuel: 5.7 wt.% (d.b.); efficiency = sum of thermal output of boiler and condenser/ energy input by the fuel based on NCV and air

Figure 6.26 also shows the rise in efficiency that can be achieved by flue gas condensation in wood chip furnaces as a comparison. Due to the higher moisture content of this fuel, the moisture content of the flue gas is accordingly higher too. This is why more condensation can take place underneath the dewpoint so that the efficiency is even more elevated by flue gas condensation in wood chip furnaces than it is in pellet furnaces. The interrelation between efficiency rise and moisture content is not of great relevance for pellets because according to the standard prEN 14961-2, pellets must have a moisture content of below 10% and fluctuations of moisture content at that range have little effect on condensation.

The O_2 content of the flue gas also has a significant impact on boiler efficiency because both boiler efficiency and dewpoint go down as the O_2 content of the flue gas rises (cf. Figure 6.27). This is why an automatic control system, as exemplified in Section 6.1.2.6, that minimises the required O_2 concentration is of great significance for efficient operation of a pellet furnace.

Measurements of pellet furnaces with and without flue gas condensation at test stands showed that by the application of flue gas condensation, efficiencies are increased by around 10 to 11% on average (cf. Figure 6.28).

It has been demonstrated above that heat recovery potential is great when flue gas condensation is applied to pellet furnaces. Whether investing in a pellet boiler with flue gas condensation is economical or not ultimately depends on the additional investment costs as well as on the achievable fuel savings by using flue gas condensation. In this regard, achieving a low as possible return temperature in order to cool the flue gas to a temperature as low as possible plays a decisive role. An economic evaluation of flue gas condensation in pellet furnaces is carried out in Chapter 8.

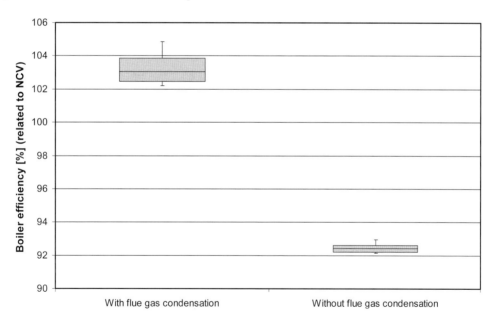

Figure 6.28: Efficiencies of pellet furnaces with and without flue gas condensation

Explanations: operating conditions of the furnace with flue gas condensation: nominal boiler capacity 20 kW; flue gas temperature 33.7°C; return temperature 22.4°C; O_2 content of flue gas 8.3 vol.% dry flue gas; box plot based on 9 measurements; operating conditions of the furnace without flue gas condensation: capacity: 18.2 kWh_{th}; flue gas temperature 144°C; return temperature 59.8°C, O_2 content of flue gas 8.6 vol.% dry flue gas; box plot based on 5 measurements; data source [282]

In addition to the achievable efficiency increase, flue gas condensation systems can also reduce particulate emissions under certain conditions [298]. A comparison of fine particulate emissions of pellet boilers with and without flue gas condensation is shown in Figure 6.29. The reduction of fine particulate emissions of the boiler with flue gas condensation is obvious but limited. A reduction of approximately 18% was achieved. Reduction of fine particulate emission always depends on the actual system and operational conditions and thus has to be evaluated individually for each case at hand.

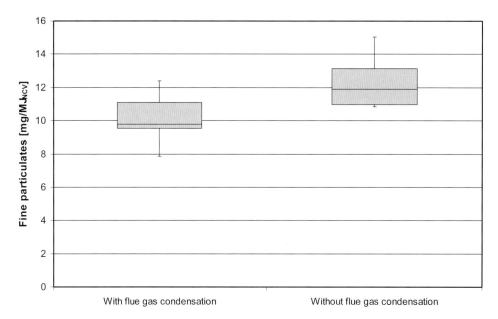

Figure 6.29: Fine particulate emissions of pellet furnaces with and without flue gas condensation

Explanations: operating conditions of the furnace with flue gas condensation: nominal boiler capacity 20 kW; box plot based on 9 measurements; operating conditions of the furnace without flue gas condensation: nominal boiler capacity 18.2 kW; box plot based on 5 measurements; data source [282]

6.1.2.9.1.2 Legal framework conditions for pellet furnaces with flue gas condensation

In practice, when pellets are used around 0.35 l/kg (w.b.)$_p$ of condensate are obtained [299]. Assuming an annual fuel consumption of 4 to 6 t (w.b.), 1,400 to 2,100 l of condensate are produced in a year. In Austria, the discharge of condensates to the sewer system is regulated in the Austrian waste water emission act [300; 301]. According to this regulation the condensate of small-scale pellet boilers with flue gas condensation can be discharged to the sewer without cleaning and neutralisation, provided that the following requirements are fulfilled:

- Fuel power input must be below 400 kW.
- Pellets according to ÖNORM M 7135 must be utilised.
- A positive type test according to [302] of the respective furnace must exist.

- Compliance with the emission limits of the Austrian waste water emission act [301] must be proven (cf. Table 6.1).
- Corrosion resistant material must be used for flue gas channels, the heat exchanger and the discharge pipe for the condensate.
- Installation, operation and maintenance have to be effected in a verifiable way according to the specifications of the type test.
- Regular inspection of the furnace and documentation thereof is required (at least every 2 years).

Neutralising the condensate is not an obligation. The corrosive properties of the condensate are met by utilisation of corrosion resistant materials.

Table 6.1 presents the limiting values for the condensate of pellet furnaces with flue gas condensation according to the Austrian waste water emission act in comparison with actual values of a furnace. It can be seen that the latter are far beneath all limiting values.

Table 6.1: Heavy metal contents of the condensate of a pellet furnace with flue gas condensation in comparison with limiting values of the Austrian waste water emission act

Explanations: all values in [mg/l]; data source [299; 300]

Parameter	Condensate	Limiting value AEV
Lead	0.017	0.5
Cadmium	0.0053	0.05
Chromium (total)	0.004	0.5
Copper	0.005	0.5
Nickel	0.003	0.5
Zinc	1.66	2.0
Tin	< 0.01	0.5

The possibility for a discharge of wood combustion condensates into a municipal waste water system is still not consistently regulated for in Germany. For heating oil and natural gas combustion, permissible waste water discharging is described in the so-called ATV-DVWK-A 251 instructions [303], but wood combustion condensates are not yet considered in that paper, although their addition is planned. However, there are regional regulations that can provide guidance. In the state of Bavaria, for example, particular instructions for wood fuel condensates apply [304]. This regulation allows the discharge of (natural) wood combustion condensates from furnaces below 50 kW without any further treatment if it is mixed with a suitable volume of sanitary sewage. For larger furnace power, controlled neutralisation is required in Bavaria.

6.1.2.9.1.3 Types of pellet furnaces with flue gas condensation

Different systems that enable flue gas condensation in small-scale pellet furnaces have entered the market in recent years. There are two fundamentally different applications available. Pellet furnaces can either be equipped with an integrated condensation heat exchanger or with an extra condensation heat exchanger module. So far there are only two pellet boilers with integrated heat recovery on the market. Additional modules for heat recovery are available on the market from four manufacturers.

If flue gas condensation is applied, flue gas channels and the chimney must fulfil specific requirements regarding fire-resistance, corrosion and moisture resistance. In addition, in case of operation with overpressure, the chimney must be pressure-tight.

A technical evaluation of all available systems is conducted in the following sections. An economic evaluation of flue gas condensation in pellet boiler technology is carried out in Chapter 8.

Despite intensive research it cannot be ruled out that other systems might exist on the market or might be in the final stages of development, however. Thus, the following sections make no claim to be complete.

6.1.2.9.1.3.1 Pellet furnace with integrated condenser

The introduction of a pellet furnace with integrated flue gas condensation in 2004 was an innovation [299; 305]. Figure 6.30 shows the scheme of such a system. It is a single stage flue gas condensation system in which the flue gas, after having passed through the boiler that generates the required feed temperature for the heating circuit, is led through a heat exchanger made of stainless steel where it gets cooled below dewpoint. Cooling medium in the stainless steel heat exchanger is the return from the heating circuit, which is thereby pre-heated. The condensate of the stainless steel heat exchanger can be discharged into the sewer without further measures. Comparing the system to a conventional pellet heating system, a gain in heat, although this is dependent on the return temperature of the heating system, can be achieved by this technology. In low temperature heating systems such as floor or wall heating systems with appropriate design, the efficiency can rise by up to 12%.

According to the manufacturer, manual cleaning of the condenser once in a year is sufficient. Because the system is integrated in the furnace it does not require much space. Operational costs rise slightly due to the marginally elevated electricity demand of the suction fan because a slightly greater pressure drop has to be overcome. This extra energy demand is very modest, however.

The pellet boiler with flue gas condensation is on offer with nominal boiler capacities of 8, 10, 15 and 20 kW.

Another pellet boiler with integrated flue gas condenser has recently been introduced into the market (cf. Figure 6.31) [306]. The heat exchanger is made of stainless steel and can be used in new pellet boilers as well as for retrofitting existing types of these pellet boilers. After the flue gas has passed through the boiler where the required feed temperature for the heating circuit is generated, it enters the flue gas condensation heat exchanger where it is cooled below the dewpoint. Cooling medium in the heat exchanger is the return from the heating circuit, which get pre-heated there. The condensate can be discharged into the sewer. The condensation heat exchanger is equipped with an automatic cleaning system to keep the heat exchanger continuously clean. The system is on offer with nominal boiler capacities of 15 and 25 kW.

Wood pellet combustion technologies

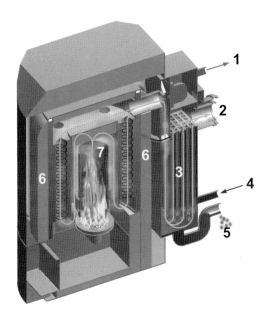

Figure 6.30: Scheme of a pellet boiler with flue gas condensation

Explanations: 1...feed; 2...flue gas; 3...stainless steel heat exchanger; 4...return; 5...condensate (to be discharged to the sewer); 6...boiler made of steel; 7...flue gas path; data source [305]

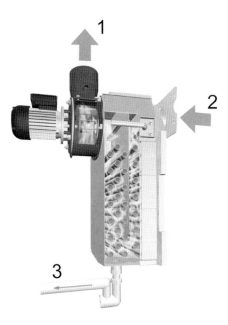

Figure 6.31: Pellet boiler with flue gas condensation

Explanations: 1...flue gas to the chimney; 2...flue gas from the boiler; 3...discharge condensate; data source [306]

6.1.2.9.1.3.2 External condensers for pellet furnaces

6.1.2.9.1.3.2.1 Racoon

The condenser that was originally developed under the name PowerCondenser [307] is now sold under the name Racoon [308] (cf. Figure 6.32) and can be integrated into new as well as already existing oil, gas or wood furnaces. Apart from the obligatory change of the granules in the neutralisation box, the system is free of maintenance.

The single stage flue gas condensation is placed downstream of the boiler. The flue gas from the boiler enters the condenser made of stainless steel. The return from the heating circuit is led through the condenser by means of plastic coated tubes, whereby the flue gas gets cooled and the return is pre-heated. The outlet temperature of the flue gas is limited due to the plastic coated tubes to 100°C. Therefore, a bypass is compulsory for skirting the condenser at higher temperatures. According to the manufacturer, efficiency rises by 15%. Such a great increase would only be possible if the outlet temperature of the flue gas was around 20°C, which is not possible in practice and thus the case is unrealistic (cf. Figure 6.27). Assertions as to the particulate matter precipitation efficiency of the system cannot be made due to lack of data.

Manual cleaning of a condenser operating in a pellet furnace should take place at least once a year. When particulate matter levels are increased, a quench unit can be placed upstream. Even though the condenser is not integrated into the furnace but is an additional module, space demand is modest. So are operational costs.

The Racoon is on offer for nominal boiler capacities of between around 25 and 1,800 kW.

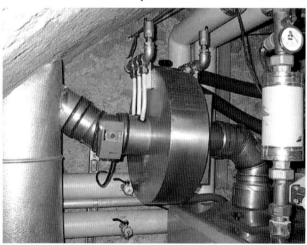

Figure 6.32: Racoon

Explanations: data source [307]

6.1.2.9.1.3.2.2 Öko-Carbonizer

The Öko-Carbonizer [309] can also be combined with new as well as existing oil, gas or wood heating systems and it is nearly maintenance free, according to the manufacturer.

The single stage flue gas condensation is installed downstream of the boiler (cf. Figure 6.33). The flue gas of the boiler passes through the Öko-Carbonizer, which is made of highly corrosion resistant carbon blocks that are impregnated with synthetic fibre.

Data on achievable efficiency increases and particulate matter precipitation efficiencies are only available for wood chip furnaces and thus cannot be assigned to pellet furnaces.

Water is injected discontinuously (hourly) to clean the condenser. The condensate could theoretically be used instead of water, but then the risk of nozzle fouling would arise. Moreover, manual cleaning of a condenser in combination with a pellet furnace should be done at least once a year. The space requirements and operational costs of the Öko-Carbonizer are moderate.

Öko-Carbonizers are on offer for nominal boiler capacities of 22 to 60 kW.

Figure 6.33: Öko-Carbonizer

Explanations: data source [309]

6.1.2.9.1.3.2.3 BOMAT Profitherm

The BOMAT Profitherm [310] (cf. Figure 6.34) can be applied in combination with new as well as existing oil, gas, wood and pellet heating systems.

The single stage condenser is installed downstream of the boiler. The flue gas leaving the boiler passes through a condenser made of ceramics and is cooled below dewpoint by a suitably cold return.

Data on achievable efficiency increases and particulate matter precipitation efficiencies are only available from combinations with wood chip furnaces [311] and thus cannot be assigned to pellet heating systems.

Water is injected hourly to clean the condenser. Moreover, manual cleaning should be carried out at least once a year. The space requirements as well as operational costs of the BOMAT Profitherm are moderate.

BOMAT Profitherms are on offer for nominal boiler capacities starting from 50 kW.

Figure 6.34: BOMAT Profitherm

<u>Explanations</u>: data source [310]

6.1.2.9.1.3.2.4 Schräder Hydrocube

The Schräder Hydrocube [312; 313; 314; 315; 316; 317] is a scrubber with upstream economiser. All the flue gas condensation systems presented up to now cool the flue gas by heat exchange between the flue gas and the return from the heating circuit. The Schräder Hydrocube combines this principle with material exchange in a flue gas scrubber (spray scrubber). Colder flue gas temperatures and thus higher efficiencies can be achieved in this way. In addition, some particulate matter precipitation takes place [315].

Figure 6.35 shows a picture of the stainless steel Schräder Hydrocube. The flue gas leaving the condenser passes through a heat exchanger first, where it is cooled by the return from the heating circuit, which itself gets pre-heated. Then, cold water is injected into the flue gas, which cools the flue gas even further, condensing the gas. The accumulating condensate is collected in a condensate box that contains a heat exchanger. Here, the condensate is cooled down to injection temperature by the cold water feed of the hot water tank, the feed of the hot water tank being pre-heated accordingly. The surplus condensate is discharged. It is clear that the condensate can only be cooled during hot water tapping. This is when efficiency reaches its maximum. If no hot water is tapped, the efficiency increase is solely achieved by the first heat exchanger in which the return from the heating circuit is made use of. In practice, efficiency increases will be somewhere in between the two extremes. Hot water demand is relatively low for small-scale users (a few 100 up to around 1,000 kWh per person and year, depending on user conduct), which is why the Schräder Hydrocube is not particularly suitable for small-scale systems. For larger users such as hotels or baths with greater hot water demand, the system could be a step in the right direction.

The system can be integrated into new as well as already existing oil, gas, wood or pellet furnaces. In retrofitting, the suitability of the chimney has to be checked (for resistance against condensate).

Figure 6.35: Schräder Hydrocube

Explanations: data source [310]

Cleaning is automatic and is integrated into the system. In small-scale furnaces (< 20 kW) additional manual cleaning is required once in a year according to the manufacturer. The Schräder Hydrocube requires quite a lot of space. Operational costs are also comparatively high. They are caused by the pump needed to inject the condensate.

R&D activities are ongoing regarding the combination of the Schräder Hydrocube with a wet electrostatic precipitator (cf. Section 12.2.1.1.3).

6.1.2.9.2 Multi fuel concepts

Many pellet boiler manufacturers offer pellet boilers that are not only dedicated to the use of wood pellets but also to other solid biomass fuels (e.g. firewood, wood chips, pellets from herbaceous biomass). Especially for wood pellet/firewood and wood pellet/wood chip combinations, proven concepts are already available on the market. However, their level of automation and the features installed vary broadly. Boilers for the combined utilisation of wood chips and pellets are usually equipped with similar features and have the same level of automation as pellet-alone applications. The simplest boilers for the combined utilisation of pellets and firewood have to be adjusted and operated manually in case of operation with firewood. Sophisticated boilers identify the fuel automatically and require no manual adjustments. Between these two boiler types several applications are available.

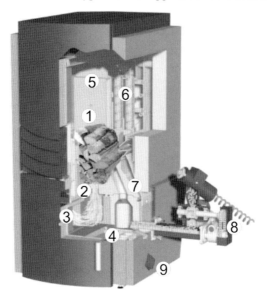

Figure 6.36: Combined boiler for the use of pellets and firewood

Explanations: 1…filling space for firewood; 2…stainless steel grate (firewood); 3…secondary combustion zone; 4…tiltable stainless steel grate (pellets); 5…heat exchanger; 6…spiral scrapers; 7…pellet combustion chamber tiled with firebricks; 8…feeding unit with rotary valve; 9…primary air valve; data source [293]

A system of that kind is displayed in Figure 6.36 [293]. The system automatically realises when there is no more firewood. This is accomplished by a temperature sensor placed downstream of the combustion chamber in a region where flue gas temperatures must be around 600°C. If this temperature stops being attained, the pellet burner will be started automatically. It is also started when the boiler's power output cannot remain steady, in spite of being required to. If firewood is inserted, this will be distinguished by the system because the downstream temperature sensor will measure a temperature above a certain value. If this is the case, the pellet burner will be turned off. If the temperature remains at the required level it will remain turned off, and if not, it will turn itself on again.

In addition, newly developed systems are already available on the market that are claimed to be suitable for the combined utilisation of wood pellets and corn or pellets made of herbaceous biomass. However, it is doubtful whether such boilers can be operated over longer periods of time without malfunctions and at low emissions, as several problems regarding the thermal utilisation of herbaceous biomass fuels in small-scale applications (e.g. fine particulate matter, NO_x and SO_x emission, deposit formation, corrosion) are still unsolved (cf. Section 3.5) [318; 319].

6.1.2.9.3 Pellet fired tiled stoves

The development of pellet fired tiled stoves is relatively recent [169]. It holds great potential as tiled stoves are very popular in Austria (around 480,000 tiled stoves are installed in Austria

[320]) and southern Bavaria. The focus is on retrofitting existing tiled stoves as well as the development of new tiled stoves that are designed and optimised for the use of pellets. Using pellet fired tiled stoves as automatically operated central heating systems by building in appropriate boilers is one more option. Test runs of pellet fired tiled stoves confirm the compatibility of modern pellet burners and tiled stoves. Automatic operation of tiled stoves can be accomplished in this way. Moreover, CO and TOC emissions could be reduced significantly, as compared to the use of firewood [321].

6.1.2.9.4 Pellet furnace and solar heating combination

The combination of pellet furnaces with solar heating systems is a current issue. It has been increasingly promoted and applied. A reasonable combination can attain a simple supply of hot water but can stretch to the support of the heating system by appropriately large collector surfaces and well designed buffer capacities. In this way, solar energy can provide a significant share of heat generation and fuel can be saved [322; 323; 324; 325].

Both state-of-the-art pellet furnaces and solar heating systems are advanced technologies and available on the market. The main challenge is combining these two technologies in an ideal way, and especially achieving the correct adjustment of their control systems. Improper adjustment of the two systems to one another can cause high heat losses in the buffer or inefficient operation of the solar heating system [326; 327; 328].

To combine wood pellet and solar heating is common, for example in Austria, Sweden, Denmark and Germany. There are three main advantages of combining pellet and solar heating. First, regarding economy, is the possibility of utilising the heat buffer storage for solar heat as well as the pellet burner. Second is the possibility of improving the efficiency of the pellet burner. Third is the possibility of having a 100% renewable heating system with a reduced need of wood fuel resources and with reduced emissions.

The combined pellet and solar heating systems on the market are commonly designed to provide heating and domestic hot water, commonly called a "combi system". The solar heat will cover 10 to 50% of the heat demand (for heating and domestic hot water) depending on type of system and heat load, while the pellet boiler covers the remaining part of the load.

Small systems for detached houses (cf. Figure 6.38 and Figure 6.39) comprise solar collectors, a well insulated heat buffer storage tank (500 to 750 litre of water, typically 75 to 100 litre per m^2 of collector area) with an integrated pellet burner or a separate pellet boiler, and an integrated or external heat exchanger for the solar circuit. Large systems for large buildings or groups of buildings comprise a separate boiler and one or several storage tanks (with 100 to 200 litres per m^2 of collector area) depending on system design and the load size.

Using solar heating will replace the summer operation of the burner, with its worse operating conditions that lead to relatively high emissions and low efficiency. This replacement results in much reduced CO emissions and higher boiler efficiency [329; 330; 331]. However, there will be heat losses from the heat buffer storage so the system efficiency can be either higher or lower depending on the insulation of the boiler as compared to the buffer store.

The main challenge in combining pellet and solar heating is having an appropriate control. A solar heating system requires a heat buffer storage in order to override the difference between the availability of solar radiation (commonly five to eight hours during daytime) and the load demand (commonly heating and hot water). Stable combustion conditions make it easier to obtain a high efficiency and reduce the emissions from processed wood burners such as pellet burners. Thus, it is an advantage in most cases to use a heat buffer storage (commonly water) to even out the difference between the actual power demand (commonly heating and hot water) and the nominal power of the burner and obtain longer burning periods and more stable burning conditions.

A typical pellet and solar heating system is illustrated in Figure 6.38. The burner can either have an on/off control or a modulating operation adjusting the combustion power to the current load situation. The on/off control is easier to adapt but the modulating operation has the potential to reduce the emissions as the number of start-ups and shutdowns can be reduced [329; 330; 331]. However, the time lag is too long using only sensors in the store for modulating operation.

In this case (cf. Figure 6.38), the burner is controlled in on/off-mode by the temperature sensors TS1 and TS2 in the buffer store. To maintain a high system efficiency it is essential that the burner and the pump P2 stop automatically when the solar collector can cover the heat demand. Therefore the burner has to be controlled by sensors placed in the store. When solar heat is enough to cover the load, the boiler can cool down and the boiler losses will be very small during the summer period. In order to further minimise the heat losses, it is also important that the boilers have a small water volume, i.e. a low thermal mass [330].

A pellet stove with water jacket connected to a small buffer store (cf. Figure 6.39) is another possible pellet and solar heating system combination that is suitable for single family houses without a boiler room. The system design can be identical to the boiler system but as the stove is usually placed in the living room, comfort levels may not be as good because the room may get too hot. Therefore it is important that the fraction of heat transferred to the water circuit is high (> 80%). [330] investigates how the systems can be designed to maintain comfort criteria. However, this means that an electric heater in the store is required to maintain the temperature.

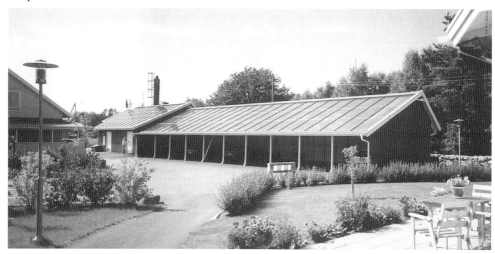

Figure 6.37: Pellet and solar heating system using roof integrated solar collectors

Explanations: roof of heating plant and carport equipped with solar collectors; plant for a small residential area with 36 residential units in nine houses in Kungsbacka, Sweden; data source [332]

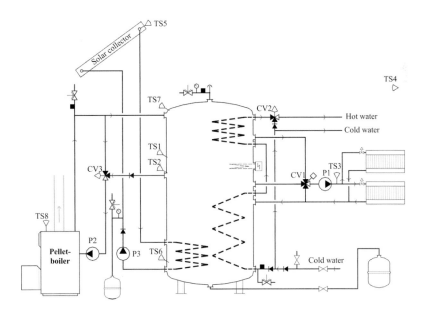

Figure 6.38: Solar and pellet heating system with pellet boiler

Explanations: TS…temperature sensor; CV…control valve; P…pump

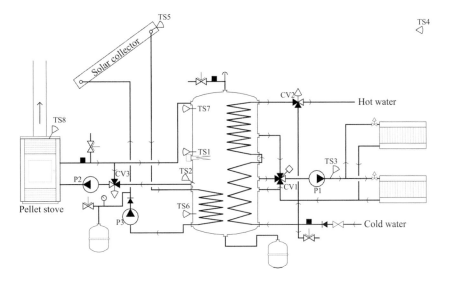

Figure 6.39: Solar and pellet heating system with a pellet stove and a small buffer store

Explanations: TS…temperature sensor; CV…control valve; P…pump

6.2 Medium-scale systems (nominal boiler capacity 100 - 1,000 kW$_{th}$)

6.2.1 Combustion technologies applied

Furnaces with a nominal boiler capacity ranging from 100 to 1,000 kWh are designated as medium-scale furnaces. In general, such furnaces are either medium-sized biomass district heating systems, commercial or industrial applications as well as biomass CHP systems. Typical applications in this power range are for instance apartment houses, caterers, schools, kindergartens, sports and other halls, baths, housing estates, churches, cloisters, castles, nurseries, hospitals or stations. Many such systems are set up as contracting systems.

With regard to medium-scale pellet furnaces, such systems play a subordinate role in Austria and Germany for instance, as wood chips or shavings are cheaper and more easily available. Nevertheless, this pellet market sector is gaining ground and some medium-scale pellet furnaces have been installed already [333]. The advantage of using pellets in this field is that less storage space is required due to the superior energy density of pellets, and fuel feeding as well as combustion technology become cheaper. Moreover, pellets of lower quality (more fines and higher ash contents) can be used without any problems, which in turn may render the utilisation of raw materials other than wood (for example bark) more interesting in pellet production.

With regard to furnace technologies and fittings, the same principles addressed in the case of small-scale furnaces (cf. Section 6.1) apply in principal in medium-sized systems, whereby furnace technologies in this range are usually able to make use of both pellets and wood chips (underfeed or grate furnaces). As concerns fuel storage, there are differences because storage capacity of an annual fuel demand is not essential in medium-sized system as these must have a supervisor available at all times in case of system failures (even when the furnace is automatically operated), who may also take pellet deliveries several times a year. Just-in-time fuel delivery with suitable long-term contracts is favoured in such systems. Storage costs are reduced thereby. Storage in silos is a good option for such applications, which are usually not applied in small-scale systems, chiefly due to aesthetic reasons. Storage rooms with automatic discharge are used too.

Fuel conveyor systems are generally built in a more robust way in medium-scale furnaces than in small-scale furnaces. Usually, flue gas recirculation is employed for combustion temperature control in medium-scale applications. In addition, medium-scale systems are usually equipped with more sophisticated measurement and control methodologies for low emission combustion and flue gas cleaning systems (especially for precipitating particulate matter). De-ashing is typically carried out automatically by a feeding screw that conveys the ash into a sufficiently large-sized container. Since medium-scale furnaces are so equipped, industrial pellets that are less durable and contain more ash than high quality pellets can be used without problem.

In contrast to small-scale furnaces, where pressurised air is generally not available, medium-scale furnaces may be equipped with pneumatic heat exchanger cleaning systems.

6.2.2 Innovative concepts

A pellet furnace that was designed for the use in apartment houses, micro-grids and other large-sized buildings with nominal boiler capacities ranging from 100 to 300 kW was developed by the company KWB (KRAFT & WÄRME AUS BIOMASSE GMBH) and brought to series production under the name "KWB TDS Powerfire 150". One key attribute of the newly developed system is its patented rotary grate that can be used with different kinds of biomass fuels next to pellets. It exhibits good charcoal burnout even when the fuel is very moist (accepting wood chips with a moisture content of up to 50 wt.% (w.b.)), which allows the operator to be flexible with regard to fuel. The heart of the furnace is a vertical cyclone combustion chamber. It has been registered for patent, was developed by means of CFD simulation and has tangentially arranged secondary air nozzles. This combustion chamber design ensures efficient flue gas burnout without streak formation and moderate combustion temperatures in order to avoid fly ash deposits and wear, as well as good precipitation of fly ash particles, and all this for a wide range of operating conditions and fuel moisture contents. Primary air is fed in from beneath the rotary grate. The ash on the grate falls through it into a screw channel underneath, where a feeding screw conveys the ash into an ash box. The ash from the heat exchanger is also fed into this ash box. Figure 6.40 displays this innovative kind of pellet furnace.

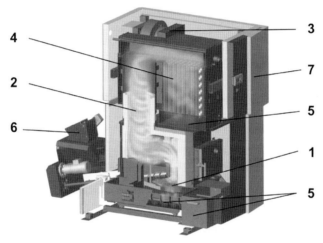

Figure 6.40: TDS Powerfire 150

Explanations: 1...rotary grate combustion system (cf. Figure 6.10 for details); 2...rotary combustion chamber; 3...lambda sensor; 4...heat exchanger; 5...de-ashing; 6...fireproof valve; 8...micro processor based control system; data source [293]

Another innovative medium-scale furnace system has a rotary combustion chamber and is from the company KÖB Holzfeuerungen GmbH, a member of the Viessmann group, introduced to the market under the name of PYROT [334]. The system is shown in Figure 6.41. The furnace was optimised for the use of dry wood fuels such as pellets or dry wood chips. In combination with a patented upstream dryer, wood chips having a moisture content of up to 60 wt.% (w.b.) are usable. The system is on offer with 6 different nominal boiler capacities of 80 to 540 kW and thus appropriate for the use in larger buildings or micro-grids. The furnace is charged by a feeding screw and the fuel is gasified under air deficiency. Combustion gases ascend towards the rotary combustion chamber where they get mixed with secondary air that is set into rotation by a fan. In this way secondary air and combustion gases are optimally mixed. Very low CO and NO_x emissions are accomplished by this system.

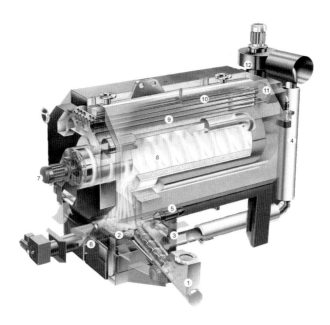

Figure 6.41: PYROT rotation furnace

<u>Explanations</u>: 1…feeding screw; 2…grate; 3…primary air supply; 4…flue gas recirculation; 5…ignition fan; 6…de-ashing system; 7…secondary air supply by rotary fan; 8…rotary combustion chamber; 9…heat exchanger; 10…safety heat exchanger; 11…pneumatic heat exchanger cleaning system; 12…suction fan; data source [335]

6.3 Large-scale systems (nominal boiler capacity > 1,000 kW$_{th}$)

6.3.1 Combustion technologies applied

Large-scale furnaces are furnaces with nominal boiler capacities of more than 1 MW$_{th}$. Such furnaces can be found for instance in the wood processing and wood working industry, in district heating plants or in CHP plants (cf. Section 6.1). In such plants the use of pellets of lower quality, which are unsuitable for use in small-scale furnaces but can be produced at low costs, can be used, though economic advantages are dependent on the price of industrial pellets.

Technologies that can be applied to this range of power output are the grate furnace, underfeed furnace and above 20 MW fluidised bed furnace. At present, plants of this scale mainly use cheap by-products of sawmills as well as waste wood. For detailed exploration of these technologies see [53; 63].

Pellet fired large-scale furnaces are also offered in container design with power outputs of up to 3 MW [336]. Container design is suitable for the use in overcoming of heating outages, temporary heating of construction sites, halls at big events but also for permanent use, for instance in boiler retrofitting. The warm air required for heating up big exposition halls or tents at large events can be supplied by making use of a flue gas-air-heat exchanger. Pellet furnaces of container design are completely pre-fitted so that only electrical and water connections have to be provided on site.

Using pellets for combustion or co-firing in retrofitted furnaces of fossil fuels, especially coal, is a potential application of pellets in large-scale plants (cf. Section 6.5).

6.3.2 Innovative concepts

In Sweden it has been common practice for many years to retrofit oil fired furnaces used for heat generation to burn wood dust (ground pellets). Pulverised fuel burners were developed for exactly the purpose of replacing old oil burners. However, requirements concerning CO and NO_x emissions became increasingly rigid over time. These increasing demands made the Swedish company TPS Termiska Processor AB of Stockholm work on the development of a new type of burner resulting in the introduction of the Bioswirl® burner (cf. Figure 6.42) [337; 338] which fulfils present requirements with regards to emission reduction.

It is a cyclone burner where the fuel particles are pyrolysed inside the cyclone combustion chamber. The rotational flow warrants complete thermal degradation of bigger particles. Flue gas is led into the secondary combustion chamber of the boiler by a nozzle. Here, secondary and tertiary air is injected and complete burnout takes place. Combustion of ground pellets with this technology allows the furnace to be controlled at a broad range of power outputs and low emissions at the same time. It was found that particle size has little influence on the system. The power range being aimed at by the Bioswirl® burner lies between 1 and 25 MW. Burners with power ranges of 1 to 3 MW are already being used successfully. In a demonstration plant equipped with a 17 MW Bioswirl® burner CO emissions could be reduced significantly in comparison with a conventional dust burner for wood [337; 338].

Figure 6.42: Bioswirl® burner

Explanations: data source [339]

The burner was originally designed for retrofitting existing oil furnaces. Due to the positive results that were achieved by the first systems, new systems on the basis of the Bioswirl® burner are on offer too.

6.4 Combined heat and power applications

Energy generation from biomass has gaining importance in recent years. The driving forces behind this development are national and international (especially in the EU) efforts aiming at the increase of the "green" share of electricity generation and reduction of CO_2 emissions. Different initiatives support and fund these aims at the EU level (e.g. White paper for a community strategy and action plan, Renewable energy directive, Green paper towards a European strategy for the security of energy supply, Directive on the promotion of co-generation based on a useful heat demand in the international energy market, etc.). On the national scale, many different instruments (e.g. feed-in tariffs, allocation of quota, bonus systems) are applied to support energy generation from biomass (e.g. in Austria, Germany, Italy, Switzerland, Finland, Belgium, Denmark, the Netherlands or Sweden) [340; 341].

For energy generation from biomass there are different technologies available, depending on the power range [342], that are exemplified and discussed in the following sections. Wood chips and bark have mainly been utilised in biomass CHP systems to date. Pellets are still playing a subordinate role in this field but a positive trend is notable here too.

6.4.1 Small-scale systems (nominal boiler capacity < 100 kW$_{th}$)

There is no proven CHP technology for small-scale furnaces below 100 kW$_{th}$ available on the market yet. However, there are plenty of R&D activities in several research institutes as well as manufacturers that are mainly focussed on two technologies, namely Stirling engines and thermoelectric generators within a power range of a few kW$_{el}$ [343; 344; 345; 346].

Stirling engines with power outputs of 35 to 75 kW$_{el}$, thus belonging to the medium-scale systems, are currently on the market (cf. Section 6.2). Stirling engines are also interesting for small-scale applications because they can be down-scaled to a few kW$_{el}$ [347; 348].

A Stirling engine, which was under development by the Stirling Power Module Energieumwandlungs GmbH in Austria, namely the SPM Stirlingpowermodule, is explained here, even though its development has been stopped for the moment [716]. A picture of the furnace with the Stirling engine at the top is shown in Figure 6.43 [349; 350]. The aim was to produce enough energy for an average household; excess green electricity could be fed into the electrical grid. As a further step, its operation independently of the electrical grid should be made possible. The Stirling engine is designed in a way that makes it possible to retrofit existing pellet furnaces with it. The Stirlingpowermodule is based on a four cylinder Stirling engine with patented gearbox with complete mass balancing. This gear box converts the totally linear movement of the two crossed-over piston rods into rotary movement for driving the generator. Air is used as the working medium. The nominal electric power output is intended to be about 1 kW$_{el}$. The engine should to be utilised in a pellet furnace with a nominal fuel power input of 16.8 kW. The electric efficiency of the system amounts to almost 6%.

The German company Sunmachine GmbH has developed a Stirling engine with an electric capacity of about 3 kW [717]. A small series of around 400 units have already been manufactured. However, technical problems currently hamper further development [718].

Figure 6.43: SPM Stirlingpowermodule

<u>Explanations</u>: data source [351]

Another way to realise combined heat and power production in small-scale furnaces is the application of thermoelectric generators [345; 346; 352; 353]. First trials with a prototype (cf. Figure 6.45) have been carried out already. The development aims to generate sufficient electric energy for the pellet furnace itself so that the system can operate independently from the electrical grid. An electric power output of < 1 kW_{el} would be sufficient for this purpose and it is hence the target value of the development. Thermoelectric generators exploit a thermoelectric effect in which a current flows in an electric circuit made of two different metals or semi-conductors as long as the electric contacts have different temperatures (the principle is shown in Figure 6.44). They enable the direct conversion of heat to electric energy. The advantages of thermoelectric generators are long operating times without maintenance and operation free of noise and without moving parts. However, their lifespan is still short and further development work needs to be done. The electric efficiency of state-of-the-art thermoelectric generators is around 5 to 6%. An increase of the electric efficiency to around 10% is expected by further improvements of the system. The main challenge is posed by integrating thermoelectric generators into pellet furnaces. The electric system efficiency that can be achieved at present (electric power output/energy content of the fuel based on NCV and air) lies at around 1.6%. Electricity generation costs amount to around 0.7 to 0.8 €/kWh_{el}. A doubling of the electric system efficiency and cutting the electricity generation costs by half could be achieved by raising the temperature level in a further development step [354]. Market introduction is not expected within the coming years. If costs for thermoelectric modules can be reduced, the technology could become of interest in the medium term.

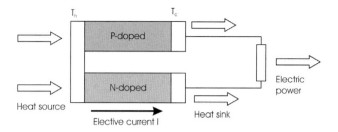

Figure 6.44: Principle of thermoelectric electricity generation

Explanations: data source [355]

Figure 6.45: Prototype of a thermoelectric generator designed for utilisation in a pellet furnace

Explanations: data source [355]

6.4.2 Medium-scale systems (nominal boiler capacity 100 - 1,000 kW$_{th}$)

6.4.2.1 Stirling engine process

CHP technology on the basis of a Stirling engine is an interesting and promising application in the field of electricity generation from biomass for power outputs below 100 kW$_{el}$. In this power range in particular there are no mature technologies available on the market yet.

Within the framework of an R&D cooperation between BIOS BIOENERGIESYSTEME GmbH, MAWERA Holzfeuerungsanlagen GmbH, BIOENERGY 2020+ GmbH and the Technical University of Denmark, a CHP technology on the basis of a Stirling engine with nominal capacities of 35 and 75 kW$_{el}$ was developed. The CHP technology on the basis of a 35 kW$_{el}$ four-cylinder Stirling engine has been tested successfully for 12,000 operating hours and already realised in the course of a commercial project (cf. Figure 6.47). The efficiency of the Stirling engine is around 25 to 27%. In test trials, an electric plant efficiency of around 12% was achieved [356; 357; 358; 359].

The Stirling engine belongs to the group of hot gas or expansion engines. In such engines, the piston is not prompted to action by expansion of a combustion gas of an internal combustion but by expansion of a sealed and thus constant amount of gas that expands due to energy supplied by an external source of heat. The power generation is thus separated from the furnace, which, in principle, can be operated with any kind of fuel and optimised with regards to emissions on its own. For a detailed description of the technology and further information, see [348; 360]. A scheme of the integration of a Stirling engine into a biomass CHP plant is displayed in Figure 6.46).

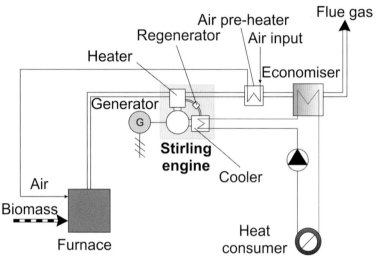

Figure 6.46: Stirling engine process – scheme of integration into a biomass CHP plant

Explanations: data source: BIOS BIOENERGIESYSTEME GmbH

The Stirling engine developed at the Technical University of Denmark uses helium as the working medium and is designed as a hermetically sealed unit. Helium is very efficient with regard to the electric efficiency but it poses high demands on seals. The generator of the CHP concept developed by the Technical University of Denmark is placed inside the pressurised crankcase, which considerably simplifies sealing the drive shaft and the piston. Only the cable connections between the generator and the grid come out of the crankcase. Inside, simpler seals may be used. In conventional Stirling engine concepts, the moving seals (especially those of the piston rods) create severe difficulties that have yet to be overcome.

Developing and designing a furnace for a CHP system that is based on a Stirling engine was a complex R&D task. In order to accomplish high electric efficiencies, flue gas temperatures have to be kept as high as possible upon entering the hot heat exchanger of the Stirling engine. The system was designed for inlet temperatures of the flue gas being approximately 1,200 to 1,300°C. Maximum flue gas temperatures of conventional biomass furnaces are around 1,000°C. The resulting higher temperatures in the combustion chamber can cause slagging of the ash that subsequently gets deposited on the inner walls of the combustion chamber leading to possible operational failures. Therefore, the target in developing the high temperature furnace was to safeguard high flue gas temperatures at the hot heat exchanger side on one hand and avoiding temperature peaks in the furnace on the other hand. Due to the high temperatures in the combustion chamber, only wood chips, sawdust and pellets with small amounts of bark may be used as a fuel. Underfeed furnaces are particularly suitable for such finely composed fuels. The development of the new furnace was supported by CFD simulations carried out by BIOS BIOENERGIESYSTEME GmbH.

Figure 6.47: Pictures of a pilot plant and the 35 kW$_{el}$ Stirling engine

Explanations: data source [359]

Furthermore, an automatic cleaning system for the hot heat exchanger was designed. The system consists of a pressurised air tank with a number of valves that are arranged at every outlet panel of the hot heat exchanger. Regularly, one valve at a time is opened and the tubes of the hot heat exchanger are cleaned by the impulse of pressurised air.

The new CHP technology is the first successful application of a Stirling engine in a biomass furnace with a power output of less than 100 kW$_{el}$ worldwide and can indeed be seen as a breakthrough with regard to the utilisation of biomass in CHP systems in the small capacity range.

Presently, several demonstration plants are in operation in order to gain long-term experience in field tests and to be able to eliminate weak points in the future.

6.4.2.2 ORC process

The ORC (Organic Rankine Cycle) process is an interesting technology for combined electricity and heat generation in decentralised biomass CHP plants of a power range of between 200 and 2,000 kW$_{el}$ (this corresponds to nominal thermal capacities of roughly 1,000 to 10,000 kW). The technology has already been proven in this range but it is not yet relevant for the power range of small-scale CHP systems. It is discussed here because expansion of the power range into the small-scale field is expected as economy-of-scale problems are overcome.

This new technology, designed especially for biomass CHP systems, was developed and realised in the course of two EU demonstration projects. The first biomass CHP plant on basis of an ORC process in the EU was put into service in the wood processing industry STIA in Admont (Austria) [361; 362]. The system has a nominal electric power output of 400 kW, has been in operation since October 1999, uses mostly wood dust, sawdust and wood chips as a fuel and is operated in heat controlled mode only. The second system was implemented in the CHP plant of Lienz (Austria) and is the enhanced technology of Admont [363]. The nominal electric power output is 1,000 kW, it has been in operation since February 2002, uses wood chips, sawdust and bark as a fuel and is also operated in heat controlled mode only. Almost 150 biomass CHP plants based on the ORC technology had been put into service up to 2010.

The principle of electricity generation by means of the ORC process is that of the conventional Rankine process. The main difference is that instead of water, an organic working medium with specific thermodynamic properties (hence the name Organic Rankine Cycle) is used. An ORC system adapted for the use in biomass CHP plants was developed by the company TURBODEN Srl in Brescia, Italy. Another provider of ORC technologies for biomass CHP plants is the German company Adoratec GmbH.

Figure 6.48 demonstrates the working principle as well as the different components of the ORC process using the example of the biomass CHP plant Lienz.

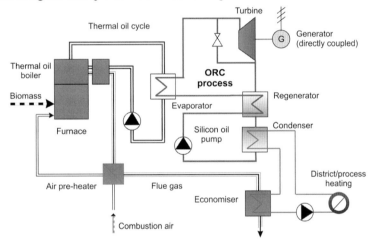

Figure 6.48: Scheme of the ORC process as integrated into the biomass CHP plant Lienz

Explanations: data source: BIOS BIOENERGIESYSTEME GmbH

The ORC process is connected to the thermal oil boiler by means of a thermal oil circuit. The ORC process is built as a closed unit and uses silicon oil as the organic working medium. The pressurised silicon oil is vaporised and also slightly superheated by the thermal oil inside the evaporator. After that it is expanded inside the axial turbine that is coupled directly to an asynchronous generator. Before entering the condenser, the expanded silicon oil is fed into a regenerator (for internal heat recovery). Condensation of the working medium takes place at a temperature level that allows utilisation of the heat for district or process heating (feed temperature between 80 and 100°C). Finally, the condensed working medium is brought to the pressure level of the hot part of the cycle and reaches the evaporator again after having flown through the regenerator. The electric plant efficiency of the system is 15 to 16%, which is much above the efficiency of systems based on the Stirling engine.

The major benefit of the ORC technology is its superb part load and load change behaviour. According to operational experience from the systems in Lienz and Admont, fully automatic operation between 10 and 100% nominal load is possible without any problems. According to measurements in Lienz, the electric efficiency at 50% part load is around 92% of the efficiency at full load, which underlines the applicability of this technology for heat controlled operation.

Due to the ORC being constructed as a closed system, there are neither liquid or gaseous emissions nor losses of working medium. The silicon oil is hardly prone to aging and does not need to be changed at any time of the lifespan of an ORC system (more than 20 years, based on experience from geothermal plants), which is why operational costs of these systems are very low. Maintenance costs (change of lubricants and seals, system check once in a year) are also very low. This is confirmed by plants in operation.

Acoustic emissions of ORC systems are moderate, with the highest emissions stemming from the capsuled generator. They are around 85 dB(A) at a distance of 1 m.

Experience with the operation of the biomass CHP systems in Lienz and Admont has shown that ORC technology is an interesting solution for small-scale biomass CHP systems from both the technical and the economic viewpoint. Due to the positive experiences of these two projects, market introduction of the ORC technology was speeded up accordingly.

The first 200 kW_{el} ORC plant was put into service in autumn 2007. This size could also be of interest for the utilisation of pellets.

For detailed information about ORC technology, see [348; 356; 361; 362; 363].

6.4.2.3 Fixed bed gasification

Fixed bed gasification is another technology suitable for biomass CHP plants. Its range of power outputs would fit medium-scale systems. In comparison with biomass CHP technologies based on biomass combustion, higher electric efficiencies are possible. However, automation and control of the gasification process as well as producer gas cleaning are much more difficult to handle. This is the reason why the technology is not on the market yet. Whether and when this will be the case is still unknown. All fixed bed gasification systems operating at present are pilot plants. Activities with regard to fixed bed gasifiers for the small- to medium-scale power range are followed especially in Germany, Switzerland and Denmark. For more information, see [348; 359; 364; 365; 366].

6.4.3 Large-scale systems (nominal boiler capacity > 1,000 kW_{th})

The utilisation of pellets is conceivable and possible in all kinds of biomass CHP systems. The CHP technologies based on biomass combustion that are relevant at the large-scale are the steam turbine process (usually applied from 2 MW_{el} upwards) and ORC process (up to 2 MW_{el}).

The benefits achievable by using pellets were discussed in Section 6.2 for the medium-scale. They are also true for large-scale biomass CHP systems. In Austria and Germany, pellets play a minor role in large-scale systems. Only a few installations have been set up to date.

6.5 Combustion and co-firing of biomass pellets in large pulverised coal fired boilers

6.5.1 Technical background

In general terms, the co-firing of biomass materials in large pulverised coal fired power plants, both as a retrofit in existing plants and, in the future, for new built applications, tends to be one of the more cost effective and energy efficient approaches to the utilisation of biomass for energy recovery. The co-firing of biomass in existing power plants can, in most cases, be implemented relatively quickly and conveniently, and normally involves relatively low levels of technical and commercial risk compared to the installation of new, dedicated biomass power plants. In most cases, maximum use is made of the existing power generation equipment, and of the civil and electrical engineering infrastructure.

There has also been increasing interest in the conversion, particularly of the smaller pulverised coal fired power plants and CHP boilers, to 100% biomass pellet firing. To date, this has been implemented in a small number of cases in Northern Europe [53], however there is evidence of increasing interest in other countries in Europe and in North America.

The quantities of biomass fired and co-fired by the electricity supply industry, particularly in Northern Europe, have increased dramatically over the past 10 years in response to the EC and member state government policies on renewable energies. This trend is also becoming more apparent worldwide as national governments are progressively introducing policy instruments aimed at the promotion of renewable energies in order to meet their international obligations to reduce CO_2 emission levels. A significant portion of the pelletised biomass materials produced worldwide is utilised in this market sector.

The principal technical options for the firing and co-firing of biomass materials in large pulverised coal fired boilers are described schematically in Figure 6.49.

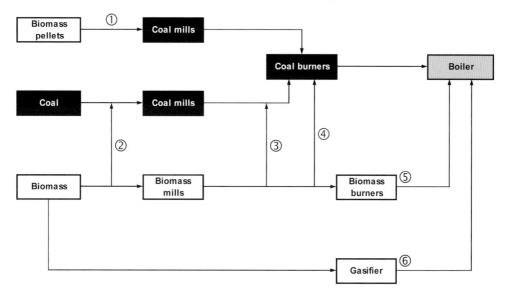

Figure 6.49: Biomass co-firing options at large pulverised coal fired power plants

Option 1 involves the milling of sawdust pellets on their own through the existing coal mills, after modification, and the combustion of the milled biomass through the existing firing system, with fairly minor modification, if required, as described in Section 6.5.2. This has been realised successfully in a small number of pulverised coal fired power plants in Northern Europe, and this approach is currently the subject of a number of feasibility studies in Europe and North America. The conversion of one or more of the mills in a boiler can be carried out, or all mills can be converted to provide 100% biomass firing.

Option 2 involves the pre-mixing of the biomass, generally in pelletised, granular or dust form, with coal in the coal handling system and at modest co-firing ratios, and the milling and firing of the mixed fuel through the existing coal firing system. This has been by far the most popular approach to co-firing as a retrofit project, as it can be implemented relatively quickly and with modest capital investment. As such, it has been the most popular option for power station operators who are embarking on biomass co-firing activities for the first time and whenever there are uncertainties associated with the security of supply of suitable biomass materials, with long-term security of the government subsidies or the other financial incentives that may be available for co-firing.

Options 3, 4 and 5 involve the direct injection of pre-milled biomass into the pulverised coal firing system, that is:

- into the pulverised coal pipework,
- into modified burners, or
- into dedicated biomass burners.

As described in more detail below, these options involve significant modifications to the installed equipment and significant levels of capital investment, but much higher co-firing ratios can be achieved than with Option 2. A number of coal fired power plants in Northern Europe have installed direct injection systems over recent years, and this is one of the more favoured options for the provision of biomass co-firing capabilities in newly built coal power plants.

Option 6 involves the gasification of the biomass, generally in chip or pellet form, in a dedicated unit that is normally based on a fluidised bed reactor, air blown and works at atmospheric pressure. The co-firing of the product syngas takes place in the pulverised coal boiler. The product gas may or may not be cleaned before firing it into the coal boiler, but the cooling and cleaning of the complex syngases produced in these systems has proved to be problematic. This approach to biomass co-firing has been adopted in a small number of plants in Northern Europe.

Overall, therefore, it is clear that a number of co-firing options are available for biomass materials, for both retrofit and newly built applications, depending on the fuels available for co-firing, and on the aspirations of the plant operator or developer. A number of these options have been implemented successfully as retrofit projects in existing pulverised coal fired boilers, mostly in Northern Europe.

A fairly wide variety of biomass materials in pelletised, chip/granular and dust forms has been utilised for co-firing in large pulverised coal fired boilers. There has, in recent years, been increasing interest in the utilisation of torrefied biomass materials for firing and co-firing in large coal boilers. Torrefaction is a low temperature thermal process that involves heating the raw biomass to temperatures in the range of 250 to 300°C at atmospheric pressure and in the absence of oxygen. The torrefied material is dry, brittle and hydrophobic (cf. Section 4.1.4.2).

The development of torrefied and pelletised material as a boiler fuel is currently in the pilot scale/demonstration phase and it is likely that this material will become increasingly available in the quantities relevant to use as a boiler fuel over the next few years.

6.5.2 The conversion of coal mills for processing sawdust pellets

The practical application of this option, i.e. Option 1 in Figure 6.49, has been demonstrated in a relatively small number of cases in Northern Europe. It has been shown that large, vertical spindle coal mills can be employed, with fairly modest modifications, to reduce dried and pelletised sawdust back to something close to the primary particle size distribution, and that the milled material can be fired successfully through the existing pulverised coal pipework systems and burners. There is no reported experience with large ball and tube coal mills.

Generally speaking, the modifications required to the milling equipment are associated with:

- the low bulk and particle densities of pellets and sawdust compared to raw coal and milled coal, and
- the high volatile content and high reactivity of the biomass. In general, the primary air temperatures when processing biomass pellets are much lower than those employed for coal for safety reasons.

Experience has shown that very little size reduction of the primary sawdust particles occurs in the mill, but provided that the product particle size distribution is suitable for combustion in pulverised fuel furnaces, this is a viable option for the firing or co-firing of biomass. There is successful, long-term experience with this approach in Scandinavia and elsewhere in Northern Europe, and it may be instructive to examine briefly one or two of the more important examples.

One of the key early applications was at Vasthamnsverket in Helsingborg, Sweden. This is a 200 MW_{th} pulverised fuel boiler, originally designed for the combustion of bituminous coals. The boiler was commissioned in 1983 and produced 82 kg/s of steam at 110 MPa and 540°C. The coal was pulverised in two large Loesche LM16.2D roller mills, with one spare mill. The original tangential firing system had three levels of coal nozzles, two levels of oil guns and one level of over-fire air nozzles.

During the period 1996 to 1998, the boiler was converted in a step-wise fashion to the firing of wood pellets, in such a way as to retain the capability to return to coal firing after a short outage. The objective was to provide at least 50%, and preferably 67%, of the heat input rate achievable with coal firing when milling the wood pellets.

The principal plant modifications included:

- The installation of new reception and covered storage facilities for the wood pellets;
- Modification of the fuel handling and bunkering system to handle the wood pellets. These modifications were mainly associated with concerns about the dust generation from pellet handling and with the associated explosion risk;
- The installation of a boiler flue gas recirculation system to reduce the oxygen concentration in the primary air supply to the mill, for mill safety reasons;
- The installation of a rotary valve on top of the mill for pellet milling to provide a seal between the mill and the bunker. This can be removed when milling coal;

- The installation of a new, adjustable louvre ring arrangement, that can be set at two positions for either pellet or coal milling;

- The installation of an adjustable inner return cone from the classifier which can be moved to two settings for pellet or coal milling.

In general, the experience at Vasthamnsverket with the milling of the wood pellets was fairly good, and the modified system was in successful commercial operation for several years.

The milling of biomass pellets, principally wood and straw, in conventional coal milling plants has also been successfully realised at Avedore Unit 2 in Denmark and in the new boiler at Amager Unit 1 in the Netherlands. In this case, the coal mills can be switched from coal to biomass pellet milling and vice versa within a day's outage. The mills are operated at a mill inlet temperature of 80 to 90°C, and the heat input from the mill group is reduced to around 70% of that achievable on coal when milling biomass pellets.

In Unit 9 at Amer Centrale in the Netherlands, two roller mills were converted in 2002/2003 to milling wood pellets, namely 300,000 tonnes of wood per annum per mill. For mill safety reasons and in order to maximise the pellet throughput, a number of internal mill modifications were made:

- An explosion detection and suppression system was installed on the mill and the associated pipework and ductwork.

- The orientation of the rollers was adjusted.

- Some holes of the upper parts of the rollers were closed.

- A baffle plate was installed in the upper part of the mill body.

This system has been in successful operation since that time.

At Hässelby in Sweden, a number of Doosan Babcock 6.3E9 mills were successfully converted by Doosan Babcock to the milling of wood pellets in the early 1990s. This system has been in commercial operation since 1993. At that time, the plant was firing around 270,000 tonnes of pellets per annum, and around 1,500 tonnes per day during the peak heating season. At best, the E mills can provide around 70% of the heat input rate from coal when firing wood pellets. Originally, the heat input to the furnace was balanced out by oil firing. More recently, hammer mills were installed to supplement the pellets being processed through the E mills already, and the plant can achieve the same full load as previously on coal when firing only wood pellets.

During the mill conversion by Doosan Babcock in the early 1990s, a number of relatively minor physical modifications of the 6.3E9 mills at Hässelby were made. The key modifications were:

- A rotary valve was installed in the coal feed pipe above the mill to provide an air seal, since it was considered that the low bulk density of the wood pellets would result in a poor sealing effect in the bunker.

- A number of baffles were inserted in the mill body above the throat to increase the local air velocities and reduce the tendency of partially milled pellets to accumulate in the mill.

- A dynamic discharge unit was installed at the outlet of the classifier return cone.

- No mill inerting, flue gas recirculation or other explosion prevention systems were installed, and the safe operation of the milling and firing system when processing wood pellets is maintained by close control of the primary air temperatures.

The wood pellet firing system in Hässelby is still in full commercial operation, with biomass pellet processing through both the coal mills via a direct firing system originally installed for coal firing, and the hammer mills, via a new bin and feeder system to the same burners.

Clearly the processing of wood pellets through conventional coal mills is a viable option for the conversion of pulverised coal fired boilers to the firing or co-firing of biomass. This option can be attractive to plant operators because of their familiarity with this type of milling equipment and the general concerns about hammer mills that are perceived as having relatively high maintenance requirements.

6.5.3 Co-firing biomass by pre-mixing with coal and co-milling

To date, the great majority of biomass co-firing in the coal-fired power plants in Northern Europe is realised by pre-mixing the biomass with the raw coal, normally in the existing coal handling and conveying system. The mixed fuel is then processed through the installed coal bunkers and mills and the installed pulverised coal firing equipment. This approach to co-firing is described as Option 2 in Figure 6.49.

The approach has been realised successfully in a large number of power stations and with a fairly wide range of biomass materials in kernel, granular, pellet and dust forms. Relatively dry biomass materials, with moisture contents of less than 20%, have been most popular for co-firing by this method, however, wet sawdust materials at moisture contents of approximately 50 to 60% have been co-fired successfully in this way.

The maximum achievable co-milling ratio and hence the level of co-firing without significant mill throughput constraints is limited and depends on the design of the coal mill, the nature of the biomass material and the plant operating regime. In most cases, the co-firing of the biomass materials up to around 10% on a heat input basis is possible, although co-firing ratios of approximately 5 to 8% are more common commercially.

Conventional coal mills generally break up the coal by a brittle fracture mechanism, but most biomass materials, including pelletised materials, tend to have relatively poor properties in this regard. There is a tendency, therefore, for the larger biomass particles to be retained within the coal mill to some extent, and this can limit the co-firing ratio achievable in this way. For instance, in vertical spindle coal mills, there may be a tendency for the primary air differential pressure and the mill power consumption to increase with increasing biomass co-firing ratio, and this may represent a limiting factor. There may also be an increase in the particle size of the mill product when co-milling biomass, due to the relatively low particle density of most biomass materials compared to coal particles. When very wet biomass materials are co-milled, there is a significant impact on the mill heat balance, and this can also be a limiting factor.

There will clearly be mill safety issues with the co-processing of biomass materials in most conventional coal mills where hot air is used to dry the coal in the mill. All biomass materials tend to release combustible volatile matter into the mill body at temperatures significantly lower than those that are applied when milling bituminous coals. It may be necessary, therefore, to modify the mill operating procedures to minimise the risks of overheating the coal-biomass mixture, thereby causing temperature and pressure excursions in the mill. The technical principles of the safe operation of conventional coal mills when co-processing biomass materials are well understood and have now been demonstrated successfully in a large number of power plants and with all of the most common types of coal mill.

Despite the potential difficulties and limitations, the co-milling and co-firing of a number of pelletised biomass materials as well as a wide range of chipped and granular materials through most of the more common designs of conventional large coal mills has been achieved successfully on a fully commercial basis in a number of coal fired power plants in Northern Europe.

6.5.4 Direct injection biomass co-firing systems

A number of the coal fired power stations in Northern Europe have installed systems for the direct injection co-firing of pre-milled biomass materials into large pulverised coal fired boilers, i.e. Options 3, 4 and 5 in Figure 6.49. All of these direct injection systems involve the by-passing of the installed coal mills and firing the pre-milled biomass material. This approach can allow operation at higher biomass co-firing ratios, potentially up to around 50% on a heat input basis. The design and operational experience with these systems in the UK and elsewhere provides the technical basis for the development of advanced co-firing systems for future retrofit and new projects.

All of the relevant technical approaches to direct injection co-firing involve the pre-milling of the biomass to a particle size distribution that will provide acceptable levels of combustion efficiency in a pulverised fuel flame, and all of the systems involve pneumatic conveying of the pre-milled biomass from the biomass handling/milling facilities to the boilers.

There are three basic direct co-firing options for the pre-milled biomass in retrofit applications, namely:

- The installation of new dedicated biomass burners with the appropriate fuel and combustion air supply systems (Option 5 in Figure 6.49);
- The injection of the biomass directly into the existing coal burners, after suitable modification (Option 4 in Figure 6.49);
- The injection of biomass into the pulverised coal pipework or at the burner, and co-firing with coal through the existing burners (Option 3 in Figure 6.49).

6.5.4.1 Dedicated biomass burners

In some circumstances, the installation of new burners dedicated to the co-firing of biomass materials as a retrofit in existing boiler plants may have some benefits. In most applications, it will be desirable to maintain the coal firing capability, which may mean that additional biomass burners are required. There will be a number of technical and commercial risk areas and significant problems to be resolved:

- New burner locations for the biomass firing, generally within the existing burner belt, may have to be identified, and significant new furnace penetrations may be required. This is expensive and it can prove difficult to find suitable locations for new burners without significant modification of the existing combustion air ductwork. A secondary air supply to the biomass burners is required, which means that significant modifications are required to the existing boiler draft plant to provide the air supply ductwork for the new biomass burners.

- The impacts of co-firing biomass through the new burners on the performance of the existing pulverised coal combustion system, and on furnace and boiler performance, may be significant, depending on the co-firing ratio and on the locations of the new burners. This is a significant risk area and will need to be assessed in some detail.

- The direct firing of biomass is relatively complex, both in terms of the mechanical interfaces and the control interfaces with the boiler, and is relatively expensive to install.

There are a number of biomass co-firing systems in Europe based on the installation of new, dedicated biomass burners, but it is fair to say that the accumulated plant experience with dedicated biomass burners is not extensive, and not all of the experience to date has been successful.

6.5.4.2   Direct injection through a modified coal burner

The direct injection of the pre-milled biomass into the existing coal burners in a wall-fired system may involve significant modification of the burners. This approach may be relatively expensive and may involve significant technical risks, but it may be necessary for some biomass materials such as cereal straws and similar materials in chopped form when there are concerns about the potential for blockage of the pulverised coal pipework system, the splitters and riffle boxes in particular and, if applicable, of the coal burners.

The cereal straw co-firing system at Studstrup Power Station in Denmark is an important example of such a system [367]. In this system the chopped straw is blown through the central core air tubes of modified Doosan Babcock Mark III Low NO_x coal burners, with the pulverised coal being fired as normal through the primary air annulus. The principal burner modifications included the relocation of the oil gun for ignition of the pulverised coal and of the flame monitor sighting tube.

6.5.4.3   Direct injection to the pulverised coal pipework

The principal alternative to injection of the biomass directly through the existing coal burners or through dedicated burners is to introduce the pre-milled biomass into the existing pulverised fuel pipework upstream the coal burners. In this case, the pulverised coal/biomass mixture is carried forward along the pulverised coal pipework and then enters the pulverised coal burners as normal. This type of approach is, in principle, equally applicable to all types of pulverised coal firing system and all burner designs.

Two potential locations for the introduction of the biomass into the pulverised coal pipework are apparent:

- The introduction of the biomass into the pulverised coal pipework just upstream of the non-return valves and next to the burners. This location is downstream the pulverised coal splitters, if there are any, and there will be one biomass delivery system for each coal burner.

- The introduction of the biomass into the mill outlet pipework and, if applicable, upstream the pulverised coal splitters. In this case there will be one biomass injection system for each mill outlet pipe.

The first of these options, i.e. the injection of the biomass stream next to the burner inlet, has a number of potential benefits:

- The point of introduction of the biomass into the pulverised fuel pipe and the associated shut-off valve, instrumentation, etc. will generally be readily accessible from the burner platforms for inspection and maintenance.

- The potential process risks associated with the introduction of a significant quantity of pre-milled biomass into the pulverised coal pipework are minimised by having the shortest possible length of pipework carrying the mixed fuel stream and avoiding the splitters in the coal pipes.

- The introduction point for the biomass into the pipework takes place at a significant distance from the coal mill, and hence the potential impacts of mill incidents and of mill vibration on the integrity and performance of the biomass conveying and injection system are reduced.

In many cases, however, the routing of the biomass pipework through the normally congested region close to the boiler front and the arrangements for supporting the biomass pipes can become complex and expensive. It should be noted also that the pulverised coal pipework next to the coal burners must move with the burners as the boiler furnace expands with increasing temperature, and sufficient flexibility of the biomass conveying pipework is essential to allow for this movement. This can add cost and complications.

For most applications, the second approach may be preferred, i.e. the introduction of the biomass stream into the mill outlet pipework just downstream the mill and upstream of any pulverised coal splitters. The mixed biomass/pulverised coal stream is then carried forward to the burners, via possible splitters in the pulverised coal pipework.

This approach is much easier to engineer and will generally be cheaper to install. In many cases, the number of biomass feeders and pneumatic conveying systems required will be significantly lower than for biomass injection downstream the splitters. The degree of movement of the pulverised coal pipework close to the mill is relatively low, and this will make the biomass injection pipework simpler to engineer.

The principal disadvantages of this system are the facts that the injection point for the biomass is closer to the coal mill, which means that there are greater risks of interference with the pulverised coal transport system and particularly at the splitters, and the impact of any mill incident on the biomass conveying system may be greater. There may also be relatively poor access to the point of introduction of the biomass for inspection and maintenance, depending on the details of the pipework layout.

In all cases, the introduction point of the biomass to the pulverised coal pipework or directly to the burner is fitted with a fast acting, actuated biomass isolation valve that allows rapid automatic isolation of the biomass system from the coal mill and the coal firing system.

If the system is engineered properly, there are a number of important advantages of the direct injection systems with biomass injection into the pulverised coal pipework, as described above, namely:

- There are no requirements for significant physical modifications of the existing coal mill, the boiler draft plant, the pulverised fuel pipework, the coal burners, etc.

- The boiler and mills are started up on coal firing as normal, and the biomass co-firing system does not start until all of the combustion and boiler systems are functioning properly.
- If there are any problems with the functioning of the biomass co-firing system on a mill group, the biomass system can be turned off quickly and in an isolated way, and the mill concerned automatically switches to coal and so boiler load can be maintained.
- If there are problems with the coal mill, e.g. coal feeder problems, a fire in the mill, etc. the biomass co-firing system can be turned off quickly and in an isolated way until the problem is resolved.
- The biomass feeder control system only communicates with the mill controls, that is to say that it is an add-on to the normal boiler, mill and burner controls, and the appropriate safety interlocks are well understood and have been successfully demonstrated in practice.
- The milled biomass is co-fired with the coal through the coal burners at up to 50% heat input, which means there are fewer risks of problems associated with combustion efficiency, burnout, flame shape and furnace heat transfer factors, etc. than for the firing of the biomass alone through dedicated burners.
- The biomass combustion is always supported by a stable pulverised coal flame. This will help to optimise the combustion efficiency of the biomass and may increase the range of biomass types and qualities that can be co-fired in this way.
- The products of combustion of biomass are always well mixed with those from coal. This means that the risks associated with striated flows in furnaces and boilers producing localised deposition and corrosion effects are minimised due to the low concentration of the products of biomass combustion.
- In the latest direct injection biomass co-firing systems, fully automatic control of the biomass feed rate allows for the same turndown capabilities of the mills that have been converted to biomass co-firing as before.

For both newly built and retrofit applications, where elevated co-firing ratios are desired, co-firing biomass will have to take place in a number of mill groups. The potential impacts on the mills and the boiler will depend largely on the nature of the biomass, the target co-firing ratio and the operational regime of the boiler plant, and this will generally need very careful consideration. In principle, the co-firing of biomass up to a co-firing ratio of 50% or so, on a heat input basis, may be possible using the direct injection method, but the range of biomass materials that can be co-fired at this ratio is limited. The possible co-firing ratio will in most cases be determined by the biomass ash content, the quality of the biomass and coal ashes, and most of all by the tendency of the mixed ashes to form troublesome deposits on boiler surfaces.

Overall, it is clear from the descriptions presented above that there are a number of viable technical options for the direct injection co-firing of pre-milled biomass materials as a retrofit to coal-fired power stations. The preferred technical option for any particular application will depend on a number of factors, these are:

- the types of biomass to be co-fired,
- the desired co-firing ratio,

- the site-specific factors, i.e. the types of coal mill, the arrangement of the installed coal firing systems, etc.
- the plant operating regime and the aspirations of the station engineers.

A number of these direct injection biomass co-firing systems have been in successful commercial operation in Northern European countries for a number of years. The design and capabilities of the direct injection systems are still being developed. For instance, one of the most advanced direct injection biomass co-firing retrofit projects currently being built in the UK will involve fully automatic control of the biomass feed rate in response to unit demand, to reinstate the turndown capability of the converted coal mills when co-firing biomass [368].

6.5.4.4   *Gasification of the raw biomass with co-firing the syngas*

The gasification of the biomass, generally in chip or pellet form, and the combustion of the product gas in the coal fired boiler, as described under Option 6 in Figure 6.49, has been applied in a couple of cases in Northern Europe. To date, there has been only limited interest in the wider replication of this approach to biomass co-firing.

6.5.5   The impacts of biomass firing and co-firing on boiler performance

The biomass firing and co-firing retrofit projects carried out to date, principally in Northern Europe, have shown that, provided the size distribution of the material supplied to the burners was acceptable (particle size of not more than around 1 mm), the combustion behaviour proved to be acceptable too, both in terms of combustion efficiency and CO emission levels. This was the case for co-firing systems involving both pre-mixing the biomass with the coal and direct injection co-firing. Biomass materials are much more reactive in combustion systems than are coals, but they do not require particle size reduction to the same level as pulverised coal.

In cases where the biomass pellets are being milled in modified coal mills prior to combustion in a pulverised fuel flame, it is common practice for the fuel purchaser to specify the primary particle size of the sawdust that is to be used to prepare the pellets, since the modified coal mill can at best only reduce the pellets back to the size of the sawdust again. A maximum particle size of around 2 mm is most common. The presence of oversize biomass particles will increase the levels of unburned materials in the furnace bottom ashes and boiler fly ashes.

In the appliances in Northern Europe, no significant changes to the basic flame shapes or the furnace heat absorption have been observed when co-firing biomass with coal, or firing 100% biomass through dedicated burners, and in the great majority of cases there was no requirement for any significant boiler modifications to permit biomass firing or co-firing.

In general terms, the non-combustion-related impacts of firing or co-firing biomass on boiler performance are associated mainly with the inorganic components of biomass and coal, particularly the ash content and ash quality, the sulphur and chlorine contents and the trace element contents. All in all, the risks of significant impacts on the boiler are apparent from the fuel specifications. The plant experience in Northern Europe indicates that the risks of excessive ash deposition on the boiler surfaces are controlled largely by the co-firing ratio, and the ash content and ash composition of the coal and biomass fuels. Usually, high grade wood pellet materials with low ash contents and modest levels of alkali metals present relatively low risks in this regard, even when fired at 100% in a coal fired boiler. At lower co-firing ratios, biomass fuels with higher ash contents and more problematic ashes can be co-fired successfully. In general, methods for the assessment of the slagging and fouling potential of coal ashes can also be employed, with some modifications, for the assessment of biomass ashes and of the mixed ashes produced by co-firing biomass with coal.

The flue gas and ash deposits from biomass firing and co-firing with coal may be more aggressive than those from coal firing with respect to their potential to cause accelerated metal wastage due to high temperature corrosion of superheater and reheater surfaces. As above, these potential risk areas can be controlled by careful consideration of the biomass quality specification, particularly in terms of the ash, sulphur and chlorine contents, the ash composition and the boiler conditions. The boiler conditions principally concern the boiler tube materials and the gas and metal temperatures.

Primary NO_x and SO_2 concentration levels when firing and co-firing most biomass materials will be lower than those when firing coal and will depend largely on the nitrogen and sulphur contents of the biomass fuel.

The characteristics of the solid products of the combustion of biomass materials are very different from those from the combustion of coal. The chemistries of the ashes are very different, as are their physical characteristics. The nature of the inorganic material in most biomass materials is such that a significant amount of sub-micron particles can be generated in the flame.

The firing and co-firing of biomass materials will generally lead to a reduction in the total fly ash dust burden compared to firing most coals due to the lower ash content of the biomass. However, the biomass ash may contain significant levels of very fine aerosol material that may present problems to conventional particulate matter precipitation systems. When electrostatic precipitators are employed, there may be an increase in the particulate emissions level from the chimney compared to the level when firing coal alone. This effect is likely to be highly site specific and will be of particular interest in retrofit projects, where the existing electrostatic precipitators were designed for coal firing alone. When fabric filters are employed for particulate collection, there may be a tendency for the very fine aerosol material to blind the fabric, resulting in difficulties in cleaning and an increased pressure drop across the system. These potential problems with the particulate emissions precipitation equipment will be dependent on the co-firing ratio, the ash content and ash composition of the biomass and a number of other site specific factors. It is generally wise to consult the supplier of the precipitation system on these matters.

The utilisation of the mixed ashes produced by the co-firing of biomass materials with coal by the cement industry is covered in BS EN 450 (2005), which was specifically modified to include the ashes from the co-firing of biomass. In most cases where the coal ashes are destined for disposal on land, the normal disposal routes are suitable for the mixed ashes from biomass co-firing as well. In general terms, the trace element and heavy metals contents of most clean biomass materials tend to be lower than those of most coals.

6.6 Summary/conclusions

In some countries, such as Austria, Germany or Italy, the primary focus lies on the utilisation of pellets in small-scale systems for residential heating in the power range of up to 100 kW_{th}. However, the use of pellets in medium-scale systems with nominal power outputs of between

100 kW$_{th}$ and 1 MW$_{th}$ is of rising significance in Austria and Germany for instance, and is well established in Sweden or Denmark. Some manufacturers already offer pellet furnaces especially adapted to this power range. Large-scale pellet applications with a power range of more than 1 MW$_{th}$ are relevant in countries such as Sweden, Denmark, Belgium and the Netherlands.

Within the area of small-scale systems, the main focus lies on pellet central heating systems (except in Italy and the USA). In this sector, high standards have been achieved in recent years with special regard to automation degree, ease of use, emission reduction and efficiency improvement. Staged air supply, micro processor control, automatic cleaning systems for the heat exchanger and automatic de-ashing systems are state-of-the-art. Looking at systems to convey the pellets from the storage space to the furnace, feeding screws and pneumatic systems are the two essentially different technologies that are available. The conveyor systems have achieved a high standard that ensures safe and trouble free system operation. Proven burn-back prevention by means of rotary valves, fireproof valves, self-initiating fire extinguishers as well as combinations of these effectively avoid fire burn-back into the storage space. Austrian furnace manufacturers play a leading role in this field and export their products to many countries worldwide.

Stoves are not an insignificant area of interest either. Here, automatic operation for a few hours to a few days has by now been achieved by means of appropriate micro processor controls and integrated pellet reservoirs. Important pellet stove markets exist in Italy and the USA.

The development and optimisation of both pellet stoves and boilers are increasingly supported by CFD simulations.

The retrofitting of existing gas or oil boilers by exchange of the gas or oil burner with pellet burners is very common in Sweden.

Massive development work is in progress within all power ranges, which is leading to innovative concepts. Flue gas condensation was recently introduced in the area of small-scale systems. The newest developments focus on small-scale furnaces with very low nominal thermal capacities in order to meet the trend towards low energy housing. Within medium-scale systems, the use of wood chip furnaces that may be fired with pellets is possible when appropriate control systems are in place. In addition, there are developments for combined wood chip and pellet furnaces in place.

The use of pellets in decentralised CHP plants is rare. Such applications are only in Scandinavian countries. However, the area will be of interest in the future as soon as small-scale biomass CHP systems are developed. Stirling engine and ORC process are interesting technologies in this respect because they are the most developed to date.

In the area of large-scale systems, it is mainly combustion and co-firing in coal fired power and CHP plants that is relevant. In general terms, the retrofitting of existing pulverised coal fired boilers with the capability to co-fire a range of biomass materials in granular or pelletised form has been reasonably successful, and there have been relatively few problems, provided that technical and other issues involved are properly addressed at the design stage. To date, the majority of the biomass co-firing activity in Europe pre-mixes the biomass with the coal in the coal handling system, and processes the mixed fuel through the installed coal mills and firing equipment. This approach permits co-firing ratios of up to around 10% on a heat input basis, and at this level the impacts of the co-firing of most biomass materials on plant operation and performance are modest. A number of recent projects involved the installation of more advanced systems with direct injection co-firing of pre-milled biomass materials, which allows operation at higher co-firing ratios. All of these systems involve the pneumatic conveying of the pre-milled biomass to the boiler and injection into the pulverised coal pipework, into modified coal burners or into dedicated biomass burners. A number of these systems are now in commercial operation, but it is fair to say that long-term operational experience with these systems is limited. The available options for the direct injection co-firing of biomass are described and discussed in some detail above. The preferred approach in most applications involves the direct injection of the pre-milled biomass into the pulverised coal pipework, and a number of such systems are in commercial operation in Europe. It is assumed that the method can be replicated in a number of coal mill groups in a large pulverised coal boiler. In principle, this approach could permit biomass co-firing ratios of up to 50% on a heat input basis to be achieved, but only with high quality pelletised materials with low ash contents. In a small number of cases, sawdust pellets were successfully processed in modified coal mills and fired through modified burners at up to 100% on a heat input basis in plants originally designed for coal firing. The co-firing of biomass by gasification of the raw biomass and just co-firing the product syngas with coal in a large coal-fired boiler was also carried out successfully in a couple of plants in Europe. The key technical issues are associated with the cooling and cleaning of the product syngas prior to co-firing. Overall, therefore, it is clear that biomass co-firing in existing coal fired boilers has become reasonably well established, particularly in Northern Europe, and that it represents an attractive means to utilise a fairly wide range of biomass materials for power generation. This is now widely recognised and the trend towards co-firing in both existing and new coal power plants is becoming more apparent on a worldwide basis.

7 Cost analysis of pellet production

In this section, a cost analysis of a typical pellet production plant with an annual pellet production capacity of approximately 40,000 tonnes using wet sawdust as a raw material is carried out under Austrian framework conditions. It should be pointed out that the results of the calculation cannot be transferred directly to a specific project in another region or country, as specific framework conditions might differ significantly. However, it can be used as a guide and gives an indication as to how the calculation should be made.

7.1 Cost calculation methodology (VDI 2067)

According to the full cost calculation based on the guideline VDI 2067, the different types of costs are divided into four cost groups. These are:
- costs based on capital (capital and maintenance costs);
- consumption costs;
- operating costs;
- other costs.

Costs based on capital consist of the annual capital and maintenance costs. The annuity (annual capital costs) can be calculated by multiplying the capital recovery factor (CRF) (cf. Equation 7.1) with the investment costs. The capital and maintenance costs are calculated for each unit of the overall pelletisation plant, taking the different wear and utilisation periods into account. Total capital and maintenance costs can be calculated by summation of these subtotals.

Equation 7.1: $$CRF = \frac{(1+i)^n \cdot i}{(1+i)^n - 1}$$

Explanations: CRF...capital recovery factor; i...real interest rate [% p.a.]; n...utilisation period [a]

Maintenance costs are calculated as a percentage of the whole investment costs on the basis of guiding values and are evenly spread over the years of the utilisation period.

All costs in connection with the manufacturing process, for example the costs of the raw material, the heat for drying and the electricity demand, are included in the group of consumption costs.

The operating costs comprise costs originating from the operation of the plant, for example personnel costs.

Other costs include insurance rates, overall dues, taxes and administration costs and are calculated as a percentage of the overall investment costs.

7.2 Economic evaluation of a state-of-the-art pellet production plant

7.2.1 General framework conditions

A cost calculation was carried out according to the full cost method of VDI 2067 for all steps of the total pelletisation process, i.e. drying, grinding, pelletisation, cooling, storage and peripheral equipment (cf. Sections 7.2.3 to 7.2.7). The personnel costs as well as the construction costs were not calculated for each step but for the whole plant (cf. Sections 7.2.2 and 7.2.8). Construction costs were directly integrated in the investment costs for storage facilities only. Finally, the raw material costs were calculated, as exemplified in Section 7.2.9. The framework conditions that are generally valid for the calculation of the base case scenario are shown in Table 7.1. They are discussed in detail in the following sections.

Table 7.1: General framework conditions for the calculation of the pellet production costs for the base case scenario

Explanations: [1]...selected according to the general trend towards continuous operation; [2]...based on information from pellet producers; [3]...average price of electricity in medium-sized enterprises (price basis 12/2008); [4]...according to VDI 2067; [5]...internal calculation guidelines; [6]...percentage of total investment costs for the plant

Parameter	Value	Unit
Number of shifts per day[1]	3	
Working days per week[1]	7	
Plant availability[2]	91	%
Annual full load operating hours	8,000	h/a
Simultaneity factor (electrical installations)[2]	85	%
Throughput (output pellets)	5.0	t (w.b.)$_p$/h
Price for electricity[3]	100	€/MWh
Utilisation period construction[4]	50	a
Service and maintenance costs construction[4]	1	% p.a.
Utilisation period infrastructure[5]	15	a
Service and maintenance costs infrastructure[5]	1	% p.a.
Utilisation period planning[5]	20	a
Interest rate[5]	6	% p.a.
Other costs (insurance, administration, etc.)[2][6]	2.8	% p.a.

The number of 8,000 annual full load operating hours is based on the assumption of a continuous plant operation on seven days per week and 24 hours per day (three shift operation), corresponding to a plant availability of 91.3%. This follows the present trend of continuously operated pellet production plants and it is an important criterion as concerns economic efficiency of the plant (cf. Section 7.2.12). The availability of 91% is based on practical experience of pellet producers and it is an achievable and realistic value.

Based on the chosen throughput of 5 t (w.b.)$_p$/h, around 40,000 tonnes of pellets are produced in one year. An average price for electricity for medium-sized enterprises in Austria amounts to about 100 €/MWh (price basis 2008). The electricity price for small-sized enterprises may be higher. The simultaneity factor for electricity demand (= electric power needed on average/nominal electric power of all units × 100) was assumed to be 85% and is based on experiences of plant operators. The selected interest rate is an average real interest rate currently achievable. The utilisation period and the maintenance costs were settled according to the guideline VDI 2067. Other costs were settled according to the experience of pellet producers and take insurance rates and administration costs into account. Utilisation periods as well as maintenance costs of infrastructure and planning were calculated on the basis of internal calculation guidelines that are also based on experience.

7.2.2 General investments

General investments subsume investments in construction, infrastructure and planning for the whole plant. Full costing of general investments of a pellet production plant is presented in Table 7.2.

It shows that specific general costs are around 1.6 €/t (w.b.)$_p$ which is equivalent to 5.7% of production and distribution costs of pellets (162.0 €/t (w.b.)$_p$ as per Table 7.20). Utilisation periods and maintenance costs were settled according to Table 7.1. Planning costs for the whole plant were settled at 10% of the investment costs. They dominate the general investment costs. In this context it has to be mentioned, that building costs are just a small share of specific production costs. This is because the utilisation period of construction is quite high at 50 years, while maintenance costs are quite low with 1% p.a. of the investment costs.

Table 7.2: Calculation of full costs for general investments of a pellet production plant

Explanations: *...related to the whole plant; framework conditions as per Table 7.1; data source: data from plants in Austria

	Investment costs €	Capital costs € p.a.	Maintenance costs € p.a.	Operating costs € p.a.	Other costs € p.a.	Total costs € p.a.	Specific costs €/t (w.b.)$_p$
Construction	140,000	8,882	1,400			10,282	0.3
Infrastructure	90,000	9,267	900			10,167	0.3
Planning*	340,300	29,669				29,669	0.7
Other costs					15,854	15,854	0.4
Total costs	570,300	47,818	2,300	0	15,854	65,972	1.6

7.2.3 Drying

Raw materials that come into consideration in pelletisation in Austria that need to be dried are shown in Table 7.3. The moisture contents of the raw materials as well as the moisture contents that have to be achieved by drying were taken from Section 3.4, whereby an average value was chosen for the moisture content of bark and forest wood chips. Sawdust and industrial wood chips out of the sawmill industry are usually available with a moisture content of 55 wt.% (w.b.) loco sawmill, which is why the higher value was chosen for calculation. The most important raw material for pellet production plants in Austria and many other countries is sawdust. Therefore, sawdust was selected as the basis for the calculation of the base case scenario.

Sawdust and industrial wood chips from the wood processing industry have significantly lower moisture contents, as the industry chiefly uses dried sawn timber as a raw material. In most cases, a water content of below 10 wt.% (w.b.) can be expected, which is why the material is usable in pelletisation without upstream drying. Even though straw and whole crops do not play any role in pelletisation in Austria, they could be used without drying as their moisture content is 15 wt.% (w.b.) on average.

Short rotation crops are similar to industrial wood chips with regard to moisture content.

Table 7.3: Moisture contents before and after drying of different raw materials for pelletisation

Explanations: *...in accordance with the usual supply chains

Raw material	Moisture content before drying wt.% (w.b.)	Moisture content after drying wt.% (w.b.)
Sawdust or industrial wood chips from the sawmill industry, short rotation forestry	55.0	10.0
Forest wood chips*	30.0	10.0
Bark	55.0	18.0

Technologies for drying were examined in Section 4.1.1.2. Costs for drying were calculated based on a full cost calculation for a belt dryer and sawdust as the raw material (and the values as in Table 7.3). Throughput was settled on the basis of dryer output to be 5 t/h with 8,000 annual operating hours. Framework conditions are presented in Table 7.4 and the full cost calculation in Table 7.5.

Table 7.4: Framework conditions for full cost calculation of drying in a belt dryer

Explanations: moisture contents as in Table 7.3; [1]...percentage per year of total investment costs of the drying system; [2]...heat price for hot water (90°C); data source: manufacturers, own research and calculations, and data from plants in Austria

Parameter	Value	Unit
Electric power demand	140	kW
Heat demand for drying (per ton evaporated water)	1,200	$kWh/t_{ev.w.}$
Utilisation period	15	a
Service and maintenance costs[1]	2.4	% p.a.
Specific heat costs[2]	35	€/MWh
Electricity consumption	952,000	kWh/a
Moisture content before drying	55	wt.% (w.b.)
Moisture content after drying	10	wt.% (w.b.)
Water evaporation rate	5.0	t/h
Heat demand for drying	48.0	GWh/a

The electric power demand of the dryer consists of the electric drives of the belt, the suction fan for exhaust vapours and a series of auxiliary units (control system, belt cleaning system, feeding screw for even distribution of the material on the belt) and is based on information from dryer manufacturers. The specific heat demand for drying, which is quantified based on tonnes of evaporated water, is also based on information from dryer manufacturers. Utilisation periods and maintenance costs are based on information from pellet producers. Water with a feed temperature of 90°C is required for operating the belt dryer. The specific heat price for that is based on the supply by a biomass fired hot water boiler. Electricity demand, amount of water to be evaporated as well as the heat demand for drying are calculated by means of the framework conditions presented in Table 7.3 and Table 7.1.

The specific drying costs of a belt dryer, which were calculated on basis of the full costing method presented in Table 7.5, are 48.1 €/t (w.b.)$_p$. The consumption costs are dominant when looking at drying costs and mainly consist of heat costs.

Thermal energy consumption for drying is around 24.5%, and electric energy consumption is around 0.5%, as based on the NCV of pellets.

Table 7.5: Full cost calculation of a belt dryer

Explanations: framework conditions as in Table 7.4; data source: manufacturers and data from plants in Austria

	Investment costs excl. Construction €	Capital costs € p.a.	Maintenance costs € p.a.	Consumption costs € p.a.	Operating costs € p.a.	Other costs € p.a.	Total costs € p.a.	Specific costs €/t (w.b.)$_p$
Dryer	950,000	97,815	23,180				120,995	3.0
Electicity costs				95,200			95,200	2.4
Heat costs				1,680,000			1,680,000	42.0
Other costs						26,410	26,410	0.7
Total costs	950,000	97,815	23,180	1,775,200	0	26,410	1,922,605	48.1

In addition to the belt dryer, there are other technologies available for drying. Tube bundle dryers and superheated steam dryers are of particular relevance (cf. Section 4.1.1.3). Detailed economic evaluation of these systems is not undertaken here, but the following framework conditions should be considered when one of these two technologies is to be employed:

- Tube bundle dryers have slightly lower and superheated steam dryers considerably higher investment costs than belt dryers.
- Both drying technologies require heating media of a raised temperature level (hot water, steam, thermal oil), and thus are more expensive. Using saturated steam for instance (e.g. saturated steam of 16 bar and 201°C can be used in both tube bundle and superheated steam dryers), heat costs of at least 40 €/MWh must be expected (this is when the heat is produced by a biomass steam boiler of one's own; if heat has to be bought, the price is even higher). Whether the investment costs that are lower than those of the belt dryer can balance out the increased heat costs for saturated steam and thus render the use of a tube bundle dryer more economical, has to be evaluated on a case-by-case basis.
- The main benefit of the superheated steam dryer is that up to 95% of the heat input can be recovered as superheated steam with two to five bar. Provided that the recovered heat can be utilised appropriately, it can be sold, which renders the superheated steam dryer attractive with regards to its economy.

7.2.4 Grinding

Pellet producers and pellet mill manufacturers make different claims as to the grinding of raw materials [147; 369; 370]. The use of wood chips, bark, straw and whole crops causes grinding to be essential. If sawdust is used exclusively, grinding does not need to be carried out. If sawdust with small amounts of wood shavings is used, grinding may also not be necessary under certain conditions. From a set amount of wood shavings onwards, grinding becomes vital, yet, many producers put sawdust into the hammer mill as well in order to make the product even more homogenous. As a rule, grinding is carried out by means of hammer mills. They are not suitable for grinding bark, however, which is why cutting mills or especially adapted hammer mills are used for grinding bark.

Table 7.6 presents the framework conditions for full cost calculation of grinding sawdust in a hammer mill. The calculation is shown in Table 7.7. The electric power demand given is the power demand of the main drive of the hammer mill. The utilisation period as well as maintenance costs are based on information from pellet producers.

Table 7.6: Framework conditions for full cost calculation of raw material grinding in a hammer mill

Explanations: [1]...percentage per year of total investment costs of the drying system; data source: manufacturers, own research and calculations, and data from plants in Austria

Parameter	Value	Unit
Electric power demand	110.0	kW
Utilisation period	15	a
Service and maintenance costs[1]	2.4	% p.a.
Electricity consumption	748,000	kWh/a

The costs for grinding are around 2.70 €/t (w.b.)$_p$, with costs for electricity being the greatest contributor. The energy expense per tonne of pellets is around 18.7 kWh or 0.38% of the energy content of the pellets (4,900 kWh/t (w.b.)$_p$).

Table 7.7: Full cost calculation of grinding in a hammer mill

Explanations: framework conditions as in Table 7.6; data source: manufacturers and data from plants in Austria

	Investment costs excl. construction	Capital costs	Maintenance costs	Consumption cost	Operating costs	Other costs	Total costs	Specific costs
	€	€ p.a.	€ p.a.	€ p.a.	€ p.a.	€ p.a.	€ p.a.	€/t (w.b.)$_p$
Hammermill	206,000	21,210	5,026				26,237	0.7
Electricity costs				74,800			74,800	1.9
Other costs						5,727	5,727	0.1
Total costs	206,000	21,210	5,026	74,800	0	5,727	106,764	2.7

7.2.5 Pelletisation

Framework conditions and full cost calculation of the pelletisation process based on ring die technology are shown in Table 7.8 and Table 7.9. The nominal electricity demand includes the demands of the main drive, the driving motor for the feeding of raw material and the mixing screw for hot water conditioning. Consumption of hot water was estimated on the basis of pellet producer's information and is negligibly low. Costs for biological additives are also derived from pellet producer's information, whereby according to prEN 14961-2, the amount of added biological additives is allowed to be not more than 2.0 wt.% (w.b.). Utilisation period as well as maintenance costs are also based on information from pellet producers. Maintenance costs arise mainly from costs for rollers and the die, which become worn. Rollers usually exhibit shorter lifetimes than dies.

Table 7.8: Framework conditions for full cost calculation of a pellet mill

Explanations: [1]...percentage per year of total investment cost of the pellet mill; [2]...price for hot water (90°C); data source: manufacturers, own research and calculations, and data from plants in Austria

Parameter	Value	Unit
Electric power demand	300.0	kW
Hot water demand for conditioning related to tons of pellets produced	1.0	%
Specific heat costs[2]	2.70	€/t
Costs for additives per ton of pellets produced	2.25	€/t (w.b.)$_p$
Utilisation period	15	a
Service and maintenance costs[1]	2.4	% p.a.
Electricity consumption	2,040,000	kWh/a

The investment costs for a pellet mill as in Table 7.9 are independent of the raw material used, because the raw material is prepared for pelletisation by the preceding process steps. Hence it already possesses the needed structure for pelletisation. The raw material does have an influence on the energy expense however. Hardwood like beech and oak calls for stronger compression forces, which increases the specific energy consumption. Investment costs of the pelletisation plant include not only the costs for the pellet mill itself but also costs for the control system and mounting of the plant as well as fittings and fixtures.

Full cost calculation of pelletisation of dry shavings or appropriately prepared raw materials yields specific costs of around 9.2 €/t (w.b.)$_p$ which is equivalent to 5.7% of production and distribution costs of pellets (162.0 €/t (w.b.)$_p$, as per Table 7.20). Energy demand is around 38 kWh/t (w.b.)$_p$ or 1.0% of the energy content of pellets (4,900 kWh/t (w.b.)$_p$). Electricity costs dominate the overall pelletisation costs, with costs for biological additives coming second. Capital bound costs of 16.2% out of total pelletisation costs are also significant.

Table 7.9: Full cost calculation of a pellet mill

Explanations: framework conditions as in Table 7.8; data source: manufacturers and data from plants in Austria

	Investment costs excl. construction €	Capital costs € p.a.	Maintenance costs € p.a.	Consumption costs € p.a.	Operating costs € p.a.	Other costs € p.a.	Total costs € p.a.	Specific costs €/t (w.b.)$_p$
Pellet mill	467,000	48,084	11,395				59,478	1.5
Electricity costs				204,000			204,000	5.1
Conditioning costs (hot water)				1,080			1,080	0.0
Additive costs				90,000			90,000	2.3
Other costs						12,983	12,983	0.3
Total costs	467,000	48,084	11,395	295,080	0	12,983	367,541	9.2

7.2.6 Cooling

Pellets coming out of the pellet mill exhibit relatively high temperature levels of up to 100°C and possibly even more due to heating up inside the pellet mill and upstream conditioning.

The temperatures vary according to the type of pre-treatment and the pelletisation technology employed. Pellets must be cooled accordingly. Cooling also makes the pellets more stable. Usually a counterflow cooler is used.

Table 7.10: Framework conditions for full cost calculation of cooling in a counterflow cooler

Explanations: [1]...percentage per year of total investment costs of the cooler; data source: manufacturers, own research and calculations, and data from plants in Austria

Parameter	Value	Unit
Electric power demand	12.0	kW
Utilisation period	15	a
Service and maintenance costs[1]	2.4	% p.a.
Electricity consumption	81,600	kWh/a

A full cost calculation of a counterflow cooler (cf. Section 4.1.3.1) was carried out in order to obtain cooling costs as shown in Table 7.10 and Table 7.11. The electricity demand includes expenses of the fan that is needed for transmitting the air for cooling as well as the conveyor system into and out of the cooler. Utilisation period and maintenance costs are based on information from pellet producers.

Investment costs include not only the costs for the cooler but also the costs for the fan as well as the costs for the cyclone needed to precipitate fines.

Cooling costs are 0.33 €/t (w.b.)$_p$ or just 0.2% out of total production and distribution costs (162.0 €/t (w.b.)$_p$, as per Table 7.20). Hence, costs for cooling are negligible (cf. Table 7.11). Energy demand is around 1.5 kWh/t (w.b.)$_p$ or 0.04% of the energy content of pellets (4,900 kWh/t (w.b.)$_p$) and thus is also negligible.

Table 7.11: Full cost calculation of a counterflow cooler

Explanations: framework conditions as in Table 7.10; data source: manufacturers and data from plants in Austria

	Investment costs excl. construction €	Capital costs € p.a.	Maintenance costs € p.a.	Consumption costs € p.a.	Operating costs € p.a.	Other costs € p.a.	Total costs € p.a.	Specific costs €/t (w.b.)$_p$
Counterflow cooler	32,000	3,295	781				4,076	0.10
Electricity costs				8,160			8,160	0.20
Other costs						890	890	0.02
Total costs	32,000	3,295	781	8,160	0	890	13,125	0.33

7.2.7 Storage and peripheral equipment

In pellet production, raw materials and pellets have to be stored appropriately. Framework conditions and the full cost calculation of the pellet and raw material storage at the producer's site are shown in Table 7.12 and Table 7.13, respectively.

Wet raw material (sawdust) that has been delivered is stored in a paved outdoor storage. As mentioned in Section 4.2.1, storage time of wet sawdust should not exceed 2 to 3 days in order to avoid physical, chemical and biological processes setting in that could lead to degradation of the dry substance. In order to ensure a certain degree of flexibility, a storage capacity of one week in an outdoor storage space of about 1.200 m² was assumed. Supposing an average storage height of 5 m, the storage capacity for wet sawdust is around 5,800 m³.

Intermediate storage of dried raw material takes place in silos with a total storage capacity of around 1,200 m³. This is equivalent to a storage capacity of 36 h and can hence serve to overcome short outages only. Since the drying as well as the pelletisation process are run on a 7 days a week and on a 24 h a day basis, greater storage capacities are not required. However, the less storage capacity is in place, the more important become transport logistics.

The pellets produced are stored in silos with a total storage capacity of 1,500 m³ or 8 days. This gives a certain degree of flexibility with regards to logistics and also signifies reasonable investment costs. The utilisation period and maintenance costs of the outdoor storage were settled according to the VDI 2067 guideline (construction). The corresponding values for the silo storage are based on internal calculation guidelines and experience.

Table 7.12: Framework conditions for full cost calculation of raw material and pellet storage at the producer's site

Explanations: [1]... percentage per year of total investment costs of storage; [2]...price basis 2008, delivered quantity of 6 t; data source: own research and calculations and data from plants in Austria

Parameter	Value	Unit
Outdoor storage before drying		
Utilisation period	50	a
Service and maintenance costs[1]	1	% p.a.
Storage capacity (in % of annual raw material demand)	1.92	%
Silo storage after drying		
Utilisation period (silo 15 years, construction 50 years)	22.29	a
Service and maintenance costs[1]	1.5	% p.a.
Storage capacity (in % of annual raw material demand)	0.41	%
Pellet storage		
Utilisation period (silo 15 years, construction 50 years)	22.19	a
Service and maintenance costs[1]	1.5	% p.a.
Storage capacity (in % of annual amount of pellet production)	2.3	%
Pellet sales price (excl. 10 % VAT and transport costs)[2]	162.84	€/t (w.b.)$_p$
Average storage filling level (in % of the storage capacity)	50	%
Average days sales outstanding	14	d

The investment costs as shown in Table 7.13 include the costs for the pavement of the outdoor storage area and the investment costs of the silos plus construction, fines precipitation and investment costs for silo discharge and loading systems.

As can be deduced from the full cost calculation above, storage costs for outdoor storage are negligibly low, whereas silo storage forms the greatest part of the storage costs. Capital costs and maintenance costs are thereby the highest. Imputed interest of pellet and raw material storage is relatively low. So are days sales outstanding (due to the gap between pellet delivery and incoming payment – two weeks after delivery on average). According to full cost calculation, storage costs are 3.8 €/t (w.b.)$_p$ in total or 2.3% of total production and distribution costs (162.0 €/t (w.b.)$_p$, as per Table 7.20).

Table 7.13: Full cost calculation of raw material and pellet storage at the producer's site

Explanations: framework conditions as in Table 7.12; data source: manufacturers and data from plants in Austria

	Investment costs excl. construction €	Capital costs € p.a.	Maintenance costs € p.a.	Consumption costs € p.a.	Operating costs € p.a.	Other costs € p.a.	Total costs € p.a.	Specific costs €/t (w.b.)$_p$
Outdoor storage	113,000	7,169	1,130				8,299	0.2
Silo raw material	580,000	47,863	8,700				56,563	1.4
Silo pellets	390,000	32,252	5,850				38,102	1.0
Imputed interest (stored goods)				20,836			20,836	0.5
Other costs						30,107	30,107	0.8
Total costs	1,083,000	87,284	15,680	20,836	0	30,107	153,907	3.8

Peripheral equipment includes investment costs and electric power demand for conveyor systems, sieving machines, fans, rotary valves, biological additive feeding system and conditioning, as long as these aggregates are not included in other calculations already. The utilisation period and maintenance costs are based on information from pellet producers. Framework conditions and full cost calculation of peripheral equipment are shown in Table 7.14 and Table 7.15, respectively.

As demonstrated in the full cost calculation in Table 7.15, specific costs of peripheral equipment are 3.5 €/t (w.b.)$_p$, with consumption costs contributing the most. With regard to production and distribution costs of pellets (162.0 €/t (w.b.)$_p$, as per Table 7.20), costs for peripheral equipment amount to 2.2%.

Table 7.14: Framework conditions for full cost calculation of peripheral equipment

Explanations: [1]... percentage per year of total investment costs of peripheral equipment; data source: own research and calculations and data from plants in Austria

Parameter	Value	Unit
Electric power demand	108.0	kW
Utilisation period	15	a
Service and maintenance costs[1]	2.4	% p.a.
Electricity consumption	734,000	kWh/a

Table 7.15: Full cost calculation for peripheral equipment

Explanations: framework conditions as in Table 7.14; data source: manufacturers and data from plants in Austria

	Investment costs excl. construction €	Capital costs € p.a.	Maintenance costs € p.a.	Consumption costs € p.a.	Operating costs € p.a.	Other costs € p.a.	Total costs € p.a.	Specific costs €/t (w.b.)$_p$
Peripheral equipment	435,000	44,789	10,614				55,403	1.4
Electricity costs				73,440			73,440	1.8
Other costs					0	12,093	12,093	0.3
Total costs	435,000	44,789	10,614	73,440	0	12,093	140,936	3.5

7.2.8 Personnel

With regard to personnel, it was assumed that one person is needed to control and operate the plant. In addition to the personnel needed per shift, a quarter person has been calculated for deputyship (holidays, illness). This means, that 1.25 persons are needed to operate the plant. Due to operation around the clock throughout the year (8,000 full load operating hours), operating staff are needed at 8,760 hours per year (corresponding to three shift operation, seven days a week), leading to 10,950 annual working hours. Based on an hourly rate of 25.0 €/h, personnel costs amount to 273,750 €/a. In addition, two full time employees are calculated for marketing and administration, with annual costs of 73,000 € leading to total annual personnel costs of 346,750 € or 8.7 € per tonne of pellets.

7.2.9 Raw material

Next to investment costs and operating costs, the kind of raw materials and raw material costs are a decisive factor for the economy of pellet production. Table 7.16 presents an overview of raw material prices. Raw materials prove to be most economical when they accumulate as a by-product or waste product in-house. If so, transport costs as they are given in Table 7.16 can be spared. Detailed discussion of raw materials was carried out in Section 3.4.

Table 7.16: Price range of possible raw materials for pellets

Explanations: prices excluding VAT and loco pellet production plant (incl. transport); bulk densities relate to softwood (spruce); [1]...out of SRC plantations, chipped; data source [147; 290; 371; 372; 373; 374; 375; 376; 377; 378] and own research (price basis 2008)

Raw material	Retail price		Share of transport costs	Moisture content	Bulk density	Specific price			
	from €/lcm	to €/lcm	€/lcm	wt.% (w.b.)	kg (w.b.)/lcm	from €/t (d.b.)	to €/t (d.b.)	from €/MWh	to €/MWh
Industrial wood chips with bark	9.49	10.99	1.49	55	378	56	65	12.8	14.8
Industrial wood chips without bark	10.69	12.99	1.49	55	378	63	76	14.4	17.5
Forest wood chips	14.05	20.71	1.13	30	250	80	118	16.4	24.2
Short rotation crops[1]	12.50	23.63	1.13	55	389	71	135	16.4	31.0
Sawdust	7.05	11.05	1.05	55	267	59	92	13.5	21.2
Bark	5.91	11.41	1.41	55	356	37	71	8.5	16.4
Straw (square bales)	6.81	17.58	0.88	15	141	57	147	12.1	31.1
Whole crops (square bales)	11.70	24.28	0.86	15	188	73	152	15.5	32.2
Wood dust	1.39	6.05	1.05	8	163	9	40	1.8	7.9
Wood shavings	10.00	12.00	1.05	10	133	83	100	16.3	19.6

The price ranges given in Table 7.16 are mean price ranges of the stated raw materials, whereby pronounced seasonal or local price fluctuations can affect some kinds of biomass. For industrial wood chips with bark, prices around 9.5 to 11.0 €/lcm are stated. Industrial wood chips without bark cost between 10.7 and 13.0 €/lcm. Forest wood chips are the most expensive raw material within the wood chip assortments costing around 14 to 21 €/lcm, whereby it should be noted that wood chips are rarely used for pelletisation. In the medium term, this sector is expected to expand though. Chipped wood from SRC plantations costs around 12.5 to 23.6 €/lcm.

Bark, straw and whole crops are not presently used for pelletisation in Austria. Bark is used to some extent for producing bark briquettes.

Wood dust and wood shavings are most probably the best suited group of raw materials for pelletisation due to their attributes (dry, small particle size) and they neither require drying nor grinding. Wood dust and shavings are used for pelletisation in Austria, whereby further expansion of pellet production capacities on the basis of these two materials is very restricted because almost all available potential is already exploited (not only but to a great extent by pelletisation; cf. Section 10.1.3).

Therefore (and also because wood chips, bark and herbaceous biomass are not as suitable for pelletisation) sawdust, which is available in great quantities, is still the most important raw material for pelletisation. Sawdust is prone to strong price fluctuations, varying between 6.0 to 10.0 €/lcm (loco sawmill). The mean price is 8.0 €/lcm (price basis October 2008). The effect of fluctuations in this field is examined by means of sensitivity analyses (cf. Section 7.2.12). Since sawdust plays an important role as a raw material in the particle board industry, there is strong competition for this raw material and this determines the price. The situation is examined in more detail in Section 10.1.3.

Figure 7.1 shows the price development of sawdust since December 2003. Beginning in March 2006, a clear price rise was noted, which correlates to the scarcity of sawdust in this period due to increased demand and reduced wood harvest at the same time. The sawdust price rose on average by almost 80% by March 2006 and reached its maximum at the beginning of 2007; looking at the period starting in November 2004 the rise was above 90%. This is why pellet production also became more costly.

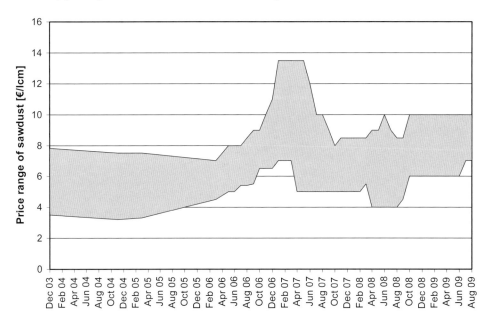

Figure 7.1: Price development of sawdust from December 2003 to August 2009

Explanations: prices loco sawmill, data source [371]

With regard to energy content, wood dust is the cheapest material on average, followed by bark, industrial wood chips with bark, industrial wood chips without bark, wood shavings, sawdust, forest wood chips, straw, SRC and whole crops, in the stated order. This order may shift when possible price fluctuations are considered.

The mean price for sawdust of the period November 2007 to October 2008 was 7.82 €/lcm (including transport costs). This price was also chosen for the calculations. Based on this raw material price and the framework conditions for the base case scenario, raw material costs are around 2,346,000 €/a or 58.7 €/t (w.b.)$_p$. Raw material costs alone thus make up 36.2% of production and distribution costs of pellets. The effects of possible price fluctuations of raw materials are discussed in Section 7.2.12.

7.2.10 Total pellets production costs

Based on the calculations and explanations as in Sections 7.2.1 to 7.2.9, total pellet production costs under the framework conditions as demonstrated and when using wet sawdust as a raw material amount to 136.6 €/t (w.b.)$_p$. Table 7.17 and Figure 7.2 show an overview of the composition of the total pellet production costs.

Table 7.17: Overview of the composition of the total pellet production costs

Explanations: data of calculations as in Sections 7.2.1 to 7.2.9

	Investment costs €	Capital costs € p.a.	Maintenance costs € p.a.	Consumption costs € p.a.	Operating costs € p.a.	Other costs € p.a.	Total costs € p.a.	Specific costs €/t (w.b.)$_p$
Drying	950,000	97,815	23,180	1,775,200		26,410	1,922,605	48.1
Grinding	206,000	21,210	5,026	74,800		5,727	106,764	2.7
Pelletisation	467,000	48,084	11,395	295,080		12,983	367,541	9.2
Cooling	32,000	3,295	781	8,160		890	13,125	0.3
Storage	1,083,000	87,284	15,680	20,836		30,107	153,907	3.8
Peripheral equipment	435,000	44,789	10,614	73,440		12,093	140,936	3.5
Personnel					346,750		346,750	8.7
Raw material				2,346,000			2,346,000	58.7
General investments	570,300	47,818	2,300			15,854	65,972	1.6
Total costs	3,743,300	350,294	68,976	4,593,516	346,750	104,064	5,463,599	136.6
Specific costs		8.8	1.7	114.8	8.7	2.6		136.6

The total pellet production costs are dominated by raw material and drying costs. These two factors constitute almost 80% of the total pellet production costs. Other important cost factors are personnel, with a share of about 6.3% and pelletisation itself with 6.7%. All other cost factors play a subordinate role, making up less than 9% on the whole. Therefore, raw material and drying hold the greatest potentials for cost reduction. Some savings potential is hinted at by the broad fluctuation of raw material costs. As concerns drying costs, using low temperature dryers operated with cheap heat (waste heat) represents a potential cost reduction. This is especially true when the drying process is combined with a biomass CHP system. Automation of the processes can reduce personnel costs and has also some potential for cutting costs.

Figure 7.3 shows the composition of the pellet production costs according to VDI 2067. Here, the specific pellet production costs are dominated by consumption costs that amount to around 84% of the total pellet production costs. Consumption costs are basically electricity, heat and raw material costs. The second largest costs are capital bound costs, including capital and maintenance cost according to VDI 2067. Operating costs include personnel costs and amount to about 6% of the total pellet production costs. Other costs of approximately 2% in total are not of great relevance.

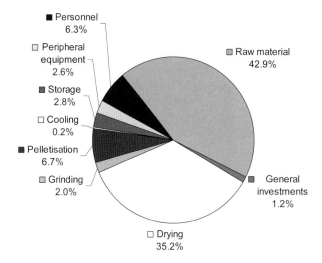

Figure 7.2: Pellet production costs and their composition according to the different cost factors when sawdust is used as raw material

Explanations: total specific pellet production costs of 136.6 €/t (w.b.)p; calculation of the specific production costs of process steps and cost factors as per Sections 7.2.1 to 7.2.9; general framework conditions: around 8,000 annual full operating hours (continuous operation); annual production of around 40.000 t (w.b.)p/a

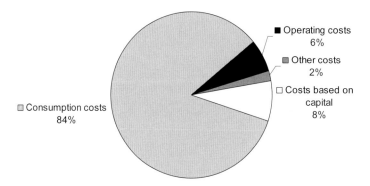

Figure 7.3: Pellet production costs and their composition according to VDI 2067 when sawdust is used as raw material

Explanations: total specific pellet production costs of 136.6 €/t (w.b.)$_p$; calculation of the specific production costs of process steps and cost factors as per Sections 7.2.1 to 7.2.9; general framework conditions: around 8,000 annual full operating hours (continuous operation); annual production of around 40.000 t (w.b.)$_p$/a

The total specific energy consumption of pellet production is 1.315 kWh/t (w.b.)$_p$ (thereof around 114 kWh$_{el}$/t (w.b.)$_p$ and 1,200 kWh$_{th}$/t (w.b.)$_p$) on basis of the framework conditions as exemplified above. Thermal energy needed for drying constitutes 93% of the energy consumption (cf. Figure 7.4). The other 7% are the electricity demands of grinding, pelletisation, cooling and peripheral equipment, whereby pelletisation makes up the greatest amount with 3.9% of the total. This explains the relatively high consumption costs and demonstrates once more the great potential for cost reduction in drying.

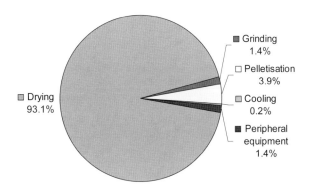

Figure 7.4: Energy consumption of pellet production when sawdust is used as raw material

Explanations: total energy consumption of pellet production of 1,315 kWh/t (w.b.)$_p$; calculation of the specific energy consumption of process steps and cost factors as per Sections 7.2.1 to 7.2.9; general framework conditions: around 8,000 annual full operating hours (continuous operation); annual production of around 40,000 t (w.b.)$_p$; electric power output 670 kW, specific heat demand for drying 1,200 kWh/t of evaporated water; drying from M55 to M10

7.2.11 Pellet distribution costs

The costs calculated in the above sections are pellet production costs loco pellet producer. In order to gain an overview of all the costs, from the raw material down to the storage of the pellets at the end user site, costs for pellet distribution were also considered. What is more, appropriate storage facilities have to be in place to overcome the difference between pellet production and delivery. Most pellet producers have only small storage capacities, which is why external storage facilities have to be hired. Such interim storage is not only necessary for balancing out the difference between pellet production and purchase but also for nationwide distribution in order to keep the transport times and distances to the end user as short as possible.

Distribution includes transport of pellets to the interim storage site (if the pellets are not delivered directly to the end user from the production site) as well as to the end user by silo truck. For the following calculations, transport from the producer to the interim storage by silo truck is also assumed. The basic data for the calculation of transport costs of silo trucks are presented in Table 7.18. The average speed of a silo truck up to a transport distance of 50 km is calculated as per Equation 7.2. For longer distances, an average speed of 60 km/h may be assumed [59].

The results of the calculation of transport costs for distribution to interim storages and end users versus the transport distance on the basis of the data of Table 7.18 and calculated according to Equation 7.2 are shown in Figure 7.5.

In Austria, it is common practice to charge a flat fill-in fee of 26.36 € per delivery (excl. VAT, price basis 12/2008). With an average amount of 6 t of pellets per delivery, a specific delivery price of 4.39 €/t (w.b.)$_p$ results. Transports of up to 70 km are covered by this (cf. Figure 7.5). Owing to the very good distribution net of pellets, it can be assumed that transport distances are usually below 70 km.

Table 7.18: Basic data for the calculation of transport costs per silo truck

Explanations: [1]...calculated as per Equation 7.2; bulk density of pellets: 625 kg (w.b.)$_p$/m³; data source [59, 379]

Parameter	Value	Unit
Capacity pellet truck	33	m³
Capacity pellet truck	20.63	t
Hourly rate truck (transport)	63.49	€/h
Hourly rate truck (time parked)	34.19	€/h
Average speed (distance up to 50 km)	[1]	km/h
Average speed (distance more than 50 km)	60	km/h
Time for pellets discharge	0.5	h

Equation 7.2: $\bar{v} = 12.95 \cdot d^{0.39}$

Explanations: data source [59]

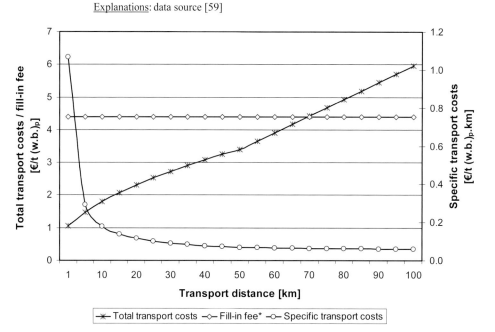

Figure 7.5: Total and specific pellet transport costs versus transport distance

Explanations: *...26.36 € per delivery (price basis 12/2008, excl. VAT); 6 t per delivery; basic data as in Table 7.18

The overall pellet distribution costs including all transport and storage costs are shown in Table 7.19. An average transport distance of 50 km was chosen for calculating transport costs from the producer to the interim storage site as well as from the interim storage site to the end user.

Looking at the supply chain of pellets from the raw material to the end user via production, storage and distribution, total costs amount to 162.0 €/t (w.b.)$_p$ (cf. Table 7.20). Total costs are thus close to the Austrian market price of pellets (167.24 €/t (w.b.)$_p$ excluding VAT, delivered; an overview of the Austrian market price development can be found in Section 8.1). If the calculation is not based on the average price of sawdust but on the present price, total costs of pellet production are slightly above the price they achieve on the market.

Table 7.19: Total costs of pellet distribution

Explanations: data source [45]; own calculations

Cost factors	Costs [€/t (w.b.)$_p$]
Transport from production site to intermediate storage	3.39
Unloading truck and loading silo	3.00
Rent for intermediate storage	7.10
Sieving before truck loading	5.50
Truck loading	3.00
Transport from intermediate storage to end user	3.39
Total costs	25.39

This shows that producing pellets with wet sawdust is in fact at the limits of economic efficiency under present general conditions. This conclusion is confirmed by information from the pellet production sector from the years 2007 and 2008, when pellets had to be sold below their total production costs in order to avoid full storage and consequential production outages.

Table 7.20: Total costs of pellet supply

Explanations: costs as per Table 7.17, Table 7.19 and Figure 7.5 for transport distances of 50 km to interim storage and end user sites; market price of pellets of 167.24 €/t (w.b.)p excl. VAT, delivered

Cost factor	Costs [€/t (w.b.)$_p$]	Share in retail price [%]
Pellet production	136.6	81.7
Intermediate storage	7.1	4.2
Transport incl. loading and unloading	18.3	10.9
Total	162.0	96.9

In this context, it should be pointed out that the total pellet production costs from raw material until end user supply of 162.0 €/t (w.b.)$_p$ are an average value for an average case under Austrian framework conditions. The used raw material, technical equipment, storage systems, client network, heat price for drying, national framework conditions in particular regarding pellet price, etc., are parameters that lead to strong changeability in pellet production costs. Transport costs for sawdust for instance can be saved by proper choice of location. Raw material price is thus reduced from 7.82 €/lcm to 6.77 €/lcm, which in turn lowers the costs of pellet production to 128.7 €/t (w.b.)$_p$. Total costs of the pellet supply would then just be 154.1 €/t (w.b.)$_p$ or 92.1% of the Austrian pellet market price. Furthermore, the price of heat has a strong influence on total production costs. If heat costs were reduced to 30 €/MWh (a realistic value under certain conditions) by optimised combination of a pellet production plant and a biomass CHP plant, pellet production costs would drop to 130.6 €/t (w.b.)$_p$. Total costs of the pellet supply chain would thus be 156.0 €/t (w.b.)$_p$ or 93.3% of market price. If both cost reduction potentials were exploited, pellet production costs would be 122.7 €/t (w.b.)$_p$ resulting in total pellet supply costs of 148.1 €/t (w.b.)$_p$ or no more than 88.6% of the Austrian market price of pellets.

Section 7.3 deals with different basic conditions. The influence they have on the economic efficiency of pellet production plants is presented and discussed.

7.2.12 Sensitivity analysis

In this section, sensitivity analyses of some important parameters are carried out in order to investigate the effect these parameters have on the total specific pellet production costs of the base case scenario. Single parameters are varied in a certain range that seems possible and reasonable. Subsequently, the total specific pellet production costs are calculated with each of these new values. By that, the extent of possible errors in the choice of parameters can be determined and cost saving potentials as well as important parameters for economic pellet production can be identified.

The specific pellet production costs of 136.6 €/t (w.b.)$_p$ that were calculated in Section 7.2.10 serve as a basis for comparison.

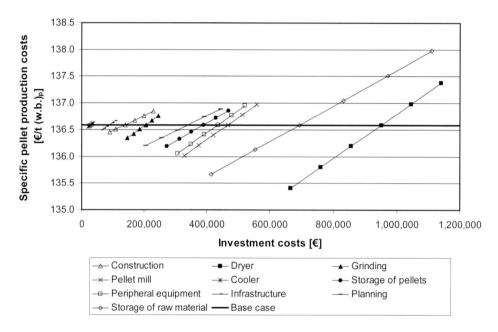

Figure 7.6: Influence of investment costs on the specific pellet production costs of the base case scenario of different plant components

Explanations: calculation of the specific pellet production costs as per Sections 7.2.1 to 7.2.9; specific pellet production costs of base case scenario: 136.6 €/t (w.b.)$_p$

Figure 7.6 shows the sensitivity analyses of investment costs for the different plant components of a pellet production plant. The slopes of the lines are proportional to the influence of the investment costs on the specific pellet production costs. It can be derived that varying the investment costs for machinery (dryer, grinder, pellet mill, cooler and peripheral equipment) has most influence on the specific pellet production costs of the base case scenario. This is because the utilisation period is comparatively short with 15 years and the maintenance costs of 2.4% of the investment costs are comparatively high. The influence of construction on investment costs is the least due to the high utilisation period of 50 years and low maintenance costs (1% p.a. of investment costs). The influence of varying the investment costs of the other units of the pellet production process lies somewhere in between.

Looking at Figure 7.6, the great influence of investment cost changes of cooler and infrastructure is put into perspective. Due to low absolute investment costs of these units, the influence of their change is limited, although their relative impact on the specific pellet production costs is significant. In contrast, investment costs of storage facilities (for raw materials as well as pellets) were found to have great absolute influence if storage capacity is large, although the relative influence is low. So, choosing the right storage capacity is of great relevance.

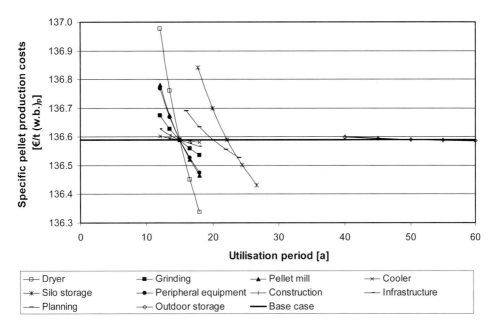

Figure 7.7: Influence of the utilisation periods of different plant components on the specific pellet production costs

Explanations: calculation of the specific pellet production costs as per Sections 7.2.1 to 7.2.9; specific pellet production costs of base case scenario: 136.6 €/t (w.b.)$_p$

The influence on the total specific pellet production costs of utilisation periods of single plant components is shown in Figure 7.7. Utilisation periods of the plant components were varied by ± 20% compared to the base case scenario. The specific pellet production costs rise with declining utilisation periods. Regarding construction, outdoor storage and cooler, little influence was found. In construction and outdoor storage this is because varying a utilisation period of 50 years hardly has an effect on capital costs and thus on specific production costs. The small influence of the cooler can be explained by its very low investment costs. Silo storage exhibits strong sensitivity as concerns utilisation period due to comparatively high investment costs. The combination of relatively short utilisation periods and comparatively high investment costs for pellet mill, dryer and peripheral equipment explain their high sensitivity. Other components are of minor relevance. Generally, reducing utilisation periods makes the specific pellet production costs of the base case scenario rise more than elevated utilisation periods decrease the costs.

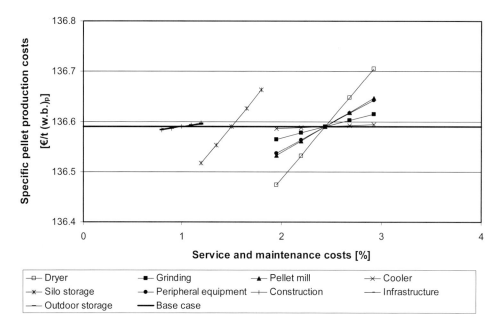

Figure 7.8: Influence of maintenance costs on the specific pellet production costs of different plant components

Explanations: calculation of the specific pellet production costs as per Sections 7.2.1 to 7.2.9; specific pellet production costs of base case scenario: 136.6 €/t (w.b.)$_p$

The influence of maintenance costs on the total specific pellet production costs of different plant components is shown in Figure 7.8. Maintenance costs of the plant components were varied by ± 20% compared to the base case scenario. In an absolute sense, drying and silo storage have the greatest influence, followed by pellet mill and peripheral equipment. Other components have minor influence. Silo storage is most sensitive due to relatively high investment costs. In total, the influence of maintenance costs on the specific pellet production costs is little.

Sensitivity analyses of plant availability and simultaneity factor are shown in Figure 7.9. In the base case scenario, a plant availability of 91.3% was assumed, which is a realistic and achievable value according to information from pellet producers and which has to be viewed as the mandatory minimum for modern pellet production plants. Reduced plant availability, be it by scheduled or unscheduled outages, raises the specific pellet production costs significantly. Increased plant availability can actually clearly decrease the specific pellet production. The simultaneity factor of electric equipment takes into account that not all of the electrical installations run on full load and not at the same time. A realistic value for the simultaneity factor is between 80 and 85%, the latter having been selected for the base case scenario. The influence of simultaneity factor variation on the specific pellet production costs is moderate.

Figure 7.10 shows the sensitivity analysis of electricity price. In the base case scenario, an electricity price of 100 €/MWh$_{el}$ was assumed, which is a realistic price for industries consuming roughly 4.5 GWh$_{el}$/a, as in this case, and under average Austrian framework conditions. The electricity price even varies strongly within Austria, depending on the federal state and the framework conditions, namely between 80 and 120 €/MWh$_{el}$. Electricity prices even outside this range might be possible in other countries. This can have a significant influence on the specific pellet production costs, as shown in Figure 7.10.

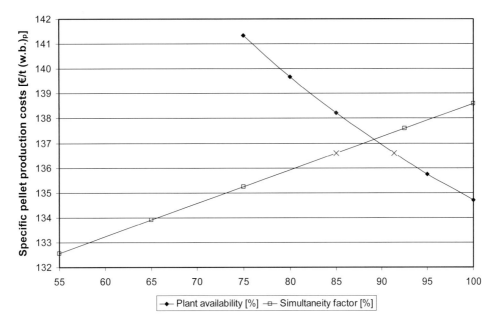

Figure 7.9: Influence of plant availability and the simultaneity factor of electric equipment on the specific pellet production costs

Explanations: x...base case scenarios; calculation of the specific pellet production costs as per Sections 7.2.1 to 7.2.9; specific pellet production costs of base case scenario: 136.6 €/t (w.b.)$_p$

A sensitivity analysis of specific heat costs for drying (based on hot water) is displayed in Figure 7.11. Variation of heat price has a strong impact on the specific pellet production costs, which demonstrates the large potential for cost reduction, as well as how high the risk of soaring costs is, in this area. In the base case scenario, a heat price of 35 €/MWh was assumed, which is a realistic and achievable average value for heat supply based on a biomass hot water boiler (of one's own). The right combination of a pellet production plant and a biomass CHP plant can reduce the heat price considerably under proper framework conditions. Realistically, heat costs can be reduced to 30 €/MWh in this case. The specific pellet production costs would thus decline by 4.4% to 130.6 €/t (w.b.)$_p$. Higher heat costs rapidly render pellet production uneconomic.

Besides making use of low temperature heat for the operation of belt dryers as in the present case, steam operated dryers (tube bundle dryers, superheated steam dryers) may be utilised as well for the drying of sawdust (cf. Section 4.1.1.2.3). Using steam as the drying medium augments specific heat costs. Since using different drying technologies also changes investment costs, their evaluation by means of a simple sensitivity analysis of the heat costs is not valid. Important framework conditions that have to be considered when one of these two technologies is to be employed are shown in Section 7.2.3.

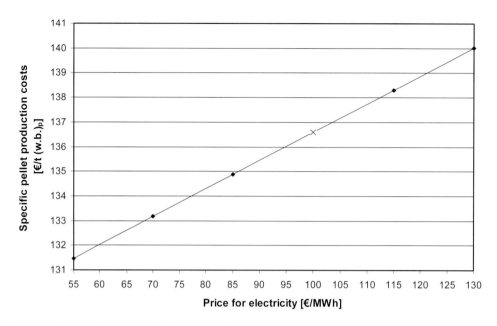

Figure 7.10: Influence of electricity price on the specific pellet production costs

Explanations: x...base case scenario; calculation of the specific pellet production costs as per Sections 7.2.1 to 7.2.9; specific pellet production costs of base case scenario: 136.6 €/t (w.b.)$_p$

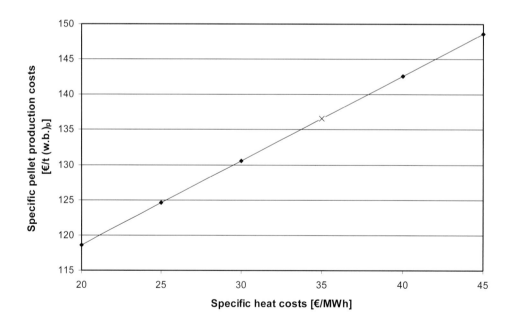

Figure 7.11: Influence of specific heat costs on the specific pellet production costs

Explanations: x...base case scenario; calculation of the specific pellet production costs as per Sections 7.2.1 to 7.2.9; specific pellet production costs of base case scenario: 136.6 €/t (w.b.)$_p$

Figure 7.12 shows a sensitivity analysis of annual full load operating hours. In the base case scenario, 8,000 h p.a. were assumed, which is based on the assumption of continuous plant operation seven days per week and 24 hours per day at a plant availability of 91.3% and represents the most economic way of operation, as illustrated in Figure 7.12. Limiting plant operation to five days a week and three shift operation would raise the specific pellet production costs by 4.4% to 142.6 €/t (w.b.)$_p$. Further reduction can therefore not be recommended. Moreover, discontinuous operation is practically impossible when dryers are used because a daily start-up and shutdown of the dryer would require far too much energy and time.

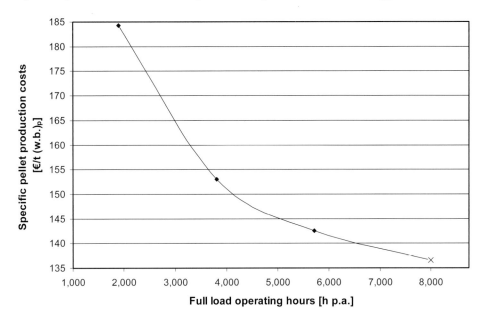

Figure 7.12: Influence of annual full load operating hours on the specific pellet production costs

Explanations: x…base case scenario; calculation of the specific pellet production costs as per Sections 7.2.1 to 7.2.9; specific pellet production costs of base case scenario: 136.6 €/t (w.b.)$_p$

Figure 7.13 shows sensitivity analyses of the parameters personnel per shift and hot water demand for conditioning.

The range of hot water consumption that was looked at, namely 0 to 0.25 wt.% (w.b.)$_p$, only causes a variation of the specific pellet production costs of less than 0.05%. Therefore, the influence is negligible.

The influences of personnel needed per shift and especially of raw materials are more prominent. In the base case scenario, it was assumed that one person is needed per shift to control and operate the plant. This is feasible and in fact state-of-the-art when appropriate automation is installed. A higher personnel demand indicates too low a degree of automation and a lesser personnel demand is not realistic at present. So, there is hardly any scope for variation in this area. Still, reduction of costs could be achieved by fully automated operation in the future by just requiring some supervising activity.

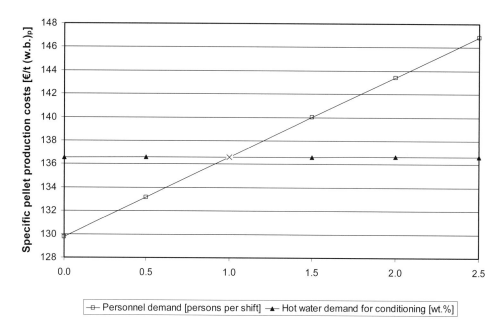

Figure 7.13: Influence of personnel and hot water demand for conditioning on the specific pellet production costs

Explanations: x...base case scenario; calculation of the specific pellet production costs as per Sections 7.2.1 to 7.2.9; specific pellet production costs of base case scenario: 136.6 €/t (w.b.)$_p$

Raw material costs have the most significant influence on pellet production costs (cf. Figure 7.14). Looking at a possible variation of sawdust prices from 7.05 to 11.05 €/lcm (cf. Table 7.16), pellet production costs are altered by almost 23%. If other raw materials are considered as well, the range of variation gets even broader in both upper and lower directions. This shows that choice of raw materials and safeguarding the costs of raw materials by means of long-term contracts are of great relevance in keeping pellet production economically efficient. Raw material prices that are too high can render pellet production uneconomic, even if other framework conditions are at an optimum. In this context, it should be noted that pellet production at the same location where sawdust is generated creates logistic advantages that make transport cost savings possible. Thus, there is cost saving potential in the right choice of location. If, for instance, the share of transport costs for sawdust from the sawmill to the pellet producer is spared as a whole, the raw material price is reduced from 7.82 €/lcm to 6.77 €/lcm (cf. Table 7.16); hence pellet production costs fall by 5.8% to 128.7 €/lcm.

Figure 7.15 shows the influence of the interest rate on the specific pellet production costs. In the base case scenario, an interest rate of 6% p.a. was chosen, which is an achievable real interest rate under present framework conditions. Possible fluctuations of the general interest rate level would not have an insignificant impact on the specific pellet production costs though.

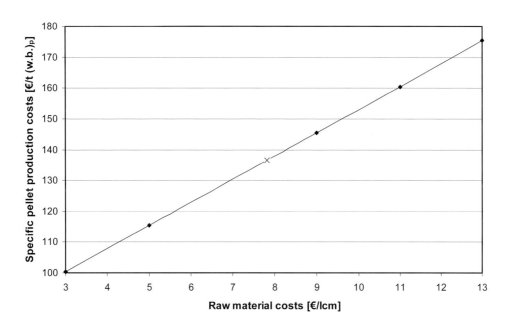

Figure 7.14: Influence of raw material costs on the specific pellet production costs

Explanations: x...base case scenario; calculation of the specific pellet production costs as per Sections 7.2.1 to 7.2.9; specific pellet production costs of base case scenario: 136.6 €/t $(w.b.)_p$

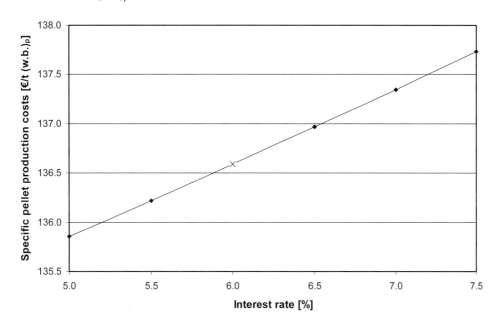

Figure 7.15: Influence of interest rate on the specific pellet production costs

Explanations: x...base case scenario; calculation of the specific pellet production costs as per Sections 7.2.1 to 7.2.9; specific pellet production costs of base case scenario: 136.6 €/t $(w.b.)_p$

Pellet throughput has a significant impact on the specific pellet production costs too, as illustrated in Figure 7.16. Even a small change of throughput causes relatively big changes of the specific pellet production costs. Optimising the throughput is thus important. It can be achieved primarily by the right choice of pellet mill and, in operation, optimisation of die geometries. The raw material also plays a significant role (e.g. softwood or hardwood). Regulation of moisture content is also very important in this respect as material that is too dry causes stronger frictional forces inside the compression channel hence leading to a reduction of throughput. A further increase of throughput can be achieved by addition of the right binding agents.

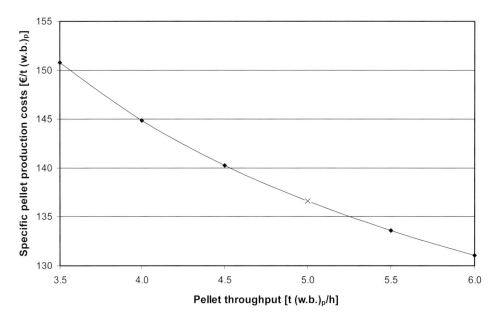

Figure 7.16: Influence of throughput on the specific pellet production costs

Explanations: x...base case scenario; calculation of the specific pellet production costs as per Sections 7.2.1 to 7.2.9; specific pellet production costs of base case scenario: 136.6 €/t $(w.b.)_p$

With regard to personnel demand for marketing and administration on the specific pellet production costs, two different causes for variation have to be examined. One are minor fluctuations by different wages, which have only moderate influence on the specific pellet production costs, which is why they may be neglected. The other, more important, cause of variation is the general marketing and administration strategy that is being pursued. Austrian pellet producers employ none to four persons for this purpose. In the first case, administrative duties are usually carried out along the way by some other employee. Marketing is outsourced as a whole. However, an increasing number of employees in this area means an increase in the specific pellet production costs, as illustrated in Figure 7.17.

Figure 7.18 presents an overview of the effects of parameter variation on the specific pellet production costs. The calculation as per Sections 7.2.1 to 7.2.9 was taken as a base case scenario. Subsequently, each parameter was altered by ± 10% and the relative change of the specific pellet production costs was calculated in percent.

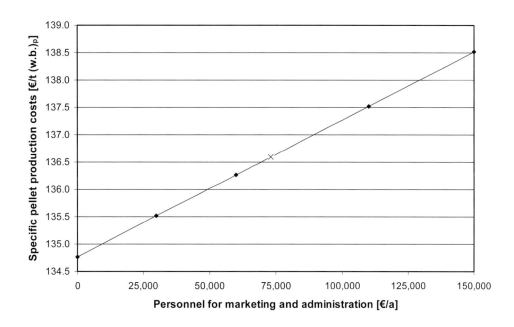

Figure 7.17: Influence of personnel demand for marketing and administration on the specific pellet production costs

Explanations: x...base case scenario; calculation of the specific pellet production costs as per Sections 7.2.1 to 7.2.9; specific pellet production costs of base case scenario: 136.6 €/t (w.b.)$_p$

Pellet throughput, plant availability and annual full load operating hours on one hand and raw material costs, specific heat costs and investment costs on the other hand, were found to be the main influencing variables (cf. also [380; 381; 382]). These areas bear the highest potential for reduction of costs as well as the highest risk of uneconomic operation. All parameters mentioned can either be optimised by appropriate planning or later, during operation. Pellet throughput can be optimised, i.e. kept at a high level, by appropriate die design, regulation of moisture content and addition of biological additives. Plant availability can be optimised by means of technical measures and choosing the right technologies. Annual full load operating hours are more or less determined by the kind of shift operation, with continuous operation being the best solution from the economic standpoint. Prices for raw material can be kept at a low as possible level by adequate supply contracts or by using raw materials accruing at one's own site. Specific heat costs can be minimised by a well designed combination of pellet production with biomass CHP plants. In particular the utilisation of low temperature heat (waste heat) in combination with low temperature drying systems is an interesting option from an economic point of view, even though investment costs of such drying technologies are higher. With regards to investment costs, storage space bears great cost saving potential. Investment costs of large storage facilities can be spared by optimised supply chains, provided that sales are safeguarded by retailers or intermediaries. Personnel costs have some cost saving potential too. Here, the production sector has to be regarded as separate from marketing and administration. In production, one person per shift is common practice. A higher personnel demand indicates too low a degree of automation. A lesser personnel demand could be achieved by further automation of the process, however, there are not a lot of possibilities for influencing this area. The choice of marketing strategy has a greater effect. A marketing department requires personnel and hence increases costs. The choice as to whether an individual marketing strategy should be pursued or an existing marketing concept should be employed, plays a decisive role in this respect.

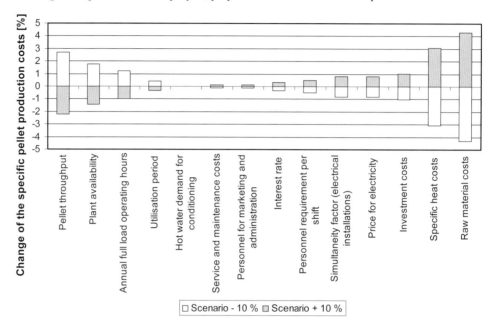

Figure 7.18: Overview of the effects of parameter changes on the specific pellet production costs

Explanations: calculation of the specific pellet production costs as per Sections 7.2.1 to 7.2.9; calculation of the relative changes of the specific pellet production costs based on the sensitivity analyses carried out above by varying each parameter by ± 10%; the parameters "utilisation period", "maintenance costs" and "investment costs" were varied for all plant components together; specific pellet production costs of base case scenario: 136.6 €/t (w.b.)p

7.3 Economic comparisons of pellet production plants under different framework conditions

As mentioned in the above sections, the pellet production costs that were calculated represent an average scenario under Austrian framework conditions. Depending on producer specific framework conditions, pellet production costs can vary greatly. In order to examine the effects of parameters on the specific pellet production costs, comprehensive sensitivity analyses were carried out and discussed in Section 7.2.12. This section takes a look at the effects producer specific framework conditions and plant sizes have on economic efficiency of pellet production.

Table 7.21 displays the most important parameters of eight scenarios that differ from the base case scenario, on the basis of which the specific pellet production costs of the different scenarios were calculated. For comparative purposes, the parameters of the base case scenario as per Sections 7.2.1 to 7.2.9 are included.

Scenario 1 is based on the assumption that wood shavings are used as a raw material instead of sawdust. In contrast to the base case scenario, only silo storage for this raw material is suitable, as outdoor storage of dry wood shavings would bear a great risk of re-humidification. The storage capacity for the dry raw material is assumed to be 0.41% of the annual demand or 36 h of storage. Such a low storage capacity as compared to the base case scenario is legitimate, if the raw material production is at the same location as the pellet production plant and the raw material storage can therefore be designed as an intermediate storage between wood shaving production and pelletisation (storage capacity being the same as storage capacity of interim storage between drying and pelletising of the base case scenario). Moreover, raw material transportation costs can be avoided in this case. Due to the use of dry raw material, the drying step can be left out. This results in reduced electricity and especially heat demand, total investment costs being reduced substantially due to the drying system being rendered unnecessary. In this scenario, the specific pellet production costs can be lowered by about 25% compared to the base case scenario. However, the cost reduction effect is partly compensated by the more expensive raw material. The price for wood shavings is around 83 €/t (d.b.), which is about 27% higher compared to the price of sawdust (around 65 €/t (d.b.)). The composition of the specific pellet production costs for scenario 1 is completely different from the composition of the specific pellet production costs of the base case scenario. When sawdust is used, 35% of the costs arise from drying and 43% from the raw material. In the case of using wood shavings, there are no drying costs and the share of raw material costs is raised to almost 73% of total pellet production costs. The other cost factors change accordingly (cf. Figure 7.19). The fact to consider in this scenario is that the raw material comes out of a wood processing plant of one's own, hence no transport costs have to be taken into account and thus the raw material is cheaper. What is more, investment costs of storage can be spared if continuous supply of raw material is given so that storage capacities need not be as great. If wood shavings were to be bought, transport costs would arise and investment costs for appropriate storage facilities would rise too.

Scenario 2 is an upscale of the base case scenario, based on a threefold annual pellet output (i.e. 120,000 $t_{pellets}$/a) and the same annual full load operating hours. Therefore, the different units of pellet production (dryer, hammer mills, pellet mills, cooler and peripheral equipment) and construction must be designed for this threefold load. The storage capacity was also adapted. An economy-of-scale effect was assumed to be between 15 and 20% for the different units. Due to this upscale, the specific pellet production costs can be decreased by about 6.4%, compared to the base case scenario. The economy-of-scale effect that can be achieved by large-scale pellet production plants of appropriate throughput becomes clear. The result confirms the economic sense of the current trend in Austria and many other countries toward erecting large-scale pellet production plants.

An example of the other end of the scale is scenario 3. It is the case of a small-scale pellet producer who produces pellets according the raw materials available in his own wood working plant. The raw material is dry wood shavings and is pelletised by a second hand pellet mill from the animal feed industry. The pellet production amounts to around 430 t (w.b.)$_p$/a. This scenario shows that economic pellet production can be achieved in small-scale plants under the right framework conditions. In this case, these are extremely low cost electricity due to the owner's having a hydropower station, existing storage facilities, dry raw material that needs neither drying nor grinding, no cooling demand owing to low throughput and extremely cheap raw material based on the maximum achievable sales price in the respective region.

Table 7.21: Key parameters of the scenarios considered in comparison to the base case scenario

Explanations: [1]…based on annual demand, annual production capacity, respectively; [2]…log wood in scm; calculation as per Sections 7.2.1 to 7.2.9 with framework conditions of this table

Parameter	Unit	Base case	Scenario 1	Scenario 2	Scenario 3	Scenario 4	Scenario 5	Scenario 6	Scenario 7	Scenario 8
General conditions										
Price for electricity	€/MWh	100	100	100	45	120	120	110	100	100
Total electricity consumption	GWh/a	4.56	3.60	12.93	0.03	0.08	0.27	0.11	5.75	7.79
Specific electricity consumption	kWh/t (w.b.)$_p$	113.9	90.1	107.8	75.3	88.4	119.4	100.1	143.7	194.7
Total investment costs	€	3,743,300	2,112,000	9,176,200	53,424	184,440	809,840	178,164	4,596,900	7,485,500
Raw material data										
Raw material		Sawdust	Wood shavings	Sawdust	Wood shavings	Wood shavings	Wood shavings	Wood shavings	FWC w. bark	Log wood
Raw material price	€/lcm[2]	7.82	9.95	7.82	1.34	9.95	9.95	9.95	11.84	40.80
Raw material storage										
Kind of storage for wet raw material		Paved outdoor storage	None	Paved outdoor storage	None	None	None	None	Paved outdoor storage	Paved outdoor storage
Storage capacity[1]	%	1.92		1.92					1.92	8.33
Kind of storage for dried raw material		Silo	Silo	Silo	Silo	Silo	Silo	Silo	Silo	Silo
Storage capacity[1]	%	0.41	0.41	0.41	existing	0.41	0.41	0.41	0.41	0.41
Drying data										
Dryer type		Belt dryer	None	3 Belt dryers	None	None	None	None	Belt dryer	Belt dryer
Required electric power	kW	140.0		420.0					140.0	140.0
Grinding / sieving data										
Unit type		Hammermill	Hammermill	Hammermill	None	None	Sieving machine	included	Hammermill	Hammermill
Required electric power	kW	110.0	110.0	330.0			2.5		110.0	110.0
Pellet mill data										
Required electric power	kW	300.0	300.0	900.0	50.0	40.0	154.0	42.0	300.0	300.0
Cooling data										
Cooler type		Counterflow cooler	Counterflow cooler	3 Counterflow coolers	None	None	None	Included	Counterflow cooler	Counterflow cooler
Required electric power	kW	12.0	12.0	36.0					12.0	12.0
Pellet storage										
Kind of storage		Silo	Silo	Silo	Storehouse	Silo	Silo	Storehouse	Silo	Silo
Storage capacity[1]	%	2.30	2.30	2.30	25.00	12.00	50.00	10.00	2.30	2.30
Peripheral equipment data (conveying systems, steel construction)										
Required electric power	kW	108.0	108.0	216.0	12.0	12.0	12.0		108.0	108.0
Pellet data										
Pellet production rate	t (w.b.)$_p$/h	5.0	5.0	15.0	0.7	0.5	1.2	0.3	5.0	5.0
Annual pellet production	t (w.b.)$_p$/a	40,000.0	40,000.0	120,000.0	430.7	952.4	2,285.7	1,142.9	40,000.0	40,000.0
Kind of shiftwork										
Shifts per day		3	3	3	1	1	1	2	3	3
Working days per week		7	7	7	1.5	5	5	5	7	7
Annual operating hours	h p.a.	8,000	8,000	8,000	615	1,905	1,905	3,810	8,000	8,000
Personnel data										
Hourly rate	€/h	25.00	25.00	25.00	8.94	25.00	25.00	25.00	25.00	25.00
Persons per shift		1.00	1.00	1.00	1.00	1.00	1.00	0.25	1.25	2.00
Persons for deputyship per shift (holidays, illness)		0.25	0.25	0.25		0.25	0.25	0.25	0.25	0.25
Personnel for marketing and administratio	€/a	73,000	73,000	73,000	3,700	2,800	8,200	1,990	73,000	73,000
Specific pellet production costs	€/t (w.b.)$_p$	136.6	102.5	127.8	50.8	180.5	162.2	135.0	148.9	196.8

The throughput of scenario 4 (0.5 t/h) is even lower than the throughput of scenario 3. The raw material is dry wood shavings again. However, due to the assumed 1-shift operation on 5 days per week, the annual pellet production is higher. Drying, grinding and cooling can be left out in this scenario too. The investment costs are based on new equipment, which makes them higher. Electricity costs as well as raw material costs are based on average Austrian prices under these framework conditions. Electricity and raw material costs are thus much above the values in scenario 3. Under these framework conditions, pellet production costs amount to approximately 181 €/t (w.b.)$_p$ which exceeds the limit for economic operation by far (which is roughly between 140 and 155 €/t (w.b.)$_p$, depending on the distribution system). A pellet mill with a throughput of 1.2 t/h as considered in scenario 5 can lower the pellet production costs under the same framework conditions as for scenario 4 to about 162 €/t (w.b.)$_p$. However, the costs are still above the limit for economic operation.

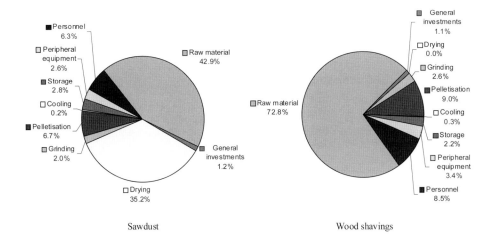

Figure 7.19: Composition of the specific pellet production costs according to different cost factors when sawdust and wood shavings are used as a raw material

Explanations: calculation of the specific pellet production costs of the cost factors as per Sections 7.2.1 to 7.2.9 with framework conditions of Table 7.21

A specific case is scenario 6. A Swedish manufacturer offers complete pellet production plants for dry raw material with throughputs of 300 kg/h, in which hammer mill, pellet mill and cooler are included [383]. The personnel demand is limited to about 2 hours per shift. Based on a two shift operation five days a week, specific pellet production costs of about 135 €/$t_{pellets}$ can be achieved. Therefore, this system is well suited to economically produce pellets in small-scale operation, if framework conditions are optimal. The system is particularly favoured by Italian pellet producers [147]. Scenarios 3 to 6 demonstrate the fact that the risk of uneconomic operation is very great in small-scale pellet production plants and only special framework conditions allow an economic operation.

Industrial wood chips (without bark) are used as a raw material in scenario 7 under the same framework conditions as in the base case scenario. This raw material is of increasing importance, as additional sawdust is scarce to allow a further expansion of pellet production in Austria (cf. Section 10.1.3). Industrial wood chips with bark would also be a potential raw material for pelletisation and they would even be cheaper. However, due to the bark content and consequently the higher ash content, a production of class A1 pellets according to prEN 14961-2 could not be carried out, which is why this option is not examined in detail. Production of pellets for industrial use, for example in power plants, would be possible though. The price for industrial wood chips without bark was settled at 11.84 €/lcm or 69.7 €/t (d.b.) (mean value of price range as in Table 7.16), hence industrial wood chips without bark are about 7% more expensive than sawdust. The investment costs of a pellet production plant able to use wood chips increase as coarse grinding (required particle size of 7 mm) before drying has to take place (additional hammer mills, construction work and peripheral equipment are necessary). Costs are slightly reduced in raw material storage due to the higher bulk density of wood chips compared to sawdust. The additional grinding unit causes additional electricity costs. In total, the specific pellet production costs under these framework conditions amount to about 149 €/t (w.b.)$_p$ or 9% more than in the base case scenario.

Due to the increasing shortage of sawdust, many pellet producers consider establishing or already have established pellet production plants able to use log wood as a raw material [111; 112]. Such a plant is the case in scenario 8. In order to produce pellets according to standard, wood without bark must be used, as in the case of industrial wood chips. The price for log wood (pulp wood) was settled as 40.8 €/scm including transport costs [371; 384]. Based on dry substance, log wood is about 50% more expensive than sawdust and 20% more expensive than wood shavings. The additional equipment already mentioned for scenario 7 is also necessary for scenario 8. Moreover, a stationary chipper is considered in scenario 8 in order to produce wood chips out of the log wood. In addition, log wood conveyor systems as well as a log wood storage have to be considered. Compared to scenario 7, the total investment costs thus increase by 63%. The electricity consumption increases by about 70%, compared to the base case scenario due to chipping and additional peripheral equipment. The specific pellet production costs amount to about 197 €/t (w.b.)$_p$, which is clearly higher then the limit for economic production. However, pellet production from log wood could be an interesting option in the future, if pellet prices increase (present pellet prices are comparatively low). Moreover, pellet production plants using log wood should be designed for higher annual outputs in order to achieve an economy-of-scale-effect. An additional cost reduction potential could be realised by utilisation of stored log wood. In this case, the natural drying effect would reduce the heat demand for drying and consequently the drying costs.

7.4 Summary/conclusions

The main cost factor of pellet production is the raw material. Therefore, raw material prices play a decisive role in the economy of pellet production. All kinds of raw materials are subject to strong seasonal and local price fluctuations. The minimum and maximum price of sawdust for instance was 4 €/lcm and 10 €/lcm (loco sawmill) between November 2007 and October 2008. Thus the difference between highest and lowest price was 150%. These numbers illustrate how important securing long-term and economic raw material availability is for pellet production. If pellet production is not located at the site where sawdust is accumulated, transport costs from sawmill to pellet producer have to be considered as well.

Specific pellet production costs for each sub-unit of pellet production (drying, grinding, pelletising, cooling, raw material and pellet storage, peripheral equipment and construction) were calculated under average Austrian framework conditions and taking personnel and raw material costs into account. The resulting specific pellet production costs are 136.6 €/t (w.b.)$_p$. They are dominated by raw material and drying costs that together make up 80% of total costs. Personnel and pelletising itself are also important cost factors that together add 13% of all costs. The remaining 9% comprise costs for general investments (mainly construction), grinding, cooling, storage and peripheral equipment. Looking at the cost groups as per VDI 2067, it is shown that the specific pellet production costs are dominated by consumption costs, which is mainly due to the heat costs for drying. This is confirmed when energy consumption is calculated, which shows that 93% of required energy is consumed by drying of the raw materials.

In order to take a holistic look at the pellet supply chain, distribution costs as well as costs for possible interim storage spaces were calculated in addition to production costs. Thus, up to 25 €/t (w.b.)$_p$ must be added to the specific pellet production costs in order to obtain specific pellet supply costs. These costs can fluctuate depending on logistics and consumer structure.

Within the framework of sensitivity analyses, raw material costs, specific heat costs, pellet throughput, plant availability, annual full load operating hours as well as investment costs were found to be the main influencing parameters (with decreasing importance according to the stated order). The greatest cost saving potential but also the greatest danger of uneconomic operation also lie within these parameters; this must be contemplated in planning as well as in plant operation.

In addition, pellet production plants operating under different framework conditions and in different scales were compared. It was shown that pellet production plants with large annual production capacities and appropriate plant utilisation, plants that dry the raw materials by means of low temperature dryers with access to low temperature heat, plants with a high degree of automation and plants that require moderate storage capacities due to intelligent logistics are the most attractive choice. By contrast, it was shown as well that economic pellet production at a very small-scale is possible too, if the framework conditions are right. Still, the danger of uneconomic operation is great in small-scale systems, as was demonstrated by two examples. Innovative concepts, such as the use of ready-to-use small-scale pelletisation plants as they are offered, for example by a Swedish manufacturer, can also make economic pellet production possible under certain framework conditions. The use of wood chips for pelletisation is an economically reasonable option too. It is only the use of log wood that is not economical under the present framework conditions with relatively modest pellet prices (and the stated pellet production of 40,000 t (w.b.)$_p$/a). Expansion of raw materials by making use of these sets of materials is needed in the medium term though to meet with the strong growth of the pellet markets. In part, this is economically reasonable already, as was shown by some examples. Larger production capacities, drying effects by log wood storage and slightly higher pellet prices would render the use of log wood for pelletisation economical too.

8 Cost analysis of pellet utilisation in the residential heating sector

In this section, a cost analysis of pellet utilisation by means of a central heating system in a typical residential house under Austrian framework conditions is provided. The use of pellets is directly compared to central heating systems based on other fuels, i.e. oil, natural gas and wood chips as well as on biomass district heat. It should be pointed out that the results cannot be directly transferred to a specific project in another region or country as specific framework conditions may differ significantly. However, they can be used as a guide and give an indication as to how the calculation should be made.

8.1 Retail prices for different fuels in the residential heating sector

Figure 8.1 shows a comparison between the average Austrian market prices of pellets, heating oil, natural gas and wood chips for 2006, 2007 and 2008, based on NCV. It follows that wood chips with a price ranging from 27.0 to 29.5 €/MWh$_{NCV}$ are the cheapest fuel, followed by pellets with 37.5 to 45.9 €/MWh$_{NCV}$, whereby the average price for pellets decreased continuously during those years. Heating oil was the most expensive fuel in 2006 and 2008. Natural gas had similar prices to heating oil in 2006 and 2007 but it did not rise in price as much as heating oil in 2008.

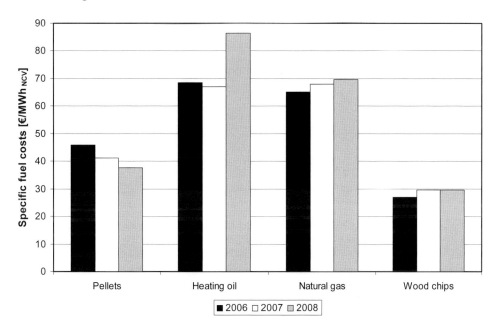

Figure 8.1: Average prices of different fuels based on NCV from 2006 to 2008

Explanations: average prices of trading products incl. VAT (10% for pellets and wood chips, 20% for heating oil and natural gas) and delivery; NCV for pellets 4.9 kWh/kg (w.b.)$_p$, heating oil 10 kWh/l, natural gas 9.60 kWh/Nm³, wood chips 3.72 kWh/kg (w.b.); average prices as according to Figure 8.2

Price development of pellets, heating oil, natural gas and wood chips since the year 1999 is shown in Figure 8.2. In this period, the natural gas price increased by 88%, heating oil price by 92% and pellet price by 48%.

The development of the different prices shows quite different characteristics. The natural gas price rose continuously without any marked ascents or downfalls. Only in November 2008 was there a steep increase of 28%, and there were price declines twice (in 2000 and 2001).

Development of the oil price is in part characterised by strong fluctuations. From June 1999 to October 2000, i.e. within 16 months, the oil price rose by 87%. After that, it dropped again, not reaching the level of 1999, however. Then the oil price remained more or less at the same level until December 2003. From then onwards, the price increased steeply again, arriving at an interim peak of 0.74 €/l in October 2005. In January 2007, 15 months later, the price was 0.57 €/l again – a quarter less than at the interim peak in October 2005. From there, the oil price rose strongly again, reaching its maximum of 1.09 €/l in July 2008 (price increase of almost 90% within a year and half, on the basis of the preceding depression). With the international financial and economic crisis, the oil price fell to the level of January 2007, or of 2005, until December 2008. The development of the oil price demonstrates that it is impossible for the end user to acquire oil at a reasonable price by means of storage logic because price development is totally unforeseeable.

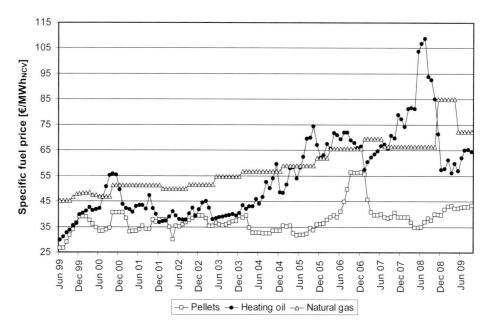

Figure 8.2: Price development of pellets, heating oil and natural gas from June 1999 to September 2009 in Austria

Explanations: NCVs according to Section 8.2; data source of fuel prices: ABEX (Austrian Biofuels Exchange), stock exchange for biomass fuels; webpage of the Austrian Energy Agency, http://www.energyagency.at; own research

The pellet trade, however, exhibited a more stable price policy, with low prices in summer and slightly raised prices in winter together with moderate increases from year to year. The strong fluctuations of the fossil fuel sector were not mirrored, which raised confidence in the pellet market by the end user. Consequently, it was possible to keep the price level of pellets low by appropriate storage strategies.

There was, however, a considerable increase in pellet prices in the year 2006 (in Austria but also in Germany and Italy). Several factors were responsible. In autumn 2005, more pellet boilers were sold than ever before, which led to unexpectedly high demands the following winter. This high demand was further increased by increased consumption during this long and harsh winter. At the same time, the long winter with high snowfall, constricted wood harvest. Sawmills produced less sawdust and thus less raw material for pelletisation. Hence the price for sawdust rose (for instance in Styria 30% on average since 2005 [371]), which in turn led to increased pellet production costs. Another outcome of this increase in demand and decrease in production were supply shortages.

The development of the pellet price in Germany is similar to that in Austria. In winter 2006/2007 the price was at its peak with up to 263 €/t (w.b.)$_p$. After that, it decreased sharply. Since the middle of May 2008, a moderate price increase with seasonal fluctuations has been noted again (cf. Figure 8.3).

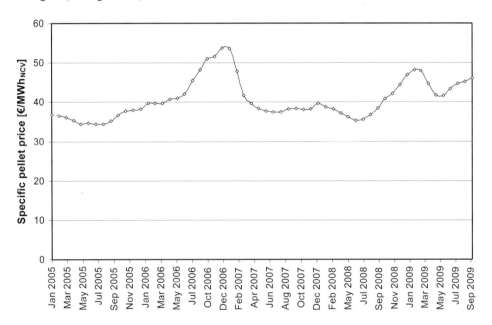

Figure 8.3: Price development of pellets in Germany

Explanations: price of pellets when purchase is > 6 t incl. delivery within distances of 100 to 200 km, all additional charges and 7% VAT; data source [385; 484]

Looking at the composition of pellet production costs, it is evident that the share of raw material costs is around 43%, when sawdust is used as a raw material (cf. Figure 7.2). Raw material price alone thus cannot account for the steep pellet price increase between May 2006 and January 2007. Export of pellets must have played a significant role. Around 490,000 t of pellets were produced in Austria in the year 2005, national consumption being around 280,000 t. In 2006, about 620,000 t were produced and around 400,000 t were consumed. So, roughly one third of Austrian pellet production is exported [386]. It clearly follows that the shortage of pellets and thus the rise of pellet price was induced also by massive exports since the pellets produced would actually have covered national demand. Pellet producers and pellet trade as a whole were able to achieve short-term profits by selling pellets on more profitable markets (such as Italy in particular, where more is paid for pellets due to the increased tax on heating oil). As a result, the prospering Austrian pellet market was at risk. This trend brings into question the two key arguments used by the pellet sector for changing from other fuels to pellets, namely stability in price and national added value. It would be desirable if the pellet sector could sustain these two arguments for the sake of sustainable pellet market development, safeguarding an appropriate price level as well as security of supply by suitable measures.

The situation was relaxed by the erection of new pellet production plants resulting in a massive increase of pellet production capacity (cf. Section 10.1.2) in 2007, lowering the price level to the value of previous years (before 2006).

Due to these developments, sales volumes of pellet heating systems broke down. Some manufacturers had reductions in sales of up to 90%. In 2007, sales volumes for pellet boilers decreased by 60% in Austria (cf. Section 10.1.4.1.1). Similar market developments were observed in many other countries (for example Germany and Sweden). It was not until 2008 end users regained their confidence in the pellet market. At that time, sales volumes of pellet boilers started to rise again.

The period in which pellet prices exhibited a high level caused uncertainty as well as resentment and lack of understanding by the end user of the price development of pellets. Nevertheless, pellet heating systems were competitive with oil heating systems in this period (even without subsidies); even though the price of pellets rose in 2006, the oil price decreasing slightly (from a high level) at the same time. Therefore, and due to the shortage of fossil fuels [387; 388] entailing the unforeseeable development of the oil price, investing in a pellet heating system surely is the right option in the long term, if no supply network for natural gas or district heat exists. Another increase in pellet prices, with the oil price remaining steady, would surely result in negative consequences however. Price development since 2006 shows that this is not to be expected.

Price development of wood chips is not shown in Figure 8.2 due to lack of data. At present (12/2008) wood chips cost around 29.5 €/MWh$_{NCV}$ and biomass district heat costs 82 €/MWh, whereby the price of biomass district heat has to be compared to the heat generation costs and not to the fuel price (cf. Section 8.2.9).

However, it is invalid to consider fuel price alone in choosing the right heating system. A holistic evaluation has to look not only at fuel costs but also at other consumption costs, costs based on capital, operating costs and other costs on the basis of full cost calculations. Full cost calculations are carried out in Section 8.2.

8.2 Economic comparison of different residential heating systems

Different residential heating systems were compared by means of a full cost calculation according to VDI 2067. Investment costs, maintenance costs, consumption costs, operating costs and other costs were considered. Investment costs are considered by means of capital costs, which were calculated by multiplying the investment costs with the CRF. The CRF is calculated according to Equation 7.1 (cf. Section 7.1). Investment costs include all costs in connection with construction of the plant, and thus not only boiler and fuel storage but also proportionate costs for furnace and storage room as well as chimney and, if needed, connection fees. Maintenance costs of all plant components are calculated as a percentage of the whole investment costs on the basis of guiding values and are evenly spread over the years of the utilisation period. Capital costs and maintenance costs are grouped together in costs based on capital. Consumption costs are all costs for fuel, respectively district heat, electric energy and costs for the imputed interest on fuel storage of pellet, wood chip and oil heating systems. Operating costs comprise costs originating from the operation of the plant, for example costs for maintenance and service on own account, chimney sweeper and meter rental. Other costs are insurance costs and administration costs, which are 0.5% p.a. of investment costs as a guiding value.

Heat costs of the following central heating systems were compared on a full cost basis:

- Pellet central heating system;
- Pellet central heating system with flue gas condensation;
- Oil central heating system;
- Oil central heating system with flue gas condensation;
- Wood chip central heating system;
- Natural gas central heating system with flue gas condensation;
- Biomass district heating.

The prices are given including VAT, in order to relate to the end user. Annual heat generation costs are calculated with the year 2008 as the basis. Prices of pellets, wood chips and heating oil include the delivery.

The framework conditions of the full cost calculations as well as the results are demonstrated in the following sections. It must be noted that the costs for the central heating systems used as a basis for the calculations are based on information from single manufacturers and installers and include installation and start-up. Therefore, actual prices may be above or beneath these prices depending on manufacturer and design. Much attention was given to choosing equivalent systems. Finally, the influence of different parameters on total costs is examined be means of sensitivity analyses (cf. Section 8.2.10).

8.2.1 General framework conditions

Table 8.1 displays the framework conditions for all the scenarios compared. The nominal boiler capacity and the annual full load operating hours were assumed to be the same for all systems in order to create a comparable basis for the calculations. They represent an average case of an Austrian detached house.

The price for electricity is an end user price including all taxes, fees and surcharges. It is the average value of all electricity suppliers in Austria based on the period December 2007 to November 2008 [389].

In order to estimate and evaluate maintenance and service effort for the end user, an hourly rate for work of one's own account was defined. Building costs for furnace and storage room are based on experience. Insurance costs and utilisation periods were settled according to the

VDI 2067 guideline. An interest rate of 6% was chosen, which represents an average value of usual real interest rates.

Table 8.1: General framework conditions for full cost calculation of different heating systems

Explanations: price basis 12/2008; data source [59; 389]; own research, VDI 2067 guideline

Parameter	Value	Unit
Nominal boiler capacity	15	kW
Full load operating hours	1,500	h p.a.
Price for electricity	173.40	€/MWh
Hourly rate (services on own account)	5.86	€/h
Construction costs of storage and furnace room (cellar)	244.16	€/m²
Insurance (in % of investment costs)	0.50	% p.a.
Utilisation period boiler (except heat transfer unit)	20	a
Utilisation period oil tank/storage room	20	a
Utilisation period construction and chimney	50	a
Service and maintenance costs construction (% of investment costs)	1.0	% p.a.
Interest rate	6.0	% p.a.

8.2.2 Pellet central heating system

Table 8.2 shows the basic data for full cost calculation of a pellet central heating system. The annual efficiency was settled according to own measurements and data from the literature [390]. Pellet consumption lies around 5.5 t (w.b.)$_p$/a under the assumed framework conditions (i.e. nominal boiler capacity, annual full load operating hours, NCV and annual efficiency).

Table 8.2: Basic data for the full cost calculation of a pellet central heating system

Explanations: price basis 12/2008; data source [59; 390; 391]; own research, VDI 2067 guideline; general framework conditions as per Table 8.1

Parameter	Value	Unit
Annual efficiency	84.0	%
Net calorific value	4.90	kWh/kg (w.b.)
Fuel demand	5,466	kg (w.b.)$_p$ p.a.
Bulk density of pellets	625	kg (w.b.)/m³
Electric power demand (in % of nominal boiler capacity)	0.70	%
Fuel price (absolute)	184	€/t (w.b.)$_p$
Fuel price (specific)	37.5	€/MWh$_{NCV}$
Maintenance and service effort (services on own account)	0.19	h/week
Chimney sweeper	143	€ p.a.
Service and maintenance furnace (% of investment costs)	2.0	% p.a.
Funding (in % of investment costs)	25	%
Funding (upper limit)	1,400	€

The average electric power demand of the system corresponds to the average value of the test protocols of the BLT Wieselburg [391]. Pellet price is the average price of pellets of the year 2008 and includes transport and delivery costs as well as all taxes and fees.

Maintenance and service by the end user require about 0.19 hours a week. This includes organising the fuel (querying prices, order, being present at delivery), emptying the ash box, being present at sweeping and annual service, storage space cleaning work and other administrative work. Costs for the chimney sweeper are based on the Styrian chimney sweep act.

Maintenance costs for the pellet central heating system were not applied according to the VDI 2067 guideline. They were assumed to be 2% instead of 1% of the annual investment costs.

Pellet central heating systems are subsidised to different extents depending on the federal states of Austria (cf. Section 10.1.4.1). A subsidy of 1,400 € was chosen, which is a possible subsidy in Styria for instance. The national subsidy of 800 € [392] between February 2008 and January 2009 was not taken into account here.

Table 8.3 displays the full cost calculation of a pellet central heating system based on the data in Table 8.1 and Table 8.2. For fuel storage, a storage space with automatic feeding system was chosen. In this way, a similar comfort of use as in oil or gas systems is achieved. The costs for boiler and storage space include costs for a hot water boiler. Apart from the costs for technical equipment, these costs also contain additional costs for installation, delivery and start-up. Furthermore, investment costs comprise building costs for the chimney, furnace and storage room as well as the needed fixtures of the storage space such as plastic baffle plate, fill-in and return pipe and sloped bottom. The storage space was designed for 1.2 times the annual fuel demand in order to be prepared for varying winter conditions.

For the calculations in Table 8.3, possible subsidies were not considered. If investment subsidies of 1,400 € were considered, specific heat generation costs would be 144.3 €/MWh; the sum of heat generation costs would decrease to 3,247 €/a. Thus, the subsidies would lower the heat generation costs by 3.6%.

Table 8.3: Full cost calculation of a pellet central heating system

Explanations: price basis 12/2008; basic data as in Table 8.1 and Table 8.2; data source [59], own research, VDI 2067 guideline; specific heat generation costs based on useful heat

	Investment costs €	Capital costs € p.a.	Maintenance costs € p.a.	Consumption costs € p.a.	Operating costs € p.a.	Other costs € p.a.	Total costs € p.a.	Specific costs €/MWh
Pellet boiler	12,600	1,099	252				1,351	60.02
Storage room discharge, storage room fixtures	2,700	235	54				289	12.9
Chimney	2,700	171	27				198	8.8
Construction costs furnace room, 3.6 m²	879	56	9				65	2.9
Construction costs storage room, 5.6 m²	1,370	87	14				101	4.5
Fuel costs				1,006			1,006	44.7
Electricity costs				27			27	1.2
Service and maintenance (services on own account)					58		58	2.6
Other costs						101	101	4.5
Imputed interest (fuel stored)				30			30	1.3
Chimney sweeper					143		143	6.4
Total costs	20,249	1,648	355	1,063	201	101	3,369	149.7
Specific costs €/MWh		73.2	15.8	47.2	8.9	4.5		149.7

A storage option that is worth mentioning is underground storage instead of a storage space in the cellar. Hereby, an underground storage tank is installed in the garden. This can save putting up a 5.6 m² storage space in the cellar with 1,370 € construction costs. In addition, costs for fixtures such as storage space door, feed-in equipment and sloped bottom are cut, saving altogether about 950 €. However, investment costs for an underground storage tank are about 7,300 € (including VAT) for a storage tank with a capacity of 10 m³ [207]. This includes the required discharge system of the tank, which is why costs for the discharge system of the storage room are spared. Assuming the above conditions, investment costs rise from 20,200 € to 23,500 €, i.e. by 16%, when underground storage is used. Specific heat generation costs rise from 149.7 €/MWh to 167.9 €/MWh, or by 12.2%. If subsidies were considered, specific heat generation costs would be 162.5 €/MWh. Despite these extra costs, underground storage is a reasonable alternative to cellar storage space when there is insufficient space and especially when the cellar is moist or no cellar is available.

Tanks made of synthetic fibre are another alternative to cellar storage spaces. A fibre tank for storing 6.7 t of pellets has a ground surface of 6.25 m². In order to set up a fibre tank in the cellar, a room of at least 9 m² is needed in order to allow access from at least two sides. Thus, in contrast to a storage room in the cellar, 3.4 m² more of surface area is required, leading to extra costs. Apart from the storage room door no other fixtures are necessary. Costs for a fibre tank amount to about 3,200 € (incl. VAT). On the whole, investment costs for this scenario rise from 20,200 € to 23,500 €, as compared to conventional cellar storage. Specific heat generation costs thus rise to 164.6 €/MWh (+10%) or, in the case of 1,400 € of subsidies, to 159.2 €/MWh.

8.2.3 Pellet central heating system with flue gas condensation

Flue gas condensation, state-of-the-art for natural gas furnaces and increasingly employed in the oil heating sector, was not introduced to the pellet furnace market until 2004, namely by the company ÖkoFEN (cf. Section 6.1.2.9.1.1). In an optimally designed system (the lower the return temperature of the heating circuit is, the more advantageous is flue gas condensation), flue gas condensation changes framework conditions, as demonstrated in Table 8.4. The annual efficiency is increased in comparison to a conventional pellet furnace. Assuming a boiler efficiency of 103% as per test protocol of the BLT Wieselburg and losses by radiation, cooling down, heat distribution and the control system of 10.4% (due to lack of data the losses were assumed to be the same as for pellet heating systems without flue gas condensation), the resulting annual efficiency is around 92.3%. The annual pellet demand then declines from 5.5 to about 5 t (w.b.)$_p$/a. Other framework conditions remain the same (cf. Table 8.2).

Investment costs for a boiler with flue gas condensation are 1,600 € more than for an equivalent boiler without flue gas condensation. Investment costs for the discharge system and the furnace room remain the same. So do investment costs for the chimney as the selected type of chimney is suitable for both conventional pellet furnaces and pellet furnaces with flue gas condensation. Investment costs for the storage space are slightly reduced due to the fuel demand and hence storage space demand is lower when this technology is used. The operating costs are just slightly elevated due to the increased electricity demand of the suction fan because a somewhat higher pressure drop has to be overcome. This extra energy demand is negligibly modest though and has therefore not been considered (cf. Section 6.1.2.9.1.3.1). On the whole, investment costs are around 21,700 €, thus 7.3% above the costs for an equivalent pellet heating system without flue gas condensation.

Fuel costs, as well as imputed interest for pellets stored, decline by around 9% due to the lowered annual fuel demand.

Table 8.4: Basic data for full cost calculation of a pellet central heating system with flue gas condensation

Explanations: price basis 12/2008; data source [59, 391], own research, VDI 2067 guideline; general framework conditions as in Table 8.2

Parameter	Value	Unit
Annual efficiency	92.3	%
Net calorific value	4.90	kWh/kg (w.b.)
Fuel demand	4,975	kg (w.b.) p.a.
Bulk density of pellets	625	kg (w.b.)/m^3
Electric power demand (in % of nominal boiler capacity)	0.70	%
Fuel price (absolute)	184	€/t (w.b.)$_p$
Fuel price (specific)	37.5	€/MWh$_{NCV}$
Maintenance and service effort (services on own account)	0.19	h/week
Chimney sweeper	143	€ p.a.
Service and maintenance furnace (in % of investment costs)	2.0	% p.a.
Funding (in % of investment costs)	25	%
Funding (upper limit)	1,400	€

Table 8.5: Full cost calculation of a pellet central heating system with flue gas condensation

Explanations: price basis 12/2008; basic data as in Table 8.1 and Table 8.2, data source [59], own research, VDI 2067 guideline; specific heat generation costs based on useful heat

	Investment costs €	Capital costs € p.a.	Maintenance costs € p.a.	Consumption costs € p.a.	Operating costs € p.a.	Other costs € p.a.	Total costs € p.a.	Specific costs €/MWh
Pellet boiler	14,200	1,238	284				1,522	67.65
Storage room discharge, storage room fixtures	2,700	235	54				289	12.9
Chimney	2,700	176	28				203	9.0
Construction costs furnace room, 3.6 m^2	879	57	9				66	2.9
Construction costs storage room, 5.1 m^2	1,247	81	13				94	4.2
Fuel costs				915			915	40.7
Electricity costs				27			27	1.2
Service and maintenance (services on own account)					58		58	2.6
Other costs						109	109	4.8
Imputed interest (fuel stored)				27			27	1.2
Chimney sweeper					143		143	6.4
Total costs	21,726	1,780	386	970	201	109	3,446	153.1
Specific costs €/MWh		79.1	17.2	43.1	8.9	4.8		153.1

All in all, specific heat generation costs of 153 €/MWh result, which is 2.3% more than for an equivalent pellet heating system without flue gas condensation. Therefore, conventional pellet heating systems are to be preferred to systems with flue gas condensation from an economic point of view, but with regards to energy efficiency, flue gas condensation clearly is the best option. It must be noted, however, that the system becomes economic once fuel prices rise because then the saved fuel costs compensate for the higher investment costs.

8.2.4 Oil central heating system

Table 8.7 shows the basic data for full cost calculation of an oil based central heating system. The annual efficiency is based on [390]. With regards to the general framework conditions, annual heating oil demand is around 2,500 l.

The average electric power demand of the system is slightly less than it is for pellet central heating systems, because fuel conveyor systems are simpler for oil than for pellets. The price of heating oil is the average price of the year 2008 on the basis of our own research. Transport and delivery costs as well as all taxes and fees are included.

Emptying the ash box is naturally not required in an oil heating system, thus maintenance and service work is reduced. Costs for chimney sweeping are based on the Styrian chimney sweep act. These costs are reduced too since heating systems based on liquid fuels need just three instead of four sweeps per year.

Maintenance costs were settled to 1% p.a. of investment costs according to VDI 2067.

There are no longer any subsidies for oil based heating systems without flue gas condensation in Austria (cf. Section 10.1.4.1).

Table 8.6: Basic data for full cost calculation of an oil central heating system

Explanations: price basis 12/2008; basic data as in Table 8.1, data source [59; 390; 393], own research, VDI 2067 guideline

Parameter	Value	Unit
Annual efficiency	90.0	%
Net calorific value	10.0	kWh/l
Fuel demand	2,500	l p.a.
Electric power demand (in % of nominal boiler capacity)	0.60	%
Fuel price (absolute)	865	€/1.000 l
Fuel price (specific)	86.5	€/MWh$_{NCV}$
Maintenance and service effort (services on own account)	0.15	h/week
Chimney sweeper	96.4	€ p.a.
Service and maintenance furnace (in % of investment costs)	1.0	% p.a.
Funding (in % of investment costs)	0.0	%
Funding (upper limit)	0	€

Table 8.7 shows the full cost calculation of an oil central heating system. The calculation is based on data from Table 8.1 and Table 8.6. Again, investment costs include not only the costs for furnace and boiler but also the costs for hot water supply as well as construction costs for the chimney, the storage and the furnace room. All other storage room equipment as well as delivery, installation and start-up costs were considered too. Storage space was designed, as in the case for pellets, for 1.2 times the annual fuel demand in order to be prepared for different winter conditions.

As mentioned already, the full cost calculation is based on the average fuel price of the year 2008, i.e. about 865 €/1,000 l (incl. 20% VAT and delivery fee, purchase quantity of 3,000 l). Heat generation costs thus are 169.5 €/MWh.

Table 8.7: Full cost calculation of an oil central heating system

Explanations: price basis 12/2008; basic data as in Table 8.1 and Table 8.6, data source [59], own research, VDI 2067 guideline; specific heat generation costs based on useful heat

	Investment costs €	Capital costs € p.a.	Maintenance costs € p.a.	Consumption costs € p.a.	Operating costs € p.a.	Other costs € p.a.	Total costs € p.a.	Specific costs €/MWh
Oil boiler	7,800	680	78				758	33.7
Oil tank and retention pond	2,600	244	28				272	12.1
Chimney	2,700	171	27				198	8.8
Construction costs furnace room, 1.9 m²	474	30	5				35	1.5
Construction costs storage room, 4.7 m²	1,148	73	11				84	3.7
Fuel costs				2,161			2,161	96.1
Electricity costs				23			23	1.0
Service and maintenance (services on own account)					46		46	2.0
Other costs						75	75	3.3
Imputed interest (fuel stored)				65			65	2.9
Chimney sweeper					96		96	4.3
Total costs	14,922	1,198	149	2,250	142	75	3,814	169.5
Specific costs €/MWh		53.3	6.6	100.0	6.3	3.3		169.5

8.2.5 Oil central heating system with flue gas condensation

As mentioned above, flue gas condensation has become increasingly utilised over recent years in the oil heating system sector. In order to make the most of this technology, a sufficiently low return temperature of the heating circuit is required. Table 8.8 shows the basic data for the full cost calculation of an oil central heating system with flue gas condensation. Compared to a conventional oil heating system, annual efficiency and fuel demand change. Oil central heating systems with flue gas condensation achieve boiler efficiencies of up to 105% (in an optimally designed system with the appropriately low return temperature). Assuming losses by radiation, cooling down, heat distribution and by the control system, the resulting annual efficiency is around 96.0% according to [390]. The annual heating oil demand declines from 2,500 to 2,350 l/a, with other framework conditions remaining the same (cf. Table 8.6).

Table 8.8: Basic data for full cost calculation of an oil central heating system with flue gas condensation

Explanations: price basis 12/2008; data source [59, 390, 393], own research, VDI 2067 guideline; general framework conditions as in Table 8.1

Parameter	Value	Unit
Annual efficiency	96.0	%
Net calorific value	10.0	kWh/l
Fuel demand	2,344	l p.a.
Electric power demand (in % of nominal boiler capacity)	0.60	%
Fuel price (absolute)	865	€/1.000 l
Fuel price (specific)	86.5	€/MWh$_{NCV}$
Maintenance and service effort (services on own account)	0.15	h/week
Chimney sweeper	96.4	€ p.a.
Service and maintenance furnace (in % of investment costs)	1.0	% p.a.
Funding (in % of investment costs)	0.0	%
Funding (upper limit)	0	€

Oil central heating systems with flue gas condensation are subsidised in some federal states of Austria if they replace older systems. Set-up in new buildings is not subsidised, even if flue gas condensation is employed. This is why subsidies were not considered with regard to the heat generation costs.

Investment costs of an oil central heating system with flue gas condensation are around 2,400 € higher than those of conventional oil central heating systems. Investment costs for chimney as well as storage and furnace room stay the same. As shown in Table 8.9, total investment costs increase to about 17,300 € and are thus 16.1% higher when flue gas condensation is employed in an oil central heating system.

Fuel costs and imputed interest on stored fuel decrease by 6.3%.

Specific heat generation costs are about 174.2 €/MWh, thus being 2.8% more than for a conventional oil central heating system. Flue gas condensation technology is thus not profitable from an economic point of view under the given framework conditions. However, strongly increasing fuel prices would render flue gas condensation more interesting as the system becomes economic once fuel prices rise, because only then can the saved fuel costs compensate for the higher investment costs. What is more important with regard to flue gas condensation is the considerable increase in efficiency that can be achieved, which is why an oil central heating system with flue gas condensation is to be preferred to a conventional oil central heating system from an ecologic point of view.

Table 8.9: Full cost calculation of an oil central heating system with flue gas condensation

Explanations: price basis 12/2008; basic data as in Table 8.1 and Table 8.8, data source [59], own research, VDI 2067 guideline; specific heat generation costs based on useful heat

	Investment costs €	Capital costs € p.a.	Maintenance costs € p.a.	Consumption costs € p.a.	Operating costs € p.a.	Other costs € p.a.	Total costs € p.a.	Specific costs €/MWh
Oil boiler	10,200	889	102				991	44.1
Oil tank and retention pond	2,600	244	28				272	12.1
Chimney	2,700	171	27				198	8.8
Construction costs heating room, 1.9 m^2	474	30	5				35	1.5
Construction costs storage room, 4.7 m^2	1,148	73	11				84	3.7
Fuel costs				2,026			2,026	90.1
Electricity costs				23			23	1.0
Service and maintenance (services on own account)					46		46	2.0
Other costs						87	87	3.8
Imputed interest (fuel stored)				61			61	2.7
Chimney sweeper					96		96	4.3
Total costs	17,322	1,408	173	2,110	142	87	3,920	174.2
Specific costs €/MWh		62.6	7.7	93.8	6.3	3.8		174.2

8.2.6 Natural gas heating system with flue gas condensation

Table 8.10 shows the basic data for the full cost calculation of a natural gas central heating system with flue gas condensation. Natural gas central heating systems with flue gas condensation achieve annual efficiencies of up to 96% [390]. The annual natural gas demand is about 2,440 Nm^3/a under the given framework conditions.

The average electric power demand of the system is 0.3% of the nominal boiler capacity, which is less than for pellet or oil heating systems. This is because no fuel conveyor systems are required. The calculation of the heat generation costs is based on the average gas price of

2008 including all taxes and fees. In contrast to pellet or oil heating systems, no storage space is required for natural gas. Since the fuel is supplied by a supply network, a gas meter has to be installed for which there is a meter rental.

Maintenance and service activities were assumed to require no more than 0.06 hours per week, which is even less than in the case of heating oil. This is because emptying the ash box is spared and the fuel does not have to be organised. Besides, there is no storage space that has to be cleaned or else serviced. As mentioned in Section 8.2.4, there are generally no longer any subsidies for heating systems based on fossil fuels in Austria (cf. Section 10.1.4.1). However, natural gas heating systems with flue gas condensation, like oil heating systems with flue gas condensation, are subsidised if they replace an older system. Set-up in new buildings is not subsidised though, even if flue gas condensation is employed. This is why subsidies were not considered with regard to the heat generation costs for this case.

Table 8.10: Basic data for full cost calculation of a natural gas heating system with flue gas condensation

Explanations: price basis 12/2008; data source [44; 59; 390], own research, VDI 2067 guideline; general framework conditions as in Table 8.1

Parameter	Value	Unit
Annual efficiency	96.0	%
Net calorific value	9.60	kWh/Nm^3
Fuel demand	2,441	Nm^3 p.a.
Electric power demand (in % of nominal boiler capacity)	0.30	%
Fuel price (absolute)	668	€/kNm^3
Fuel price (specific)	69.6	€/MWh$_{NCV}$
Meter rental	57.6	€ p.a.
Maintenance and service effort (services on own account)	0.06	h/week
Chimney sweeper	43.5	€ p.a.
Service and maintenance furnace (in % of investment costs)	1.0	% p.a.
Funding (in % of investment costs)	0.0	%
Funding (upper limit)	0	€

Costs for chimney sweeping are again based on the Styrian chimney sweep act. These costs are the lowest of all scenarios, except biomass district heating (as there is no chimney required), since heating systems based on natural gas need just one sweep per year. Maintenance costs were settled at 1% of annual investment costs according to VDI 2067.

Table 8.11 shows the full cost calculation of a natural gas central heating system with flue gas condensation. Again, investment costs include not only the costs for furnace and boiler but also the costs for hot water supply as well as all costs for delivery, installation and start-up. Storage facilities are not needed. However, there is a connection fee of 1,900 € (this value may be different depending on federal state and gas supplier). A separate furnace room is also not needed in natural gas central heating systems with flue gas condensation, hence no costs are incurred by this element.

Since the fuel is obtained from a supply network, a gas meter has to be installed and rent has to be paid. Since there is no fuel storage, there is no imputed interest on it.

Table 8.11: Full cost calculation of a natural gas central heating system with flue gas condensation

Explanations: price basis 12/2008; basic data as in Table 8.1 and Table 8.10, data source [59], own research, VDI 2067 guideline; specific heat generation costs based on useful heat

	Investment costs €	Capital costs € p.a.	Maintenance costs € p.a.	Consumption costs € p.a.	Operating costs € p.a.	Other costs € p.a.	Total costs € p.a.	Specific costs €/MWh
Natural gas condensing boiler	8,800	767	72				840	37.3
Connection fee	1,900	166	16				181	8.1
Chimney	2,700	171	27				198	8.8
Fuel costs				1,632			1,632	72.5
Meter rental					58		58	2.6
Electricity costs				12			12	0.5
Service and maintenance (services on own account)					18		18	0.8
Other costs						67	67	3.0
Chimney sweeper					43		43	1.9
Total costs	13,400	1,104	115	1,643	119	67	3,049	135.5
Specific costs €/MWh		49.1	5.1	73.0	5.3	3.0		135.5

Heat generation costs as calculated by means of a full cost calculation amount to 135.5 €/MWh for a natural gas central heating system with flue gas condensation.

8.2.7 Wood chips central heating system

Table 8.12 shows the basic data for the full cost calculation of a wood chip central heating system. Wood chip central heating systems achieve annual efficiencies of around 80% on average [390]. The annual wood chips demand is about 7.6 t (w.b.)/a under the given framework conditions.

Table 8.12: Basic data for full cost calculation of a wood chip central heating system

Explanations: price basis 12/2008; data source [44; 59; 390], own research, VDI 2067 guideline; general framework conditions as in Table 8.1

Parameter	Value	Unit
Annual efficiency	80.0	%
Net calorifc value (M25)	3.72	kWh/kg (w.b.)
Fuel demand	7,560	kg p.a.
Bulk density of wood chips	233	kg (w.b.)/m^3
Electric power demand (in % of nominal boiler capacity)	0.80	%
Fuel price (absolute)	109.7	€/t (w.b.)
Fuel price (specific)	29.5	€/MWh$_{NCV}$
Maintenance and service effort (services on own account)	0.27	h/week
Chimney sweeper	143	€ p.a.
Service and maintenance furnace (in % of investment costs)	2.0	% p.a.
Funding (in % of investment costs)	25	%
Funding (upper limit)	1,400	€

The average electric power demand of the system is 0.8% of the nominal boiler capacity, which is more than in pellet fired furnaces due to the conveyor systems that have to be built in a more robust way. The price of wood chips is around 24.0 €/lcm (price basis 2008, incl. VAT and delivery). With a moisture content of 25 wt.% (w.b.), this is equivalent to 110 €/t (w.b.).

Maintenance and service activities were assumed to require 0.27 hours per week, which is more than for pellet furnaces. This is because emptying the ash box is needed more frequently and fuel acquisition is not as simple since wood chips have a lower energy density than pellets, which demands fuel delivery at least twice a year. Chimney sweeping costs are almost the same as for pellet central heating systems, namely 143 €/a, based on the Styrian chimney sweep act.

In this case, maintenance costs were not put down according to the VDI 2067 guideline. They were assumed to be 2% instead of 1% p.a. of the investment costs; hence twice as high as for oil or gas central heating systems.

Wood chip central heating systems are subsidised, like pellet central heating systems, to different extents depending on the federal states of Austria (cf. Section 10.1.4.1). Specific heat generation costs were calculated according to the Styrian framework conditions with maximum subsidies of 25% or 1,400 €. The national subsidy of 400 € [392] that was available from February 2008 to January 2009 was not taken into account here.

The full cost calculation of a wood chip central heating system is presented in Table 8.13.

Table 8.13: Full cost calculation of a wood chip central heating system

Explanations: price basis 12/2008; basic data as in Table 8.1 and Table 8.12, data source [44; 59], own research, VDI 2067 guideline; specific heat generation costs based on useful heat

	Investment costs €	Capital costs € p.a.	Maintenance costs € p.a.	Consumption costs € p.a.	Operating costs € p.a.	Other costs € p.a.	Total costs € p.a.	Specific costs €/MWh
Wood chip boiler	18,600	1,622	372				1,994	88.6
Storage room discharge, storage room fixtures	3,200	279	64				343	15.2
Chimney	2,700	171	27				198	8.8
Construction costs furnace room, 6.4 m²	1,563	99	16				115	5.1
Construction costs storage room, 12.5 m²	3,052	194	31				224	10.0
Fuel costs				829			829	36.9
Electricity costs				31			31	1.4
Service and maintenance (services on own					82		82	3.7
Other costs						146	146	6.5
Imputed interest (fuel stored)				25			25	1.1
Chimney sweeper					143		143	6.4
Total costs	29,115	2,365	509	886	226	146	4,131	183.6
Specific costs €/MWh		105.1	22.6	39.4	10.0	6.5		183.6

The storage room was not designed in a way that 1.2 times the annual fuel demand can be stored since the low energy density of wood chips would demand an extensive storage volume thus leading to significantly increased investment costs. Installing such a large storage space is not realistic, so a storage space for about half the annual fuel demand was chosen. The increased service effort caused by this is taken into account by the personnel demand of 0.27 hours per week. The reason why investment costs are so much higher than for an equivalent pellet heating system is that not only is a feeding screw needed for fuel discharge but also an agitator inside the storage room. In addition, the investment costs of furnace and boiler are higher. Once more, investment costs comprise costs for furnace and boiler as well as hot water supply, chimney, delivery, installation and start-up costs, as well as all the furnace and storage room equipment needed.

Specific heat generation costs of the wood chip central heating system as presented and on the basis of a full cost calculation amount to 183.6 €/MWh. If subsidies of 1,400 € are considered, the costs would decrease to 178.2 €/MWh. The total heat generation costs would be reduced by around 3.0% from 4,131 €/a to 4,009 €/a.

8.2.8 Biomass district heating

Table 8.14 shows the basic data for the full cost calculation of a biomass district heating system. The annual efficiency of a biomass district heating system is stated to be 96% [394], where only distribution losses in the house to be heated are considered since all other losses occur during the conversion process in the district heating plant. Heat demand is around 23.4 MWh/a under the given framework conditions.

A biomass district heating system has no electricity demand of its own because there is neither a fuel conveyor system nor a suction fan (the pump for the heating circuit in the house to be heated was not taken into account in any of the scenarios here). The heat price is based on an average price of an actual Austrian biomass district heating plant.

The use of a biomass district heating system hardly requires any maintenance or service. Servicing by users, fuel organisation and de-ashing are not necessary. Usually there are no separate maintenance costs because maintenance is normally carried out by the system provider and is included in the heat price. Costs for the chimney sweep are also spared, however, requirement to be present during maintenance, plus the demands of administrative activities, were considered.

Table 8.14: Basic data for the full cost calculation of biomass district heating

Explanations: price basis 12/2008; data source [59], own research, VDI 2067 guideline; general framework conditions as in Table 8.1

Parameter	Value	Unit
Annual efficiency	96.0	%
Electric power demand (in % of nominal boiler capacity)	0.00	%
Heat price	82.1	€/MWh
Meter rental	86.3	€ p.a.
Maintenance and service effort (services on own account)	0.04	h/week
Utilisation period heat transfer unit	30	a
Maintenance heat transfer unit	0.0	% p.a.
Funding (upper limit)	1,200	€

The extent to which biomass district heating is subsidised varies by federal state. Calculations were based on the subsidy that is achievable in the federal state of Upper Austria, i.e. 1,200 €.

As concerns investment costs, connection to a biomass district heating network also demonstrates numerous benefits in comparison to other heating systems. The costs of the heat transfer station merely include costs for the boiler for hot water supply, i.e. costs for the boiler itself as well as delivery, installation and start-up. Storage and furnace rooms are not required, so construction costs are avoided. So are costs for a chimney. Despite connection costs of 1,900 €, total investment costs are rather low, amounting to around 8,800 €. Heat generation costs were calculated according to Table 8.15 without consideration of investment subsidies. They are 120.2 €/MWh. Taking investment subsidies into account, specific heat generation costs are lowered to 116.4 €/MWh and annual heat costs are reduced from 2,705 to 2,618 €.

Table 8.15: Full cost calculation of biomass district heating

Explanations: price basis 12/2008; basic data as in Table 8.1 and Table 8.14, data source [44; 59], own research, VDI 2067 guideline; specific heat generation costs based on useful heat

	Investment costs €	Capital costs € p.a.	Maintenance costs € p.a.	Consumption costs € p.a.	Operating costs € p.a.	Other costs € p.a.	Total costs € p.a.	Specific costs €/MWh
Heat transfer unit	6,900	501	0				501	22.3
Connection fee	1,900	138	0				138	6.1
Heat costs				1,924			1,924	85.5
Meter rental					86		86	3.8
Other costs						44	44	2.0
Total costs	8,800	639	0	1,924	98	44	2,705	120.2
Specific costs €/MWh		28.4	0.0	85.5	4.4	2.0		120.2

8.2.9 Comparison of the different systems

In this section, the heating systems discussed in Sections 8.2.2 to 8.2.8 are compared with regard to investment costs, fuel, respectively heat costs and heat generation costs, both with and without possible subsidies. The price basis for the comparison of the systems is December 2008. All prices include VAT as well as all other taxes and fees.

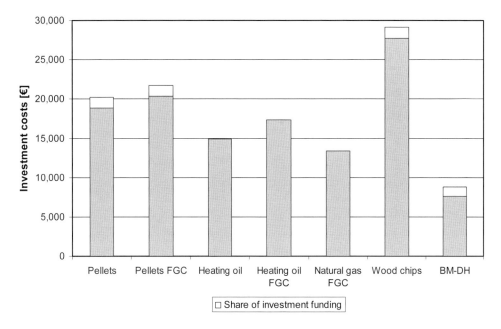

Figure 8.4: Comparison of investment costs for different heating systems

Explanations: nominal boiler capacity 15 kW; price basis 12/2008; possible subsidies: 1,400 € for pellet and wood chip furnaces; 1,200 € for connection to a biomass district heating network (guiding values that are different for each Austrian federal state)

A comparison of the investment costs of the different heating systems is shown in Figure 8.4. The investment costs of all scenarios include costs for delivery, installation and start-up. Costs for hot water supply as well as chimney, furnace and storage room construction were considered for all systems too, if applicable. Investment costs for heat distribution equipment in the house to be heated were not taken into account.

It must be noted that actual prices may be above or beneath the prices stated here, depending on manufacturer and design. The comparison is based on the case of one distinct, and great care has been taken to select equivalent systems with regard to power output and design criteria.

A central heating system based on district heat has the lowest investment costs, followed by gas and oil heating systems. Oil heating systems with flue gas condensation are 16% more expensive. Investment costs for pellet and wood chip furnaces are comparatively high. The investment costs for wood chip furnaces are especially high due to the agitator that is required for storage discharge. Pellet systems with flue gas condensation are also more costly – namely around 7.3% more than the costs of a conventional system.

Looking at oil and pellet heating systems in direct comparison, it can be noted that pellet heating systems are around 36% (conventional system) and 25% (system with flue gas condensation) more expensive. If investment subsidies are taken into account, these values are reduced to 26% and 17%, respectively.

Figure 8.5 displays a comparison between annual fuel and heat costs. The dissimilar fuel demands are considered by the different annual efficiencies of the systems, which were calculated according to Equation 8.1.

Equation 8.1: $$C_{F,a} = \frac{P_N \cdot t_f}{\eta_a \cdot NCV} \cdot C_F \cdot 100$$

Explanations: $C_{F,a}$ in €/a; P_N in kW; t_f in h p.a.; η_a in %; NCV in kWh/amount; C_F in €/amount

Fuel costs are lowest for wood chips as these exhibit the lowest prices. Pellet heating systems also show inexpensive fuel prices that are reduced by a further 9% when the system uses flue gas condensation. Fuel costs are higher for fossil fuels. Conventional oil heating systems have the highest annual fuel costs. Fuel costs for heating oil can be reduced by 6.3% when flue gas condensation is employed. Natural gas heating systems show the lowest fuel costs within the group of fossil fuels systems. Comparing heat costs of biomass district heating directly to fuel costs of the other heating systems, the heat costs are slightly less than the fuel costs for oil heating systems. However, this comparison is not actually permissible since the heat price includes the total heat generation costs of the biomass district heating plant as well as the service and maintenance costs related to the heat transfer station. This is why the only valid evaluation is the comparison of heat generation costs based on full cost calculations, which is carried out below (in principle this is also true of the other systems).

Figure 8.6 displays the results of the full cost calculations of the different heating systems by means of the specific heat generation costs.

The two network bound heating systems, i.e. natural gas heating system with flue gas condensation and biomass district heating, have the lowest specific heat generation costs (with and without possible subsidies for biomass systems) with clear advantages of biomass district heating. Application of these systems implies the presence of a supply network. The wood chip heating systems exhibit the highest heat generation costs (with and without possible subsidies). Heat generation costs of pellet and oil heating systems are in between but pellet heating systems have clear advantages. If possible subsidies for pellet heating systems are considered, these advantages are even more prominent. Heat generation costs are slightly raised by the use of flue gas condensation in both pellet and oil heating systems. The advantage of lowered fuel demands is more than counterbalanced by the elevated investment costs. Yet, from an ecologic point of view, using flue gas condensation is recommended because of the efficiency increase.

The reason for the low costs of natural gas heating systems and biomass district heating systems are their comparatively low investment costs. A biomass district heating system has the further advantage that no chimney is needed. Usually there are no maintenance costs since maintenance is normally carried out by the system provider and thus the costs are included in the heat price. The elevated heat generation costs of wood chips furnaces are mainly caused by the increased investment costs. The cheap fuel only compensates for this in part. The higher investment costs of pellet systems as compared to heating oil systems are compensated for by the pellet price being far beneath the price for heating oil. So pellet heating systems have clear advantages in this respect.

If subsidies of 1,400 € for pellet and wood chip heating systems and 1,200 € for biomass district heating connection (subsidies being varied among Austrian federal states) are taken into account, the cost–benefit ratio is shifted more towards biomass district heating. The above order of heat generation costs is not influenced by this, however, it can be one more incentive compared to oil or gas heating systems.

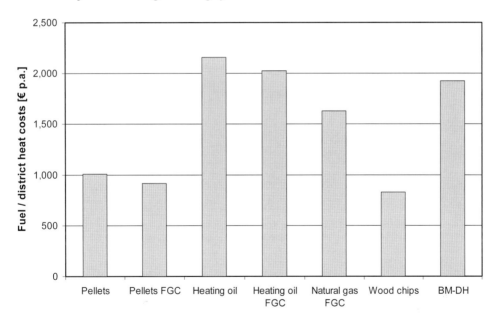

Figure 8.5: Comparison of annual fuel and heat costs

Explanations: bases 12/2008; values incl. VAT; basic data according to Sections 8.2.1 to 8.2.8; calculation according to Equation 8.1

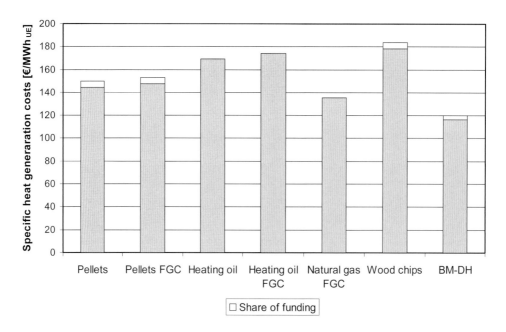

Figure 8.6: Comparison of specific heat generation costs of different heating systems

Explanations: nominal boiler capacity 15 kW; price basis 12/2008; values incl. VAT; 1,500 annual full load operating hours; interest rate 6% p.a.; further basic data as in Sections 8.2.1 to 8.2.8; specific heat generation costs based on useful heat

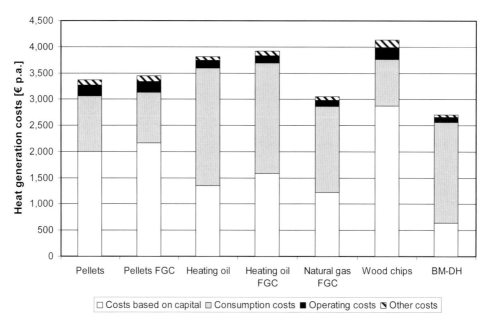

Figure 8.7: Comparison of annual heat generation costs broken down into costs based on capital, consumption costs, operating costs and other costs

Explanations: nominal boiler capacity 15 kW; price basis 12/2008; 1,500 annual full load operating hours; interest rate 6% p.a.; further basic data as in Sections 8.2.1 to 8.2.8

Total annual costs of heat generation split up into costs based on capital, consumption, operating and other costs are shown in Figure 8.7.

It can be seen that operating costs and other costs play a subordinate role. Most cost is generated by costs based on capital and consumption costs. Within these two groups of costs there are pronounced differences with regard to the heating systems.

The costs based on capital comprise capital costs (investment costs) and maintenance costs. Maintenance costs make up around 18% of costs based on capital in pellet and wood chip central heating systems. This share is around 11% for oil heating systems, 9% for natural gas heating systems and zero for biomass district heating. Thus costs based on capital mainly consist of investment costs. Costs based on capital make up 59 to 70% of total costs in pellet and wood chip heating systems. In oil and natural gas heating systems, consumption costs make up 54 to 59% of total costs. Consumption costs are therefore the main costs in fossil fuel systems and they are dominated by the fuel costs.

8.2.10 Sensitivity analysis

The foremost cost factors for specific heat generation costs were found to be fuel and investment costs. Sensitivity analyses were carried out for both parameters as shown in Figure 8.8 and Figure 8.9. They display the correlation between specific heat generation costs and fuel price (heat price in the case of biomass district heating), respectively investment costs.

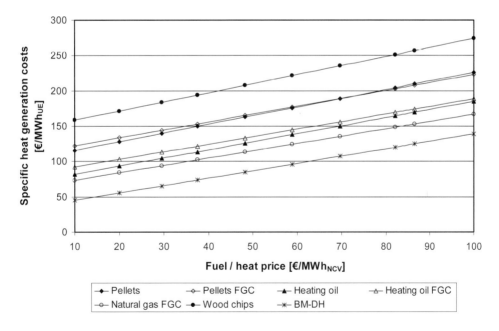

Figure 8.8: Influence of fuel or heat price on specific heat generation costs

Explanations: calculation of specific heat generation costs according to Sections 8.2.1 to 8.2.8

Changing fuel or heat price by 10% has the strongest effect on specific heat generation costs, which rise by 7% in the case of biomass district heating. This can be explained by the dominant role heat costs play in this system. In oil and natural gas heating systems, the specific heat generation costs would change by 5.3 to 5.8%. The effect would be rather small in pellet heating systems both with and without flue gas condensation, i.e. 3.1% and 2.7%, respectively. Specific heat generation costs of wood chips systems would change the least at 2.1%.

The picture is changed when investment costs (furnace and storage only) are varied by ± 10% (cf. Figure 8.9). This results in a very slight change of specific heat generation costs of biomass district heating (less than 2%). Heating systems on the basis of natural gas and heating oil also prove to be low in sensitivity as concerns investment costs (2.8 to 3.4%). The greatest change is found in wood chips furnaces with 5.9%, while pellet heating systems also display rather large changes of 5.1 to 5.5%. This shows a great potential for cost reduction, especially for pellet heating systems since a price reduction of these systems can be expected if the production capacities of manufacturers keep rising.

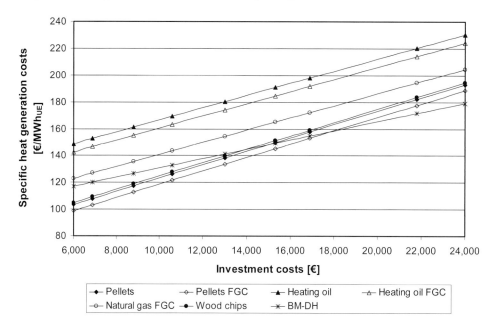

Figure 8.9: Influence of investment costs on specific heat generation costs

Explanations: calculation of specific heat generation costs according to Sections 8.2.1 to 8.2.8

There is some uncertainty in the calculation concerning the determination of annual efficiencies. The annual efficiency does not only take conversion losses (boiler efficiency) into account but also the losses due to heat distribution at the end user site, start-up, shutdown, radiation, cooling down and to the control system, which means that it partly depends on user behaviour. Individual annual efficiencies may therefore differ strongly from the average annual efficiency. The efficiencies were determined according to available data in the literature and, in the case of pellet heating systems with flue gas condensation, according to test stand measurements together with average losses (cf. 8.2.2 to 8.2.8). Figure 8.10 shows the dependence of the heating systems in this respect. A decrease of annual efficiency in particular can lead to pronounced increases of specific heat generation costs. Therefore, proper system design and installation of efficient control systems are of great significance.

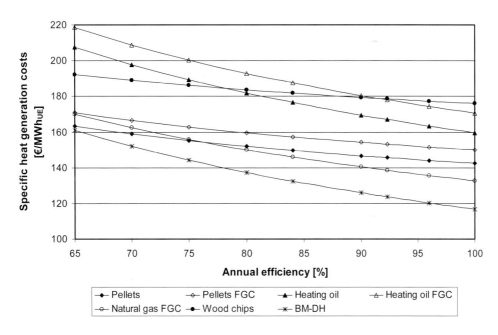

Figure 8.10: Influence of annual efficiencies on specific heat generation costs

Explanations: calculation of specific heat generation costs according to Sections 8.2.1 to 8.2.8

The effects of different scenarios with regard to the development of fuel and heat prices as well as investment costs are looked at in detail below. The effect of subsidies in biomass heating systems is also discussed. An overview of the scenarios and their effects is displayed in Table 8.16.

The specific heat generation costs of the pellet furnace (without flue gas condensation) are less than for the oil furnace without flue gas condensation under the framework conditions. If the oil price was lowered to 0.692 €/l (- 20%), the specific heat generation costs of the heating oil furnace would be the same as for the pellet furnace. This outcome could also be achieved with flue gas condensation and an oil price of 0.668 €/l (- 22.7%). In this context it has to be noted that the fuel prices taken into account for the calculations are average prices of 2008, because calculation with a price derived from a single point in time is not valid due to strong fluctuations in fuel prices. If the maximum oil price of July 2008 is chosen as the calculation basis, specific heat generation costs would be raised to 195.0 €/MWh (without flue gas condensation) and 198.1 €/MWh (with flue gas condensation) – values that are far above those of the calculation based on average prices. The same is true for pellets when looking at the price maximum of December 2006. A calculation based on this price results in specific heat generation costs of 172.7 €/MWh (without flue gas condensation) or 174.1 €/MWh (with flue gas condensation) – costs that are still notably beneath the costs of oil heating systems based on the maximum price of oil.

The following investigations are based on the average prices of heating oil and pellets that were taken as a basis for the full cost calculation of the previous sections.

Table 8.16: Different scenarios and their effects on specific heat generation costs

Explanations: price basis 12/2008; framework conditions, basic data and full cost calculation as in Sections 8.2.1 to 8.2.8

Heating system	Scenario	Value	Change	SHGC [€/MWh]	Comments
Heating oil	Oil price decrease	0,692 €/l	- 20,0%	149.7	equal to pellet heating systems
Heating oil-FGC	Oil price decrease	0,668 €/l	- 22,7%	153.1	equal to pellet heating systems with FGC
Heating oil	Oil price increase	1.087 €/l	+ 25,7%	195.0	>> pellet heating systems (oil price maximum in July 2008)
Heating oil-FGC	Oil price increase	1.087 €/l	+ 25,7%	198.1	>> pellet heating systems (oil price maximum in July 2008)
Pellets	Investment costs reduction (boiler and fuel feeding system)		- 18,7%	135.5	equal to natural gas heating system with FGC
Pellets FGC	Investment costs reduction (boiler and fuel feeding system)		- 20,9%	135.5	equal to natural gas heating system with FGC
Pellets	Pellet price increase	263,0 €/t (w.b.)$_p$	+ 42,9%	174.2	equal to heating oil systems (pellet price maximum in December 2006: 275,8 €/t (w.b.)$_p$)
Pellets FGC	Pellet price increase	276,5 €/t (w.b.)$_p$	+ 50,3%	174.2	equal to heating oil systems with FGC (pellet price maximum in December 2006: 275,8 €/t
Pellets	Pellet price decrease	127,1 €/t (w.b.)$_p$	- 30,9%	135.5	equal to natural gas heating system with FGC
Pellets FGC	Pellet price decrease	106,5 €/t (w.b.)$_p$	- 42,1%	135.5	equal to natural gas heating system with FGC
Pellets	Taking investment funding into account (federal state level)	1,400 €		144.3	6.5% higher than natural gas heating system with FGC
Pellets FGC	Taking investment funding into account (federal state level)	1,400 €		147.7	9% higher than natural gas heating system with FGC
Pellets	Taking investment funding into account (national level)	2,200 €		141.2	4.2% higher than natural gas heating system with FGC
Pellets FGC	Taking investment funding into account (national level)	2,200 €		144.6	6.7% higher than natural gas heating system with FGC
Natural gas FGC	Gas price increase	982 €/kNm³	+ 49,9%	169.5	equal to heating oil systems
Natural gas FGC	Gas price increase	1.025 €/kNm³	+ 53,4%	174.2	equal to heating oil systems with FGC
Natural gas FGC	Gas price increase	800 €/kNm³	+ 19,6%	149.7	equal to pellet heating systems
Natural gas FGC	Gas price increase	831 €/kNm³	+ 24,3%	153.1	equal to pellet heating systems with FGC
Wood chips	Investment costs reduction (boiler and fuel feeding system)		- 31,1%	149.7	equal to pellet heating systems
Wood chips	Investment costs reduction (boiler and fuel feeding system)		- 28,0%	153.1	equal to pellet heating systems with FGC
Wood chips	Investment costs reduction (boiler and fuel feeding system)		- 13,0%	169.5	equal to heating oil systems
Wood chips	Investment costs reduction (boiler and fuel feeding system)		- 8,6%	174.2	equal to heating oil systems with FGC
Wood chips	Taking investment funding into account (federal state level)	1,400 €		178.2	> than all other systems
Wood chips	Taking investment funding into account (national level)	1,800 €		176.6	> than all other systems
Wood chips	Wood chip price decrease (M25)	11.9 €/t (w.b.)	- 89,1%	149.7	equal to pellet heating systems
Wood chips	Wood chip price decrease (M25)	21.8 €/t (w.b.)	- 80,2%	153.1	equal to pellet heating systems with FGC
Wood chips	Wood chip price decrease (M25)	69 €/t (w.b.)	- 37,1%	169.5	equal to heating oil systems
Wood chips	Wood chip price decrease (M25)	82.7 €/t (w.b.)	- 24,7%	174.2	equal to heating oil systems with FGC
BM-DH	Taking investment funding into account	1,200 €		116.4	increases the difference to the other systems
BM-DH	BM-DH price increase	129,4 €/MWh	+ 57,6%	169.5	equal to heating oil systems
BM-DH	BM-DH price increase	133,9 €/MWh	+ 63,2%	174.2	equal to heating oil systems with FGC
BM-DH	BM-DH price increase	110,4 €/MWh	+ 34,5%	149.7	equal to pellet heating systems
BM-DH	BM-DH price increase	113,7 €/MWh	+ 38,5%	153.1	equal to pellet heating systems with FGC
Heating oil	Basis: current oil price (December 2008)	0,574 €/l	- 33,6%	136.3	< pellet heating systems
Heating oil-FGC	Basis: current oil price (December 2008)	0,574 €/l	- 33,6%	143.1	< pellet heating systems
Natural gas FGC	Basis: current gas price (December 2008)	817 €/kNm³	+ 22,2%	151.6	similar to pellet heating systems
Pellets	Basis: current pellet price (December 2008)	193,8 €/t (w.b.)$_p$	+ 5,4%	152.2	slight increase
Pellets FGC	Basis: aktueller Pelletspreis (Dezember 2008)	193,8 €/t (w.b.)$_p$	+ 5,4%	155.4	slight increase

A decrease in the investment costs of pellet boilers and storage space discharge system by 18.7% would put the specific heat generation costs at the same level as natural gas heating systems with flue gas condensation. A pellet boiler with flue gas condensation would require a decrease in investment costs by 20.9%. Investment costs for pellet heating systems are actually expected to decline due to increasing sales volumes of pellet heating systems and hence bigger production capacities. As great a decrease as mentioned above is not expected, however.

For pellet heating systems to accomplish the same specific heat generation costs as natural gas systems, the price of pellets would have to fall by an unrealistic 30.9% (without flue gas condensation) or even 42.1% (with flue gas condensation).

If subsidies on federal state level of 1,400 € are taken into account, specific heat generation costs would be around 144.3 €/MWh. A system with flue gas condensation would have specific heat generation costs of 147.7 €/MWh. If the 800 € national subsidy that was available until January 2009 were taken into account, specific heat generation costs would be reduced to 141.2 €/MWh (without flue gas condensation) or 144.6 €/MWh (with flue gas condensation).

For a natural gas heating system, the gas price would have to rise by 49.9 to 53.4% as compared to the average value of 2008 in order to have the same specific heat generation costs as an oil heating system. For a gas heating system to reach the level of specific heat generation costs of the pellet heating system, an increase by 19.6 to 24.3% would be enough. A gas price of this level was actually reached in November 2008. With that, specific heat generation costs of natural gas heating systems have measured up to those of pellets.

In order to lower the specific heat generation costs of a wood chip heating system, investment costs would have to decrease by 9 to 13% or the price of wood chips would have to decrease by around 25 to 37%. For the specific heat generation costs of wood chips heating systems to reach the level of pellet heating systems, investment costs would need to decrease by around 28 to 31% or wood chip prices by 80 to 90%. Both cases are unrealistic.

Taking both the 1,400 € possible federal state subsidy and the 400 € possible national subsidy into account, the specific heat generation costs would decline to 178.2 € /MWh or 176.6 €/MWh. Still, the wood chips heating system would have the highest specific heat generation costs of all systems. The use of wood chip furnaces seems reasonable in a sustainable economy when used by farmers who hold a forest of their own. Lowering the price of the wood chips in this case is not permissible, however, because the market price that could be achieved is still the same. Therefore, transport costs alone may be omitted.

Making use of biomass district heating is very economical. Considering a possible investment subsidy of 1,200 €, specific heat generation costs of just 120.2 €/MWh are reduced even further to 116.4 €/MWh. Heat price would have to rise by 58 to 63% for the specific heat generation costs of biomass district heating to reach the level of oil heating systems. In order to attain the level of pellet heating systems, the heat price would need to rise by 35 to 39%.

8.3 External costs of residential heating based on different heating systems

All the costs that have been discussed up to this point are internal costs and thus costs that are covered directly by the end user of the actual heating system. Looking at the national economy, it seems reasonable to include external costs in the calculation of specific heat generation costs. External costs refer to costs caused by environmental impacts such as health damage, damage to flora and fauna and damage to buildings as well as climate and safety risks (major accidents, waste disposal, etc.). These external costs are not included in the market price and are thus covered by the general public. This is why effects such as these cannot be evaluated by means of market prices, however, many different methods for monetary evaluation of external effects have been developed. Two basic methods can be distinguished: determination of damage costs and determination of prevention costs [395]. Damage costs are directly determined by means of market prices, provided that the damage is reversible and can be removed by appropriate "repairing". Damages of an immaterial nature or irreversible damages such as loss of human life, extinction of species or evaluating the risk of major accidents cannot be carried out in such a simple way and are thus difficult to determine. Methods can be employed for this, which derive monetary values for relevant kinds of damage from the preferences of the affected. Determination of prevention costs aims to determine the costs needed to avoid any such damaging effect by preventive measures. A damage prevention approach alone is not concerned with the extent of anticipated damages at all. Thus, damage and prevention costs can in fact not be balanced out with the aim of overall optimisation.

An "exact" determination of external costs is legitimately impossible due to the above reasons. Including external costs in economic considerations would be a step into the right direction from the national economy point of view since the environment is not a reproducible good that can be consumed and manufactured again. Not taking external costs into account means that they are designated a value of zero, which surely is a false approach.

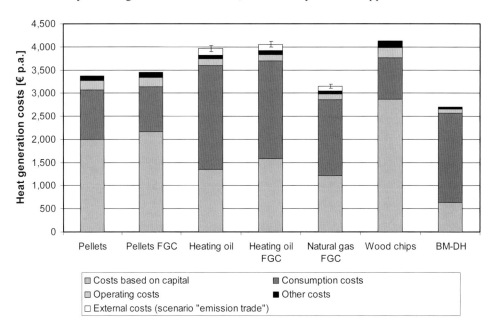

Figure 8.11: Specific heat generation costs of central heating systems with external costs for the scenario "emission trade"

Explanations: consideration of CO_2 emissions on the basis of prices for emission certificates of 9.3 to 31.9 €/t CO_2 (minimum and maximum value between January 2008 and January 2009 [396]) for the fictional case of small-scale furnaces being included in the emission trade

In this work two approaches were chosen to estimate the external costs of the heating systems.

In the first approach, the effects the CO_2 emission trade would have on the economy of small-scale furnaces, if these were considered (which is not the case at present), were examined. Costs for CO_2 emissions were set to range from 9.3 to 31.9 €/t CO_2 (minimum and maximum value between January 2008 and January 2009 [396]). Specific heat generation costs of oil and natural gas heating systems would rise by about 1.8 to 5.6% in this case (depending on the scenario, cf. Figure 8.11). The economy of biomass based systems is not altered as biomass is CO_2 neutral. Thus systems based on fossil fuels become more expensive. The order of the systems with regard to specific heat generation costs is unchanged, however.

In the second approach, data for monetary evaluation of pollutant emissions were drawn from different studies in this field [59; 395; 397]. Apart from external costs for CO_2 emissions, emissions of CO, C_xH_y, NO_x, SO_2 and particulate matter are also considered. The data used are not derived from a pure prevention costs approach. Damage costs and costs for risk assessment are also included. Since data of this kind vary considerably, external costs for pollutant emissions were calculated on the basis of minimum, maximum and average values of external costs. Thereby, some idea of the range of variation of these costs can also be gained.

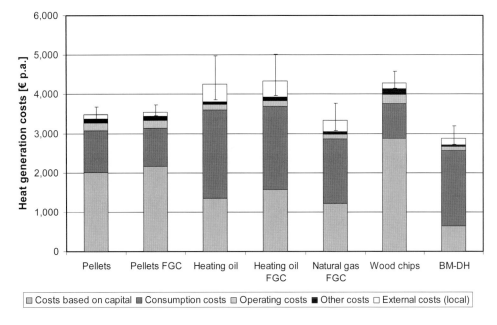

Figure 8.12: Specific heat generation costs of central heating systems with external costs based on local emission prognoses

Explanations: monetary evaluation of emissions on the basis of data available from studies in the field [59; 395; 397]; calculation of external costs based on local emission prognoses, i.e. considering emissions directly caused by the combustion of the different fuels

The economic efficiency of the systems is clearly changed when external costs are considered. If the calculation of external costs is based on a local emission prognosis, i.e. on emissions directly caused by the combustion of the different fuels, then total costs of the systems based on fossil fuels rise clearly more than total costs of systems based on biomass fuels (cf. Figure 8.12). In the average case, this leads to heating oil and wood chip based systems having the highest specific heat generation costs. Natural gas heating systems and district heating remain the cheapest options.

If the calculation considers a global emission prognosis, i.e. taking emissions alongside the fuel and auxiliary energy supply chain into account, the effects already found for local emission prognoses are enhanced. Natural gas heating systems with flue gas condensation would be more expensive than all biomass based systems if the calculation is based on the maximum value of external costs. Fossil fuel based systems would have the highest specific heat generation costs. Biomass district heating is the cheapest option in this case again (cf. Figure 8.13).

External costs have a pronounced impact on total heat generation costs but they cannot be determined exactly. With regards to the national economy, not taking external costs into account is surely incorrect. Considering external costs based on average values and in a global sense is thus recommendable. Pellet central heating systems are already more economical than oil heating systems, even without external costs being considered. If average external costs are considered, the use of pellet furnaces seems to be more reasonable in both ecological and national economy terms. Therefore, and also considering an expected scarcity of fossil fuels, pellet heating systems and biomass district heating are to be given preference.

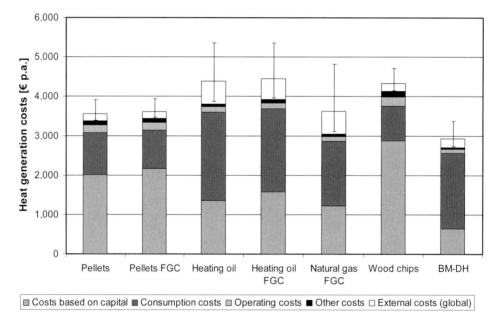

Figure 8.13: Specific heat generation costs of central heating systems with external costs based on global emission prognoses

Explanations: monetary evaluation of emissions on the basis of data available from studies in the field [59; 395; 397]; calculation of external costs based on global emission prognoses, i.e. considering emissions directly caused by the combustion of the different fuels and emissions alongside the fuel and auxiliary energy supply chain

8.4 Summary/conclusions

With regard to energetic utilisation of pellets, their prices and price development of previous years were compared to end user prices of other energy carriers (heating oil, natural gas, wood chips). It was found that the pellet trade exhibited continuity in price policy over time, with low summer prices, slightly higher winter prices, and moderate annual price increases. The strong fluctuations in fossil energy prices (above all heating oil) were not mirrored, which improved user confidence in the pellet market. Therefore, very low pellet prices could be achieved by end users employing appropriate storage strategies.

There were price rises of pellets in 2006, however (in Austria but also in Germany and Italy). Several factors were responsible: in autumn 2005 more pellet boilers were sold than ever before and this led to an unexpected rise in fuel demand the following winter. In addition, consumption and hence demand were boosted by the long and harsh winter of that year. At the same time, the snowy winter inhibited the wood harvest, which is the reason why sawmills produced less sawdust – the main raw material for pellets. As a consequence, the sawdust price rose and pellet production costs increased substantially. Increased demand and lower production led to shortages in pellet supply. The situation was improved in 2007 by the establishment of new pelletisation plants in Austria and many other countries worldwide and a thus massive increase in production capacities. The price of pellets fell to the low levels seen in previous years (before 2006).

The price fluctuation put the growing pellet market in Austria in risk, however. Sales volumes of pellet boilers dropped by 60% in Austria. Similar declines in boiler sales were noted in many other countries too. It was not until 2008 that things improved, but the two key arguments underpinning the change from other fuels to pellets, namely stability in price and national added value, were undermined. The pellet sector needs to return its attention to these two arguments for the sake of the sustainable development of the pellet market, safeguarding an appropriate price level as well as security of supply by suitable measures.

In order to evaluate the economy of different heating systems in a comprehensive way, full cost calculations of pellet, oil, natural gas and wood chip central heating systems as well as biomass district heating were carried out. The cheapest alternatives for supplying an average detached house with a nominal heat load of 15 kW with heat from a central heating system were found to be the network based systems, i.e. biomass district heating and natural gas heating with flue gas condensation, with biomass district heating being most inexpensive and also most ecological owing to the use of renewable energy. The limited availability of biomass district heat is one drawback, however.

Naturally, systems requiring house based fuel storage space are more cost intensive, whereby the wood chip furnace is the most expensive option due to high investment costs and great storage demand. However, from the ecological standpoint, wood chips, like pellets, are a reasonable alternative to fossil fuels.

Owing to low oil prices, oil heating systems had long been the cheapest option for residential heating up to around 2003. The situation changed because of both increased heating oil prices and lowered pellet prices. Pellet central heating systems are cheaper than oil heating systems under present framework conditions (2008), even without taking investment subsidies into account. This gain will shift further in the direction of pellet central heating systems by further increasing oil prices caused by the scarcity of fossil fuels. Even short-term price drops of heating oil, for example caused by the current worldwide financial and economic crisis, cannot obscure this trend. Moreover, the higher security of supply of pellets due to their domestic production must be highlighted as a major advantage of pellets as compared to oil or gas. The insecurity of gas supply became evident in January 2009 when Russia disrupted gas supplies to Western Europe.

The full cost calculations based on internal costs concerning the end user were expanded by a consideration of external costs (costs caused by environmental impacts such as health damage, damage to flora and fauna and damage to buildings as well as climate and safety risks) in order to evaluate each heating system from a national economy point of view. Since an "exact" determination of external costs is impossible, different scenarios were looked at to calculate external costs. These scenarios were based on the incorporation of small-scale furnaces in the emission trade (which is not the case at present) on one hand and on local (emissions from the furnace only) as well as global (emissions alongside the fuel and auxiliary energy supply chain as well) emission prognoses. Monetary evaluation of single pollutant emissions was based on data from studies in the field. In this way, a possible range of external costs as well as effect tendencies on the different heating systems could be determined. It was found that external costs have a significant impact on specific heat generation costs. They tend to burden heating systems based on fossil fuels more than biomass heating systems (depending on the scenario). Although external costs cannot be exactly determined, not taking them into account is surely incorrect with regards to the national economy. Considering average external costs based on global emission prognoses is thus recommendable. Pellet central heating systems are already more economic than oil heating systems, even without considering external costs. If average external costs are considered, the use of pellet furnaces seems to be even more reasonable from both ecological and national economy cases. This evidence, in conjunction with the expected fossil fuel scarcity, means that pellet heating systems and biomass district heating should be given preference.

9 Environmental evaluation when using pellets for residential heating compared to other energy carriers

9.1 Introduction

In this section, the ecological evaluation of the utilisation of pellets in the residential heating sector is carried out on the basis of emission factors, considering the emissions that occur along the process chain from fuel pre-treatment and supply to auxiliary energy supply to thermal utilisation in the furnace. It is assumed that production and disposal of the furnace itself play a subordinate role, so a detailed investigation of this is left out (cf. for instance [393; 398; 399]).

In addition, an ecological comparison to the heating systems that were economically assessed in Chapter 8 is carried out (central heating systems based on pellets and heating oil, each with and without flue gas condensation, natural gas with flue gas condensation, wood chips and district heat).

The emission factors are given for each emission in mg/MJ final energy (FE), i.e. relating to the net calorific value of the fuel. For district heat, the final energy relates to the supplied heat at the heat transfer station of individual houses. Finally, the emission factors are converted to useful energy (UE) under consideration of the annual efficiencies of the different systems, which takes conversion and distribution losses at the end user site into account, making the emissions related to useful energy directly comparable.

The ecological evaluation is based on Austrian framework conditions concerning fuel supply, distribution and utilisation. The most important influencing steps in the whole supply chain, i.e. fuel production and utilisation, will not or will only be slightly influenced if another country is taken as a base case. Differences might occur concerning the electricity consumption in production and utilisation of fuels (auxiliary energy demand). However, these factors are in general of minor relevance and a change in the electricity mix in another country will therefore have a very low absolute influence. Another difference to be taken into account is the international or even intercontinental trade of pellets. The Austrian base case in this chapter is related to local pellet production and utilisation with limited transport distances. Long distance transport by truck, train or, in the case of intercontinental trade, by ocean vessel would significantly influence the emissions caused by fuel supply. Possible deviations from the Austrian case when examining other countries and their impact on the emission factors are discussed in the relevant sub-sections.

9.2 Pollutants considered for the evaluation

Solid as well as gaseous reaction products emerge from the fuel and the combustion air during combustion and are released to the environment by the flue gas. The main components of the flue gas are nitrogen (N_2), water vapour (H_2O), carbon dioxide (CO_2), oxygen (O_2), carbon monoxide (CO), organic compounds (C_xH_y), nitrogen oxides (NO_x), sulphur dioxide (SO_2) and particulate matter. The ecological evaluation is carried out on the basis of the classic pollutants CO, C_xH_y (sum of hydrocarbons), NO_x, SO_2 and particulate matter as well as on the greenhouse gas CO_2. Trace elements in the flue gas are not taken into account. In this regard, the literature of the field is referenced [393; 400; 401].

Fine particulate emissions are dealt with in Section 9.9 owing to their prominence in public debate.

9.3 Fuel/heat supply

The emission factors of fuel supply include all emissions in the steps that relate directly to the supply of the fuel, i.e. fuel extraction and/or production, raw material and fuel transport as well as storage and combustion at the end user site. Thus, depending on the fuel, system boundaries have to be drawn in different ways.

Looking at pellets, the supply chain starts with the raw materials "wood shavings" or "sawdust" since these are by-products of the wood industry and hence production is assigned to the wood industry. The supply chain ends at the storage space of the end user. All interim steps, from raw material transport to the pellet producer, to pellet production including all process steps to pellet transport to the end user site, are considered.

The basic data for the calculation of the emission factors along the pellet supply chain are given in Table 9.1. Table 9.2 presents the calculated emission factors.

Table 9.1: Basic data for the calculation of the emission factors along the pellet supply chain

Explanations: data source [59; 393; 402; 403]

		Emission factor					
		CO_2	CO	C_xH_y	NO_x	SO_2	Dust
Transport	mg/t.km	75,800	240	123	960	24	53
Electricity supply	mg/MJ_{el}	70,000	67	290	67	77	6
Heat supply	mg/MJ_{th}	4,585	71	26	138	17	28

An average transport distance of 50 km was assumed for both raw material and pellet transport. The emission factors for the electricity supply for pellet production are based on the average annual Austrian electricity mix, including hydropower. The emission factors of district heat from wet wood chips were taken as a basis for the heat supply needed for drying. The emission factors for raw material supply as well as pellet transport, as in Table 9.2, follow from that, whereby raw material supply solely includes raw material transport to the pellet producer and not production, since raw material production is assigned to the wood industry, as mentioned above.

The framework conditions of the base case scenario and scenario 1 according to Table 7.21 and as separated by process steps according to Table 9.2 were adopted for the calculation of the emission factors of pellet production from sawdust and wood shavings. Thus the electricity demand for the production of pellets from wood shavings is about 90.1 kWh/t (w.b.)$_p$ and for the production of pellets from sawdust is about 113.9 kWh/t (w.b.)$_p$. The production of pellets from sawdust requires additional heat, amounting to 1200.8 kWh/t (w.b.)$_p$.

The emission factors related to final energy of pellet supply result from these framework conditions for pellets made of wood shavings and of sawdust, as shown in Table 9.3.

The emission factors of fuel supply for heating oil, natural gas and wood chips as well as the emission factors for the supply of district heat are shown in Table 9.4, which presents data derived only from the literature.

The emission factors for the supply of heating oil and natural gas include the whole supply chain, i.e. from extraction (domestic and foreign, according to the import shares) and all of transport up to the furnace at the end user site. The supply chain of wood chips begins at wood chipping. Emissions from harvest and forwarding are not taken into account because first, the energy demand of the wood harvest is very low and second, the wood harvest is not carried out for the main purpose of delivering wood chips. So emission factors include emissions from chipping and from transport to the end user site. The emission factors for the supply of district heat include all emissions caused by the fuel supply for the use in the district heating plant, by combustion, by the auxiliary energy demand of the furnace, by losses of the district heating network and the auxiliary electric energy demand of the district heating network.

Table 9.2: Energy consumption of the pellet production process steps for pelletisation of wood shavings and sawdust

Explanations: values in kWh/t (w.b.)$_p$; framework conditions as in Sections 7.2.1 to 7.2.10

	Wood shavings		Sawdust	
Production step	Electricity demand	Heat demand	Electricity demand	Heat demand
Drying			23.8	1,200.0
Grinding	18.7		18.7	
Pelletisation	51.0		51.0	0.8
Cooling	2.0		2.0	
Peripheral equipment	18.4		18.4	
Total	90.1	0.0	113.9	1,200.8

Table 9.3: Emission factors of the pellet supply chain

Explanations: [1]...same emission factors for wood shavings and sawdust, as long as they have the same bulk densities; [2]...made of wood shavings; [3]...made of sawdust; average transport distance of 50 km for raw material and pellet transport; specific electricity and heat demand for pellet production as per Table 9.2; data source [59; 393; 402; 403], own calculations

	Emission factor [mg/MJ$_{FE}$]					
	CO_2	CO	C_xH_y	NO_x	SO_2	Dust
Raw material supply[1]	716	2.27	1.16	9.07	0.23	0.50
Pellet production[2]	1,287	1.23	5.33	1.23	1.42	0.11
Pellet production[3]	2,751	18.94	13.01	35.47	6.06	6.98
Pellet transport	321	1.01	0.52	4.06	0.10	0.22
Pellet supply[2]	2,324	4.51	7.01	14.36	1.74	0.84
Pellet supply[3]	3,787	22.22	14.69	48.60	6.39	7.70

Table 9.4: Emission factors of the supply of heating oil, natural gas and wood chips as well as the supply of district heat

Explanations: data source [59; 393; 403]

	Emission factor [mg/MJ$_{FE}$]					
	CO_2	CO	C_xH_y	NO_x	SO_2	Dust
Heating oil	7,000	27	42	54	29	4
Natural gas	3,300	93	490	12	5	1
Wood chips	1,900	8	7	23	3	2
BM-DH	5,583	72	30	139	19	28

Deviations related to the pellet supply in other countries might on the one hand occur due to different transport distances of raw materials to pellet producers or of pellets to end users and on the other hand due to a different electricity mix. An evaluation is conducted in Section 9.6.

9.4 Auxiliary energy demand for the operation of the central heating system

Apart from emissions caused by fuel supply and by their thermal utilisation in the furnace, emissions that arise from the auxiliary energy needed for the operation of the central heating system have to be considered. The auxiliary energy demand is different depending on the heating system that is used and it can also vary for specific systems to some degree according to the fuel conveyor system and the control system used. The data that were assumed for the full cost calculations of the different systems concerning their auxiliary energy demand in Sections 8.2.2 to 8.2.8 were taken as a basis for the calculation of the emission factors of the auxiliary energy demand. Then, the emission factors were calculated based on literature data and converted to the specific auxiliary energy demand of the different systems, as shown in Table 9.5.

The emission factors for the electricity supply for auxiliary energy are again based on the average annual Austrian electricity mix, including hydropower. Different electricity mixes in different countries would consequently influence the emission factors of auxiliary energy use during combustion. An evaluation is conducted in Section 9.6.

Table 9.5: Emission factors of auxiliary energy use during operation of central heating systems

Explanations: auxiliary energy demand (average electric power demand in percent of nominal boiler capacity according to Section 8.2.2 to 8.2.8); data source [393]

Heating system	AED	Emission factor [mg/MJ$_{FE}$]					
		CO_2	CO	C_xH_y	NO_x	SO_2	Dust
Pellets	0.7%	699	0.66	2.86	0.66	0.76	0.06
Heating oil	0.6%	599	0.57	2.45	0.57	0.65	0.05
Natural gas	0.3%	300	0.28	1.22	0.28	0.33	0.02
Wood chips	0.8%	799	0.76	3.26	0.76	0.87	0.06
BM-DH	0.0%	0	0.00	0.00	0.00	0.00	0.00

9.5 Utilisation of different energy carriers in different heating systems for the residential heating sector

9.5.1 Emission factors from field measurements

Emission factors are the amount of emissions of a certain fuel and furnace combination related to the energy content of the fuel. The emission factors of the investigated heating systems based on literature data as assumed for the ecological evaluation are presented in Table 9.6 and Figure 9.1. With regard to wood chip heating systems, old and new systems were differentiated because there are clear differences owing to the technological developments of the last years. The emission reductions in modern pellet and wood chip furnaces are discussed in detail later (cf. Section 9.5.2). Table 9.6 also presents the limiting values for automatically fed furnaces based on biomass fuels at nominal load for up to 300 kW [29]. However, it must be noted in this respect that the direct comparison of emission factors from field measurements that also take different load conditions into account, with limiting values valid for nominal load, is given as an indication only. An exceeding of the limiting value in the legal sense (valid for nominal load) cannot thus be derived from emission factors that are above the limiting value.

Second to water, CO_2 is the main product of any combustion process. However, CO_2 emissions from the combustion of biomass can be regarded as neutral emissions since the CO_2 that is emitted during combustion is taken up again by plants during growth, providing sustainable forestry. Thus emissions from biomass combustion are climate neutral. As long as the sustainable use of biomass is maintained, which is the case in Austria as more biomass is growing than exploited, the CO_2 emissions from biomass combustion can be set at zero.

Emission factors of district heat are all equal to zero since no emissions emerge at the end user site. The emissions along the supply chain of district heat were taken into account in Section 9.3 already.

Table 9.6: Emission factors of different central heating systems based on field measurements

Explanations: [1]...old systems from before 1998; [2]...new systems from 2000 onwards; [3]...no emissions at the end user site; [4]...valid for automatically fed furnaces based on biomass fuels of up to 300 kW at nominal load; [5]...value for SO_2 based on old systems due to lack of field measurements; data source [29; 292; 393; 400; 401; 404; 405; 406; 407; 408; 409; 410]

Heating system	\multicolumn{6}{c}{Emission factor [mg/MJ$_{FE}$]}					
	CO_2	CO	C_xH_y	NO_x	SO_2	Dust
Pellets	0	102	8	100	11	24
Heating oil EL	75,000	18	6	39	45	2
Natural gas	55,000	19	6	15	0	0
Wood chips[1]	0	1,720	88	183	11	54
Wood chips[2] [5]	0	717	18	132	11	35
BM-DH[3]	0	0	0	0	0	0
Limiting value[4]	-	500	40	150	-	60

CO is a good indicator of combustion quality. It shows that CO emissions from pellet and wood chip furnaces are generally higher than those of systems based on heating oil or natural gas. Wood chip furnaces even pass the limiting value of 500 mg/MJ$_{NCV}$ to a notable extent, whereby old systems have an even more pronounced CO emission factor. Pellet furnaces exhibit CO emissions that are clearly underneath the limiting value, which underlines the benefits of this homogenous and dry fuel.

Hydrocarbon emissions are products of incomplete combustion, such as CO. Elevated hydrocarbon emissions can be caused by too low combustion temperatures, too short residence times of the flue gases in the combustion zone or lack of oxygen. Although hydrocarbon emissions of pellet and new wood chip furnaces are slightly above those of furnaces based on fossil fuels, they are notably below the Austrian limiting value. It is just old wood chip systems that exceed the limiting value for hydrocarbon emissions.

NO$_x$ emissions of biomass furnaces are mainly the product of partial oxidation of the nitrogen in the fuel. Moreover, at temperatures exceeding about 1,300°C atmospheric nitrogen can react with oxygen radicals to form NO. Since the temperatures in biomass furnaces are usually lower than that, NO$_x$ formation by atmospheric nitrogen is of almost no relevance. NO$_x$ emissions of biomass furnaces are slightly higher than those of oil or gas furnaces but they still clearly adhere to the prescribed limiting values (except in old wood chip systems).

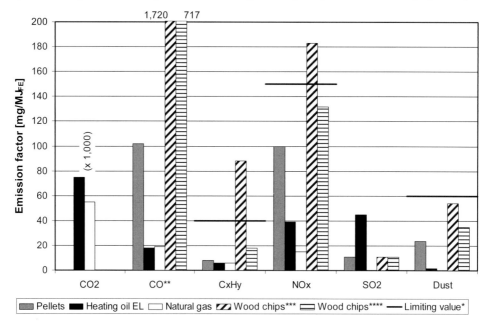

Figure 9.1: Emission factors of different central heating systems based on field measurements

Explanations: *...valid for automatically fed furnaces based on biomass fuels of up to 300 kW at nominal load; **...limiting value 500 mg/MJ; ***... old systems from before 1998; ****...new systems from 2000 onwards; data source [29; 292; 393; 400; 401; 411]

SO$_2$ emissions of biomass furnaces are about 11 mg/MJ$_{NCV}$ and thus between the comparatively high SO$_2$ emissions of oil furnaces (owing to the comparatively high sulphur content of heating oil) and the relatively low SO_2 emissions of natural gas furnaces (owing to the very low sulphur content of natural gas).

Fine particulate emissions of biomass furnaces are also higher than those of fossil fuel based furnaces. The emissions of modern pellet and wood chip furnaces are notably below the limiting values in Austria however, and even old wood chip furnaces adhere to the required values.

9.5.2 Emission factors from test stand measurements

As mentioned above, emission factors are the amount of emissions of a certain fuel and furnace combination as related to the energy content of the fuel. In addition to the fuel and the furnace used there are other parameters that influence emissions, however. These include:

- Age of the furnace;
- Changing load conditions;
- Maintenance and service of the system;
- Hot water supply by the furnace or external (e.g. by a solar heating system);
- Chimney design;
- Moisture, kind and particle size of the solid fuel;
- User behaviour in system operation.

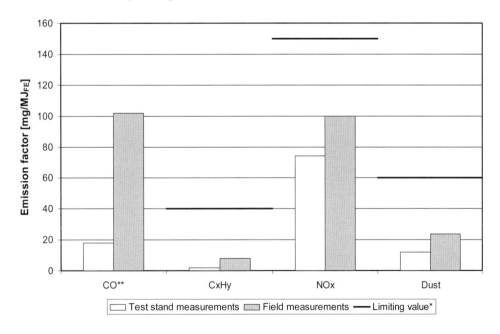

Figure 9.2: Comparison of test stand and field measurements of Austrian pellet furnaces

Explanations: *...valid for automatically fed furnaces based on biomass fuels of up to 300 kW at nominal load; **...limiting value 500 mg/MJ; data source [29; 292; 391; 393; 400; 401; 404; 407; 408]

The emission factors used in Section 9.5.1 on the basis of field measurements thus represent average emission factors for specific fuel and furnace combinations. Test stand measurements, however, keep the aforementioned influencing parameters at a constant level. The emissions determined in this way are hence solely dependent on the used fuel and furnace combination, and lower than those of the field measurements since the measurements are taken at nominal load and ideal conditions, as shown in Figure 9.2 for CO, C_xH_y, NO_x and fine particulate matter. In any case, the emissions of Austrian pellet furnaces based on the test stand and field measurements that were carried out are clearly below the limiting values of the ÖNORM EN 303-5.

The test stand measurements of the BLT Wieselburg show one interesting aspect [391; 412]. Looking at CO and particulate matter emissions measured from 1996 to 2008 (cf. Figure 9.3 and Figure 9.4), a decreasing tendency can be found in newer systems (cf. [413] in this respect). Concerning emissions of hydrocarbons and NO_x, no correlation between year of manufacture and emissions could be established. With regard to hydrocarbon emissions, this is because they are very low (around the detection limit). However, the share of measurements below the detection limit rose from 68% at the beginning (in the period 1999–2001) to almost 82% (2005–2008), demonstrating another positive trend. NO_x emissions of pellet furnaces are mainly dependent on the nitrogen content of the fuel (hardly any formation of thermal NO_x), which is why NO_x emissions are independent of the year of manufacture. Possible primary measures for NO_x emission reduction are appropriate air staging in combination with sufficient residence time and sufficient flue gas temperature in the primary combustion zone.

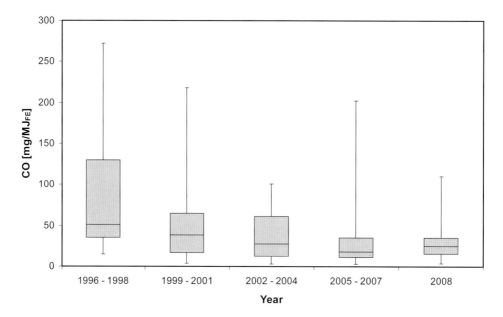

Figure 9.3: Development of CO emissions from Austrian pellet furnaces from 1996 to 2008

Explanations: data source [391; 412]

The results based on test stand measurements show that clear reductions of particulate matter and CO emissions could be achieved within 1996 to 2008 by technological improvements of pellet furnaces. This leads to the conclusion that modern pellet and wood chip furnaces will show lower average emission factors in field measurements too. However, there is no validation of this by a sufficient number of field measurements based on a statistical sampling plan.

The following evaluations of the total emission factors are based on the emission factors as determined by the field measurements.

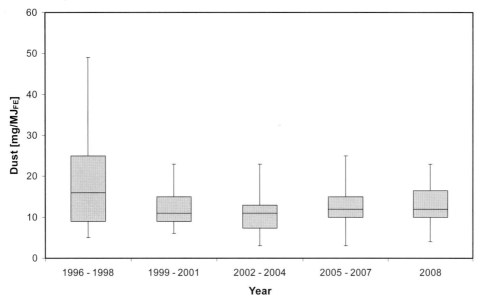

Figure 9.4: Development of particulate emissions from Austrian pellet furnaces from 1996 to 2008

Explanations: data source [391; 412]

9.6 Total emission factors for the final energy supply for room heating

Table 9.7 and Figure 9.5 present an overview of the total emission factors of final energy supply of residential heating by different heating systems.

The CO_2 emissions of systems based on renewable fuels are notably lower than those of systems based on fossil fuels. The reason for this is that CO_2 emissions from biomass combustion can be regarded as climate neutral and thus set at zero. In the overall evaluation, it is the CO_2 emissions of fuel supply (by the use of electric energy and the consumption of fossil fuels in transport) and system operation (auxiliary electric energy) that are taken into account.

With regard to greenhouse gas emissions, CO_2 is the most important parameter for furnaces. Thanks to the climate neutrality of CO_2 emissions from biomass furnaces, CO_2 emissions can be avoided by the use of biomass furnaces, which would be emitted by the use of furnaces based on fossil fuels. The replacement of an oil furnace by a pellet furnace (utilisation of pellets made of sawdust) thus saves CO_2 emissions to an extent of about 78,000 mg/MJ$_{NCV}$. If a gas heating system is replaced by the same pellet furnace, about 54,000 mg/MJ$_{NCV}$ of CO_2 emissions are saved.

Table 9.7: Emission factors of the final energy supply of different heating systems in order of supply steps

Explanations: sum of data from Sections 9.3, 9.4 and 9.5; data source [59; 292; 391; 393; 400; 401; 402; 403; 404; 405; 406; 407; 408; 409; 411]

Central heating system based on	Emission factor [mg/MJ$_{FE}$]					
	CO_2	CO	C_xH_y	NO_x	SO_2	Dust
Pellets produced from wood shavings						
Fuel supply	2,324	4.5	7.0	14.4	1.7	0.8
Auxiliary energy supply	699	0.7	2.9	0.7	0.8	0.1
Thermal utilisation	0	101.9	7.9	100.0	11.0	23.6
Total	**3,023**	**107.1**	**17.8**	**115.0**	**13.5**	**24.5**
Pellets produced from sawdust						
Fuel supply	3,787	22.2	14.7	48.6	6.4	7.7
Auxiliary energy supply	699	0.7	2.9	0.7	0.8	0.1
Thermal utilisation	0	101.9	7.9	100.0	11.0	23.6
Total	**4,487**	**124.8**	**25.5**	**149.3**	**18.2**	**31.4**
Heating oil						
Fuel supply	7,000	27.0	42.0	54.0	29.0	4.0
Auxiliary energy supply	599	0.6	2.4	0.6	0.7	< 0.05
Thermal utilisation	75,000	18.0	6.0	39.0	45.0	1.6
Total	**82,599**	**45.6**	**50.4**	**93.6**	**74.7**	**5.6**
Natural gas						
Fuel supply	3,300	93.0	490.0	12.0	5.0	1.0
Auxiliary energy supply	300	0.3	1.2	0.3	0.3	< 0.05
Thermal utilisation	55,000	19.0	6.0	15.0	0.0	0.0
Total	**58,600**	**112.3**	**497.2**	**27.3**	**5.3**	**1.0**
Wood chips (old units till 1998)						
Fuel supply	1,900	8.0	7.0	23.0	3.0	2.0
Auxiliary energy supply	799	0.8	3.3	0.8	0.9	0.1
Thermal utilisation	0	1,720.0	88.0	183.0	11.0	54.0
Total	**2,699**	**1,728.8**	**98.3**	**206.8**	**14.9**	**56.1**
Wood chips (new units since 2000)						
Fuel supply	1,900	8.0	7.0	23.0	3.0	2.0
Auxiliary energy supply	799	0.8	3.3	0.8	0.9	0.1
Thermal utilisation	0	717.0	18.0	132.0	11.0	35.0
Total	**2,699**	**725.8**	**28.3**	**155.8**	**14.9**	**37.1**
Biomass district heating						
Heat supply	5,583	71.8	29.6	139.4	19.4	27.9
Auxiliary energy supply	0	0.0	0.0	0.0	0.0	0.0
Total	**5,583**	**71.8**	**29.6**	**139.4**	**19.4**	**27.9**

Assuming a distribution of oil to gas furnaces in Austria of 46.4% to 53.6% (based on 2006) [467] and assuming that the exchange takes place according to this ratio, an average of 65,000 mg/MJ$_{NCV}$ of CO_2 could be saved by the use of a pellet furnace instead of a fossil fuel based system.

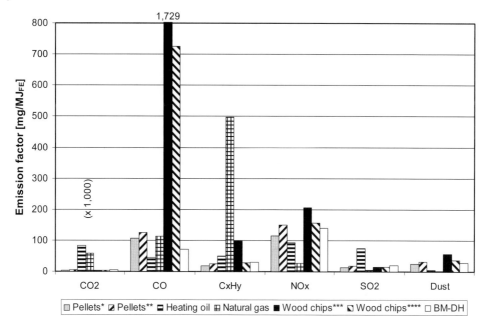

Figure 9.5: Emission factors of final energy supply of different heating systems

Explanations: *...made of wood shavings; **...made of sawdust; ***...old systems before 1998; ****...new systems from 2000 onwards; sum of data from Sections 9.3, 9.4 and 9.5; data source [59; 292; 393; 400; 401; 402; 403; 404; 405; 406; 407; 408; 409; 391; 411]

Based on the pellet consumption of 500,000 t (w.b.)$_p$ of 2008 in Austria, which is equivalent to 8.8 PJ related to NCV, and assuming that it is just oil and gas heating systems that get replaced by pellet heating systems, the use of pellets would save about 575,000 t/a of CO_2. Pellet consumption is forecast to be up to 1.3 million t/a until the year 2010. Under the stated framework conditions, this would be equivalent to saving almost 1.5 million t of CO_2/a. Pellet furnaces also replace heating systems based on other fuels (e.g. coal or wood) that in part are also CO_2 neutral energy carriers and hence the actually saved amount of CO_2 is less. The numbers, however, do underline the possible contribution of pellets to reducing greenhouse gas emissions in Austria. In addition, pellets are able to make a substantial contribution to climate protection on a European level due to current developments of the pellet market in many European countries (cf. Chapter 10).

Heating systems based on heating oil show the lowest CO emissions and heating systems based on wood chips the highest. Systems based on natural gas, pellets and district heating systems are of a similar level in between.

Hydrocarbon emissions lie in a relatively narrow range of between 18 and 50 mg/MJ$_{NCV}$ for pellet, oil and new wood chip heating systems as well as for district heating systems. The emission factor of old wood chip systems is about twice as high (owing to the poor combustion quality of old systems) and that of natural gas is ten times as high, which is mainly caused by the fuel supply.

NO_x emissions of all examined systems (except natural gas) are in a relatively narrow range of 94 to 207 mg/MJ_{NCV}. NO_x emissions of natural gas are lower with around 27 mg/MJ_{NCV}.

The SO_2 emissions of biomass based systems are on roughly the same level. The emission factor for SO_2 of oil furnaces is above this level and the emission factor of gas furnaces below it.

With regard to fine particulate emissions, the systems based on fossil energy carriers exhibit clear benefits, which is mainly due to near non-existent ash in these fuels. Biomass fuels can have an ash content of between about 0.1 wt.% (d.b.) and 6.0 wt.% (d.b.) depending on the kind of biomass, and even more when the fuel is minerally contaminated. The fact that fine particulate emissions are, despite these high ash contents, clearly underneath the strict limiting values in Austria, was shown in Section 9.1.

Figure 9.6 shows the emission factors of different heating systems in order of origin (fuel supply, auxiliary energy supply and thermal utilisation). Displaying the emission factors of biomass district heating systems is abstained from here since they are almost exclusively caused by the supply of district heat and no emissions arise at the end user site.

It is shown for all compared systems that the emission factors caused by auxiliary energy supply are negligible. The low CO_2 emissions of biomass based systems are solely due to the fuel supply because the CO_2 emissions from combustion can be regarded as climate neutral. In contrast, the CO_2 emissions of fossil fuels based heating systems originate almost completely from combustion.

The main share of CO emissions is caused by the combustion in biomass based systems. CO emissions from the combustion of pellets made of sawdust are the same as from the combustion of pellets made of wood shavings. Total emissions of the former increase, however, owing to upstream emissions caused by the more complex pre-treatment of the fuel, especially the required drying step. As concerns heating oil and especially natural gas, CO emissions of fuel supply prevail.

Hydrocarbon emissions are generated to the most part during combustion in biomass furnaces (except for pellets made of sawdust). With regard to pellets made of sawdust and fossil energy carriers, it is the hydrocarbon emissions of fuel supply that are dominant, whereby there are extremely high hydrocarbon emissions from natural gas due to leakages during extraction and transport.

Except for oil heating systems, NO_x emissions in most part originate from combustion in all furnaces.

SO_2 emissions of biomass furnaces are relatively low and mainly caused by combustion. Systems based on heating oil have higher SO_2 emissions that are also dominated by combustion. The SO_2 emissions from natural gas use are negligible.

Fine particulate emissions play a subordinate role for fossil fuels, with fuel supply being their main origin. As concerns pellets and wood chips, most fine particulate matter is emitted during combustion. Regarding pellets made of sawdust, an additional amount of fine particulate emissions comes from the heat supply for raw material drying, whereby a biomass furnace was also assumed for this heat supply.

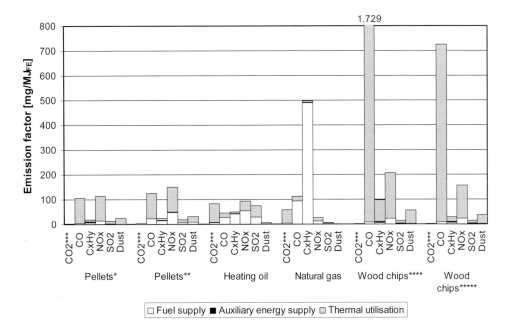

Figure 9.6: Emission factors of final energy supply for different heating systems as well as their composition

Explanations: *...made of wood shavings; **...made of sawdust; ***...× 1,000; ****...old systems before 1998; *****...new systems from 2000 onwards; sum of data from Sections 9.3, 9.4 and 9.5; data source [59; 292; 391; 393; 400; 401; 402; 403; 404; 405; 406; 407; 408; 409; 411]

On the whole, the use of sawdust as a raw material increases upstream emissions to some extent by the more sophisticated fuel processing that is required. Emissions originating from utilisation of wood chips or pellets made of wood shavings are dominated by combustion emissions due to the comparably low pre-treatment effort.

As already mentioned, the ecological evaluation shown in this section is based on Austrian framework conditions concerning fuel supply, distribution and utilisation. Under framework conditions in other countries or by the import of pellets from other countries, differences with regard to the ecological evaluation might occur concerning the electricity mix in the respective country (with influence on the auxiliary energy supply both of production and utilisation of pellets) and concerning different transport distances for raw materials and in particular pellets. Such an evaluation has been done for the CO_2 emissions, the most important greenhouse gas emission, also constituting the main advantage of pellets due to their CO_2 neutrality.

During production and utilisation of pellets produced from sawdust, CO_2 emissions of 2,326 mg/MJ$_{FE}$ are caused, which are allocated to electricity consumption. This represents about 52% of the total CO_2 emissions of 4,487 mg/MJ$_{FE}$ during production and utilisation (cf. Table 9.7). A variation of the emission factor for CO_2 by ± 10% due to a varying electricity mix would result in a variation of the share of CO_2 emissions caused from electricity consumption by ± 2.5%. A varying transport distance of raw materials and pellets per truck would change the total CO_2 emissions by 20.7 mg/MJ per kilometre. For instance a doubled average transport distance for both raw materials and pellets from 50 to 100 km would result in an increase of CO_2 emissions from 4,487 mg/MJ_{FE} to 5,523 mg/MJ_{FE} or by about 23%. Intercontinental pellet transport by ocean vessels would increase the CO_2 emissions by up to 4,200 mg/MJ_{FE} or more than 90% (cf. Table 10.11 in Section 10.11).

9.7 Conversion efficiencies

The emission factors presented and discussed in the previous sections are based on the fuels' net calorific value or, in the case of district heating systems, on the supplied district heat. In order to relate the emission factors to the useful energy, the emission factors of the final energy supply have to be related to the annual efficiency of the heating system in hand. The annual efficiency takes not only conversion losses (boiler efficiency) into account but also losses due to heat distribution at the end user site, start-up and shutdown losses, losses due to radiation, losses due to cooling down, as well as losses due to the control system. It is defined according to Equation 9.1 as the ratio of the annual UE of the end user for keeping the desired room temperature and hot water supply to the FE fed into the furnace or conversion system [393]. The annual efficiencies of the different heating systems for this environmental evaluation were used according to Sections 8.2.2 to 8.2.8. An annual efficiency of 69.0% was assumed for old wood chip systems according to [393]. The annual efficiencies and corresponding boiler efficiencies are shown in Figure 9.7. For biomass district heat, the "boiler efficiency" is indicated to be 100% because the heat transferred via the heat transfer station to the end user is measured as the basis. The annual efficiency of biomass district heating thus only includes losses due to heat distribution at the end user site.

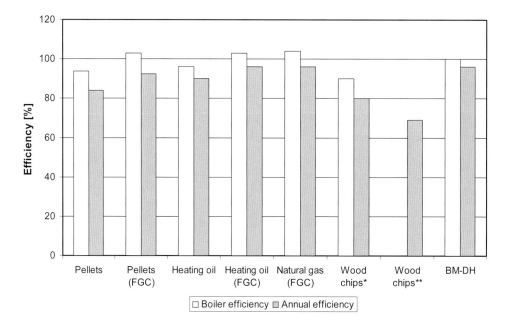

Figure 9.7: Comparison of boiler and annual efficiencies of the systems compared

Explanations: *…new systems from 2000 onwards; **… old systems before 1998 (no corresponding boiler efficiency available)

Equation 9.1: $\quad \eta_a = \dfrac{UE}{FE} \cdot 100\, [\%]$

Explanations: data source [393]

The annual efficiencies of pellet heating systems that are achievable in practice require evaluation at greater depth. The annual efficiency taken as a basis for the economic evaluations in Chapter 8 as well as for the environmental evaluation in this section, i.e. 84.0%, can be achieved under good framework conditions. The target for annual efficiencies of pellet boilers should be above 90% [414]. However, annual efficiencies in the range of 69.9 to 80.4% were measured in field tests (cf. Figure 9.8) [292; 415]. The following points have been identified to be the main reasons for these lower values:

- Often oversized boilers are installed in residential houses. What is less problematic for oil or gas heating systems can lead to pronounced negative effects for biomass boilers in general and in particular for pellet boilers. If the actual heat demand of the building is lower than the nominal thermal capacity of the heating system, on–off operation of the heating system is often necessary, which in turn leads to increased start-up and shutdown losses, losses due to cooling down and higher auxiliary electricity consumption due to more frequent automatic ignition. Up to 7.2% auxiliary electricity demand related to the useful heat output has been measured [415].

- The installation of pellet boilers without heat buffer storage also leads to an increased number of start-ups and shutdowns with the same effects as mentioned above.

- A combination of oversized pellet boilers and their installation without heat buffer storage makes the situation even worse.

- A problem inherent to the system of pellet boilers is their thermal inertia. Pellet boilers are often designed as massive steel constructions and their combustion chamber is often made of fireclay. Their weight is typically between 300 and 430 kg and the boiler water volume is usually in the range between 30 and 115 l. Modern wall-mounted oil or gas heating systems usually have weights between 40 and 50 kg and water volumes between 1.5 and 3.0 l only. This fact makes pellet boilers slow with regard to load changes. A reduction of the heat demand thus often leads to shutdowns of the boiler because the maximum water temperature is reached even though the furnace is operating at part load.

- The problem can partly be overcome by appropriate control systems, but many pellet boilers are unable or largely unable to cope with this situation (although this is a typical situation in residential households).

- Insufficient maintenance of the boiler leads to fouling at boiler surfaces and consequently to reduced efficiencies.

- It has been observed that screw conveying systems have a lower electricity demand than pneumatic feeding systems. However, pneumatic feeding systems are often when the installation of a feeding screw would have been possible. This fact should be considered when the feeding system is selected.

Apart from the correct dimensioning of pellet boilers, which is an important basic precondition, the load control strategy seems to be a key issue regarding annual efficiency. Moreover, the integration of the pellet heating system in the hydraulic system of the building plays a major role. In this context research and development is underway (cf. Section 12.2.3).

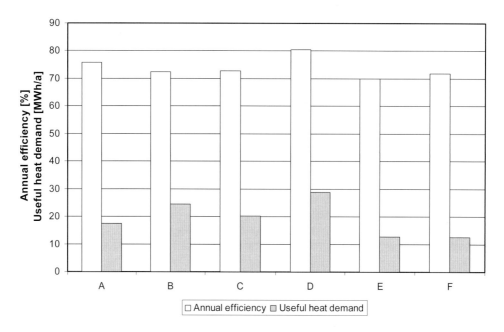

Figure 9.8: Annual efficiencies and useful heat demands of pellet boilers based on field measurements

Explanations: A to F...different pellet boilers with a nominal load below 15 kW; data source: adapted from [415]

9.8 Total emission factors of useful energy supply for room heating

The emission factors of useful energy supply for all compared systems are presented in Figure 9.9.

When emission factors are related to useful energy instead to final energy, the annual efficiency becomes relevant. Therefore, systems with flue gas condensation need to be looked at separately (natural gas heating systems without flue gas condensation are not examined since their numbers are low, especially with regard to new installations), whereby the higher efficiencies of systems with flue gas condensation result in lower emission factors. Due to the higher annual efficiencies of natural gas and oil heating systems with flue gas condensation as well as biomass district heating systems, the relative change of emission factors due to the conversion to useful energy is lowest in these systems and highest in wood chip systems (both old and new) due to relatively low annual efficiencies. Comparing systems based on heating oil and natural gas to pellet furnaces, the relation of the emission factors to useful energy results in a slight shift in favour of the fossil fuel systems due to their slightly higher annual efficiencies.

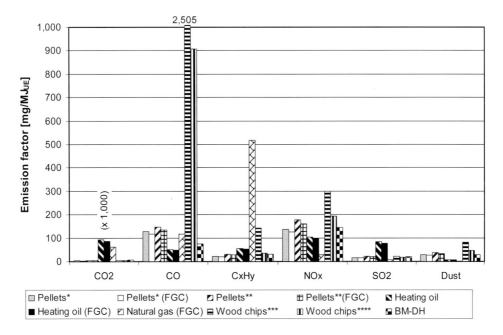

Figure 9.9: Emission factors of useful energy supply for different heating systems

Explanations: basic data for emission factors of final energy supply as in Table 9.7; conversion to emission factor related to useful energy supply on the basis of annual efficiencies according to Sections 8.2.2 to 8.2.8; [*]…made of wood shavings; [**]…made of sawdust; [***]…old systems before 1998; [****]…new systems from 2000 onwards

9.9 Basics of ash formation and ash fractions in biomass combustion systems

This section briefly describes the basic principles of ash formation during biomass combustion, which should provide a basis for understanding the following sections on fine particulate emissions and solid residues.

Biomass fuels contain varying quantities of ash forming elements in addition to their main organic constituents (C, H, O, N). The most important elements in this respect are Si, Ca, Mg, K, Na, P, S, Cl as well as heavy metals such as Zn and Pb.

There are generally two sources for inorganic ash forming matter in biomass fuels. First, ash forming elements can originate from the plant itself, as they are part of the structure of the fibres (e.g. Si, Ca) or macro or micro plant nutrients (e.g. K, P, Mg, S, Zn). Second, inorganic matter in biomass fuels can come from contamination with soil, sand or stones, while coatings, paints, glass pieces and metal parts are major sources of contamination in waste wood.

During the combustion of solid biomass fuels, the behaviour of ash forming elements follows a general scheme, which is depicted in Figure 9.10. Upon entering the combustion unit, the fuel is first dried, followed by devolatilisation of the volatile organic matter. Subsequently, the remaining fixed carbon is oxidised during heterogeneous gas–solid reactions, which is called charcoal combustion. During these steps the ash forming elements behave in two different ways depending on their volatility. Non-volatile compounds such as Si, Ca, Mg, Fe and Al are engaged in ash fusion as well as coagulation processes. Once the organic matter has been released or oxidised, these elements remain as coarse ash constituents. Easily volatile species such as K, Na, S, Cl, Zn and Pb generally behave differently. A considerable proportion of these elements is released to the gas phase due to the high temperatures during combustion. There they undergo homogeneous gas phase reactions and later, due to supersaturation in the gas phase, these ash forming vapours start to nucleate (formation of submicron aerosol particles) or condense on and react with the surfaces of existing particles, or they directly condense on heat exchanger surfaces. The submicron particles, so-called aerosols, form one important fraction of the fly ashes. The second fly ash fraction consists of small coarse ash particles entrained from the fuel bed with the flue gas. Depending on particle size, they are either precipitated from the flue gas in the furnace or boiler or are entrained with the flue gas, forming coarse fly ash emissions. Consequently, the most relevant difference between coarse fly ashes and aerosols is that coarse fly ashes always remain in the solid phase while aerosols undergo phase changes during their formation process (release to the gas phase and gas-to-particle conversion). The easily volatile elements such as K, Na, S, Cl, Zn and Pb are finally enriched in the fine fly ash (aerosols).

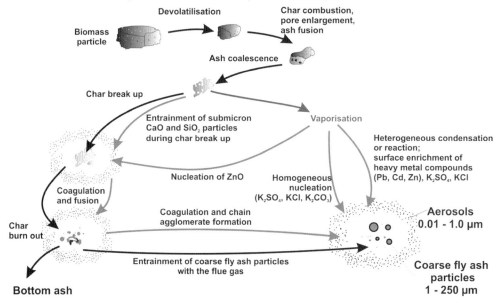

Figure 9.10: Ash formation during biomass combustion

<u>Explanations</u>: modified from [416]

According to this ash formation scheme, ashes formed during biomass combustion can generally be divided into:

- bottom ashes;
- coarse fly ashes;
- aerosols (fine fly ash).

The bottom ash is the ash fraction remaining in the furnace after combustion of the fuel and is then removed by the de-ashing system. Coarse fly ash particles, which are entrained from the fuel bed with the flue gas, are partly precipitated on their way through the furnace and the boiler by inertial impaction, gravitational and centrifugal forces and therefore form the so-called furnace or boiler ash. Particles that are small enough to follow the flue gas on its way through the furnace and the boiler finally form the coarse fly ash emission at the boiler outlet.

As already mentioned, aerosols are formed by gas-to-particle conversion processes in the furnace and in the boiler. Some of the aerosol particles coagulate with coarse fly ashes due to collisions. Moreover, a smaller part of aerosols is also precipitated in the boiler and therefore forms part of the boiler ash, while the major part of this fraction is emitted with the flue gas at boiler outlet (typical particle size significantly <1 μm (ae.d.)).

In small-scale pellet furnaces and boilers, the main ash fraction is bottom ash. Furnace and boiler ash form the major share of coarse fly ash, which is usually precipitated and mixed with the bottom ash. A small amount of course fly ash is emitted with the flue gas. Around 95% of the ash is collected in the furnace and the boiler, while about 5% is typically emitted. In large-scale plants, the coarse fly ash fraction is usually higher and particulate matter precipitation systems are installed (e.g. cyclones for coarse fly ash precipitation, electrostatic precipitators or baghouse filters for aerosol precipitation).

The easily volatile heavy metals Zn, Cd and Pb are enriched in the filter fly ash. Therefore, only bottom and coarse fly ashes from biomass combustion are usually used on soils as a fertilising and liming agent, while filter fly ash is usually disposed of (cf. Section 9.11).

9.10 Fine particulate emissions

Most recently, fine particulate pollution gave rise to debate in numerous European countries as concentration limits were overstepped more frequently and more clearly than in the past. Urban areas were especially affected. In 2006 in Austria, for instance, the fine particulate concentration limit of 50 μg/m³ ambient air (daily mean value) was exceeded on more than the allowed 30 days at 71 measuring points. The highest number of excesses, namely 120, was observed at the Don Bosco measuring point in Graz [417]. The situation has become less dramatic since 2006 (cf. Figure 9.11). However, the reduction of fine particulate concentrations from 2006 to 2007 can primarily be explained by distinct climatic conditions (with regard to the reason for the ongoing reductions in the year 2008 there are no evaluations available yet) and trends in fine particulate concentrations cannot be derived to date [418]. High levels of fine particulate concentrations and surpassing of the limiting values were found mainly under adverse distribution conditions, whereby conditions in winter are especially relevant. Adverse conditions are characterised by frequent areas of high pressure weather conditions in middle and Eastern Europe, rare weather conditions with inflow of air masses from the west, the frequent inflow of air masses from the east already with some particulate matter level, as well as low wind velocities. Weather conditions such as these were present in 2006, which caused the fine particulate pollution. 2007, by contrast, was characterised by frequent low pressure, western and northern weather conditions, accounting for the low fine particulate concentration levels of that year.

Industry, traffic, agriculture and residential heating prove chiefly responsible for high fine particulate emission levels. Thus pellet central heating systems are perceived to be related to the surpassing of the limits, especially by manufacturers and sellers of heating systems based on fossil fuels. Such arguments are often based on data attained by poorly exercised research or unrepresentative data from biomass furnaces. Emission values from poorly controlled old systems are often taken as the basis for assumptions and they in no way reflect the situation of modern biomass furnace technologies.

Operators of biomass central heating systems and potential new customers are often alienated by such arguments and media statements. The significant technological developments of biomass furnaces over recent years with regard to fine particulate emissions are often not considered. Modern pellet furnaces especially exhibit by far lower fine particulate emission levels than old and poorly controlled small-scale biomass furnaces.

In the following sections, the definition of fine particulate matters, their formation and effects as well as the newest findings in the field of fine particulate emissions of pellet central heating systems are examined in more detail in order to bring the discussion back to actual facts.

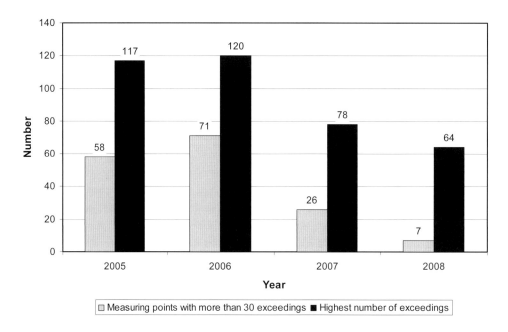

Figure 9.11: Excesses of the fine particulate emission limit in Austria from 2005 to 2008

Explanations: limiting value for fine particulate concentrations (daily mean value): 50 µg/m³; 30 excesses are allowed per year; data source [417]

9.10.1 Definition of fine particulates

All particles in the ambient air are called total dust or total suspended particulate matter (TSP). TSP is of little relevance since there is no limiting value for it since 01/01/2005 [419] and it hardly has any adverse health effects (cf. Section 9.10.2). Fine particulate matter, being one part of total particulate matter, is far more relevant. Dust particles with aerodynamic diameters (ae.d) of less than 10 µm (PM10) are called fine particulate matter. The aerodynamic diameter is used because the airborne particles neither have a uniform shape nor density. In order to determine the aerodynamic diameter, a spherical particle with a density of 1 g/cm³ is assumed. Then, the diameter this particle would have to have in order to sink in air as fast as the particle in question is calculated.

In order to be able to comprise and describe different fractions of particles, the terms PM2.5 (particulate matter with an aerodynamic diameter of < 2.5 µm) and PM1.0 (particulate matter with an aerodynamic diameter of < 1.0 µm) are used.

9.10.2 Health effects of fine particulates

The concentration of fine particulate matter in air is of particular importance as fine particulate matter is the thoracic fraction out of total particulate matter, i.e. the fraction that can pass the larynx and reach the lung (because fine particulate matters are not sufficiently filtered by the nose and the bronchia). Particles < 10 µm pass into the trachea, while particles < 2 to 3 µm can get into the pulmonary alveoli [420].

In order to evaluate health effects of air pollutants, a number of studies from different disciplines should be taken into account. This includes studies of personal exposure, toxicological studies including animal testing, controlled particulate exposure experiments and in-vitro studies, whereby each of these approaches demonstrates specific strengths and weaknesses [421].

There is strong epidemiologic indication that particulate matter in air has serious adverse health effects. Epidemiologic studies of recent years yield clear indications of effects on the cardiovascular system. A number of relevant physiological effects were found to be associated with fine particulate pollution [418; 420].

Between 2001 and 2003, the World Health Organization (WHO) carried out a review of health aspects in relation to ambient air quality [422; 423; 424]. According to the review, the correlation between fine particulate exposition and health effects is stronger than previously thought. PM2.5 shows stronger correlation with some serious health effects than the coarse fraction of fine particulates (PM10 minus PM2.5). However, there are indications that the coarse fraction of fine particulates is also related to certain adverse health effects. Thresholds for concentrations that do not cause any adverse health effects could not be derived to date. Some studies indicate that contents of certain metals, organic compounds, ultrafine particles (< 100 nm) and endotoxins are toxicologically active. Furthermore, some studies show a correlation between a reduction of fine particulate pollution and reduction of health effects. Chronic exposure to fine particulate matter shortens life expectancy of the population by one year on average, according to current model calculations [419].

Moreover, the first results of research activities in this field show that the health relevance of fine particulate matter seems to strongly depend on the concentration of carbonic particles in the fine particulate matter. Finnish investigations where lung cells were exposed to fine particulate samples from wood combustion (so-called in-vitro tests) showed stronger reactions as well as more dead cells when the cells were exposed to fine particulate matter from incomplete combustion than when the cells were exposed to fine particulate matter sampled from plants operated under ideal conditions [425]. In-vivo studies (inhalation tests) carried out in Germany, where rats were exposed to particles from complete combustion of biomass, showed that these particles did not have a negative effect on the respiratory system [426]. This is especially relevant since fine particulate emissions of poorly controlled, old furnaces are dominated by carbonic particles, whereas fine particulate emissions of modern small-scale biomass furnaces are dominated by inorganic particles from complete combustion (cf. Figure 9.13 in Section 9.10.3).

9.10.3 Fine particulate emissions from biomass furnaces

In contrast to heating oil or natural gas, as already mentioned in Section 9.9, biomass contains a considerable amount of ash, which inevitably leads to fly ash emissions during combustion. Fly ash emissions of complete biomass combustion consist mainly of potassium sulphates, potassium chlorides and potassium carbonates, i.e. salts, and are related to the PM1.0 fraction to a great extent (> 90%) [427]. Concerning the exact formation mechanisms of coarse fly ash and aerosols, see also [65; 428].

It is important to be aware of the fact that hardly any coarse fly ash arises in modern small-scale pellet furnaces (in contrast to medium- and large-scale biomass furnaces where fly ash concentrations of a few g/m³ are possible at the boiler outlet but precipitation of coarse fly ash can be achieved by using cyclones or multi-cyclones without problems). The low concentrations of coarse fly ash are the result of the low ash content of wood pellets, the relatively low flue gas velocities in the combustion chamber and the boiler as well as the calm nature of combustion (as compared to the moving grates of medium- and large-scale furnaces). So total suspended particulate matter emissions of pellet furnaces can chiefly (> 90%) be assigned to the PM1 fraction (thus to aerosols) [65].

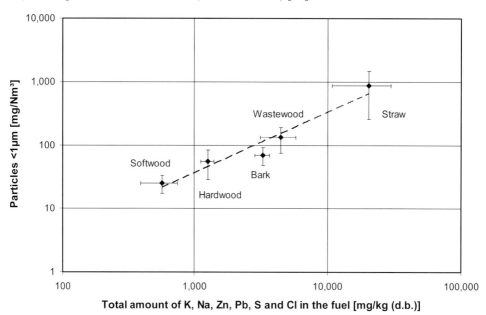

Figure 9.12: Aerosol emissions from medium- and large-scale biomass furnaces compared to aerosol forming elements in the fuel

<u>Explanations</u>: emissions related to dry flue gas and 13 vol.% O_2; results of measurements in grate furnaces with power outputs between 400 kW$_{th}$ and 50 MW$_{th}$; data source [319; 429; 430]

With regard to aerosols, organic and inorganic aerosols have to be distinguished. The formation of inorganic aerosols cannot be influenced significantly by operational or control measures in state-of-the-art biomass furnaces. Since inorganic aerosols are formed by easily volatile inorganic components that are released into gaseous phase during combustion, it is the chemical composition of the fuel and the release behaviour of the aerosol forming elements in the fuel that are the decisive factors in aerosol formation. Figure 9.12 shows measurement results of aerosol emissions from medium- and large-scale biomass furnaces in correlation with the content of easily volatile inorganic components (K, Na, S, Cl, Zn, Pb) in the fuel. It has to be noted as a principle that the content of aerosol forming elements in the fuel rises notably, beginning with softwood, followed by hardwood and with the highest content in bark and waste wood, as shown in the figure. Straw and whole crops have even higher contents of aerosol forming elements than waste wood. It is mainly the K, S and Cl contents that are decisive for aerosol formation in chemically untreated biomass fuels, whereas heavy metals such as Zn and Pb gain relevance in chemically treated fuels (waste wood).

In general, softwood pellets have relatively low contents of K, Na, S, Cl and easily volatile heavy metals (Zn, Pb, Cd) and thus low quantities of aerosols are formed during combustion. However, Figure 9.12 also shows that with the use of pellets with certain hardwood, bark or straw contents, the contents of aerosol forming elements rise significantly and thus aerosol emissions are increased. Owing to the high K, S and Cl concentrations in straw and whole crops, increased aerosol emissions can be expected when making use of these fuels.

In contrast to the formation of inorganic aerosols, the formation of carbonic aerosols that are a product of incomplete combustion consisting of elementary carbon (soot) or condensed hydrocarbon compounds (organic aerosols) can be significantly influenced by technical measures concerning combustion and control system. The more complete the combustion is (flue gas burnout), the less organic carbon compounds are available for condensation in the heat exchanger. In modern medium- and large-scale furnaces for instance, which are operated at CO concentrations in the flue gas of < 100 mg/Nm³ and organic carbon emissions of less than 10 mg/Nm³, only very low concentrations of carbonic aerosols could be found. The influence of burnout quality on aerosol emissions can be shown in a comparative investigation of emissions from poorly controlled or improperly operated biomass furnaces and modern small-scale automatic biomass furnaces. Emissions from poorly controlled or improperly operated furnaces are dominated by carbonic particles, whereas emissions from modern automatic biomass furnaces are dominated by organic salts (cf. Figure 9.13) [405]. Therefore, it can be stated that all measures to reduce CO and organic carbon emissions also minimise aerosol formation. Examples of such measures are thorough mixing of combustion air and flue gas in the combustion chamber as well as long enough residence times of the flue gas at sufficiently high combustion chamber temperatures.

Aerosols can lead to massive problems in boilers with regard to deposit formation. In addition, there is an increased risk of corrosion because of the high chlorine contents that are often present. In addition to these problems in the furnace, aerosols also cause substantial emission problems, which can generally only be counteracted by means of sophisticated precipitation technologies, such as electrostatic precipitators or baghouse filters. At present, these are only economic in medium- or large-scale furnaces, resulting in the need for R&D in the field of small-scale systems for residential heating. Current activities in this respect are dealt with in Section 12.2.1. Two international workshops in March 2005 and January 2008 in Graz also engaged in this topic [428].

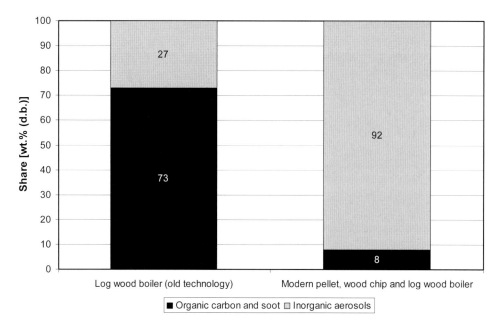

Figure 9.13: Composition of fine particulate emissions from old and modern small-scale biomass furnaces at nominal load

Explanations: data source [405]

9.10.4 Fine particulate emissions from pellet furnaces in comparison to the total fine particulate emissions of Austria

Emission factors of pellet furnaces were discussed in Section 9.5. On the basis of field measurements, an average emission factor for fine particulates of pellet central heating systems of 24 mg/MJ$_{NCV}$ can be assumed. The emission factor for fine particulates of pellet stoves is 54 mg/MJ$_{NCV}$ [408; 419]. Since 95% of pellets are used in pellet central heating systems and the rest in stoves (on the basis of existing systems in 2006; cf. Section 10.1.4.1.1), the average emission factor for fine particulate matter is 25.4 mg/MJ$_{NCV}$. Around 400,000 t of pellets were used in Austria in 2006. With a NCV of 4.9 kWh/kg (w.b.)$_p$ (equivalent to 17.64 MJ/kg (w.b.)$_p$), fine particulate emissions from pellet furnaces amounted to around 179.0 t in 2006.

Total fine particulate emissions in Austria were 43,500 t in the year 2006. Fine particulate emissions of domestic heating were about 7,900 t [431]. If fine particulate emissions of pellet heating systems on the basis of the pellet consumption in 2006 are put in relation to total fine particulate emissions and fine particulate emissions of domestic heating, pellet heating systems are found to cause 0.41% of the total fine particulate emissions and 2.27% of the fine particulate emissions from domestic heating. These facts are also shown in Figure 9.14, which displays fine particulate emissions in Austria as arranged by the sector of origin.

It can be seen in Figure 9.14 that 18.2% of fine particulate emissions can be assigned to domestic heating. The main share, namely 89.8%, of those emissions originates from wood furnaces (without pellet systems) and coal furnaces (7.0%). The share of fine particulate emissions from oil central heating systems is negligibly low (1.0%). Natural gas heating systems hardly produce any fine particulate emissions.

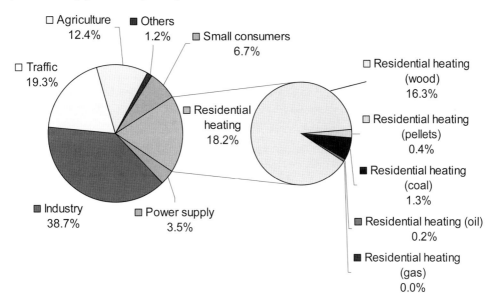

Figure 9.14: Fine particulate emissions in Austria according to sources

Explanations: data source [406; 408; 419; 431]; fine particulate emissions of pellet heating systems based on pellet consumption in 2006 (cf. Section 10.1.4.2)

Even an increase in pellet consumption to 1.3 million t/a, as it is expected for 2010, would elevate the share of fine particulate emissions from pellet heating systems to just 1.39% of the total.

These statements are not supposed to trivialise fine particulate emissions of pellet heating systems. On the contrary, they must be taken serious and all measures that can keep fine particulate emission factors of pellet heating systems low and further reduce the emissions must be taken. However, the figures clearly show that key fine particulate emission sources are to be searched for in other sectors with accordingly greater emission reduction potential (industry, traffic and agriculture).

However, there is potential for significant reductions in the residential heating sector, which can be demonstrated by means of the following theoretical scenarios. If all wood and coal heating systems were replaced by pellet heating systems, fine particulate emissions caused by residential heating could be reduced from 7,900 to 2,140 t/a (- 72.9%). If all wood and coal heating systems were replaced by oil or gas heating systems, fine particulate emissions could be reduced to 330 or 250 t/a, respectively (- 95.8% or - 96.8%, respectively). The efficiency increase that can be achieved by replacing old heating systems with modern pellet, oil or gas central heating systems was not considered in these theoretical scenarios. It would lead to a further reduction of fine particulate emissions.

The main conclusion that can be drawn is that old wood and coal heating systems especially must be retrofitted in order to reduce fine particulate emissions caused by residential heating. Looking at the reduction of fine particulate emissions only, the increased use of natural gas or oil heating systems would be the most reasonable option. However, this would lead to a massive increase of CO_2 emissions from the residential heating sector. Due to legally binding obligations to reduce CO_2 emissions in Austria, this would be counterproductive. An increased use of pellet heating systems would result in significant advantages for both areas of concern. The use of modern pellet heating systems could reduce fine particulate emissions and CO_2 emissions at the same time. A total changeover of the whole residential heating sector to pellet heating systems would lead to a reduction of fine particulate emissions caused by residential heating by 28% (owing to the great reduction of fine particulate emissions of wood and coal furnaces, which would more than compensate for the higher fine particulate emissions of pellet heating systems as compared to oil or gas heating systems).

For the above reasons, ambitions to prevent the expanding distribution of pellet systems because of the fine particulate emissions they cause, by cutting investment subsidies for pellet heating systems or even introducing subsidies for systems based on fossil fuels, must be seen as unreasonable and counterproductive. On the contrary, the reinforced use of pellets can indeed lead to a reduction of greenhouse gas emissions as well as fine particulate emissions.

9.11 Solid residues (ash)

If standardised pellets are used, as usually in the case of small-scale systems, small amounts of ash can be expected. If the upper limiting value for the ash content of 0.5 wt.% (d.b.) is assumed, 4.5 kg of ash per tonne of pellets will accrue. Assuming a pellet consumption of 6 t (w.b.)$_p$/a, a maximum of 27 kg of ash will add up over one year, whereby actual ash contents and thus ash amounts are slightly lower.

In Austria, for instance, wood ash is not hazardous waste (according to the waste catalogue ordinance 2008 [432]). Waste disposal can take place together with residual waste or biodegradable waste.

However, ash from biomass combustion contains significant amounts of nutrients. Typical nutrient contents of biomass ashes are given in Table 9.8. It is shown that biomass ashes mainly consist of calcium, magnesium and potassium oxides, which makes the ashes an interesting option for fertilisation and liming in the garden. Nitrogen as a nutrient is completely absent in biomass ashes, as almost all of it leaves the furnace via the flue gas. Thus it must be added when wood ash is used as a fertiliser. With regard to heavy metal contents, no problems are to be expected as long as chemically untreated natural biomass is burned, which is the case when standardised pellets are used (cf. Section 3.2.6). Hence, the use of ash from pellet heating systems in gardens as a secondary raw material with fertilising and liming properties is both tolerable and reasonable.

Table 9.8: Typical Ca and nutrient contents of different biomass ashes

Explanations: data source [433]

Fuel	CaO	MgO	K_2O	P_2O_5
Bark	32.12	5.35	4.45	1.77
Wood chips	38.70	4.68	6.54	3.57
Sawdust	28.06	5.37	7.56	2.35

Guiding values for dosing of 0.1 kg ash/m² in gardens and 0.075 kg ash/m² on meadows should be adhered to [63]. Assuming an average pellet consumption of 6 t (w.b.)$_p$/a, a 270 m² garden or 360 m² meadow can be fertilised, i.e. in most cases, all the ash of a pellet central heating system in a detached house can be utilised in the furnace owner's garden.

9.12 Summary/conclusions/recommendations

Emission factors that take emissions along the supply of useful energy into account (fuel supply, auxiliary energy supply and thermal utilisation) were used as the basis for the ecological comparison of central heating systems for residential heating based on the biomass fuels pellets and wood chips, the fossil fuels natural gas and heating oil and on district heat. For central heating systems based on pellets and heating oil, flue gas condensation was also considered, which is an innovation in the area of these two fuels. Systems without flue gas condensation were not taken into account in natural gas heating systems since this technology already dominates gas heating systems, with conventional systems playing a subordinate role in new installations.

Regarding CO_2 emissions and climate protection in consequence, the heating systems based on biomass fuels have clear advantages over systems operated with fossil fuels. If an oil furnace is replaced by a pellet furnace, CO_2 emissions of about 78,000 mg/MJ$_{NCV}$ can be saved. If a gas heating system is replaced by the same pellet furnace, about 54,000 mg/MJ$_{NCV}$ of CO_2 emissions are saved. Based on the pellet consumption of 500,000 t (w.b.)$_p$ in 2008 in Austria and assuming that it is just oil and gas heating systems that get replaced by pellet heating systems, the use of pellets would make it possible to save about 575,000 t/a of CO_2. Under the stated framework conditions, almost 1.5 million t of CO_2 could be saved in 2010 (with pellet consumption forecast to be 1.3 million t/a). In reality, pellet furnaces also replace heating systems based on other, partly CO_2 neutral, energy carriers (e.g. wood chips or firewood), hence the actually saved amount of CO_2 is less. However, the numbers do underline the importance of pellets as a contributor to the reduction of greenhouse gas emissions in Austria.

Regarding CO as well as particulate emissions, the biomass heating systems show clear disadvantages as compared to the fossil energy carriers. It is only pellet heating systems based on pellets made of wood shavings that have slightly lower CO emission factors than natural gas heating systems. However, particulate emissions of new pellet furnaces that mainly come from the comparatively high ash content when using biomass fuels and also CO emissions could be very much reduced due to recent technological developments. It must be noted in this respect that the reduced emissions due to these technological developments have yet not been included in the emission factors for CO and particulate matter. So CO and particulate emissions are overvalued in the current national inventory report of Austria since current emission factors still relate to old and in part poorly controlled systems. For an adequate consideration of technological developments of recent years, new emission factors for CO and particulate matter of modern biomass furnaces (modern pellet, wood chip and firewood furnaces) should be taken into account for national CO and particulate matter inventory reports.

Hydrocarbon emission factors are much higher in natural gas based systems than in any other system, which is a consequence of the emissions along the fuel supply chain. Old wood chip furnaces also exhibit comparatively high hydrocarbon emissions, which in this case is a consequence of high emissions during combustion caused by poor combustion control and hence poor burnout.

With regard to NO_x emissions, systems based on natural gas have very low emissions, which is mainly due to low emissions during combustion. The other systems show NO_x emissions of about the same level. As concerns SO_2 emissions, heating oil based systems stand out for their comparatively high emissions, caused by emissions during combustion due to the comparatively high sulphur content of heating oil. The SO_2 emissions of natural gas based systems are comparatively low, owing to low emissions during combustion. The SO_2 emissions of the other systems are on a similar level.

If the emissions factors are related to useful energy, which is carried out by taking the annual efficiencies into account, there is least effect on heating oil and natural gas systems with flue gas condensation as well as on district heating systems owing to the high annual efficiencies. The effect is greatest on old as well as new wood chip furnaces owing to the comparatively low annual efficiencies. The relation of the emission factors to useful energy results in a slight shift in favour of the fossil fuel systems due to their slightly higher annual efficiencies when pellets, heating oil and natural gas heating systems are compared.

Due to higher annual efficiencies when using flue gas condensation (about 6 to 8% higher), the systems in which this technology is applied show slightly lower emission factors. Flue gas condensation has been the state-of-the-art for many years in natural gas heating systems and it is currently being introduced to pellet and heating oil systems. With regard to the resulting efficiency rises and thus emission reductions, flue gas condensation is to be preferred over conventional furnace technologies.

Fine particulate and aerosol emissions are a special problem due to their adverse health effects. Therefore, the issue is dealt with in a number of national and international R&D projects (cf. Section 12.2.1). The findings suggest that there are three key approaches to reduce fine particulate emissions by small-scale biomass furnaces.

The use of modern combustion technologies should be supported in new buildings by appropriate subsidies. Modern pellet furnaces should especially be endorsed in this respect. They burn wood pellets under ideal conditions, which results in lower emissions of elementary carbon (soot) and organic hydrocarbons, leading in turn to lower fine particulate emissions. The reduction of soot and hydrocarbon emissions is of special relevance with regard to the health effects of the emitted fine particulate matter since the adverse health effects of fine particulate emissions seem to be reduced when less soot matter and organic hydrocarbon compounds are present (cf. Section 12.2.1 and [425; 426; 434; 435]). Flue gas condensation in pellet furnaces could be of relevance in this respect since emissions could be further reduced by making use of this technology (cf. Sections 6.1.2.9.1, 8.2.3 and 9.7).

For the above reasons, owners of old wood furnaces should be encouraged to change over to modern wood or pellet furnaces by subsidies programmes or boiler exchange campaigns. Old wood furnaces in particular have been proved to have high emissions of elementary carbon (soot) and organic hydrocarbons and thus high fine particulate emissions, with the entailed increase of adverse health effects (see above). A significant contribution to solving the fine particulate problem could be achieved by the right legal measures in this field.

While the formation of carbonic aerosols can be notably reduced by technical measures concerning combustion and control technology, the formation of organic aerosols cannot be significantly prevented in the same way.

In order to efficiently reduce inorganic aerosols, installation of appropriate fine particulate precipitation systems is an additional measure. Such systems are being tested and developed by current R&D projects [434; 435]. Before introduction on the market, further R&D needs to be done.

Regarding solid residues of pellet furnaces, i.e. bottom ash, its utilisation as a fertiliser in the garden is recommended since ashes from biomass combustion consist of considerable amounts of nutrients and soil improving substances. They are a valuable secondary raw material with fertilising and liming attributes. If this is not possible, the ash may follow the path of residual or biodegradable waste.

10 Current international market overview and projections

This chapter gives an overview of pellet market developments in the world. It makes no claim to be complete as further activities that are not publicly available might be ongoing in many regions of the world. However, the chapter covers all markets in the world that are well developed and known to be emerging pellet markets. The main focus is on European countries, as pellet production and utilisation is concentrated here. The most important countries are described on an individual basis, for example Austria, Germany, Sweden and others. Another important pellet market in the world is located in North America and therefore this market is also described in detail. Activities in other countries of the world are summarised in a separate sub-section.

The market descriptions comprise – as far as they are available – data concerning historical development and present situation of pellet production, consumption, import and export as well as of possible raw materials that are used and available. Moreover, the different sectors of pellet utilisation are described.

The chapter is supplemented by sub-sections with an international overview of pellet production potentials, the international and intercontinental trade of pellets as well as socio-economic aspects of production and utilisation of pellets.

10.1 Austria

10.1.1 Pellet associations

In 1997 Pelletsverbrand Austria (PVA), an Austrian pellet association, was founded. The PVA was involved in technical coordination projects between pellet production and the boiler industry. The technical standard for pellets as a fuel was steered and influenced to a great extent by the PVA and the first pellet logistics standard of Europe was created. The PVA represented Austria in negotiations with the European Union concerning European standardisation.

The PVA developed a standard for its members, which partly set down stricter values than did the ÖNORM M 7135 and it also comprised parameters not covered by the ÖNORM (in particular limiting values for some heavy metals). In addition, an official quality certification was developed, which set down strict quality criteria for pellet furnaces and pellets. In addition, it also regulates pellet transport. According to this certificate, the pellets have to be sieved prior to loading onto the lorry in order to safeguard a maximum of 1% of fines. Moreover, the standard warrants interim storage in closed warehouses or closed storage spaces such as silos as well as the delivery of pellets in special pellet silo trucks.

In order to appropriately safeguard the high quality set down for the members of the PVA, an innovative quality control and assurance system was introduced in March 2002. For this purpose, coded pieces of wood were added to the pellets. These coded pieces of wood make it possible to find out who produced the pellets, where and when the pellets were produced, where they were stored temporarily, if applicable, and who was in charge of transport. Thus, the pellets are completely traceable. In addition to the coding, the pellets are subject to unannounced checks by an independent and nationally accredited institution four times a year.

Thus, the PVA must be regarded as a pioneer concerning regulations for the transport of pellets and traceability along the pellet supply chain. Pellet transport and storage regulations were later covered by two separate national standards (cf. Section 2.5) and are currently being implemented in the new certification system ENplus on a European level. Moreover, a system for complete traceability along the pellet supply chain will be implemented under ENplus (cf. Section 2.6).

The PVA no longer exists in its original form. In 2005 it was changed from a limited liability company to a new organisational from, namely a registered association. Owners of heating systems, installers and technicians are its members. The functions of the newly organised PVA comprise giving information for and protection of pellet users, regular information service for all members, information and help for installers, publication of price comparisons within the pellet sector but also relating to oil and gas, and networking of different market actors within the renewable energy sector [436].

In March 2005 proPellets Austria [437], another Austrian pellet association, was founded. The network proPellets Austria brings together key market actors from different fields such as furnace, boiler and storage system manufacturers, pellet production and trade and energy suppliers. The aim of proPellets Austria is to promote the importance of pellets as a revolutionary way of heating and achieve greater awareness among consumers [437]. ProPellets has set its focus on informing the general public about the importance of pellets as an extraordinarily environmentally friendly and renewable fuel. Standards such as those developed by the PVA are not developed by proPellets Austria. The existing standards (cf. Chapter 2) are regarded as sufficient.

10.1.2 Pellet production, production capacity, import and export

The framework conditions for pellet production are beneficial in Austria, as there is still a growing demand for pellets due to the large number of new pellet furnace installations every year, mainly in the residential heating sector. Numerous projects for the establishment of new pellet production plants with appropriate drying units for sawdust confirm this trend. With regard to the most important raw material at present, namely sawdust, the long expected shortage is now a reality, not least owing to the competition with the particle board industry.

In Austria, 19 pellet producers are active at 29 production sites (as per March 2010). Figure 10.1 presents an overview of locations and sizes of these production plants. It shows that around the half of all production sites have production capacities of 10,000 to 50,000 t (w.b.)$_p$/a. The share of production sites with lower capacities is 17%, that with higher capacities 31%.

Figure 10.2 displays the development of pellet production capacity in Austria since the beginning of pellet production in 1996. Up to 2000, strong annual growth rates of production capacities of up to 115% could be noted and 12 pellet producers had a capacity of 200,000 t (w.b.)$_p$/a. After that a phase of no new production capacity development but increased domestic consumption followed. From 2004 onwards, there were strong increases in production again. In 2009, pellet production capacity reached about 1.1 million t (w.b.)$_p$/a owing to the erection of new production sites.

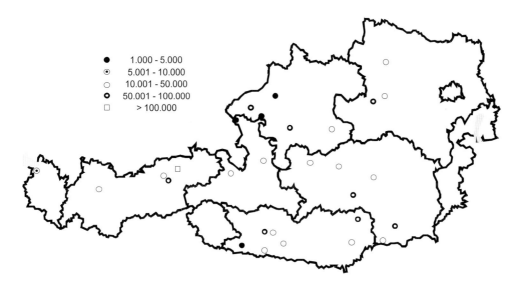

Figure 10.1: Pellet production sites in Austria and their capacities

Explanations: status March 2010; values in t (w.b.)$_p$/a, data source: own research

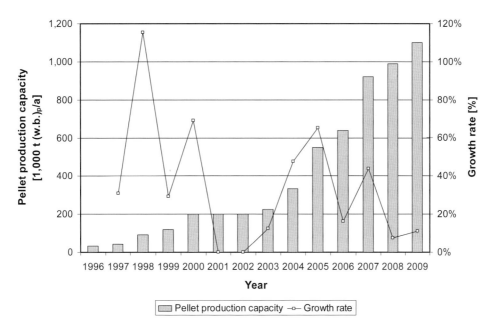

Figure 10.2: Development of Austrian pellet production capacities from 1996 to 2009

Explanations: data source [170; 437; 438; 439; 440; 441; 442; 443], own research

Figure 10.3 shows the development of pellet production in Austria since 1995 as compared to the installed production capacity, the annual demand and export. In 1995, pellet production began with the production of 2,500 t (w.b.)$_p$/a and grew to 695,000 t (w.b.)$_p$/a by 2009, with annual growth rates of 10 to 500% (though from 2007 to 2008 a decline in pellet production by about 9% occurred). In comparison, Austrian pellet production capacity grew from 2,500 t (w.b.)$_p$/a to 1.1 million t (w.b.)$_p$/a within the same period. From 1996 to 2002, production capacity was significantly higher than actual production. The average utilisation rate was just 48% during these years. It was not until 2003 that pellet production almost neared production capacities with the average utilisation rate being around 93% between 2003 and 2006. Since 2007, the average utilisation rate declined due to pronounced increases in production capacities but only moderate rises in consumption.

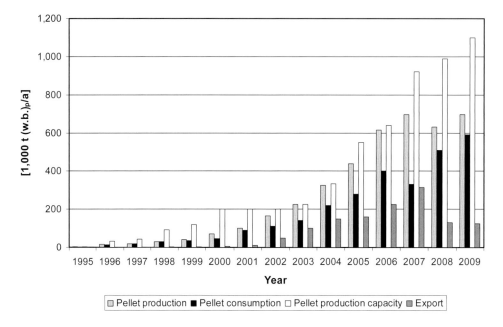

Figure 10.3: Development of Austrian pellet production, consumption and export from 1995 to 2009

Explanations: data source [170; 437; 439; 440; 441; 442; 443; 444; 445], own research

From the beginnings of pellet production in Austria, consumption has been below production. In the first years (around until 1998) of the industry, consumption was 95% of production, so that the two were almost balanced. This share decreased from 1999 onwards with some ups and downs to a minimum in 2007 of 47%. In 2008 and 2009, it increased again to almost 85%. There is insufficient information available as to where the surplus pellets ended up. It is known that from 1995 to 2000 a part of production was exported [170], mainly to northern Italy and southern Germany. Minor quantities were exported to Switzerland too. At the same time, imports were observed from 1998, which mainly came from Eastern European countries. From 2001 onwards, export quantities grew strongly due to expanding markets in Italy, Germany and partly in Switzerland, reaching their maximum at 313,000 t (w.b.)$_p$/a in 2007 [446]. A part of surplus production was probably stored for security of supply reasons, as such storages were first established in 2000 [333; 447]. Although there are no reliable data available, it is assumed that imports from Eastern Europe have further increased since then.

Prognoses concerning pellet market development are difficult and thus should be evaluated carefully as they have always had to be amended upwards in the past. This is demonstrated when looking at the pellet production prognosis of [170] in 2000. Then, pellet production in

Austria of 200,000 t (w.b.)$_p$/a was forecast for 2010. This value was surpassed in 2003 with a production of 225,000 t (w.b.)$_p$/a. Therefore, detailed prognoses of future developments of the pellet markets are abstained from at this point. Additional increases of production, production capacities and consumption are to be expected in any case. It is only the extent to which this will happen that is in question.

Whether and to what extent the required raw material potential is available for further increases of pellet production are discussed in the next section.

10.1.3 Pellet production potential

Several woody biomass fractions can potentially be used as raw materials for pelletisation, namely wood shavings, sawdust, industrial and forest wood chips (either with or without bark), short rotation crops and log wood.

The wood shavings potential in Austria is difficult to estimate because the amount of wood shavings that accumulates in the wood processing industry is not recorded. The total amount is stated to be somewhere within the very broad range of 0.8 to 2.4 million t (w.b.)/a [448], about 100,000 t (w.b.)/a of which are already used for pellet production. The majority of wood shavings is used in-house (thermal utilisation for the most part) and a smaller part is used for briquette production. So, under present framework conditions, an expansion of pellet production on the basis of this raw material does not seem possible since practically all the available quantities are being used already. Only production increases in the wood processing industry could create additional potential in this respect [60].

A greater resource than wood shavings is sawdust that is usually available with a moisture content of around 55 wt.% (w.b.). The amount of sawdust that accumulates in Austria is asserted to be up to 2.1 million t (w.b.)/a [449]. Large quantities of sawdust are used for in-house process heat generation and also most recently in CHP plants. In addition, the raw material is mainly used in the particle board industry. It is difficult to estimate actual quantities available for pelletisation owing to the competitive situation with the particle board industry and unreliable data concerning the actual total amounts present. In the past, most estimations assumed a potential of about 1 million t (w.b.)/a [59; 450], of which roughly 500,000 t/a of pellets could be produced. However, based on pellet production in 2009, i.e. the year with the highest production up to now of around 700,000 t, and taking into account the pellets made of wood shavings, it is shown that the available sawdust quantity for pelletisation must be more than thought or there must have been a shift of sawdust from the particle board industry to the pellet industry. Overall, the supply of the two raw materials of wood shavings and sawdust seems to be largely exhausted.

Some expansion of pellet production on the basis of sawdust could be possible under certain circumstances by a further shift of sawdust flows from the particle board to the pellet industry. In the short term, however, an expansion on the basis of more expensive raw materials will take place. Industrial wood chips without bark would be a suitable option. However, they are a much valued raw material of the paper and pulp industry. The amount that will actually be available for pellet production is difficult to estimate at present since this will mainly depend on the prices the different industries will be able to pay.

Another option would be bark containing biomass fractions, i.e. industrial wood chips with bark, forest wood chips and short rotation crops, whereby the bark content of these materials is their main disadvantage (cf. Chapter 3).

At present, industrial wood chips with bark are predominantly used thermally in biomass heating plants and would have to be substituted by other fuels in order to create potential for pelletisation. The question as to the extent to which this will take place will again be answered according to the price level of this material. The next group of materials that can be used for pelletisation are forest wood chips. Since forest wood chips usually contain bark, there are problems in using this raw material for the production of standardised pellets as the required ash content for class A1 pellets according to prEN 14961-2 cannot normally be adhered to. Although the elevated ash content can be handled by technical measures, for instance appropriate de-ashing systems in pellet heating systems, standardised pellets for the small-scale systems market cannot be produced. However, such pellets could be used in medium- and large-scale furnaces, where standardised pellets are currently often used. They could replace standardised pellets that in turn would be available for the market of small-scale systems. Forest wood chips are mostly thermally utilised at present. Realistically, a shift of this raw material to pelletisation will not be possible. According to [59], however, an additional potential for forest wood chips of 2.5 million tonnes (d.b.)/a could come from increased thinning. This would make the production of additional amounts of pellets out of this fraction possible. Another possibility for raw materials for pelletisation is the production of wood from SRC plantations (e.g. willow, poplar). A major drawback when using SRC, as is the case for all groups of wood chips containing bark, is their high ash as well as nitrogen, chlorine and sulphur contents. Such pellets would not comply with the top pellet quality class A1 according to prEN 14961-2.

The theoretical potential for pellets from industrial and forest wood chips is estimated to be more than 4 million tonnes per year [60]. Taking the SRC potential into account, an additional 2.7 million tonnes of pellets could be produced, totalling about 6.7 million tonnes of pellets [60].

One more alternative is the production of pellets using log wood. Separation of the bark before processing is possible but would add extra costs to production [111] (cf. Section 7.3). Some activities moving in this direction are already underway by some pellet producers in Austria [109; 111; 112]. Internationally, it is already common practice to produce pellets from log wood. The potential of this kind of raw material is not investigated in depth here, however. It suffices to mention that the annual increment of Austrian forests amounts to about 30 million scm, of which only about 20 million scm are currently used. The difference of 10 million scm per year can be regarded as the theoretical potential for pellet production on a sustainable basis. How much of it could actually be utilised in practice needs to be investigated.

Another aspect must not be neglected when looking at available potential in a holistic way. The amount of available sawdust varies according to wood processing capacities. Until recently, an increase of wood processing capacities was expected in the coming years [419] and so an increase of available sawdust was expected too. Due to the ongoing financial and economic crisis, wood processing capacities were cut, leading to a reduction in available sawdust. The alternative raw materials discussed above, such as wood chips or log wood, gain more relevance within this context.

The pellet industry is currently nearing the depletion of the sawdust potential. The situation is made even more dramatic by reduced amounts from wood processing. A possible shift of sawdust from the particle board industry to the pellet industry could relieve the situation and postpone the point of depletion somewhat. Exact prognoses in this respect are not possible at present, but taking into account the additional pellet production potential out of the biomass fractions discussed above of about 6.7 million tonnes per year and the unused increment of about 10 million scm in Austrian forests, it can be concluded that sufficient raw material is available for the coming decades, even for further pronounced growth of the pellet market.

However, it can be assumed that only a certain amount of the available theoretical potential will actually be exploitable by the pellet industry. If the pellet market needs to find new supplies, the importance of imports will probably also grow. The import of pellets from Eastern Europe and Russia could gain special relevance. Developments in this direction are already being noted [451; 452; 453]. Growing imports will mean that the Austrian and most likely the whole European markets will need to inform pellet producers in exporting countries of the required quality demands and standards, as well as needing to prevent low quality pellets from being used in small-scale furnaces by taking appropriate measures.

10.1.4 Pellet utilisation

10.1.4.1 General framework conditions

In Austria, investment subsidies are granted for new installations as well as for changeovers of residential heating system to biomass fuels by means of non-repayable allowances. The guidelines for these subsidies are different according to the federal states, which is why no general statement about the extent of subsidies can be made. Wood pellet furnaces are subsidised by 25 to 30% or a maximum of approximately 1,000 to 5,500 €. Many of the subsidies are restricted to certain periods, however, funding programmes were often prolonged without major changes to the amount or type of subsidy.

In addition, extra subsidies in addition to basic housing or renovation subsidies in the form of increased credit lines are granted when renewable energy systems are installed, the extent of which, again, varies according to the federal state.

Besides federal state subsidies, many Austrian municipalities offer the possibility to acquire subsidies for the installation of alternative energy systems (e.g. heat pump, solar heating system, wood chip heating system, pellet heating system, photovoltaic system). The amount of such subsidies depends on the municipality.

In addition to investment subsidies for pellet furnaces, pellets also have advantages regarding taxes when compared to fossil fuels. An energy tax on electricity and gas was introduced in Austria on 1 June 1996 and has since been amended twice. The extent of the tax is 7.92 €ct/Nm³ natural gas (equivalent to 0.713 €ct/kWh) at present (March 2010). The energy tax on electricity is 1.8 €ct/kWh (both tax values include 20% VAT). Energy taxes such as these are not claimed when biomass fuels are used. Moreover, VAT is only 10% for pellets. All fossil fuels and district heat command 20% VAT.

10.1.4.1.1 Small-scale users

Within the field of pellet central heating systems, a pronounced increase in new installations was noted from 1997 to 2006, with the exception of 2002, when there was a slight decrease. In Austria, 10,500 new pellet central heating systems and 5,640 pellet stoves were installed in 2006 [454; 455; 456]. One reason for the notable increase in pellet heating systems was the aforementioned investment subsidies that are granted by the different federal states for new installations of pellet heating systems or the replacement of old systems (cf. Section 10.1.4.1). In addition, information and marketing initiatives by various market players such as proPellets Austria also made their contribution to the growth in pellet heating systems. Another major reason was the rise in gas and oil prices over previous years. However, in 2007 there was a clear reduction in the numbers of newly installed pellet central heating systems as well as pellet stoves owing to the massive price rise of pellets and partial supply shortages (cf. Section 8.1) that took place in the preceding period, which caused uncertainties in consumers regarding price stability and security of supply. Pellets price could again be reduced by the expansion of production capacities, and thus user confidence regarding price stability and security of supply could be restored. In 2008, increases of all types of newly installed systems were apparent.

Figure 10.4 presents the numbers of installed pellet stoves since 2001. Before 2001, hardly any pellet stoves were likely to have been installed. On the whole, more than 20,000 pellet stoves were installed in Austria by the end of 2008.

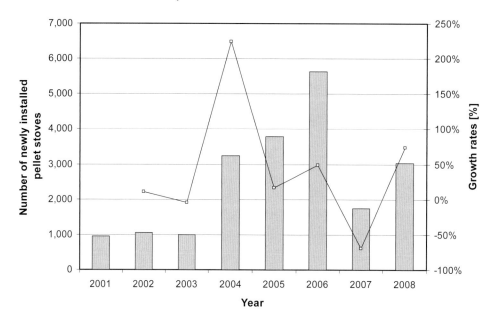

Figure 10.4: Development of pellet stoves in Austria from 2001 to 2008

Explanations: data source [454; 455; 456; 457]

The development of annual new installations of pellet central heating systems as well as annual growth rates are presented in Figure 10.5. Figure 10.6 presents the total nominal power installed as well as its growth rates. Around 62,400 pellet central heating systems with a nominal power of almost 1,200 MW$_{th}$ were installed in Austria by the end of 2008. Ignoring 2002 and 2007, when there were reductions, the growth rates in newly installed systems were between 16% and 211%. After the significant decrease in 2007, a strong increase was noted in 2008, whereby even the peak of 2006 was surpassed. For 2009, a decrease in new installations by about 23% is expected. The strong increase in numbers of installed pellet furnaces is not least due to the pronounced increases in the price of heating oil.

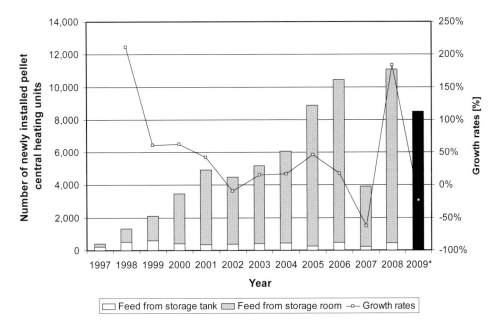

Figure 10.5: Development of pellet central heating systems in Austria from 1997 to 2009

Explanations: *...prognosis; data source [443; 455; 456; 458]

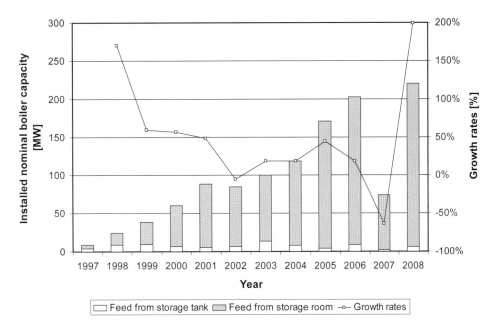

Figure 10.6: Development of annually installed nominal boiler capacity of pellet central heating systems in Austria from 1997 to 2008

Explanations: data source [454; 455; 457]

Figure 10.6 also shows that pellet furnaces that are equipped with an integrated pellet reservoir that has to be filled with packaged pellets by hand play a subordinate role. On the whole, only 6.6% of pellet central heating systems are equipped with such pellet storage boxes.

10.1.4.1.2 Medium- and large-scale users

Within the area of medium- and large-scale systems, pellet furnaces play a subordinate role in Austria. Here, wood chip furnaces or plants using mixtures of wood chips and sawmill by-products (including bark) are predominantly in use. The development of medium- and large-scale plants is presented in Figure 10.7. A decline in newly installed medium-scale systems up to 1993 was followed by an increase until 1998. In 1999 and 2002, clear decreases in new installations were noted and from 2003 to 2006 there were strong increases again. Like with pellet heating systems, there were far fewer new installations of medium-scale systems in 2007. The reason was the pronounced rise in fuel price because while the price of pellets was high, the price of forest wood chips and sawmill by-products was also high. New installations of large-scale systems increased continuously to about 50 installations per year until 1998. During 1999 to 2004 there were between 26 and 54 new installations per year. In 2005, new installations almost doubled. Since then, new installations have maintained this strong growth rate reaching a new peak of 88 new installations in 2007.

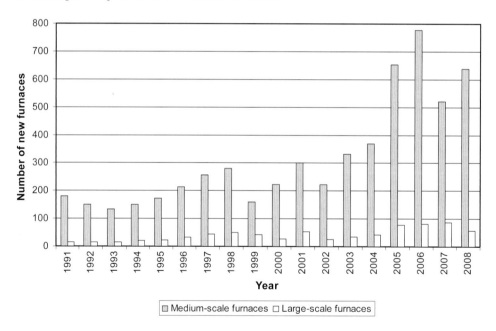

Figure 10.7: Development of medium- and large-scale wood chip furnaces in Austria from 1997 to 2008

Explanations: medium-scale systems: nominal boiler capacities of 100 to 1,000 kW; large-scale systems: nominal boiler capacities of more than 1 MW; data source [455]

There are only a few pellet furnaces with nominal boiler capacities of more than 100 kW operating to date, even though this power range in becoming increasingly significant. In the coming years, increased use of pellets is expected in apartment buildings in particular. The utilisation of pellets in this power range is particularly interesting because pellets require less storage space due to their high energy density. In addition, industrial pellets can be used as well owing to the fact that such systems can deal with pellets of a lower quality because large particle size and high dust or ash contents tend not to cause problems in larger furnaces. Moreover, the combined use of pellets and wood chips is usually possible, which makes the system flexible with regard to fuel.

10.1.4.2 Pellet consumption

Figure 10.8 presents the development of pellet consumption in Austria since 1996. Around 590,000 t of pellets were used by 2009. Pellet consumption has thus increased by a factor of 42 since 1996, the year in which a noteworthy amount of pellets was consumed for the first time. Within this, annual growth rates were between 16 and 100%. Only once, in 2007, was there a reduction of 18%. This reduction was the consequence of the extremely mild winter of 2006/2007, which led to decreased consumption and hence decreased demand. Pellet consumption in 2015 is forecast to be about 1.5 million tonnes of pellets [212]. In total about 150,000 new small-scale pellet central heating systems with an average consumption of about 6 t (w.b.)$_p$/a would be necessary to consume this amount of pellets. Consequently, approximately 25,000 new pellet central heating systems would have to be installed every year until 2015.

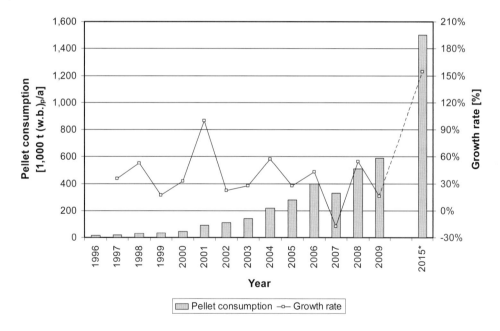

Figure 10.8: Development of pellet consumption in Austria from 1997 to 2008

Explanations: *...prognosis; data source [212; 437; 439; 441; 442; 443; 447; 459; 460; 461; 462; 463]

The gross domestic consumption of renewable energy sources (without hydropower) was around 230 PJ in 2007 in Austria. The share of pellets was 3.7% or 8.6 PJ, as shown in Figure 10.9. The share of pellets as related to total primary energy consumption of 1,421 PJ was thus

0.61% in 2007 in Austria. This share increased to around 0.73% in 2009 (approximately, as founded on pellet production in 2009 and related to total primary energy consumption in 2007). The share of bioenergy (renewable energy according to Figure 10.9 without ambient heat, wind and photovoltaics) out of total use of primary energy is around 15% (213 PJ).

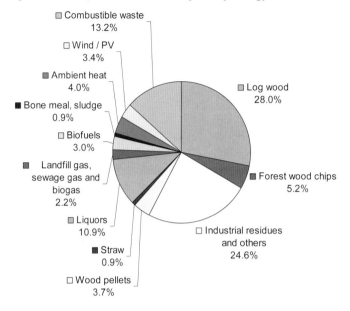

Figure 10.9: Gross domestic consumption of renewable fuels (without hydropower) in Austria (2007)

Explanations: total consumption 230 PJ; ambient heat: solar collectors, heat pump, geothermal energy; data source [464]

10.1.4.3 Pellet consumption potential

10.1.4.3.1 Framework conditions needed for further market growth

Annual costs, operational reliability, investment costs and in-house air quality were found to be the main criteria in purchasing a new heating system according to a questionnaire study of 1,500 Swedish house owners (in order of decreasing importance). Moreover, environmental aspects were found to play a minor role in decision making [465]. Although these results from Sweden are applicable to Austria to a limited extent only, key conclusions can be derived. It can be assumed that the three most important decision criteria of Swedish house owners are the same as for Austrian house owners but in a different order. In Sweden it is mainly pellet burners that are used to change over to pellets without having to replace the existing boiler. In Austria, pellet furnaces as a complete system are almost exclusively used, which are more expensive due to the fact that they are a complete system as well as the high state-of-the-art of this furnace technology. Therefore, it can be assumed that the high investment costs are the main hindrance for choosing a pellet furnace (despite available investment subsidies). This is repeatedly confirmed by many house owners who are about to take a decision in this respect.

The increased use of pellets as a fuel was inhibited for a long time by the low price for oil. The pellet market experienced a fundamental change with the introduced investment subsidies and price increases in oil, which is reflected in the growing number of new pellet central heating installations every year.

High investment costs continue to restrain pellet furnace installation to some extent. The investment costs of a pellet central heating system with automatic storage discharge remain above 9,000 € (boiler and storage discharge). The investment costs of oil or gas central heating systems are much less. In order to push the pellet market forward, investment costs would have to decrease or subsidies would have to be increased. An increase of subsidies is unlikely due to limited and even dwindling funding budgets. The price of pellet furnaces has not fallen over recent years despite increased sales volumes. Compared to oil or gas heating systems, the sales volumes of pellet furnaces are still low but an increase in production could quite possibly result in lower investment costs. Owing to the more sophisticated technology of pellet furnaces, the investment costs will finally remain above those for oil or gas furnaces. Another incentive to use pellets will arise from a sustained high level in the oil price and a stable pellet price at the same time. The oil price is forecast to rise again by many experts.

After some troubles in the early years, security of supply and a constant quality are no longer problems for the Austrian pellet sector since a sufficient number of producers and traders safeguard the supply of pellets, while the standards ÖNORM M 7135, M 7136 and M 7137 set down high quality requirements for pellets that are regularly enforced by independent inspection bodies. These standards will soon be replaced by the European standard EN 14961-2 and the certification system ENplus. Imports present some danger to pellet quality, however. Traders and also end users should carefully check whether these pellets are actually certified according to the standard and whether this is confirmed by the delivery documents.

10.1.4.3.2 Small-scale applications

Of the total 3.51 million homes in Austria, 19.4% are heated by biomass (in 2006). As shown in Figure 10.10, this share was greater in the 1980s and in the early 1990s (up to 21.5%). The main reason for the decline in wood furnaces was that old wood boilers were partly replaced by new oil or gas heating systems. This shows a tendency to favour high operational comfort that could not be provided by wood heating systems at that time. In 2000, the number of wood heated homes reached its lowest level at around 14.3%. The level remained steady with a slight rise until 2003. Since 2004 the share has been clearly increasing. The trend reversal in 2001 seems to have been due to the increased use of modern log wood, wood chip and especially pellet boilers. Thanks to the high user comfort of modern pellet heating systems and the unstable price of oil and gas there is good potential to increase the share of biomass heated homes in Austria by means of pellet central heating systems.

The share of oil heated homes remained around 27 to 28% for a long time but it declined to 24.7% in 2006. The share of electric heating has been falling since 1995. District heating (and other) shows steady increases except for a few slight decreases. The share rose from 4.3% in 1980 to 23.9% in 2006. The share of gas heating systems has also increased steadily since 1980, reaching its maximum of almost 30% in 2005 straight after a slight decrease in 2004. It declined again in 2005 and 2006. In 2000, the share of natural gas heating systems was more than that of oil heating systems for the first time. Coal was in second place after oil in 1980. The share of coal heating systems in Austrian homes continuously fell from then on though, reaching its absolute low in 2006 with 1.1%, thus having the lowest share now.

The amount of homes in total increased steadily except in 1986, 1987 and 2003, reaching 3.51 million in 2006. This constitutes great potential for pellet heating systems, whereby the use of pellets is not only possible in classic central heating but also in district heating plants (starting in theory from micro grids up to large-scale district heating networks). An additional increase in pellet heating systems can take place by means of installing pellet heating systems in newly built houses as well as by exchanging old systems. The installation of pellet heating is always possible, that is, installation is possible regardless of the system it is replacing (except for electric heating systems where there is no chimney, necessitating chimney installation).

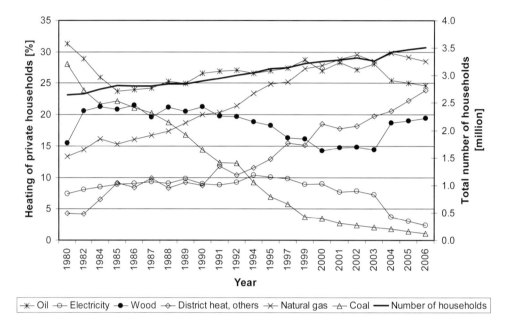

Figure 10.10: Heating systems in Austrian homes from 1980 to 2006
Explanations: data source [466; 467; 468]

Figure 10.11 shows the development of the heating market in Austria since 1997. It follows that more than 90,000 boilers were installed annually until 1999; in 1999 the number was above 97,000. However, from 2000 to 2002 a reduction was noted and in 2002 only 76,000 systems were installed. This signifies a decrease of 22% in relation to 1999. It can be derived from these numbers that required investments were not carried out. The reason probably lies with consumer anxiety caused by the price of oil. Since the exchange of many old systems did not take place then, new installations were expected to rise again, which they have done since 2003. In 2007, there was a notably decline in the entire boiler market of almost 16% as compared to the previous year, which was mainly due to the collapse of the pellet boiler market by almost 63%. This means that new installations that would have taken place by means of pellet heating systems, had the market developed as normal, did not take place and not were systems installed based on other fuels. In 2008, the decrease of 2007 was almost fully cancelled out by a rise in new installations. In order to push forward the trend towards using renewable energy sources, all market players as well as politicians need to take appropriate measures in order to re-enforce the trend and prevent increases in the use of fossil fuels in small-scale applications owing to misplaced incentives.

Looking at annual new installations by fuel, a different picture arises. Gas boilers are by far the most frequently installed applications, even though there were decreases in this area in 2006 and 2007 of 5.2% and 8.5%, respectively. Around 40,000 gas boilers are installed every year in Austria. A dramatic collapse occurred in oil boiler installations. Starting from 1999, when around 31,500 systems were installed, new installations decreased to 3,900 by 2008. This means a reduction of almost 88%. Heat pumps are gaining increasing relevance. In 2008, around 12,000 new heat pump systems were installed. Biomass boilers were on the increase until 2001 with a slight reduction in new installations in 2002 and subsequent increase by 2008, disrupted by a massive market decline in 2007. The rise originates mainly from pellet boilers because new installations of other biomass boilers (without pellet boilers) increase notably slower than pellet boilers. As shown in Figure 10.9, log wood still makes up 37% of renewable fuels (without hydropower). Many firewood boilers are old systems though that will have to be replaced within the coming years. It is important to take measures to avoid replacing these furnaces by systems based on fossil fuels, as often happened in the past.

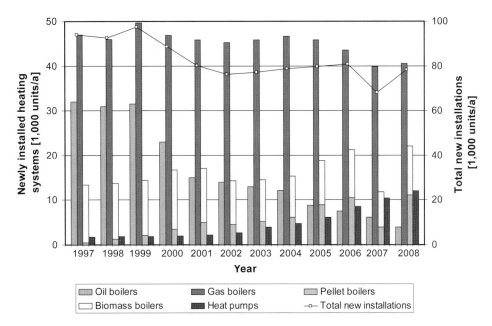

Figure 10.11: Annual boiler installations in Austria from 1997 to 2008

Explanations: systems up to 100 kW; data source [457]

10.1.4.3.3 Medium- and large-scale applications

The share of Austrian houses and flats that are supplied with district heating is also climbing, but only a small part of district heat originates from renewable fuels [170].

In Austria there are about 1,500 biomass district heating and CHP plants in operation [469] (at the end of 2008), and a steady increase in the number of these systems was noted in the preceding years. The nominal boiler capacity of Austrian biomass district heating and CHP plants is about 1,400 MW$_{th}$ in total. It is mainly by-products of sawmills, bark and forest wood chips that serve as fuels, with a trend away from bark and sawmill by-products to forest wood chips. The future supply of forest wood chips for fuel poses no problems but they are more expensive than sawmill by-products and bark. The difference in price as compared to pellets is reduced in this way. Pellets could gain significance in this area of application in the future having the advantages of uniform quality, less storage space demands, less material wear and less ash content. The use of pellets in medium- and large-scale systems would be interesting at a low pellet price, which could be achieved by the use of industrial pellets. Attempts to establish industrial pellets in Austria have been unsuccessful to date, probably owing to other, cheaper biomass fuels being available in sufficient quantities.

10.2 Germany

10.2.1 Pellet associations

The German pellet association DEPV (Deutscher Energie-Pellet-Verband e.V., since September 2009 Deutscher Energieholz- und Pellet-Verband e.V.) was founded in 2001 [484]. The association advocates the interests of its members in the economy and politics and towards end users. Its members are important actors in the pellet sector such as pellet producers, boiler and furnace manufacturers, retailers and manufacturers of components. The main target of the association is to increase the share of pellet heating systems in the heating sector.

10.2.2 Pellet production, production capacity, import and export

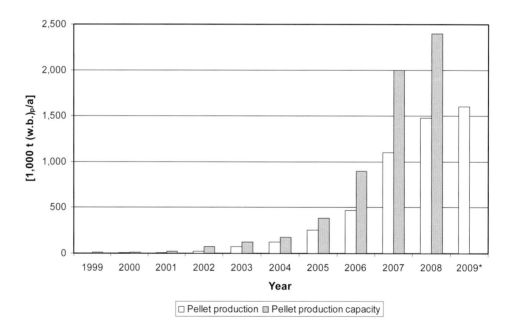

Figure 10.12: Pellet production and production capacities in Germany from 1999 to 2008

Explanations: *…prognosis; data source [470; 471; 472; 473; 474; 475]

In the early years of the industry, pellets were mainly imported from other countries, especially Austria. However, numerous pellet production plants have been erected in recent years, so national production capacities and production have increased and imports have fallen (cf. Figure 10.12). In addition, pellets have also been exported to neighbouring countries [476]. In 2009, around 1.6 million t (w.b.)$_p$/a were produced from about 60 companies at 75 sites in Germany. Production capacity is around 2.5 million t (w.b.)$_p$/a. In 2010, production is expected to increase to about 1.7 million t (w.b.)$_p$/a [477; 478; 475].

Data concerning import and export of pellets from and to Germany are scarce. An estimated amount of about 560,000 tonnes of pellets were exported in 2008, mainly industrial pellets. Industrial pellets were exported to Scandinavia, Belgium and the Netherlands. DIN$_{plus}$ pellets were exported to France, Austria, Italy and Switzerland. The total amount of DIN$_{plus}$ pellets exported is less than 2% of the total volume. Some small amounts of imported pellets are known from Austria, Eastern European countries and Sweden [480].

10.2.3   Production potential

The potential of wood shavings and sawdust will be depleted in the near future in Germany, as in Austria (in the winter of 2005/2006 there were some shortfalls in some regions). Therefore, several pellet producers are focussing on the expansion of the possible raw materials to wood chips, log wood or SRC, for example [107; 108; 109; 111; 112; 116; 479; 480].

10.2.4   Pellet utilisation

10.2.4.1   General framework conditions

Nationwide supply is safeguarded in Germany, as in Austria. The highest density of pellet traders is still found in the south but pellets are available everywhere in the north too [475; 481].

In Germany there is strong political support for renewable energies, including the use of pellets. One of the targets of Germany is to increase the share of heat generation from renewables from the current 7% (2009) to 14% in 2020. A major part of this share will have to be achieved in the residential heating sector mainly by an increased use of biomass and solar heating systems. In order to reach the target, a renewable heat law was introduced in 2009, which obliges home owners to provide a certain share of their heat demand from renewable energy sources in new buildings. The required share is different depending on the technology applied. In Baden-Württemberg, one of the federal states of Germany, a renewable heat law is in force since 2010, where even in existing buildings at least 10% of heat demand must be provided from renewable energy sources as soon as the heating system is replaced by a new one [482].

In the German renewable heat law, the continuation of the market incentive programme (MAP), which provides subsidies for the installation of renewable heating systems in new and old buildings is also included. Subsidies are provided by means of non-repayable allowances and are dependent on the kind of heating system (pellets, wood chips, firewood), its nominal power output and whether it is installed in a new or an existing building. Bonuses on the basic funding are granted for combinations with solar heating systems, for buildings with a higher insulation standard and for innovative installations such as flue gas condensation or dust precipitation systems [483]. MAP has been one of the main drivers for the fast development of the German pellet market in recent years.

For pellets, wood chips and firewood, a reduced VAT rate of only 7% applies. All fossil fuels command 19% VAT.

The pellet furnace market has to date been restricted mainly to the area of small-scale systems (central heating systems). However, the area of medium-scale systems with power outputs of between 50 and 1,000 kW is now gaining increasing significance. In the early stage of pellet market development in Germany, many Austrian pellet boiler manufacturers exported their products to Germany. Now there are about 200 pellet furnace manufacturers located in Germany, including many Austrian pellet boiler manufacturers with branches in Germany.

10.2.4.1.1 Small-scale users

Pellet heating systems were mainly imported to Germany from Denmark, Sweden and Austria in the early years. Since then, the number of pellet furnace manufacturers in Germany has grown. The players in this field are mainly small- and medium-sized enterprises [472].

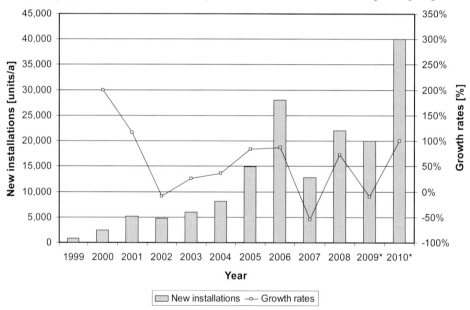

Figure 10.13: Development of pellet central heating systems in Germany from 1999 to 2010

Explanations: pellet central heating systems below 50 kW$_{th}$; *...prognoses; data source [476; 477; 470; 472; 475]

The development of pellet heating systems in Germany is shown in Figure 10.13. In contrast to Austria, the development started two years later, namely in 1999. Since then, annual new installations have risen by 14 to 200%, with the exception of 2002 due to the general weak economic situation of the building sector. In 2007, there was a notable reduction in new installations by more than 50% that, as in Austria, was caused by the massive rise of the pellet price and some supply shortages. The situation eased in 2007 due to the massive expansion of pellet production capacities. In 2008, a strong increase in newly installed pellet furnaces was observed again. Due to the current worldwide economic and financial crisis, a slight decrease of about 9% is expected for 2009. However, further increasing installation numbers are expected from 2010 onwards. Overall, the market has expanded rapidly and in 2001, annual new installations exceeded those of Austria. By the end of 2008, more than 100,000 pellet furnaces were installed in Germany. The number is expected to rise to 125,000 heating systems by 2009 and to 165,000 units by 2010 [477]. The prognoses for the cumulated number of installed pellet heating systems are that there will be more than 600,000 units by 2015 and more than 1 million units by 2020 [477; 475].

In Germany, to date, pellet heating systems have mainly been installed in new buildings. The purchase of a pellet boiler when exchanging an existing boiler has been hesitant. However, great market potential lies in this area. Regional distribution is striking in Germany as the market is presently concentrated on the southern federal states Bavaria and Baden Württemberg, where about 62% of all pellet boilers are sold [475].

10.2.4.1.2 Medium- and large-scale users

Similar to Austria, the German pellet market is concentrated in the residential heating sector with small-scale systems below 50 kW$_{th}$. The development of the medium- and large-scale sector has been slow in recent years. However, since 2008 a steep increase of new installations in this sector has been observed. The largest pellet furnace with a nominal thermal power output of 3.8 MW was put into operation in October 2009. The total number of pellet furnaces with nominal power capacities above 50 kW$_{th}$ is estimated to be about 5,000 (as per the beginning of 2009). A further increase in this sector is expected [484].

10.2.4.2 Pellet consumption

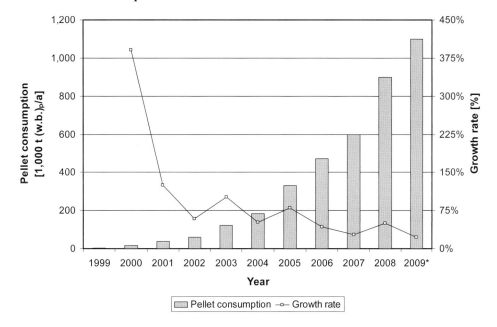

Figure 10.14: Pellet consumption in Germany from 1999 to 2008

Explanations: *…prognosis; data source [442; 476; 477; 470; 471; 472; 481]

Figure 10.14 shows the development of pellet consumption in Germany. As mentioned already, the main development of the German pellet market started in 1999. Since then, growth rates of annual consumption of up to 400% have been achieved. The most moderate increase was around 28% in 2007, when the demand was lower due to the mild winter and owing to the collapse in new installations of pellet boilers. The growth rate in 2009 is expected to be around 22%.

10.2.4.3 Pellet consumption potential

In Germany around 17 million heating systems are installed in private households. Their final energy consumption amounts to about 536 TWh in total (in 2008), around 13% of which are covered by using renewable energy sources. Around 1.0% of this final energy consumption or about 7.5% of the renewable energy sources are provided by pellets (about 5.1 TWh); for firewood these figures are 9.8% and 77.2%, respectively. About 18% of all heating systems are more than 24 years old, with nominal efficiencies below 65%. These systems should be replaced within the coming years, providing a large potential for pellet heating systems [475; 485]. As in Austria, it is important to take measures to avoid replacing these furnaces by systems based on fossil fuels, as often happened in the past, and to support pellet heating systems in Germany.

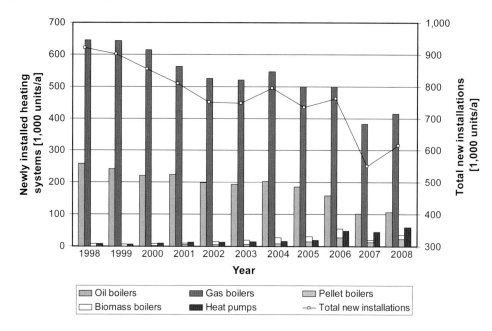

Figure 10.15:   Annual boiler installations in Germany from 1998 to 2008
Explanations: data source [475]

Figure 10.15 shows the development of the German heating market since 1998. It follows that boiler installations declined almost every year from about 920,000 new installations in 1998 to only about 550,000 units in 2007. In 2008, the number increased to 616,000 units. The long lasting decrease until 2007 is the reason for the above mentioned obsolescence of the heating systems, and modernisation of the oldest systems in particular is urgently required to increase energy efficiency as well as to reduce gaseous and particulate emissions. Again, this necessity for retrofitting old and inefficient units shows the great potential pellet heating systems have in the residential heating sector.

Gas boilers are by far the most frequently installed applications, even though their number has decreased since 1999 (except in 2004 and 2008). In 2008, around 415,000 gas boilers were installed in Germany, accounting for more than 67% of all new installations. Oil boiler installations were between about 200,000 and 260,000 units annually between 1998 and 2004. Then a steep decrease to slightly above 100,000 units per year occurred. This means a reduction of more than 60% from the highest number in 1997 to the lowest in 2007. Heat pumps are gaining increasing relevance, as in Austria. In 2008, around 60,000 new heat pump systems were installed. Biomass boilers were on the increase until 2006 followed by a massive market collapse in 2007. Pellet boilers contributed significantly to the rise, but also other biomass boilers were installed.

10.3 Italy

In South Tyrol the installation of automatically fed and controlled wood chip and pellet furnaces is subsidised with up to 30% of investment costs [486]. Some funding schemes are known in other Italian provinces but the subsidies are not as high as in South Tyrol. In many provinces there are no such subsidies. On a national level, the purchase of pellet boilers is supported by a income tax reduction [487].

The Italian pellet market began to develop in the early 1990s, but was weak for many years. It was not until recently that there was a significant increase in pellet production and use. Pellet consumption increased from 150,000 t (w.b.)$_p$ in 2001 to 1.2 million t (w.b.)$_p$ in 2008. Around 750,000 t of pellets are produced (2008) by around 75 producers (cf. Figure 10.16). The rest is imported. Italy has always been a pellet importing country because production has always been lower than consumption (cf. Figure 10.17). Imports mainly come from Austria, Germany and Eastern European countries, but imports from China and Brazil have also been reported. The large majority of production sites, about 80%, have production capacities below 5,000 t (w.b.)$_p$/a. The raw material used for pellet production is mainly sawdust with a share of 65%, followed by wood shavings with 19%. The rest are other raw materials such as wood chips and other residues. Currently there are no additional raw material sources available for a further increase in pellet production. Therefore, pellet producers are increasingly importing raw materials from other countries. Around 90% of the pellets used in Italy are packaged in bags, typically 15 kg bags [487; 488]. This large share of bagged pellets is due to the high number of pellet stoves (740,000, cf. Figure 10.17), mainly in northern Italy. Only a few pellet central heating systems are installed in Italy. Their number is estimated to be around 1,000 [487]. Part of consumption is in district heating plants [489]. In 2003, 41 small-scale district heating networks were in operation that used pellets to some extent. In addition, the wood working and processing industry partly use pellets for in-house heat production. Exact figures for this are not available, however. It can be assumed that other biomass fuels such as bark and wood chips are used in such systems too. An expanding area of pellet use is the application in micro grids. Public buildings such as schools or sports halls use such micro grids, whereby the nominal boiler capacity of these systems is usually between 600 and 1,000 kW [490]. Systems with a nominal boiler capacity of up to 400 kW, that were originally designed for the firing of wood chips, are often used in South Tyrolean hotels and increasingly often changed to use pellets.

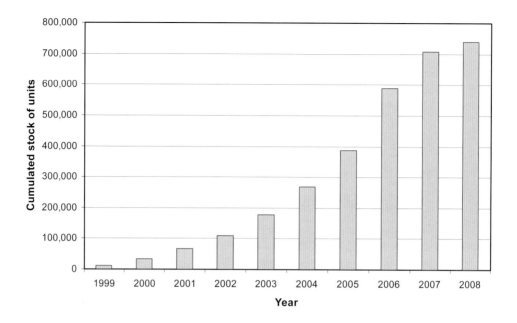

Figure 10.16: Development of pellet stoves in Italy from 2002 to 2008

Explanations: data source [491; 492; 493; 494; 495]

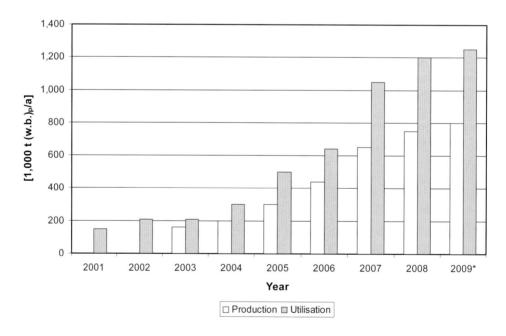

Figure 10.17: Pellet production and use in Italy from 2001 to 2009

Explanations: *...prognosis; data source [488; 489; 490; 492; 495; 496; 497; 498; 499; 500]

Numerous medium-sized enterprises are active in Italy that manufacture pellet central heating systems and pellet stoves with a predominant focus on local markets. In addition, there are five larger companies manufacturing pellet central heating systems that are active throughout the Italian market [490].

The Italian pellet market was restrained by the low quality of its pellets for a long time. System failures and operation problems in pellet furnaces leads to a poor image. The introduction of quality guidelines (as a recommendation) in 2003 solved this problem [490]. In 2006, the so-called Pellet Gold standard was established by AIEL (Associazione Italiana Energie Agriforestali). However, it is not a real certificate but rather a "label", and there is no independent certification agency in Italy, which causes some mistrust among end users and consequently hampers market development. Moreover, there is no national organisation representing the Italian pellet sector. An attempt to found such an organisation under the name Propellet Italia has not been successful [487].

Another difficulty especially with regard to central heating systems is posed by the lack of suitable silo trucks. There remain hardly any pellet producer that capable of delivering pellets by silo truck and pneumatically feeding them into the end user's storage space. The insecurity of supply renders selling pellet central heating systems a difficult task [141; 488].

Similar to Austria and Germany, there was a pellet shortage in the winter 2005/2006, not least caused by a boom in pellet stoves [501].

10.4 Switzerland

In Switzerland there are different guidelines concerning subsidies for pellet furnaces depending on the canton. In principle both investments in new installations and exchanging systems are subsidised [502]. In addition, the installation of a few hundred pellet furnaces was subsidised by a national subsidies programme from 2000 to 2003 [503; 504].

The Swiss pellet market began to develop in 1998 with the installation of 170 pellet boilers and pellet stoves (cf. Figure 10.18). By 2008, the number of installed systems had risen to about 14,300, pellet boilers having a share of 59% out of the total [505; 506; 510]. Similar to Germany and Austria, pellets are mostly used in small-scale applications in the residential heating sector. The main share of pellets is distributed loosely by tank trucks. Delivery in bags plays a role in the southern part of Switzerland, where, similar to Italy, many stoves are in operation. In Switzerland, pellet prices also rose significantly in 2006, which led to a steep decrease in new installations in 2007 and 2008.

Pellet consumption and production are balanced in Switzerland, both being around 90,000 t (w.b.)$_p$/a in 2007 (cf. Figure 10.19). Pellet production is dominated by small-scale producers with production capacities typically of between 1,000 and 12,000 t (w.b.)$_p$/a. Only 2 pellet producers (out of 14 in total) have higher production capacities. The dominant raw material for pellet production is sawdust. However, pellet production from log wood has already been started by 2 smaller pellet producers. To date, more pellets were consumed than produced, except in 2004. Thus pellets were imported, mainly from Austria and Germany, in the past. However, pellets were also exported to Italy to some degree [442; 507].

In total around 800,000 oil heating systems are installed in Switzerland of which every sixth could be replaced by a pellet heating system between 2016 and 2021. The total potential for the Swiss pellet consumption is indicated to be around 3.5 million tonnes per year [508].

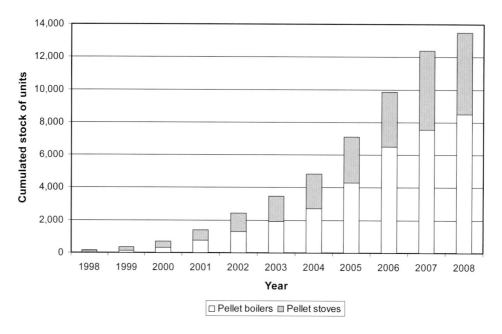

Figure 10.18: Cumulated pellet furnace installations in Switzerland

Explanations: data source [503; 505; 509; 510]

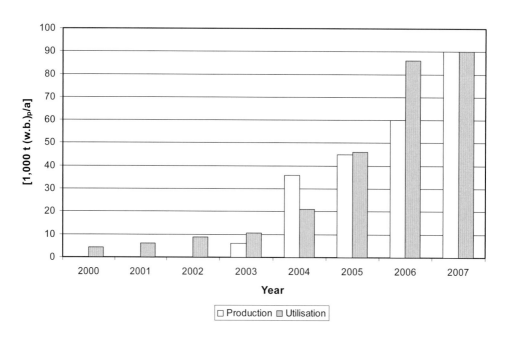

Figure 10.19: Pellet production and use in Switzerland from 2000 to 2007

Explanations: data source [489; 503; 505]

10.5 Sweden

10.5.1 Pellet production, production capacity, import and export

Swedish data concerning pellet production, import and export are shown in Figure 10.20. Imports have always been higher than exports, so Sweden is a net pellet importing country. Moreover, imports are expected to increase further as consumption cannot be covered by domestic production. Imports come mainly from Finland and Canada as well as the Baltic states. Pellets are also imported from Russia and Poland.

Pellet production was around 1.6 million t (w.b.)$_p$/a in 2009 in Sweden [521]. There were 83 pellet producers in operation at the beginning of 2009 [511].

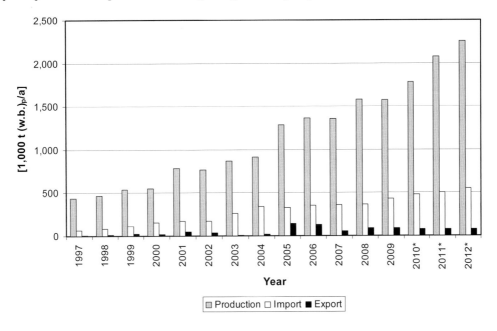

Figure 10.20: Pellet production, import and export in Sweden from 1997 to 2012

Explanations: *...prognoses; data source [521]

10.5.2 Pellet utilisation

10.5.2.1 General framework conditions

The market for and production of pellets in Sweden have grown rapidly in the last 15 years. There are several factors that explain the rapid growth of the pellet market in Sweden [519].

In 1980, oil was the dominant fuel in the district heating sector accounting for 112 PJ, compared to the 1.1 PJ of biofuels. The situation was reversed by 1999, when oil accounted for only 18 PJ and wood fuels for 57 PJ in the district heating sector. The growth of the Swedish biofuel market in this period was mainly driven by increased demand in the district heating sector. District heating networks are well distributed in Sweden and are one sector that promoted the use of biofuels [519].

In 2002, about 50% of 1.6 million residential house owners used electric or oil based systems for space heating. However, annual installations and sales of pellet burners in the market sector of small-scale applications boomed from 1994 to 2006. In 2007 the number of newly installed small-scale applications dropped significantly, as in many other European countries.

The reason for changing from oil combustion to the combustion of biofuels in large-scale facilities can foremost be explained by the tax on carbon dioxide introduced in 1991 that made fossil fuels more expensive. The conversion to pellets increased the demand, and several new production plants for pellets have been built [519].

In Sweden, there are about 1,750,000 one family houses [512]. About 60% of these were built before 1970, about 24% in the 1970s and about 16% after 1980. Heating is done by a boiler or heat pump connected to a water based heat distribution system or by electrical radiators. Some boilers can use more than one energy source, normally fuel oil or biomass in combination with electricity. The percentage of boilers that make use of oil to some extent is 9%, or about 140,000 units. About 60%, or 1,050,000 units, make use of electricity to some extent. Biomass, like firewood or pellets, is used in about 660,000 houses, i.e. 38%. Out of the houses using electricity for direct heating, 55% are estimated to have a water based distribution system, and the remaining 45% use electrical radiators.

In [513], the number of furnace installations in one family houses reported in the mandatory fire safety control system is given (cf. Table 10.1). These include not only the one family houses and premises.

Table 10.1: Cumulated number of combustion equipment in the residential sector

Explanations: based on year 2005; data source [513]

Type of installation	Units
Pellet boilers and burners	84,000
Pellet stoves	11,280
Firewood boilers	265,000
Firewood roomheaters, cookers, etc.	1,450,000

In Table 10.2, the average sales volumes of different heating systems in the residential heating sector is given for the period 2003 to 2007.

Table 10.2: Average sales of some combustion equipment between 2003 and 2007

Explanations: data source [514]

Type of equipment	Units/a
Pellet burners/boilers	17,200
Firewood boilers	6,200
Electrical boilers	8,600
Fuel oil/electrical boilers	1,800
Oil burners	3,700
Gas boilers	250

Within the area of small-scale systems it is mainly so-called pellet burners that are used in Sweden. They are usually made by Swedish manufacturers and enable a changeover to pellets without having to replace the existing boiler. Low investment costs are the main advantage. The disadvantage is that burner and boiler are not ideally adjusted to each other and hence disadvantages concerning the furnace technology and also the environment have to be accepted (lower efficiency, shorter cleaning intervals, higher emissions). There are no investment subsidies for small-scale pellet furnaces in Sweden but there are CO_2 and energy taxes that disadvantage fossil fuels. The VAT rate in Sweden is 25% for all fuels.

Figure 10.21 shows the development of small-scale installations, namely pellet burners and pellet stoves. The very small number of pellet boilers installed in Sweden is included in the number of burners. In 2007, 131,300 systems were installed in Sweden.

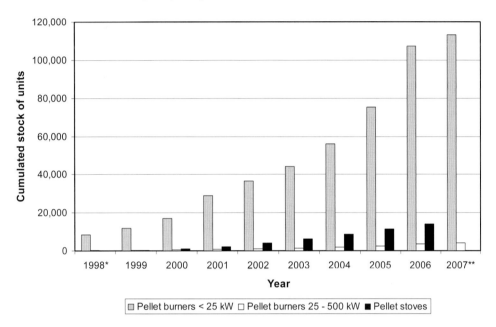

Figure 10.21: Cumulated pellet central heating and pellet stove installations in Sweden from 1998 to 2007

Explanations: *...accumulated number since 1994; **...no data for pellet stoves available; data source [492; 515; 516; 517; 518; 519]

10.5.2.2 Pellet consumption

10.5.2.2.1 Small-scale users

The share of pellets used in the residential heating sector played a subordinate role in Sweden for a long time. However, it did rise continuously for some years reaching its maximum at 37% of total consumption in 2007. Since then this share has been decreasing and is expected to decrease further in the coming year due to strong increases in overall consumption (cf. Figure 10.22).

The total use of wood pellets in detached and semi-detached houses in Sweden 2007 was approximately 461,000 tonnes (2.2 TWh) [520]. The total use of wood pellets in detached and semi-detached houses in Sweden during the period 1999 to 2007 is shown in Figure 10.23.

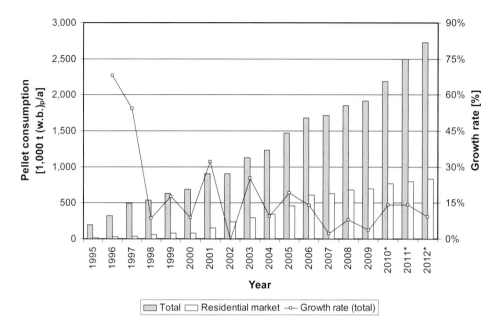

Figure 10.22: Development of pellet consumption in Sweden from 1995 to 2012

Explanations: *...prognoses; data source [518; 521; 522]

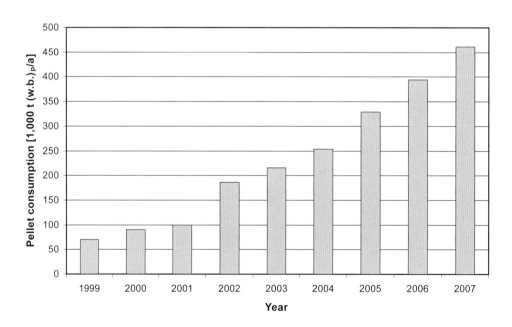

Figure 10.23: Total use of wood pellets in detached and semi-detached houses in Sweden from 1999 to 2007

Explanations: data source [520]

10.5.2.2.2 Medium- and large-scale users

In Sweden, pellets are chiefly used in large-scale furnaces. The CHP plant Hässelby, which supplies Stockholm with district heat, alone requires around 300,000 t of pellets per year. Large-scale furnaces are normally equipped with pulverised fuel burners, which is why the pellets need to be ground before use. In this case, pelletisation has the sole purpose of reducing transport and storage costs.

Pellet consumption was above 1.7 million t (w.b.)$_p$/a in 2007 (cf. Figure 10.22). Sweden is thus the largest pellet consumer worldwide. An additional increase of consumption to 2.25 million t (w.b.)$_p$/a is expected by 2010. Starting from 1995, annual increases have been between 2.1 and 68%, except for a slight decrease in 2002 [521].

10.6 Denmark

Denmark had long been the second largest pellet consumer in Europe (after Sweden). However, large-scale power plants such as those in the Netherlands now consume more pellets (either in co-firing or in power plants that were retrofitted for the sole use of pellets). A positive development in the area of small-scale systems was enabled in part by investment subsidies. However, investment subsidies ceased in November 2001. Due to higher taxes on fossil fuels (CO_2 and energy tax) there is still a strong incentive to change over to pellets [523].

Owing to the long tradition of using pellets, there are numerous pellet furnace manufacturers and manufacturers of pellet production plants in Denmark.

No primary data exist for the number of installed residential heating systems based on wood pellets. Pellet stoves are in use to a very small extent in Denmark. So for market estimations, one may assume that all residential wood pellet use takes place in small-scale boilers. To estimate the number of residential boilers, we assume an annual unit consumption of 6 tonnes per installation (Figure 10.24). It should be noted that the primary data may be problematic, as illustrated by the last three years, when the market has been stable and not falling, as the data indicate [524].

In medium-scale applications, wood pellets were first used in the district heating sector in Denmark in the late 1980s when coal fired heating plants were retrofitted to use wood pellets. The number of plants quickly rose to about 30 and has been stable for almost two decades. Industry and public service buildings such as schools, sports centres and hotels provide a stable intermediate market, totalling about 100,000 tonnes per year. Pellets are especially suitable for businesses that experience high costs for heating based on natural gas or oil.

The vast majority of large-scale consumption presently (2008) takes place in one power plant, the Avedøreværket Unit no. 2 near Copenhagen, which since 2003 has used 100,000 to 355,000 tonnes of pellets per year. In 2009, Herningværket and Amagerværket Unit no. 1 will also commence wood pellet firing at a relatively large scale, and for the coming years more power plants, including Avedøreværket Unit no. 1, are expected to convert from coal to wood pellets.

The development of pellet consumption in Denmark since 2001 is shown in Figure 10.25. An increase from 88,000 t (w.b.)$_p$/a in 1990 to 1.06 million t (w.b.)$_p$/a in 2008 can be seen with annual growth rates of between 5 and 63%. Only in 1993 can a slight reduction be noted. The consumer structure in Denmark is characterised by utilisation of pellets in all areas of energy generation. So in 2008, around 44.4% of pellets were used in small-scale systems for residential heating, 45.5% in CHP and district heating plants, 5.1% in industry and 5.0% in public services [525].Two large-scale plants were put into operation in 2003. At full load, the power plant of Avedøre uses around 300,000 t of wood pellets a year and the power plant of Amager (Unit no. 2) around 150,000 t of straw pellets [523]. However, the actual consumption of straw pellets at Amager Unit no. 2 was considerably lower between 2004 and 2008 (around 50,000 t/a) [526]. From 2010 onwards the new Amager Unit no. 1 will take over biomass utilisation from Unit no. 2. The use of straw pellets is expected to increase.

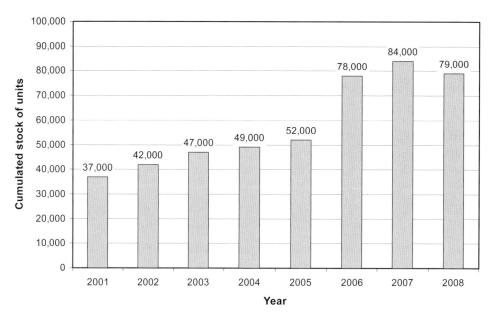

Figure 10.24: Cumulative number of residential pellet boiler installations

Explanations: data source: FORCE Technology

The high demand for pellets cannot be covered by Danish production plants, even less so since pellet production decreased in 2006 due to technical problems, resulting in the production of just 134,000 t (w.b.)$_p$/a in 2008 [494; 496; 525]. Therefore, imported quantities increased, reaching more than 900,000 t (w.b.)$_p$/a in 2008 in order to cover the national demand. Imports come mainly from the other Scandinavian countries, the Baltic states and North America.

No studies are available to accurately determine the amount of raw material available for pellet production in Denmark. Traditionally, domestic pellet production uses dry shavings and sawdust from the furniture industry and other wood industries working with dry wood. Due to the economic crisis in this industry sector, the availability of this raw material has decreased significantly during recent years. Industry representatives estimate that only about 60,000 to 90,000 tonnes are currently (2008) available.

During recent years, wet raw materials have been increasingly used for wood pellet production at facilities where wood drying is now part of the process. As such raw materials can, and are, easily imported to the country as logs or chips, there is no hindrance to the domestic production of such pellets. However, a general logistic assessment highlight the more practical option of producing pellets from wet raw material at the origin of the wood, and then importing the product.

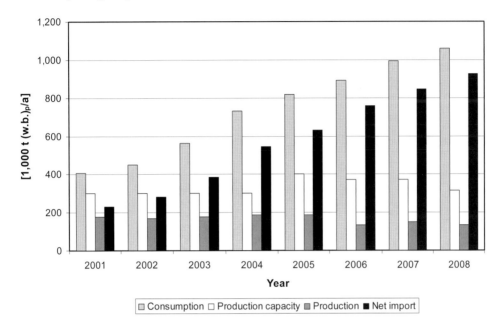

Figure 10.25: Development of pellet consumption, production capacity, production and net import in Denmark from 2001 to 2008

Explanations: data source [120; 441; 496; 523; 527; 528]

Concerning estimations of future consumption potential, the Danish wood pellet market will grow significantly in the years to come. The demand in the residential sector can be expected to increase due to high fossil fuel prices and high energy taxes. However, in the long term, the total market for residential heating outside the district heating networks is limited. A 10% annual increase is likely for the coming years. In the medium-scale sector the district heating plants present a consumption of around 100,000 t/a, which can be expected to remain at this level, or even slightly decline as wood chips or straw fuels take over this market. Some of the development will take place in the commercial and industrial sector that has only recently shown any significant interest in renewable energy fuels. The present large-scale consumption level of approximately 350,000 tonnes (2008) is expected to increase steeply in the coming years. This assumption is based on the increased biomass obligation (700,000 tonnes more) on power companies to meet Denmark's need to significantly reduce CO_2 emissions. The plans of the major market actors, i.e. DONG Energy and Vattenfall, are not yet (May 2010) confirmed, but one likely scenario includes the giving of priority to pellets above other biomass fuels, which may lead to an increase in annual consumption in this sector in the order of 1 to 3 million tonnes per year within a few years time.

10.7 Other European countries

Figure 10.26 presents an overview of the production and use of pellets in other European countries. The markets of the these countries are examined below.

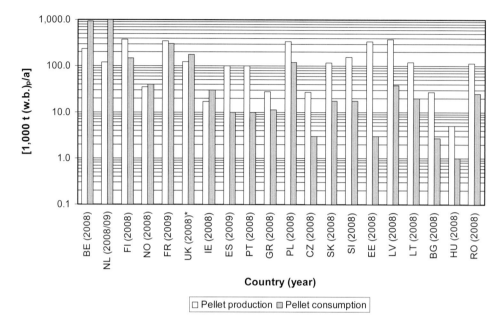

Figure 10.26: Pellet production and utilisation in selected European countries

Explanations: *...estimates for consumption vary greatly, up to 800,000 t (w.b.)$_p$/a may be possible; data source [532; 535; 543; 560; 562; 563; 565; 568; 569; 572; 573; 574; 575; 576; 582; 577; 578]

- Belgium

1.3% or 54,000 homes were heated with wood in Belgium in 2001, of which 1.1% or 600 were central heating systems [529]. At the end of 2006, there were 5,500 pellet stoves and 800 pellet boilers installed in the country with a pellet consumption of around 35,000 t (w.b.)$_p$/a [530]. The cumulated power of installed small-scale devices in the Walloon region is about 96.9 MW$_{th}$, which represents a consumption of about 30,000 t of pellets a year. This amount is smaller than the 65,000 t produced in the Walloon region for that purpose and makes the region self-sufficient in pellets for small-scale appliances. Small-scale pellet use in the Flemish region has been slow, about 100 pellet boilers (< 40 kW) and 1,000 stoves are known in the residential heating sector [534]. For 2008, pellet consumption in the residential heating sector is estimated to be about 120,000 t (w.b.)$_p$/a [532]. Reliable data on installed small-scale appliances in 2008 are not available.

The pellet stoves and central heating systems available on the Belgian market are mainly imported. There are very few Belgian manufacturers of pellet stoves. Only in Wallonia are investment subsidies granted for automatically fed pellet boilers, namely up to 3,500 €, depending on the nominal power output of the system. At the national level, wood heating systems are supported by a tax reduction. Income tax on private households can be reduced by 40% of the investment costs (up to 3,440 € in 2009).

By far the largest share of pellets is used in two power plants for electricity generation. The power plant of Les Awirs (80 MW$_{el}$) was retrofitted from coal to pellet use and needs around 350,000 t of pellets per year as the only fuel (cf. Section 11.10). The pellets are either produced in Belgium or imported [531]. In addition, pellets are co-combusted in the power plant of Rodenhuize (75 MW$_{el}$) to a degree of 25%, thus 300,000 t (w.b.)$_p$/a are used. On the whole, about 800,000 t (w.b.)$_p$/a of pellets are used in these two large Belgian power plants and some smaller industrial systems [532].

A noteworthy amount of pellets was first produced in Belgium in 2005 and started to increase substantially in 2007, for both industrial and small-scale use. In 2008, production dedicated to small-scale uses remained steady, while industrial pellet production continued to grow strongly. In the Walloon region, six pellet producers can be counted, with an installed capacity of 421,000 t/a. Actual production reaches 213,000 t/a. This production is mainly dedicated to industrial use, for electricity production. Nevertheless, producers indicate that 62,000 t were sold for domestic use in 2008 [533]. The Flemish pellet market is still very young and information about production capacity is scarce. About 20,000 t/a are produced by four producers; the main raw material is wood sawdust [534]. Additional production plants are being planned. The rest of the national demand is satisfied by imports.

- Netherlands

The main use of wood pellets in the Netherlands is co-firing in coal fired power plants. Here, quality standards are agreed on by producers and suppliers on a bilateral basis (the Netherlands takes part in the international standardisation process on solid biomass within the framework of CEN, however, there is more interest in the country in other biomass types such as wood chips). Currently, approximately 1 million tonnes of biomass pellets are co-fired, and this is expected to increase to approximately 5 million tonnes in 2020. The biomass required for this purpose will mainly be imported from overseas (Canada, Brazil, South Africa, Baltic states, etc.). The AMER power plant of Essent in Geertruidenberg alone currently consumes approximately 600,000 tonnes of wood pellets (cf. Section 11.11). In the EON Maasvlakte power plant, pellets are produced from a mixture of several biomass waste products, before they are dumped on the coal conveyor.

The main reason for the large amounts of pellets co-fired in Dutch power plants was the MEP subsidy scheme (to enhance environmental quality of electricity production) between 2003 and 2006, which was a system of feed-in premiums. The determination of the height of the premium was rather complicated and depended on the capacity of the power plant, the type of biomass used, the period when the electricity was produced and the point when the first request for subsidy was received. The premiums are valid for up to ten years, which means that from 2013 onwards, when the first contracts from 2003 will be terminated, a decrease of pellet co-firing must be expected, unless a new subsidy scheme is put in place. While most co-firing schemes are supported until 2012, when the previous support scheme MEP phases out, the current subsidy scheme for the production of renewable electricity, i.e. the SDE scheme, no longer supports the use of wood pellets for large-scale co-firing [535].

In the medium-scale power range, an estimated number of 30 to 50 industrial companies have switched to the use of pellets in the last 10 years for heat supply of industrial users such as poultry farms and cattle breeding farms. In the utility sector, only a few examples exist to date.

Small-scale pellet furnaces are not at all common in the Netherlands. This is due to the fact that almost all consumers have been connected to the natural gas network since the 1970s. As a result, most consumers have lost affiliation with handling solid or liquid fuels, and payback periods are relatively long compared to countries where heating oil is usually replaced. In addition, the space in which to construct a pellet boiler house with storage room is usually limited. In the absence of a mature pellet market, the delivery of wood pellets is usually arranged by the boiler supplier. Quality standards are not well maintained.

With estimated 150,000 to 200,000 t, the current production capacity for wood pellets in the Netherlands is relatively small compared to other countries. Actual production is lower at the moment – in 2008 it was about 120,000 t (w.b.)$_p$/a. Raw materials currently originate mainly from wood processing industries. According to a recent study [536], there are approximately 600,000 t available per year. Of this amount, some 150,000 t are currently used in the wood processing industries for heat production, the remainder is used externally for the production of fibreboard, energy production or for the production of pellets. A large fraction of the biomass pellets produced is exported to Germany, where prices are higher. One producer (production of 80,000 t/a) uses waste wood as raw material and exports to Sweden.

A new factory with a capacity of 100,000 t of product based on fresh wood from landscape maintenance is planned to start operation in 2011.

The Dutch biomass pellet market will grow significantly in the years to come, if the coal power sector agrees with the government on a new support mechanism or an obligation to co-fire or reduce CO_2. For 2020, different authors estimate potential consumption to be between 5 and 10 million tonnes.

- Luxembourg

In Luxembourg the use of pellets is confined to the residential heating sector, where approximately 2,000 pellet heating systems are installed (2008). Their typical nominal thermal capacity is between 30 and 50 kW. Up to 30% investment funding (maximum 4,000 €) is granted to households from the state.

- Finland

Finnish pellet production began in 1997. With an annual production of 373,000 t (w.b.)$_p$/a (2008) Finland has become one of the largest pellet producers in Europe. In total, 16 pellet producers are operating at 24 locations. Around 149,000 t (w.b.)$_p$/a are consumed domestically, whereby the domestic market is dominated by small-scale applications [537]. The rest is exported to Italy, Belgium, the Netherlands, Germany, the UK, Denmark, Sweden and Baltic countries [441; 538; 539; 540]. The pellets produced in Finland are mainly made of dry raw materials (about 50% each). The potential to use dry woody raw materials is already fully exploited. Two pellet producers have dryers in order to use wet raw materials [541]. Furnaces are mainly imported from the USA, Sweden and Austria. Pellet stoves are not fabricated in Finland. Some manufacturers do produce pellet central heating systems though [542]. The use of pellets is indirectly supported by CO_2 and energy taxes on fossil fuels.

There are many different kinds of "stoker burners" on the market in Finland. All of them are multi-fuel burners, so customers can use many kind of fuels and choose those that are available for a good price in the neighbourhood. Dedicated wood pellet burners and boilers are available only up to a size of about 50 kW.

- Norway

In Norway, the pellet market is still in the early stages of development. In 1998, 10,000 t of pellets were produced. Production increased to 51,000 t (w.b.)$_p$/a in 2006 and decreased again to 35,000 t (w.b.)$_p$/a in 2008, mainly due to a lack of raw materials [543]. Production capacity was 164,000 t (w.b.)$_p$/a in 2008. Owing to the small sales volumes in Norway, a great share of production is exported, mainly to Sweden. Exports made up 57% of domestic production in 2006 [544; 545]. After 2006, exports decreased and in 2008, Norway became a net pellet importer due to increased pellet sales and reduced domestic production [543]. Domestic consumption rose by over 50% in 2006 as compared to the year before due to increased sales volumes of pellet stoves (around 10,000 pellet stoves were installed by the end of 2006). In 2008, domestic consumption reached almost 40,000 t (w.b.)$_p$/a, however, domestic consumption in Norway is still relatively low in comparison with other countries. With current production, there is an abundance of raw material. District heating plants are not very common in Norway, which is mainly due to the low density of settlements. Therefore, this area is not of great relevance to the pellet market. Only 12% of houses are equipped with a central heating system in Norway; 75% of the houses are heated electrically. So, the market potential for pellets of this area is also low. What is widespread though is the combined use of different heating systems. A great part of houses and flats that are electrically heated possess a stove that is usually fired with wood. These houses and flats as well as those that are heated electrically and are equipped with a chimney can be adjusted to the use of pellets without large investments by installation of pellet stoves. The potential for such an exchange exists in two thirds of all Norwegian homes (equivalent to about 1.2 million homes) [170; 546; 547]. In recent years, increasing electricity prices have become evident; increasing energy consumption cannot be accommodated by further hydropower plants. Therefore, new buildings are often equipped with water based central heating systems and district heating systems are developing, which could form the basis for an emerging Norwegian pellet market [543].

The consumption of pellets in Norway is mainly in pellet stoves and smaller pellet boilers up to 25 kW. The sale of pellets as bulk was about 59% in 2008 [543].

The raw materials are mainly by-products from the forest industries. As these raw materials are fully utilised, new capacity will be based on log wood, mainly from pine [545].

A large pellet production plant was built in the southern part of Norway with a production capacity of 450,000 t (w.b.)$_p$/a. The start-up of this plant was in June 2010 and makes Norway a large pellet exporter [543].

- France

In addition to Sweden and the USA, France is one of the pioneers of pellet production. In the early 1980s, 12 pellet producers were operating in France. From these early operations, wide ranging experience with the production of wood and straw pellets was gained. Owing to a lack of political support and competition from cheaper fossil fuels, the market almost totally collapsed in the 1990s. In 2009, around 60 small- and medium-scale pellet producers were operating again, producing around 345,000 t (w.b.)$_p$/a. The French pellet market is confined to the residential heating sector and in total, around 87,000 pellet stoves and about 20,000 pellet boilers are in place in France, with a consumption of about 305,000 t (w.b.)$_p$/a (2009). Both import and export takes place in France with trade flows to and from Germany, Spain, Italy, the UK and other countries. Net exports amounted to about 40,000 t (w.b.)$_p$ in 2009 [496; 548; 549; 550; 551; 552]. Market potential in France is assumed to be very large. At present, around 35 million t of wood are used for energy generation in France (around 4% out of total energy consumption). Around 6.5 million households are in possession of a wood heating system, some of which are already old. Many owners of such heating systems are likely to change over to automatic gas or oil heating systems. Pellets could be an alternative to fossil fuels in this area. This would require intensified marketing activities and political will, however. With such a background, the targeted expansion of the pellet market to 1 million t (w.b.)$_p$/a seems an ambitious but realistic objective. It would only represent just 3% of the current wood energy market.

- UK

For the UK pellet market of small-scale systems only rough estimations exist. There are estimated numbers of around 130 pellet stoves and around 400 pellet boilers (of which more than 93% have a nominal thermal capacity below 100 kW). Pellets are usually sold in small or large bags; loose delivery is almost unknown [553; 554]. Pellet consumption in this sector is stated to be about 6,200 t (w.b.)$_p$/a [496]. Although development is slow, the potential for pellets is assumed to be great. The competitiveness with oil and gas could be an incentive for the increased use of pellets, even though the higher investment costs are a major hindrance.

It is common practice in the UK to use pellets for co-firing in coal fired power plants for electricity generation. However, only rough estimates of consumption are available for this sector and they differ greatly from one another, namely between 176,000 and 800,000 t of pellets in 2008 [553; 554]. Not only wood pellets but also pellets made of miscanthus or olive stones are used in this sector. Pellet imports mainly come from Russia, the Baltic states, Finland and Canada [553]. Minor import quantities come from France, Germany, USA and Argentina. Pellet exports to Ireland and Italy are also reported.

Domestic production was around 125,000 t (w.b.)$_p$/a in 2008, with a production capacity of 218,000 t (w.b.)$_p$/a. There are 13 active pellet producers in the UK. There are no national standards for pellets, however, many producers and retailers described their pellets according to the relevant European technical standards (even though they are not certified accordingly) [554].

- Ireland

The Irish pellet market is dominated by domestic and small commercial users with an estimated number of 4,000 pellet stoves and boilers installed (status 2009). No applications for electricity generation from pellets are known. Ireland produces about 17,000 t (w.b.)$_p$/a (2 producers) and consumes around 30,000 t (w.b.)$_p$/a. Similar to the UK, Irish pellet producers describe their pellets according to foreign standards, in this case usually the national German and Austrian standards, even though they are not certified accordingly. As domestic production cannot cover demand, pellets are imported from Latvia, Finland, Canada, Germany, Sweden and France [555].

- Spain

Pellet production in Spain is around 100,000 t of pellets per year (2009). Only a small part of this, around 10,000 t of pellets, is used in pellet stoves in Spain itself (based on 2009). The rest is exported to Italy, German speaking countries, Portugal, France, Ireland and the UK [556; 557; 558; 559].

The Spanish pellet market has encountered several problems. Due to competition with the particle board industry for the raw material, there was insufficient raw material available for wood pellet production and a number of pellet producers had to cease production. Also there was little public awareness of renewable energy, and especially pellets, and little financial support from public authorities. Moreover, there is no national standard for pellets and standards from other European countries are not applied either. Only one pellet producer is known to be certified according to DIN_{plus}. Therefore, confidence of potential users in pellets is very low [441; 560; 561]. The number of installed pellet boilers is thus not high, being slightly above 1,000 by the end of 2009 [559]. Present data on production quantities show that the problem of raw material supply has been solved. Moreover, there are national subsidies of 20 to 30% for pellet boilers now in place [556]. Pellet boiler installations are expected to rise.

- Portugal

The Portuguese pellet market is similar to the Spanish market with an annual pellet production of about 100,000 t and a consumption of about 10,000 t in 2008. The difference, namely approximately 90,000 tonnes per year, is exported, mainly to Northern European countries. Total pellet production capacity amounts to about 400,000 t per year by six plants currently in operation. New plants are planned, however, which will further increase production capacity. Domestic consumption is confined to pellet stoves and boilers in the residential heating sector [562].

- Greece

According to [563], around 27,800 t $(w.b.)_p$/a are produced in Greece by five producers and about 11,100 t $(w.b.)_p$/a are consumed (based on 2008). The use of pellets is mainly in industrial applications; there is no use in households. The difference between domestic production and domestic consumption is exported, mainly to Italy and usually in small or big bags. Imports of pellets to Greece are unknown. There are no pellet quality standards in Greece and European or national standards from other countries are not applied.

- Eastern Europe

Recently a new pellet market has become established in Eastern Europe. There are market players in Estonia, Latvia, Lithuania, Poland, Slovakia, the Czech Republic, Slovenia, Bulgaria, Hungary and Romania. Together these countries already produce more than 1.6 million tonnes of pellets per year (2008). Utilisation in these countries is low and amounts in total to about 250,000 t of pellets per year (2008). The main share of the pellets produced is exported. Only Poland and Lithuania that have noteworthy domestic consumption [564; 565; 566; 567; 568; 569; 570; 571; 572; 573].

10.8 North America

In North America, market development started in 1984 when pellet stoves were on offer for the first time and around 200 t of pellets were produced. Then and now most pellets are sold packaged in bags and mainly to pellet stove owners. The development of pellet consumption in North America since 1995 is shown in Figure 10.55. Except for a slight decrease in the year 1999, consumption rose by 1.9 to 61.2% every year, reaching 2.1 million t $(w.b.)_p$/a in 2007. In the USA alone, about 1.2 million pellet stoves (2008) are in place. Whereas around 55,000 pellet stoves were sold in 2007, sales volumes reached 140,000 in 2008 [574]. Interest is also growing in pellet boilers [499; 575; 576]. The growing demand for pellet stoves and pellet boilers in 2008 was mainly a consequence of the high oil price. Especially the north eastern USA is characterised by a high number of oil heated houses and it was there that the strongest increases were noted.

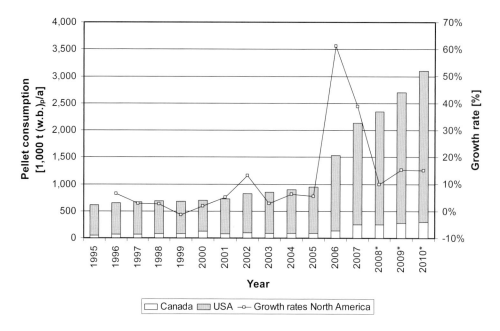

Figure 10.27: Development of pellet consumption in North America from 1995 to 2010

Explanations: *…prognoses; data source [486; 499; 577;578; 579; 580]

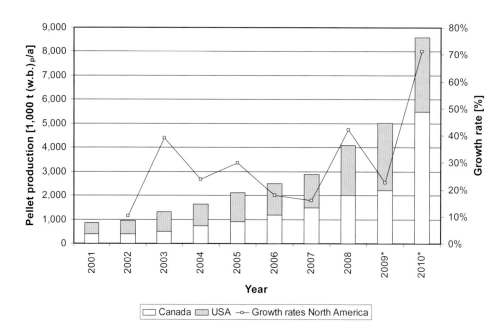

Figure 10.28: Development of pellet production in North America from 1995 to 2010

Explanations: *…prognoses; data source [499; 574; 575; 577; 580; 581]

Canada is a large pellet exporter that exports mainly to European countries (Belgium, Denmark, Sweden, the Netherlands and the UK) as well as to Japan, where there is also a developing pellet market. Pellets are also exported to Europe from the USA and the USA import pellets from Canada at the same time. In Canada, around 30 producers made 2.0 million t (w.b.)$_p$/a in 2008. In the USA, the number of pellet producers grew from 21 in 2004 to 90 in 2009 [582]. In total, they produced around 2.1 million t (w.b.)$_p$/a in 2008. Figure 10.28 shows the development of production in the USA and Canada since 2001 as well as prognoses until 2010. Canada, especially, has great potential for growth in this area. Pellet production of up to 5.5 million t (w.b.)$_p$/a is forecast for 2010. The prognosis for the USA for 2010 is 3.1 million t (w.b.)$_p$/a. Thus, the North American market could supply almost 9 million t (w.b.)$_p$/a in 2010.

10.9 Other international markets

There are pellet production activities and/or consumption evident in many countries worldwide.

China has the ambitious goal of reaching a pellet consumption of 50 million t (w.b.)$_p$/a by 2020 [545; 583]. Pellet production is currently just beginning, however [584].

Pellet production in Japan is moderate at about 8,600 t (w.b.)$_p$/a [14; 585] (2006). There are no data available concerning residential pellet use. However, initiatives are being carried out, for example by the Pellet Club Japan, to foster market development. In addition, pellets have been imported from Canada from 2008 onwards for co-firing in a coal fired power plant. In 2008, about 110,000 t of pellets were imported from western Canada and this amount is expected to increase in the coming years.

A potential study concerning pellets made of agricultural biomass was carried out in Turkey [586], but as yet, there is no actual pellet market there. This also applies to Mongolia, but initiatives to establish pellets are in progress [587].

South Africa also produces pellets that are exported to Europe via ocean vessels [197].

In South America, markets are developing in Brazil, Argentina and Chile. In Brazil there are pellet production plants with production capacities of around 60,000 t (w.b.)$_p$/a, in Argentina of around 36,000 t (w.b.)$_p$/a and in Chile, three production plants with a capacity of 85,000 t (w.b.)$_p$/a are in operation. In Chile further production capacities of around 300,000 t (w.b.)$_p$/a are planned by 2011. In 2007, Brazil, Argentina and Chile produced around 25,000, 18,000 and 20,000 t (w.b.)$_p$/a, respectively. There are almost no domestic pellet markets, so almost all production was exported [545; 588; 589].

An important pellet production market has been established in Russia. Pellet production in 2009 was 1 million t (w.b.)$_p$/a. Production capacity amounts to about 1.7 million tonnes per year and is expected to be increased to approximately 3 million tonnes per year in 2010 (not least by one large-scale plant with an annual pellet production capacity of almost 1 million tonnes of pellets per year to be started in 2010) [590]. Around 90% of production is exported to Scandinavian countries and the Benelux region owing to lack of sales possibilities in Russia. Only 50,000 t (w.b.)$_p$ were used in Russia [591].

Pellet production in the Ukraine amounted to about 120,000 t (w.b.)$_p$/a, of which more than 90% is exported [592].

South Korea recently put its first biomass power plant into operation. Around 145 t of wood chips and pellets are used in this plant daily [545]. Approximately 800 pellet boilers are installed with nominal thermal capacities below 30 kW (2009). The pellet boilers are produced by domestic manufacturers and are comparatively cheap. Only a few imported pellet boilers are in operation. Further pellet market development is mainly expected in industrial applications. Pellet production takes place in four plants with a total production capacity of 40,000 tonnes of pellets per year. However, all of them are still in the start-up phase and have not yet reached their full production capacity [593].

A pellet market is developing in New Zealand. Pellet boilers are still considered innovative in New Zealand but they could rapidly gain relevance. A total of five pellet production plants made about 20,000 t (w.b.)$_p$/a in 2006 at a pellet production capacity of 100,000 t (w.b.)$_p$/a.

In 2009, the first pellet production plant began operations in Australia, with a capacity of 125,000 t (w.b.)$_p$/a. In total, six plants should be erected by 2012 with a cumulated production of 1.5 million tonnes of pellets per year. The pellets produced are designated for the export to Europe and to be utilised in power plants. Long-term contracts are already in place to this end [594].

10.10 International overview of pellet production potentials

In order to actually achieve further increases of pellet production, appropriate raw materials supply have first to be in place, and second, production sites have to be erected. This section is concerned with evaluations in this regard.

Possible raw materials for pellet production were evaluated in Chapter 3. Evaluations and considerations of raw material potential for pellet production on a national basis are discussed in the first sub-sections of this chapter for countries where data are available. The evaluations in this section are concerned with European and global potentials. Only woody biomass is evaluated because technological problems with the use of herbaceous biomass for pelletisation as well as their thermal use in small-scale biomass furnaces have up to now not been solved to a sufficient extent. Ongoing R&D activities in this area are dealt with in Chapter 12. Herbaceous biomass is not significant for pelletisation at present but the situation may well change in the coming years.

Three different evaluations concerning pellet production potentials are shown in the following sections. In Section 10.10.1 the distribution of pellet production plants in Europe is evaluated and areas with high and low concentrations of pellet production plants are identified. For Austria, Sweden and Finland, the evaluation is more detailed. An evaluation of alternative raw material potential in Europe, such as forest logging residues, energy crops and short rotation woody biomass is in Section 10.10.2. Finally, an evaluation of the worldwide sawdust potential available for pellet production is in Section 10.10.3.

10.10.1 Pellet production plants in Europe

10.10.1.1 Distribution of pellet production plants and market areas

The pellet market and supply structures in Europe and North America are currently undergoing rapid development. As pellet markets develop, the supply side is also growing constantly. In some countries, the supply side is growing faster than domestic use, while others need to import pellets to satisfy growing demand. The growing demand for pellets has naturally increased supply in terms of increased number of pellet production plants and total production capacities [595; 596].

The world's 10 largest pellet producing countries together produced approximately 8.5 million tonnes of pellets in 2007, and this figure has slightly increased in recent years. In Europe, the leading pellet production and consumption countries are Sweden, Germany and Austria. Other countries, such as Denmark and Italy, are large consumers but their production is small, so they are dependent on pellet imports. Finland, Poland and Russia have low domestic consumption and pellet markets are export oriented. The pellet trade in Europe has increased steadily and growing demand has also increased imports from Canada, which exported about 765,000 t of pellets in 2007. The largest flows of pellets are from Austria, Finland, Germany, Poland and Russia to Sweden, Denmark and Italy [596; 597].

The location and distribution of the pellet production plants in Europe is grouped around certain hot spots with a very high density of pellet production plants. One of the possible ways to analyse the distribution of pellet production plants is by using geospatial kernels that help to identify the areas with highest production or market core areas. The kernel analysis is a non-parametric method for the estimation of the spatial distribution of probabilities, based on a pool of observed events. For a region, a continuous grid is first created, and the probability of occurrence of a specific event is calculated – in this case the existence of a pellet factory and its production capacity. This calculation is made according to the observed events, creating a density function according to the frequency of the pellet production plants. The function of density is subsequently calculated for all the points on the grid, which results in a continuous distribution of the frequencies for all the territory.

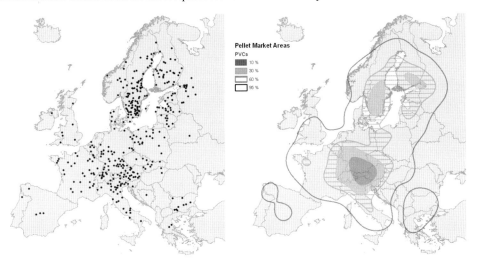

Figure 10.29: Location of the pellet production plants in Europe (left) and market analysis using percent volume contours (right)

Explanations: PVCs...percent volume contours; market analysis resulting from kernel estimations of the location of existing plants, weighted by the production capacity; the line contains 95% of the total pellet production in Europe; dark areas correspond to the locations with the highest concentration of pellet production

The maps in Figure 10.29 show the result of the application of this methodology to the current locations of pellet production plants in Europe. The search radius needed for defining the kernel curves was based on Worton's reference value [598]. The maps resulting show standardised isopleths, based on percent volume contours (PVC) in order to locate the areas with higher pellet production intensity. The PVCs represent a defined percentage of pellet producers in the smallest possible area. For instance, the isopleths containing the 10th percentile area shows the areas with the highest density of pellet production plants since it represents the smallest possible area to contain 10% of all the pellet production plants in Europe, i.e. the market core areas. The 95th percentile area represents the lowest density since it contains almost the total number of pellet production plants, and therefore defines the area of pellet production in Europe.

The analysis shows the high density of pellet production in Austria, Bavaria and middle Sweden. In these areas, the potential to further expand production capacity is very limited due to a lack of raw material and increasing competition between existing pellet producers. Other areas have sparse and low pellet production, such as the UK, northern Germany, western Poland, France and Spain. However, in order to identify areas with a high potential for a further increase in the production and use of pellets, it is essential to know the market areas that are already close to saturation both in terms of pellet production and utilisation.

10.10.1.2 The development of the Austrian market

Austria presents one of the highest densities of pellet production plants in Europe, with an annual production of about 700,000 t. Pellet production plants cover almost all the territory, and are increasingly close to each other. This results in increasing competition between the plants, which can be a symptom of market saturation.

In these cases, one possible way to analyse the market evolution of pellet production is by considering adoption curves. Studies on the adoption pattern of new technologies have been based on the initial works of [599]. This approach uses sigmoidal curves to define the aggregate number of adopters, evolution in time and the final saturation. When a new product or technology, such as wood pellets, is introduced to the market, there are very few entrepreneurs willing to invest in what is perceived as a high risk enterprise. As time passes, more and more entrepreneurs are convinced of the potential benefits of the new product, until the least risk takers finally join and maximum production is reached, which fits with current demand. In general, for each spatial unit of aggregation, a maximum ceiling is defined that is assumed to be a function of the socio-economic context of the area (cf. Section 10.11.7).

These concepts can also be used to explain the market development of pellet production plants in countries such as Austria where existing pellet plants are spread evenly across the country and, as a result, new plants only increase competition. According to this approach, the Austrian pellet production potential might reach a maximum around 2015, with a saturation productivity of approximately 1,351,000 t annually (Figure 10.30). However, this ceiling can be affected by policy and institutional measures introduced to encourage the use of wood pellets, which are exogenous variables, making future predictions uncertain. In particular, the use of raw materials other than wood shavings and sawdust, namely wood chips, log wood and short rotation crops, which are not considered in this prognosis, extends the raw material basis and allows for greater pellet production. This trend is already ongoing in Austria. Many pellet producers are planning to build or already operate plants able to use these raw materials, and therefore the competition for sawdust will be lessened as alternatives are available (cf. also Section 10.1).

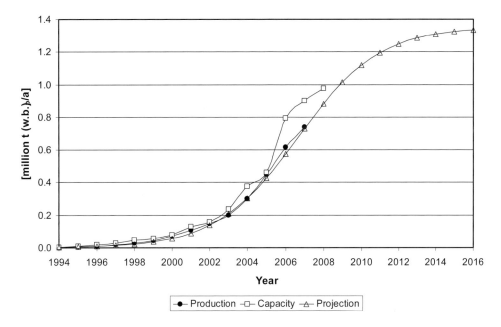

Figure 10.30: Pellet production and production capacity in Austria for the period 1994 to 2006 and projection of production until 2016

Explanations: projection based on assuming a sigmoidal curve

Nevertheless, this pattern can serve as an indicator of future potentials for pellet production in Austria as well as other European countries in similar conditions of market saturation. As more data on pellet market trends become available in different European and North American countries, a more concrete estimate of the actual potentials in the future can be made. In the case of Austria, the results show that the potential for increase is limited to the next 5 to 10 year period, unless there are significant socio-economic or policy related changes that affect this upper ceiling and as long as only traditional raw materials (wood shavings and sawdust) are taken into account. This suggests that a very similar evolution can also be expected in other countries or regions with a high concentration of pellet production plants. Consequently, their maximum production potentials based on wood shavings and sawdust from a market competition perspective could be reached in the short term if the socio-economic and political framework remain the same. However, as already indicated, this prognosis is restricted to pellet production with traditional raw materials. In many countries worldwide, it is already common practice to produce pellets from wood chips and log wood. If wood chips without bark are available or if the bark is separated from the log wood, even class A1 pellets according to prEN 14961-2 can be produced, which is particularly relevant for the residential heating sector. Raw materials containing bark can be used to produce industrial pellets for large-scale applications. Another strong increase can be expected when looking at the demand side, as the residential heating sector of many countries is still dominated by firewood, oil and gas heating systems, which could potentially be replaced by modern pellet furnaces in the coming decades. Moreover, further increasing oil and gas prices must be expected, which will also support this trend. In large-scale applications there is a trend towards firing and co-firing pellets due to financial support mechanisms in different countries.

10.10.1.3 Production in Sweden and Finland

Sweden is one of the biggest producers as well as consumers of wood pellets in the world [595; 596]. In Sweden, three factors affecting the fast development of the pellet industry have been identified: good availability of raw materials, a taxation system favourable towards biofuels and extended district heating networks [600; 601]. The pellet market is highly developed and pellet use covers all customer sectors at small, medium and large scales.

Total estimated production capacity is about 2,344,000 t/a (2007). Of the plants studied, six have an annual capacity equal or above 100,000 t, while 15 plants have an annual capacity of between 50,000 to 100,000 t. There are around 50 small-scale pellet producers whose production capacity is from a few hundred tonnes to several thousand tonnes a year. The total combined capacity of the small-scale producers (equal or below 5,000 t/a) is about 150,000 t/a, which is around 6% of the total production capacity of the pellet industry in Sweden. In the coming years, a new pellet production plant with a capacity of 160,000 t is expected to go online, which would raise the country's total pellet production capacity to over 2.5 million tonnes (cf. Section 10.5).

In Finland, the pellet market has been export oriented from the beginning. However, recently domestic consumption has started to increase. About 58% of total pellet production was exported in 2007, when one year earlier this was 75%. Nevertheless, there is still unutilised market potential [596; 602]. In the last ten years, the number of producers has increased to 24, raising total production capacity to approximately 750,000 t by the end of 2008. There are six plants with a capacity of over 50,000 t, of which one is 100,000 t, four small-scale producers (annual capacity under 5,000 t) and one in between. Five new plants are being planned and, once operative, total production capacity could reach up to 1.16 million tonnes (cf. Section 10.7).

In both countries, raw materials are mainly domestic, by-products of the wood processing industry, mostly wood chips and sawdust. This has determined the location of the pellet production plants, which are linked to the sawmilling industry as the main provider of raw materials. In almost all the cases, there is a sawmill in the proximity of every pellet production plant (Figure 10.31). In addition, there is a clear relationship to sawmilling capacity when both are aggregated in areas around 60 to 80 km of distance (Figure 10.32). According to this figure, around 12% of the sawmill capacity is linked to pellet production in the nearby area. Additional sources that are not included in the estimates come from other wood working industries, such as the furniture industry. In some cases, small amounts of sawdust were imported to Finland from Russia and to Sweden from Finland [597; 600] but these imports may cease in the future.

The geographical method performed can be used to estimate potential for pellet supply in countries with a less developed pellet market. Raw material availability sets clear limits on further development and therefore it is essential to investigate existing raw material supply structures. Furthermore, the study highlights the areas that are currently out of pellet production. In order to increase pellet production potential in these areas, new raw materials for pelletising or innovative supply structures need to be investigated.

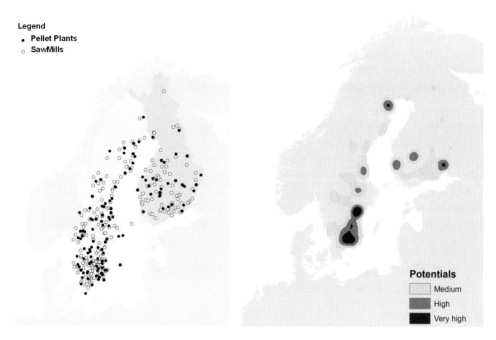

Figure 10.31: Location of sawmills and pellet production plants in Sweden and Finland

Explanations: based on 2007; right map: the darker the area, the greater the difference between sawmill production capacity and pellet production capacity in t/a in an aggregated area with a radius of 80 km; darker areas thus show the highest potential for pellet production with regard to raw material

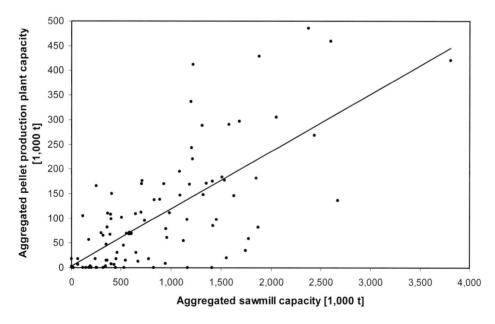

Figure 10.32: Correlation between sawmill and pellet production plant capacity aggregated using 80 km radius in Sweden and Finland

Explanations: coefficient of correlation $r^2 = 0.51$ (highly significant statistical correlation)

10.10.2 Evaluation of alternative raw material potentials in Europe

The potential development of pellet production will be determined by the availability of raw materials. There is increasing competition for good quality raw material between pellet producers and other forest based industries, such as the board industry.

The studies in Sweden and Finland confirm the suitability of using the production of the sawmill industry as a proxy for the evaluation of raw material supply. Accordingly, Sweden, the current leader in production, with a well established pellet infrastructure, has a pellet capacity proportionally higher than the European average with respect to sawlog production (Figure 10.33). In general, a shortage of raw materials has been observed, and as a result many pellet production plants are not using their full production capacities. During recent years, the high demand of raw materials has increased prices, including the price for pellets. Pellets are also considered one of the main competitors for raw material by heat and CHP plants, which have traditionally been using the same raw materials [600].

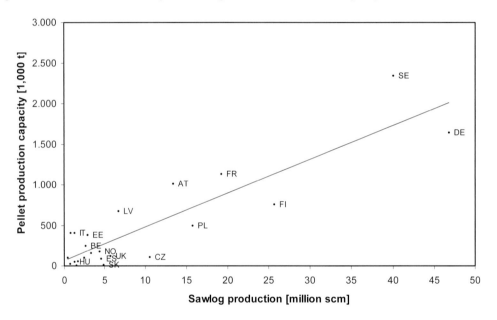

Figure 10.33: Estimated pellet production capacity of several European countries compared to the annual sawlog production.

Explanations: 1 scm ≈ 0.4 t (d.b.)

Figure 10.33 also shows the underused potential in countries such as Germany, Finland and Poland, where pellet production could be increased. However, development is influenced by changes in the sawmilling and pulp and paper industries as well as the socio-economic and policy framework in these countries. As an example, Russian wood tariffs on timber exports are reducing the amount of log wood coming to sawmills in nearby areas (e.g. Finland) and thus also reducing the raw material supply for pellets [603; 616]. At the end of 2008, the economic situation had reduced the output from sawmills throughout Finland, which could affect trends in pellet production, at least in the short term.

In order to increase production, the areas with a shortage of raw materials will have to either rely on imports or alternative raw materials. Although research into alternative sources is being initiated in many countries, the challenge is that many of the sources are not suitable for small- or medium-scale producers since drying costs are high and productivity is too low to be economically feasible [382]. However, in the near future, the shortage of traditional raw materials, particularly sawdust, in those areas with well established infrastructure for pellet production and an increasing demand, will be a driving force in the development of adapted technologies for the more intensive use of these alternatives.

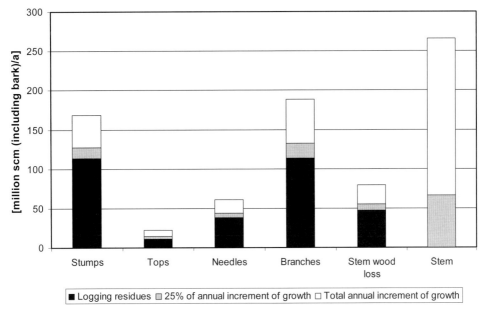

Figure 10.34: Theoretical forest fuel potential for the EU27 from logging residues and potential sustainable surplus of commercial growing stock (annual change rate)

Explanations: 1 scm ≈ 0.4 t (d.b.); data source [604]

Forest logging residues are considered as a raw material with great potential (cf. Figure 10.34). The theoretical annual fuel potential provided by tops, needles and branches in the EU27 would be 22.5, 61.1 and 188.1 million scm, respectively [604]. The total potential including stumps, coarse roots of trees and stem wood losses was estimated to be 785 million scm annually, including the potential from the surplus of commercial growing stock. However, the handling and drying of these residues would need to be optimal in order to decrease the corrosive agents that can cause problems, particularly in small-scale boilers [605]. Additionally, bark could be another possible alternative for pellets, although it is usually utilised at the place of debarking, typically in pulp mills, or used for landscaping or gardening purposes [606]. Furthermore, due to the high ash content of bark, at the moment it can only be used in larger boilers [605; 607]. With regard to needles, it must be pointed out that they are a potential source of nutrients for the soil and they should therefore not be removed from the forests. Moreover, needles contain many problematic elements with regard to combustion behaviour and would cause problems in furnaces, which is a another important reason not to use them as a fuel.

Finally, an additional potential source of raw material is energy crops such as reed canary grass and short rotation woody plantations. Their use, mostly as a fuel for district heating plants, is still under development in Europe. Currently, Finland has around 20,000 ha of reed canary grass under production. Sweden is the leader in commercial plantations for bioenergy purposes in Europe, with approximately 16,000 ha of short rotation willow plantations – about 0.5% of the total arable land in the country. In Sweden, short rotation coppice is currently mainly combusted in large-scale boilers as wood chips, although possibilities for its pelletisation are under research, presenting broad potential for the biofuel trade in Central Europe [608].

Table 10.3: Estimates for theoretical forest fuel and short rotation coppice potential

Explanations: n.d.…no data available; 1 scm ≈ 0.4 t (d.b.); forest fuel potential from both logging residues plus 25% of potential surplus from the annual increment of growth [604] and expected potential for short rotation coppice production by 2010 in harvestable annual oven dry tonnes, assuming a plantation area equal to 1% and 5% of the countries' total arable land, compared to current (2007) annual pellet production capacity

Country	Code	Current pellet production capacity [1,000 t/a]	Forest fuel potential [1,000 scm/a]	SRC 1% [1,000 t/a]	SRC 5% [1,000 t/a]
Austria	AT	1,011	22,000	228	1,142
Belgium	BE	250	4,500	146	729
Bulgaria	BG	n.d.	4,500	95	473
Czech Republic	CZ	111	19,700	220	1,101
Denmark	DK	410	1,100	270	1,349
Estonia	EE	385	6,300	58	289
Finland	FI	755	63,000	173	865
France	FR	1,136	42,000	2,338	11,688
Germany	DE	1,646	73,700	1,504	7,522
Greece	GR	25	n.d.	204	1,021
Hungary	HU	53	2,600	216	1,079
Ireland	IE	n.d.	3,500	633	3,166
Italy	IT	405	12,900	711	3,553
Latvia	LV	674	12,000	91	453
Lithuania	LT	162	6,500	128	638
Luxembourg	LU	n.d.	300	11	56
Netherlands	NL	100	900	172	861
Norway	NO	178	n.d.	86	431
Poland	PO	498	32,200	913	4,564
Portugal	PT	100	8,700	129	643
Romania	RO	n.d.	8,500	473	2,365
Slovakia	SK	12	7,400	28	138
Slovenia	SI	55	3,900	88	439
Spain	ES	88	16,800	1,444	7,222
Sweden	SE	2,344	75,600	301	1,506
United Kingdom	UK	123	11,100	1,958	9,790
Total		10,521	439,700	12,618	63,083

The potential from alternative raw materials is large, considering current pellet production capacity. Table 10.3 includes the potential supply of wood from short rotation coppicing, using 1% and 5% of the arable land, assuming proper tending and good management practices, based on the models for willow provided by Swedish experience [609; 610]. These cultivations could contribute to the development of pellet production in areas with limited forest resources but large agricultural areas, such as Spain or the UK. It can also result in additional sources for further pellet production in many other countries. Nevertheless, these scenarios will depend on the market development of the pellet sector. Research on adequate varieties and management practices, as well as technologies adapted to these raw materials, will be needed in order to fully develop their potential.

10.10.3 Evaluation of the worldwide sawdust potential available for pellet production

10.10.3.1 Forest biomass resources and wood use in forest industry

Wood pellets can technically be produced from almost all kinds of wood materials. Until recently, wood pellet production was based on the forest industry's by-products, the major raw materials being dry and fine grained by-products from the carpentry industry, and sawdust. The aim of this section is to give an overview of forest biomass resources and their use in the forest industry at a global level and to consider the transformation of log wood into forest products and by-products within the forest industry.

10.10.3.1.1 An overview of forest biomass resources and mechanical wood processing

Forest biomass is the major raw material of the forest industry and has an important role as a source of bioenergy. It has been estimated that there are 3,870 Mha of forest worldwide. Forest covers 30% of the Earth's land area, of which about 95% are natural forests and 5% are plantations. Tropical and subtropical forests comprise 61% of the world's forests, while temperate and boreal forests account for 38%. The average area of forest and wooded land per inhabitant varies regionally. The area varies between 6.6 ha in Oceania, 0.2 ha in Asia and 1.4 ha in Europe. The world's total above ground biomass in forests is 420,000 million t. The worldwide average of above ground woody biomass is 109 tonnes/ha. Brazil (114,000 million t), Russia (47,000 million t) and USA (24,000 million t) have the largest biomass resources in their forests.

The current rate of the utilisation of forest resources varies between the world regions. Deforestation, poor forest management and overuse of wood resources are serious problems in several areas, but in many parts of the word, the sustainable utilisation of forest resources can be increased. Estimates by the Food and Agriculture Organization of the United Nations (FAO) show that the global production of industrial log wood and wood fuel reached a total of 3,350 million scm in 2000 [611] (1 scm is equivalent to approximately 400 kg (d.b.) of wood). As much as 53% of this was wood fuel and about 90% of wood fuel is currently produced and consumed in developing countries [612]. In the statistics of FAO, industrial log wood is classified into three different groups: sawlogs and veneer logs, pulpwood and other industrial log wood. In 2004, the total consumption of industrial log wood was as follows (figures do not include bark) [613]:

- Sawlogs and veneer logs: 992 million scm;
- Pulpwood: 505 million scm;
- Other industrial log wood: 146 million scm;
- In total: 1,643 million scm.

Logs are mainly used as raw material in the manufacturing of sawn timber and plywood, whereas smaller diameter pulp wood is consumed in wood pulp production. For pellet production from sawdust, sawmills are the major sources of raw material. Furthermore, plywood mills generate sawdust and similar fractions as by-products that are a potential raw material of pellets. A review of the production of logs, sawn timber and plywood gives a preliminary view on global sawdust resources from the forest industry (cf. Table 10.4 and Table 10.5). North and Central America and Europe are the largest consumers of logs, and the USA, Canada and Russia are the largest producers of sawn timber.

Table 10.4: World production of industrial log wood, logs, sawn timber and plywood

Explanations: data in million scm; 1 scm ≈ 0.4 t (d.b.); by continent in 2004; data source [613]

Continent	Industrial log wood	Pulp wood	Production of sawn timber	Production of plywood
Africa	70	27	9	0.7
Asia	229	150	72	38.5
Europe	504	284	138	6.9
North & Central America	628	425	159	17.5
Oceania	48	24	9	0.7
South America	164	83	35	3.8
World	1,643	993	422	68.1

Table 10.5: World top 15 countries in the production of logs, sawn timber and plywood in 2004

Explanations: data in million scm; 1 scm ≈ 0.4 t (d.b.); data source [613]

Position	Production of logs		Production of sawn timber		Production of plywood	
1	USA	248.0	USA	93.1	China	21.0
2	Canada	167.1	Canada	61.0	USA	14.8
3	Russia	67.9	Russia	21.4	Malaysia	5.0
4	Brazil	54.9	Brazil	21.2	Indonesia	4.5
5	China	52.2	Germany	19.5	Japan	3.1
6	Sweden	35.4	India	17.5	Brazil	2.9
7	Germany	32.2	Sweden	16.9	Canada	2.3
8	Indonesia	26.0	Japan	13.6	Russia	2.2
9	Finland	24.3	Finland	13.5	India	1.9
10	Malaysia	22.0	China	11.3	Finland	1.4
11	France	19.9	Austria	11.1	Taiwan	0.8
12	India	18.4	France	9.8	South-Korea	0.8
13	Chile	15.9	Chile	8.0	Chile	0.5
14	Poland	13.0	Turkey	6.2	Italy	0.5
15	Australia	12.2	Malaysia	5.6	France	0.4

Figure 10.35 presents trends in the world's consumption of logs and production of sawn timber and plywood from 1985 to 2004. There were no remarkable changes during the reviewed period in the production of sawn timber and plywood. Considering sawn timber production at the country level between 1990 and 2004, the most remarkable increases in the annual production occurred in Canada (10.5 million scm), India (9.6 million scm), China (4.9 million scm) and Germany (3.2 million scm). In some of the largest countries in sawn timber

production, annual production declined from 1990 to 2004. Examples include Japan (- 3.5 million scm) and Brazil (- 1.9 million scm).

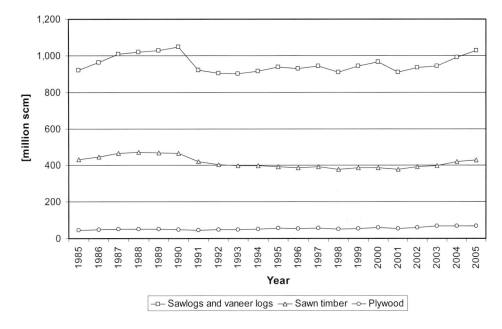

Figure 10.35: The consumption of logs and the production of sawn timber and plywood from 1985 to 2004

Explanations: 1 scm ≈ 0.4 t (d.b.); the volumes of log wood do not include bark; data source [613]

Table 10.6: World top 15 countries in the production of particle board and fibreboard in 2004

Explanations: data in million scm; 1 scm ≈ 0.4 t (d.b.); data source [613]

Position	Production of particle board		Production of fibreboard	
1	USA	21.8	China, Mainland	15.3
2	Canada	11.3	USA	7.5
3	Germany	10.6	Germany	5.1
4	China, Mainland	6.4	Canada	2.1
5	France	4.4	Poland	1.9
6	Poland	4.1	Korea, Republic of	1.6
7	Italy	3.7	France	1.3
8	Russian Federation	3.6	Spain	1.3
9	Spain	3.2	Malaysia	1.2
10	Turkey	2.7	Russian Federation	1.2
11	United Kingdom	2.7	Italy	1.1
12	Austria	2.4	Turkey	1.0
13	Belgium	2.2	Brazil	1.0
14	Brazil	1.8	Japan	0.9
15	Japan	1.2	New Zealand	0.9
	World	97.5	World	52.9

Particle board and fibreboard mills utilise by-products from sawmills and plywood mills as a raw material. The world's largest producers of particle board and fibreboard are depicted in Table 10.6.

10.10.3.1.2 Use of wood as raw material and energy in forest industry

In addition to log wood, some by-products such as pulp chips and sawdust are important raw materials for the forest industry (recovered fibres, i.e. recycled paper products, have become an important raw material for the forest industry, however, they are not considered in this study; in 2004, the total production of recycled paper in the world was 159 million t [613]). The raw material use of these by-products improves the efficiency of wood conversion into products. On average, 40–60% of log wood can be converted into forest products by the forest industry, the remainder being by-products such as black liquor, bark, sawdust and chips that have no feasible raw material use within the forest industry. The conversion efficiency varies between the production processes of different products. Also the level of technology applied and the integration of the production processes affect the conversion efficiency. Mechanical wood processing can convert wooden raw material into products more efficiently than chemical pulp making, for instance.

10.10.3.1.2.1   Forest industry's solid by-products

The majority of the solid by-product fuels in the forest industry consist of bark. The bark content of log wood is 10–22% of the total volume of wood with bark, depending on tree size and species [614]. There is no established market for bark as a raw material within the forest industry.

Sawmills produce large quantities of by-products, which are a suitable raw material for other processes within the industry. Pulp chips are equal to pulpwood as a raw material in pulp making and are the most important by-product of sawmills in many regions. For example in Finland, the sales of pulp chips to pulp mills improve the economic situation of sawmills because pulp mills pay a higher price for pulp chips than energy producers. The other by-product that pulp mills can use as a raw material is sawdust, but its quality is lower when compared to pulpwood. Instead, sawdust is an important raw material for particle board and fibreboard mills. Plywood mills produce bark, peeler cores, sawdust, veneer chippings, panel trim and sander dust as by-products. From these, sawdust and sander dust can easily be exploited in wood pellet manufacturing. The market situation of forest industry by-products varies considerably between regions. The amount of wood chips and sawdust produced in sawmills usually varies between 30 and 40% of the total amount of log wood utilised.

The by-products of sawmills (bark and sawdust) are usually utilised in heat production for timber drying, and the rest is sold as fuel to other heating and power plants or pellet production plants. In many cases, sawmills are located on the same site as paper and pulp mills, which allows the efficient utilisation of raw material. In such cases, sawmill by-product fuels are utilised in-house for heat and power production.

10.10.3.1.2.2   Forest industry's liquid by-products (black liquor)

Black liquor is the most important by-product in energy production at forest industry mills. Energy production from black liquor is part of the chemical pulping process. The wood material consists of two primary components, i.e. cellulose and lignin in approximately equal quantities. Lignin is a kind of glue that holds wood fibres together. In the chemical pulping process, a chipped wood material is cooked in a lye solution that dissolves the lignin and leaves behind the cellulose. The cooking solution consists of cooking chemicals, and the lignin is burned in a recovery boiler for gathering the cooking chemicals and for utilising the energy of the dissolved wood material.

10.10.3.1.2.3 By-products in energy production

Generally, forest biomass has been a marginal source of energy in industrial applications, but in some countries with a large forest industry sector, such as Sweden, Finland and Austria, forest biomass has huge importance. In Finland, for instance, renewable energy sources cover approximately 25% of the total primary energy consumption, and over 80% of renewable energy is derived from wood. Nearly 80% of wood energy is generated from the processing residues of the forest industry [615]. Figure 10.36 illustrates wood streams in the Finnish forest industry. In 2007, the Finnish forest industry was able to convert approximately 60% of the total raw wood consumption into products.

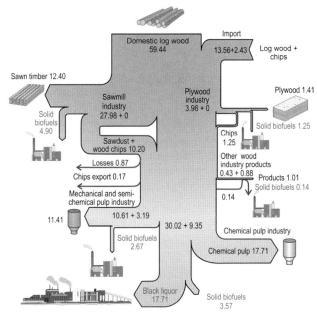

Figure 10.36: Wood streams in the Finnish forest industry in 2007

Explanations: data in million scm related to log wood equivalents; 1 scm ≈ 0.4 t (d.b.); X + Y...X stands for log wood, Y for by-products; log wood includes bark; particle board and fibreboard mills were included in the other wood products industry; data source [616]

Worldwide, and especially in industrialised countries, the by-products of the forest industry are one of the major sources of bioenergy. A preliminary calculation shows that the total volume of by-products of the forest industry is 700 to 1,100 million scm/a (equalling 4.4 to 6.9 EJ; the initial data and assumptions of the calculation were: the conversion factor of log wood into by-products 40 to 60%, the total use of industrial log wood 1,600 million scm/a excluding bark, average bark content 12%, average calorific value of wood 6.3 GJ/scm based on an average moisture content of 55 wt.% (w.b.)). In comparison, the total use of bioenergy in industrialised countries was an estimated 15 EJ in 2002 according to the International Energy Agency (IEA) [612].

10.10.3.2 Evaluation of global raw material potential for wood pellets from sawdust

10.10.3.2.1 Modelling the wood streams of forest industry at country level

An MS-Excel based spreadsheet model was developed to investigate the wood streams of the forest industry at the country level and to identify the countries that have the largest sawdust resources for energy purposes. The main objective of the model was to evaluate the excess volumes of the forest industry's solid by-products, taking into account the raw material use of by-products in the forest industry. The model uses country specific data on the production of industrial log wood and forest products and the trade of raw wood as the initial data. In this case, the data was obtained from the forestry data base of FAO. The conversion factors of bark free wood into forest products and by-products were sourced from the literature [614]. The structure of the model and the conversion factors applied are depicted below (Figure 10.37). The model uses universal conversion factors over the world regions and countries.

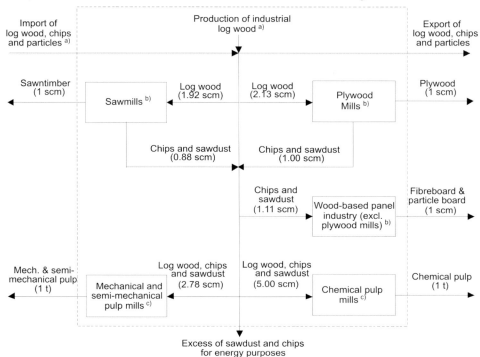

Figure 10.37: Illustration of the wood stream model and its main parameters

Explanations: data in scm related to log wood equivalents; 1 scm ≈ 0.4 t (d.b.); none of the wood streams presented include bark; material losses are not shown; a)...the net consumption of raw wood was calculated as follows: production of industrial log wood minus production of other industrial log wood minus export of industrial log wood plus import of industrial log wood minus export of chips plus import of chips; b)...the conversion factors are derived from [614]; c)...1 tonne of pulp was assumed to equal 2.5 scm wood, the pulp yield from wood was assumed to be 50% for chemical pulp and 90% for mechanical and semi-mechanical pulps

By means of the model, the total global volume of by-products (excluding bark) from sawmills and plywood mills was an estimated 440 million scm/a. Approximately 70% of the total volume consists of wood chips of sufficient quality to be used as a raw material in pulp production. The total volume of sawdust from sawn timber production was estimated at 120 to 130 million scm, which equals 53 to 58 million t (w.b.) of wood pellets (the density of wood was assumed to be 400 kg (d.b.)/scm). However, the prevailing use of by-products should be taken into account in the consideration of the availability of by-products for energy purposes. The countries with the largest production of sawn timber do not automatically have the best availability or the largest excess of sawdust or other by-products.

10.10.3.2.2 Sawdust excess from forest industry

The largest producers of by-products are presented in Figure 10.38. The USA is by far the largest producer of by-products from mechanical wood processing, producing over a fifth of the by-products under review.

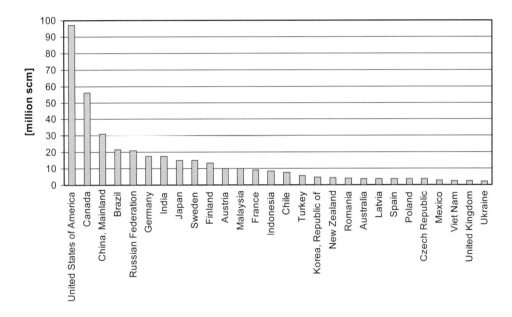

Figure 10.38:   The largest producers of by-products from sawmills and plywood mills

Explanations: 1 scm ≈ 0.4 t (d.b.); bark is excluded from the volumes; the global total volume is 440 million scm

In the next phase, the demand for raw material of particle board and fibreboard mills was subtracted from the total volume of by-products (Figure 10.39). The USA, Canada and Germany are the largest producers of particle boards, and China, the USA and Germany are the leading countries in the production of fibreboards (Table 10.6). The calculations showed that the demand for raw material from the particle board and fibreboard industry exceeds the theoretical volume of by-products in Spain, the UK and Poland. The potential reasons for this could be the import of raw materials by the wood panel industry or the actual conversion of wood into products at sawmills and plywood mills is less efficient than the model assumed.

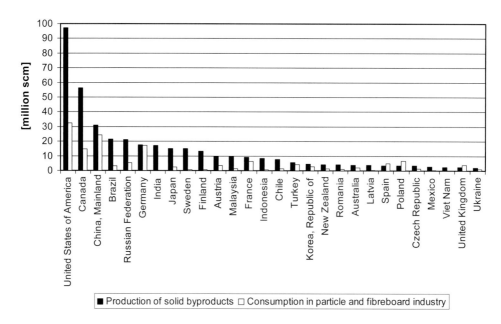

Figure 10.39: Comparison of the production of solid by-products in the sawmill and plywood industry and the demand for raw material in the particle board and fibreboard industry

Explanations: 1 scm ≈ 0.4 t (d.b.)

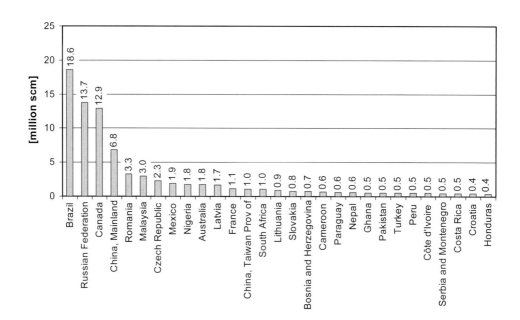

Figure 10.40: Theoretical excess of solid by-products from the mechanical wood processing industry

Explanations: 1 scm ≈ 0.4 t (d.b.)

As mentioned above, a total of about 70% of bark free by-products from sawmills and plywood mills are suitable raw material for pulp manufacturing. Figure 10.40 shows the theoretical excess of by-products in various countries after the by-products were allocated to the forest industry according to the calculated demand. The countries are presented in the figure according to the theoretical surplus of solid by-products. The USA, Canada, Finland, Sweden, Japan, Brazil and Russia are the world's largest producers of wood pulp, and in these countries, the pulp industry is the most important user of by-products from mechanical wood processing but Brazil, Russia and Canada seem to produce an excess of by-products that could be utilised for other purposes.

In total, the excess of solid by-products available for pellet production amounts to about 83.4 million scm (also including countries not shown in Figure 10.40). This amount would be equivalent to a global pellet production potential of approximately 37 million tonnes (w.b.) from forest industry by-products.

When interpreting the results, one should bear in mind that the calculations were made by means of a model using universal conversion factors for wood into forest products and by-products. Furthermore, the current situation regarding local utilisation of by-products was excluded from the scope of this study. The areas with sawmills but no local demand for by-products as a raw material or as a fuel are the most favourable for constructing new wood pellet production capacity, and feasible options for wood pellet production can also be found in countries that are not mentioned in Figure 10.40. Furthermore, several other factors, such as the market for biofuels and logistics, have an effect on the feasibility of utilising by-products in pellet production. It is clear that country specific studies will be needed to obtain more comprehensive data on the commercial possibilities of wood pellet production.

10.11 International pellet trade

10.11.1 Main global trade flows

Today, wood pellets are one of the most traded solid biomass commodities used specifically for energy purposes. In terms of traded volume, approximately three to four million tonnes are traded annually over a border, which is in a similar range as the amount of traded biodiesel or bio-ethanol [617]. This is probably due to the fact that wood pellets have relatively favourable qualities for long distance transportation, i.e. low moisture content and a relatively high energy density (17 GJ/t). While handling wood pellets still requires care (no exposure to moisture, potential dusting during handling), the advantages over other solid biomass types such as wood chips or agricultural residues are long-term storability and relatively easy handling.

The success of wood pellets as an energy carrier has in many cases been linked to international trade. While some traditional markets such as Sweden or Austria are largely self-sufficient, other markets depend on the import of wood pellets to a very large extent (e.g. the Netherlands, Belgium, Denmark and Italy), and for many producing countries (Canada being the prime example, but also other areas such as the Baltic countries, northwestern Russia and the Western Balkan area) the pellet production sector largely depends on export opportunities.

Obtaining accurate figures on production, consumption and trade figures of wood pellets is a challenge. In many countries, wood pellet production plants were built and capacities expanded rapidly, making it difficult to obtain accurate figures on production capacities and actual production. Similarly, in many countries, no accurate data are available on the sales of small-scale pellet boilers, so consumption for these countries can only be roughly estimated. Only in some countries, such as Austria, Canada, Finland, Germany and Sweden, are good data available, mainly due to the presence of well organised pellet industry associations. Regarding international trade, the traded volumes of common commodities are registered under CN (combined nomenclature) codes in the EU, and globally using HS codes. However, up until the end of 2008, no such code was available for the trading of wood pellets, so wood pellets were often traded under categories such as wood waste, making it difficult to derive trade flows from international trade statistics. In 2009, a specific CN code (4401 30 20) for the wood pellet trade in the European Union was introduced [618], while a globally accepted HS code for wood pellets should become operational in 2012, and thus more detailed trade data will be available in the future.

All figures presented in this chapter should be seen as (best) estimates. Nevertheless, they provide a reasonably accurate picture of the overall production, trade and consumption patterns.

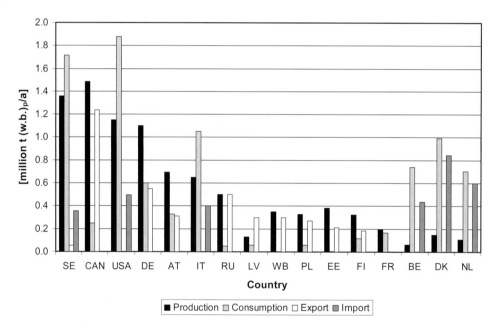

Figure 10.41: Overview of pellet production, consumption and trade flows for the most important pellet markets in 2007

Explanations: data source [496]

A quantitative overview of pellet production, consumption and trade flows for the most important pellet markets in 2007 is shown in Figure 10.41 (in many cases, these volumes are estimated, and should hence be considered with caution). Sweden, Canada, the USA and Germany all produced (well) over 1 million tonnes of wood pellets. The biggest wood pellet consumers in 2007 were Sweden, the USA, Denmark, Italy, Belgium and the Netherlands, consuming roughly two thirds of global pellet production. Figure 10.42 provides an overview of the main pellet trade routes in Europe for the period 2006 to 2008, based on [619] and updated with data from [620]. As can be seen from this figure, pellets are either imported or exported from almost every country in the EU and beyond. In general, it can be stated that the largest part of international wood pellet trade is carried out by means of ocean vessels, coasters and river barges, but that another substantial part is transported by truck. In some cases, anecdotal data suggest that trucks transport pellets up to several hundred kilometres.

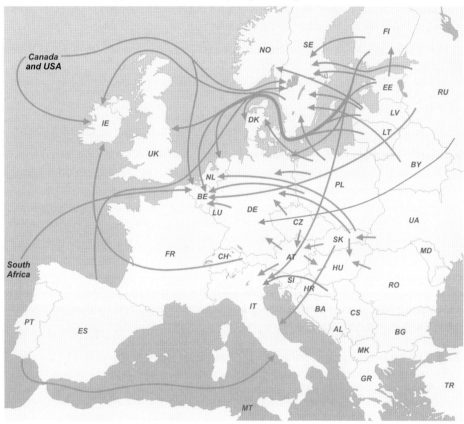

Figure 10.42: Overview of main wood pellet trade flows in and towards Europe

Explanations: data related to trade flows between 2006 and 2008, adapted from the EUBIONETII project, see [619]

Major ports and destinations for wood pellets in the period 2007 to 2008 include:

Vancouver to Antwerp-Rotterdam-Sweden: In 2007, Canada produced 1,485,000 t of wood pellets in 26 plants. 495,000 t were exported to the USA, primarily by train, and 740,000 t were exported by large ocean vessels to Europe; 500,000 t to Belgium, 100,000 t to the Netherlands, 130,000 t to Sweden, and a small amount to Denmark.

Vancouver to Japan: Canadian pellet producers shipped 110,000 t of pellets to Japan in 2008, and expect exports to reach 400,000 t. Owing to a shortage of coal for power generation in order to support the tremendous growth in Chinese industry, many private Chinese energy companies are interested in securing pellets as an alternative fuel.

Panama City (Florida, USA) to Antwerp-Rotterdam: In 2006, the USA had 60 pellet plants producing 800,000 tonnes of pellets, essentially for domestic use. New pellet plants are now being built in south east USA for production for export. Green Circle Bio Energy completed a 560,000 t mill in Cottondale Florida and began production in May 2008. Green Circle plans to build more such plants, but only if pellet prices are likely to be high enough because the feedstock is log wood, which is more costly than the sawdust originally envisioned. It exports by the deepwater port of Panama City, Florida. Customer data are confidential, but in all likelihood the main exports go to Belgium and the Netherlands.

St. Petersburg and Archangel to Swedish ports: Data for Russia are unreliable but it supposedly produced 500,000 t of pellets in 2007, for which 47,000 t were exported to Sweden via St. Petersburg. However, despite its potential as a pellet producer, there are bottlenecks. Russian ports have not received the investment required to support efficient loading and plants are also under-resourced to keep costs competitive. Russia will not become a significant pellet exporter unless drastic changes occur in these basic conditions.

Finnish ports to Sweden-Antwerp-Rotterdam: Wood pellet production in Finland started in 1998, founded on exports (supplying pellets to Sweden, where the pellet market was developing rapidly at the time). Since then, pellet production has increased steadily, climbing to 376,000 tonnes in 2008. Of these, 227,000 tonnes were exported to Sweden (45%), Denmark (31%), the UK (10%) and Belgium (8%). In 2008, the import of wood pellets to Finland was statistically recorded for the first time. Imported pellets came most probably from Russia and the Baltic states.

Latvia-Estonia-Lithuania to EU: In 2006, these three countries exported approximately 620,000 tonnes of pellets, of which 150,000 went to Sweden. These countries are suffering from a reduced supply of wood (mainly due to increasing export taxes on Russian wood), and exports of pellets will fall accordingly.

10.11.2 The history of intercontinental wood pellet trade – the case of Canada

The heart of the long distance wood pellet trade lies in British Columbia, Canada. Due to the vastness of its forest resources, in the mid 1990s, Canada was one of the dominant pulp, paper and lumber producers in the world. In 1997, Canada produced 25 million tonnes of paper grade pulp, and exported 10.2 million tonnes. This also included large quantities of residues, such as sawdust, but also so-called heritage piles, i.e. piles of bark that were left beside mills for 10 to 30 years in Ontario, Quebec and other provinces. This led to the development of a minor wood pellet production sector, and in 1997, the production of pellets in Canada was 173,000 t, of which roughly two thirds were exported to the US market, mostly the Seattle area in the west and the New England region in the east. However, in the latter half of the 1990s, the Seattle natural gas grid was extended and the pellet market declined sharply – pellet producers in British Columbia had to find new markets. One possibility was the district heating and CHP sector in Sweden, where fossil fuel use for heating was heavily taxed, and biomass enjoyed exemption from these taxes. The Öresundskraft CHP plant in Sweden had direct access to a harbour, allowing pellets to be delivered competitively, and thus, the first intercontinental shipment occurred in April 1998, when the "Mandarin Moon" brought a shipment of 15,000 tonnes from Vancouver (British Columbia) through the Panama Canal and across the Atlantic to Helsingborg in Sweden. Since then, intercontinental exports have increased rapidly. By 2007, total pellet exports grew to over 1.2 million tonnes or 85% of production, but more importantly, European markets grew from zero to 63% of Canadian pellet exports in 10 years, displacing the USA as the major trade partner. Markets include large power companies in Belgium, the Netherlands and Sweden, and also large district heating companies in Sweden.

Canada has succeeded in becoming one of the world's leaders in wood pellet production and trade in the past ten years. The factors contributing to this success story include a surplus of low cost mill residue, excess pellet production capacity due to a decline in USA pellet demand in the late 1990s, policies in Europe promoting biomass use, and the will of entrepreneurs to ship pellets over 15,000 km to Europe. The Canadian case can be seen as a prime example of successful and sustained intercontinental wood pellet (and in general bioenergy) trade. It managed to bring together abundant feedstock resources on the one hand and on the other hand, meet the increasing demand for clean and easy-to-handle high density solid biomass fuels for large-scale heating, CHP production and substitution of coal for electricity production. Despite severe logistical challenges, fluctuating oil prices and subsidies for wood pellet use in Europe, increasing freight rates and many more obstacles, Canada has managed to continuously increase its exports and build up one of the largest pellet industries in the world. It has demonstrated convincingly that biomass energy carriers can be developed into a commodity that can be traded internationally and even intercontinentally. Recent developments such as the decrease in oil prices, the increasing strength of the Canadian dollar, the break-down of the USA housing market and not least the global economic crisis have posed large barriers to overcome. However, the diversification of exports to Asian markets, falling freight rates, the possible development of torrefied pellets (made from torrefied wood with an approximate energy density of 23 GJ/t) and utilisation of new and abundant feedstocks such as forestry residues and wood damaged by the mountain pine beetle may also provide new opportunities for Canada's future exports [621].

The Canadian example also shows the main prerequisites for long distance trade, namely abundant availability of cheap feedstocks in some world regions, high demand in other (resource scarce) regions, and cost efficient logistics.

10.11.3 Wood pellet shipping prices, shipping requirements and standards

Even though the international trade in wood pellets has been going on for several years, statistics on harbour prices are scarce. Figure 10.43 depicts the CIF ARA (cost insurance and freight delivered to the Amsterdam/Rotterdam/Antwerp area) as collected by the Pellets@las project. Recently, pellet price indices were published by different organisations, for example proPellets Austria [437], FOEX [622], ENDEX/APX [623] or Argus [624]. Long-term contracts based on such indices have already been concluded.

While pellet prices are in general determined by factors such as raw material availability and fluctuating demand, especially for internationally traded biomass, freight rates can be a major contributor to the total wood pellet prices. As the historical development of freight rates shows, the charter market is highly competitive. Freight rates can change dramatically over a short period of time. A shortage of transport capacity causes freight rates to rise (and in times when no further sea transport capacity is available freight rates skyrocket), and an oversupply of ships causes freight rates to plummet.

Whereas intercontinental trade is carried out using compartments of large dry-bulk Panamax and Capesize carriers, shipments within Europe are carried out by means of smaller Handysize vessels. An overview of vessel specifications is shown in Table 10.7.

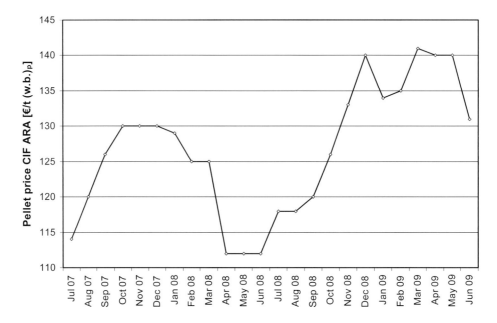

Figure 10.43: Wood pellet spot prices CIF ARA

Explanations: excl. VAT; for bulk delivery of 5,000 tonnes of wood pellets by ocean vessel; data source [496]

Table 10.7: Overview of vessel specifications

Vessel type	Maximum deadweigth [tonnes]	Main limiting factor for dimensions
Handysize	15,000 - 35,000	Typical bulk terminal restrictions of smaller ports
Handymax Panamax	35,000 - 58,000	Typical bulk terminal restrictions Just fitting through the Panama canal (max length 294 m)
Capemax	Not limited	Too large to transit the Suez Canal, hence they have to pass either the Cape of Good Hope or Cape Horn

Shipping rates for dry bulk stayed fairly constant from 1999 to 2002. As a result, few ships were coming online – only enough to replace scrapped capacity. From 2002 to 2005, shipping demand from the booming Chinese manufacturing sector drew considerable capacity from other routes. By 2006/2007 much manufacturing that had formerly taken place in the developed countries had moved to China and India. Since goods that were formerly made in the countries where they were consumed were now manufactured offshore, more shipping was required to bring these goods to the major consuming regions. Orders for ships were placed, but because ship building takes years, ships could not come off the line fast enough to match demand. The shipping shortage caused prices to rise considerably. For example, the Capesize rate rose from 4,000 US$ per day in 2004 to 20,000 US$ per day in 2008.

The current world financial crisis, which has lead to falling demand and over-supply of inventories, as well as collapsing oil prices, has caused a fall in shipping prices. At the moment, the pacific Panamax rates have slumped and are showing little chance of a quick revival as few cargoes enter the market and the tonnage list gets longer by the day. Referring to [625] the Baltic Exchange, the average time charter rate on 23 March 2009 was near 11,900 US$. Table 10.8 gives the five year maximum and minimum of the charter rates for the following types of vessels: Capesize, Panamax, Supramax and Handymax.

Table 10.8: Maximum and minimum charter rates

Explanations: data source [626]

Type of ship	Charter rate 5 year high [US$/day]	Date	Charter rate 5 year low [US$/day]	Date	Movement of prices [%]
Capesize	233,988	05 June 2008	2,316	03 December 2008	-99.0
Panamax	94,997	30 October 2007	3,537	12 December 2008	-96.3
Handymax	72,729	30 October 2007	4,065	19 December 2008	-94.4
Handysize	49,397	22 May 2008	3,948	12 November 2008	-92.0

The Baltic Dry Index (BDI) measures the costs to transport dry bulk materials by sea. The index is a weighted average of different routes and vessel sizes; it is a composite of the Capesize, Panamax and Handymax Indices. It came into operation in 1999 and is the successor to the Baltic Freight Index (BFI). In this index inflation is not taken into account. The BDI is the most important price index for dry bulk cargoes.

The cost of transport is also highly dependent on whether a port is on a common route or not. For example, a considerable volume of pellets is moved 14,000 km from Vancouver through the Panama Canal to the major European ports of Antwerp and Rotterdam. A much smaller volume of pellets is moved from Halifax to Europe, yet the costs of the two routes are almost the same. Halifax to Europe is not a common route, and ships are considerably smaller. In order for biomass to be shipped long distances at low cost, it must be on an existing route with many ships, or there must be sufficient volume of biomass to warrant establishing a major route of its own.

The following is a typical calculation for estimating the freight rates of shipping bulk cargo from A to B, reflecting market conditions as of spring 2009 (cf. Table 10.9). The example is based on a pellet transport from Indonesia to Italy, overall costs are 40 US$/t. However, it must be pointed out that every single position in the calculation could be different for all other locations and/or dates.

Table 10.9: Sample calculation for estimating the freight rates for 22,000 t pellets by bulk cargo for a shipment from Indonesia to Italy through the Suez canal

Explanations: [1]...price for the travel of the empty vessel to the harbour for loading (in the best case this price is zero, if the vessel can be loaded at the same place where the last unloading took place, which is usually not the case); [2]...loading and unloading capacity and hence duration depend on the equipment available at the harbour and the actors involved, and can differ in a broad range; [3]...optional (relevant in case of partial shipments unloaded in two different harbours); data source [627]

Parameter	Unit	Value
General data		
Time charter: daily rate of ship (charter)	US$/day	10,000
Cargo	t	22,000
Loading[2]	t/day	3,500
Unloading[2]	t/day	2,500
Duration/travel time		
Ballast voyage[1] (travelling to A)	days	3.0
Duration (travelling from A to B)	days	21.0
Duration loading[2]	days	6.3
Duration unloading[2]	days	8.8
Overall duration	days	39.1
Costs of bunker fuel		
IFO 380 - fuel oil	US$/t	200
MDO - marine diesel	US$/t	350
Overall bunker fuel requirements		
IFO 380	t	879
MDO	t	169
Costs (US$)		
Harbour dues	US$	60,000
Costs of 1st unloading	US$	60,000
Costs of 2nd unloading[3]	US$	0
Bunker fuel costs IFO	US$	175,800
Bunker fuel costs MDO	US$	59,150
Fees for Suez Canal	US$	120,000
Assurance	US$	3,740
In lieu of holds cleaning	US$	3,080
Miscellaneous	US$	2,200
Commission	US$	9,771
Overall charter rate	US$	391,000
Overall costs US$	US$	884,741
Freight rate US$	US$/t	40.2
Exchange rate	US$/€	1.27
Overall costs €	€	696,646
Freight rate €	€/t	31.7

10.11.4 Prices and logistic requirements for truck transport

While intercontinental or long distance international transport of wood pellets is mainly by vessel transportation, local distribution is carried out mostly by means of trucks (cf. Section 4.2.2). Generally wood pellets are transported either in bulk or bagged. The bulk wood pellets are delivered from the producer to the wholesale trader or middle- to large-scale consumer. The wholesale traders and some of the producers themselves employ a fleet of special tank trucks to load approximately 15 to 24 tonnes of pellets, complete with the equipment to directly blow the pellets into the in-house storage of the individual buildings.

Dump trucks, either with hydraulic unloading or as walking floor trucks will load approximately 23 tonnes. A truck will cost – depending on country – approximately 1 to 1.2 €/km of transport. A transport distance of 200 kilometres hence results in approximate costs of 10 €/t. However, freight can vary depending on route and situation in the freight market. Freights from east to west can be negotiated in certain periods due to quite significant amounts of empty room in return trucks.

Standard trucks are employed when wood pellets are packed before delivery. The standard is the 15 kg bag that, stapled on a pallet (cf. Figure 10.44) and wrapped by shrink foil, can load complete standard trucks but can be merged into mixed parcels as well. A truck will hold up to 23 t – depending on destination and truck type – which, again depending on packing, can be up to 40 pallets.

Typically, 15 kg bags contain mostly high quality (standardised) wood pellets. Lower grades are sometimes packed in big bags of 700 to 1,200 kg each. This mostly serves to utilise cheap return freights from eastern countries into Western Europe. Although handling of bulk wood pellets in big bags does increase costs by approximately 8 to 10 €/t, it is economic for certain destinations as the cost differential between dump truck and trailer truck may be higher.

Figure 10.44: Pellets stapled on a pallet

Explanations: data source [628]

Regarding technical standards for internationally traded wood pellets, to date no technical standard is widely used. In general, industrial wood pellets for co-firing or use in large district heating plants do have lower quality requirements than wood pellets for stoves. For example, the ash content is far less critical, as other fuels utilised (e.g. coal) can also have high ash contents, and the boilers are built to deal with large ash quantities. The most important criterion for large-scale users is net calorific value.

Overall, it can be stated that significant technological development has occurred over recent years to optimise wood pellet logistics, both transported in bulk and in bags. Under the right circumstances, even long distance transport distances are not prohibitive. Nevertheless, logistical challenges remain to be solved, as discussed in section 10.11.6.5.

10.11.5 Future trade routes

With fluctuating oil prices, subsidy schemes and the global economic climate in disarray, it is hard to tell which new wood pellet market will open up and how international wood pellet streams will develop. Yet, based on current policy initiatives, announced investments in production capacities and so on, a number of trends for the future supply and demand of wood pellets can be identified.

The following countries will likely be able to increase wood pellet production and supply (for export in the near future; see also Figure 10.45 for an overview of expert opinions):

- North America is expected to remain a large-scale producer and exporter of wood pellets in the coming years. The eastern provinces of Canada have plans to further increase production capacity for both domestic use and export. Investors in the province of Ontario have signed agreements to build six new pellet plants to produce 1 million tonnes of pellets by 2011. While Ontario Power Generation is moving forward with plans to co-fire biomass in its coal plants, pellet prices are higher in Europe and so the initial markets will probably be there. Similarly, it is anticipated that Quebec will also see 1 million tonnes of pellet production by 2012. The domestic market for pellets will continue to grow, but only slowly. Some quantities may be sold to the New England states, but with bioenergy incentives in the EU, it is entirely possible that 1.5 million tonnes will be destined for the EU. However, in the long term, it is also well possible that the US will develop a major domestic market, and that exports may be reduced to meet this new demand.

- With the construction of several extremely large wood pellet production plants in the southeastern states of Alabama and Florida, the USA is also destined to become a major exporter of wood pellets, probably to European power plants.

- (Northwestern) Russia certainly has the feedstock resources to produce and export additional wood pellets. The installed capacity and actual production will further increase due to several pellet production plants being planned or under construction (one of which will be the world's largest pellet production plant with an annual production capacity of about 900,000 tonnes of pellets, located in Vyborg, north of St. Petersburg, close to the Finnish border [629]). The scope of this increase will largely depend on the demand market and the development of logistics strategies.

- Chile is a major softwood pulp producer and has plans to become a pellet supplier. There is the potential for 2 million tonnes of pellets for export [630]. Given its geographical location, exports are likely to be destined for China or Japan. However, specialised handling and storage facilities will require investment.

- In South Africa, one plant is in operation in Sabie, Mpumalanga Province with a capacity of 80,000 t per year, using sawdust and offcuts from surrounding sawmills. The sawmill residues stem from forests that are sustainably managed under Forest Stewardship Council (FSC) certification. The pellets are then transported to Maputo, Mozambique, from where they are shipped to Europe and Japan.

- Australia has recently become a pellet producer and exporter. In May 2009, a manufacturer based in Albany, western Australia, announced that it had signed a supply agreement with utilities in Belgium and the Netherlands, where the first wood pellet manufacturing facility is in operation.

- Japan is the first country in the ASEAN region to utilise wood pellets on a large-scale. Since 2008, wood pellets have been shipped from British Columbia to Japan for co-firing in a coal power plant of a large Japanese utility. South Korea has so far only imported minute quantities from China and Canada, but has very ambitious plans for using wood pellets both for residential heating and power plants, with projected demand reaching 750,000 tonnes in 2012. To secure the pellet supply, South Korea signed contracts with Indonesia and Cambodia to produce wood for pellet production on a scale of 200,000 hectares each.

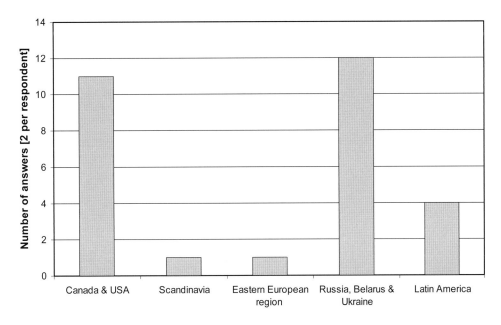

Figure 10.45: Expectations for the main growth in wood pellet production in the coming five years by wood pellet experts.

Explanations: June 2008 [631]

Increasing demand (cf. Figure 10.46) can be expected in the following regions:

- The UK has not been a major importer of wood pellets, but incentives by the UK government to develop renewable power have caused major power companies both to increase co-firing, but also to develop plans for 100% biomass power plants to use a number of different biomass feedstocks. Announced capacities are around 300 MW electrical output, which would equal a demand of more than a million tonnes of wood pellets per year.
- Also, traditional markets for the small-scale use of pellets may continue to grow in the future. In Austria, drivers for increasing consumption may be sustained subsidy programmes for pellet boilers and stoves, linked to special training for pellet boiler installers, and the development of nationwide storage concepts to ensure supply security. In Germany, the per capita wood pellet consumption in 2008 amounted to around 11 kg per person – almost a factor of four lower than the Austrian per capita consumption. Thus, there also seems to be considerable potential for further market growth.

- Japan received its first wood pellet shipment in 2008 from western Canada, and this route is likely to expand in the future.

- China has not been a pellet importer to date. However, due to the huge demands for power in its burgeoning economy, private power producers, which have had difficulty acquiring consistent shipments of coal, are now considering pellet imports, not so much as a renewable energy source, but as an available energy source.

- Finally, while the USA is expected to become an important exporter of wood pellets, domestic demand may also strongly increase, although this could probably be covered largely by domestic capacity.

- Greece is a producer with very little domestic consumption. A strong increase of consumption is not expected. Italy has a well developed pellet market, mainly concentrated on the small-scale sector of pellet stoves. A further increase in this market segment as well as in the field of pellet boilers is expected. Pellet market development in France is in its early stages, with huge potential for further growth.

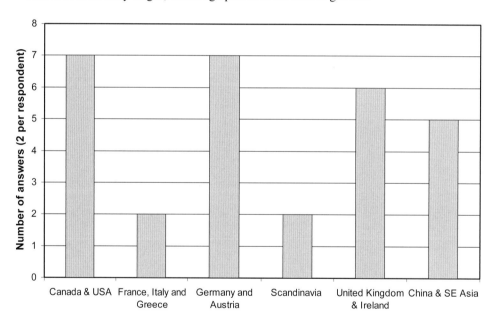

Figure 10.46: Expectations for the main growth in wood pellet demand in the coming five years by wood pellet experts

Explanations: June 2008 [631]

In general, in Europe there is a trend toward utilities converting existing power plants to handle solid biomass (mainly wood pellets) and increasingly building new power plants that are able to handle a wider variety of solid biofuels. A rising price of CO_2 and possibly increasing price of coal could be major drivers to push up the demand for wood pellets in many European countries (and to a certain extent also globally). More drivers and barriers for international wood pellet trade are discussed in the following section.

When it comes to estimating the future market and trade flows for wood pellets, it is mainly a question of what fuels wood pellets can replace. Assuming that roughly 75 million tonnes of fuel oil currently used for heating in Europe were replaced, this would represent a demand for 150 million tonnes of wood pellets. If wood pellets were to be co-fired with (or to fully replace) coal in current electricity plants, this number would probably be far above 150 million tonnes. In addition, with the advent of second generation biofuels, lignocellulosic biomass would be in even higher demand. Thus, in theory there are tremendous growth markets for wood pellets in Europe. [632] demonstrates that by simply extrapolating current growth, varying between 18 and 25% per year, wood pellet demand could be between 130 and 170 million tonnes per year by 2020, which equals about 2.5 EJ, or 0.5% of the current global primary energy consumption.

10.11.6 Opportunities and barriers for international pellet trade

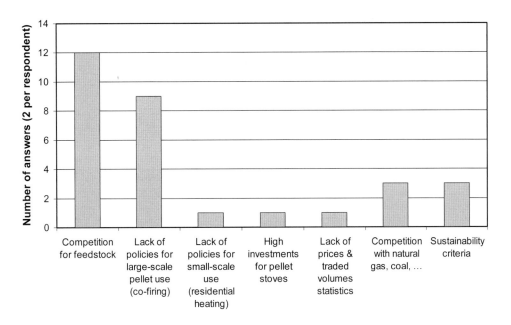

Figure 10.47: Main barriers for international wood pellet trade in the coming five years as stated by wood pellet experts

Explanations: June 2008 [631]

As already briefly illustrated in the previous sections, the international wood pellet trade has experienced exponential growth over the last decade, but it is also facing significant challenges in the near future. In this section, a number of drivers and barriers for the international trade of wood pellets are highlighted. Interestingly, in many cases, factors can both be seen as drivers and barriers, for example oil prices (high or low), financial policy support (sufficient or otherwise), sustainability criteria and certification (guaranteeing sustainable production or administrative hassle and additional costs). The following section is based partially on a workshop in Utrecht, in the Netherlands in June 2008 within the framework of the Pellets@las project, where more than 40 pellet traders, large-scale users and scientists were present, including members of IEA Bioenergy Task 40 on sustainable international bioenergy trade. The participants were asked to fill in a short questionnaire regarding opportunities and barriers to the international pellet trade. An overview of the main drivers and barriers identified are shown in Figure 10.47 and Figure 10.48. Based on the results of the questionnaire, a number of opportunities and barriers for the international wood pellet trade are discussed below.

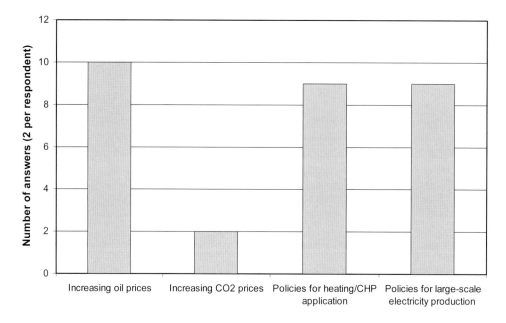

Figure 10.48: Main drivers for international wood pellet trade in the coming five years as stated by wood pellet experts

Explanations: June 2008 [631]

10.11.6.1 Fossil fuel prices

Due to the different end uses of wood pellets, competition occurs with all main fossil energy carriers to replace natural gas and oil for heating purposes, and substitute coal for electricity production. While coal prices had been increasing up until 2008, the price per GJ has always remained below that of wood pellets. Also, the value of avoided greenhouse gas emissions obtained by substituting coal by wood pellets is not likely to be sufficient to allow wood pellets to compete directly with coal for electricity generation, and thus additional policy support will probably be necessary in the coming years. However, with many countries having ambitious renewable electricity targets in place, increasing numbers of power plants with flexible fuel capacities and only limited domestic (biomass) resources, the demand for (and thus import of) wood pellets is likely to increase further over the coming years.

Regarding the replacement of oil, with oil prices peaking around 140 US$ per barrel in 2008 only to drop to about 40 US$ in 2009, one can only guess at how oil prices will develop in the future. However, based on the net calorific value of both fuels, even at low oil prices, wood pellets are often cheaper than heating oil. While investment costs for wood pellet boilers are still considerably higher than those of oil boilers, in some cases (especially when an old boiler needs to be replaced), wood pellet boilers can be economically competitive without subsidies. Especially due to possibly increasing (and likely fluctuating) oil prices, substituting heating oil by pellets is likely to be a trend independent of policy measures.

Finally, the increasing dependence of Central and Eastern European countries on natural gas imports for heating is probably an additional driver for policy makers to support the transition to a more diversified fuel portfolio.

10.11.6.2 Policy support measures

Policy support measures are often the main driver for increasing wood pellet demand, even though the type of policy instrument may vary widely:

- The Netherlands use a system of feed-in premiums for renewable electricity, differentiated by conversion technology and feedstock utilised. While the system has been modified frequently over the past five years, in general, co-firing of clean woody biomass with coal has been economically attractive, leading to large-scale imports of wood pellets since 2002.
- In Belgium, a quota system of green certificates is in place, in which each supplier of electricity has to reach a certain share of renewable electricity. This has led to the (partial and full) conversion of coal power plants to wood pellets. A similar system is operating in the UK, where producers can obtain renewable obligation certificates for electricity produced from biomass. Over the coming years, this may be a strong driver for the import of wood pellets.
- Sweden has been taxing fossil fuel use for heating for a long time, giving biomass fuels an advantage, which has lead (among other factors) to the continuously increasing use of wood pellets for district heating and CHP production. Although the majority of wood pellets are produced and consumed domestically, imports of wood pellets have been continuously growing along with the increasing demand.
- In addition to the promotion of wood pellets for the small- and large-scale production of electricity and heat, in many countries (such as Austria, Belgium, Germany, Italy and the USA), investment grants have been in place to reduce the investment costs of small-scale pellet boilers for residential heating. Especially in northern Italy, this has lead to increasing imports of wood pellets to satisfy demand.

In addition, (failing) policy support is also frequently mentioned as a barrier. Especially when policy support schemes are changed frequently (or cancelled altogether), this can lead to serious market disruptions and price fluctuations.

10.11.6.3 Feedstock availability and costs

As can be seen in Figure 10.47 most participants deemed rising feedstock costs as the most important barrier to near future development. Especially in many Western and Central European countries, there is a limited supply of wood by-products such as sawdust and wood shavings, and as utilization for wood pellet production is increasing, they are becoming increasingly scarce. In times of high demand for wood pellets, the tripling of raw material costs for sawdust has been reported anecdotally. Also, other industries such as the wood panel manufacturing sector are worried that pellets will claim substantial quantities of their raw material supply, and pellet producers will be able to pay higher prices due to the subsidies given for the end use of pellets (e.g. policy support for green electricity production or investment grants for pellet boilers). Another factor contributing to the shortage of woody by-products is the declining demand for timber products in North America. Due to the collapse of the housing market in the USA, less building of wooden houses has led to a downfall in lumber production and resulting sawdust and shavings by-products. The effect is a reduced supply of raw materials for the pellet producers.

These developments may lead, on the one hand, to the utilisation of other by-products, such as bark and reject wood, and other woody biomass of minor economic value such as wood from early thinnings and forestry residues. On the other hand, other feedstocks are increasingly being utilised, such as wood chips and even prime log wood. Especially in south eastern USA, several very large pellet production plants were built during 2008 to utilise, among others, timber from southern pine plantations as feedstock.

Ultimately, increasing demand for wood pellets may become a driver for more international trade and utilisation of so far untapped resources, as the Canadian case has shown (see also Section 10.10 on raw material potentials). Major other regions with abundant raw material could be parts of Latin America (e.g. Brazil), Russia and parts of Sub-Saharan Africa. Efficient logistics will be pivotal to access these resources.

10.11.6.4 Sustainability criteria, certified production and traceable chain management

In recent years, liquid biofuels for transportation have been especially alleged in the media to increase food prices, cause the deforestation of rain forest, and to not lead to actual greenhouse gas (GHG) reductions. Wood pellets have so far been excluded from this discussion since they mainly utilise forest residues and by-products. However, with the increasing utilisation of higher value feedstocks and long-distance trade, issues such as sustainable forest management and overall GHG performance are becoming increasingly important [633; 634]. Following the case of liquid biofuels, it is possible that a guarantee for sustainable production of solid biomass fuels (including wood pellets) will be implemented in the EU, for example. One example of a case where certified production and traceable chain management has already been put in place is GDF-SUEZ/Electrabel in Belgium.

As many other countries, Belgium has ambitious policy objectives to increase the use of renewables and, among others, the country aims at utilising biomass for electricity production. To this end, quotas for renewable electricity production were set. As producing wood for energy is not an objective of Belgian forest policy, forest by-products are becoming increasingly scarce, and it can be expected that Belgium will import significant quantities of biomass to meet the renewable energy targets [635].

Since 2002, the Belgian utility GDF-SUEZ/Electrabel has been carrying out co-firing of different biomass resources in its pulverised coal power plants. In 2005, Electrabel retrofitted two pulverised coal power plants for firing wood pellets instead of coal or for co-firing wood pellets with coal. Rodenhuize power plant generates electricity with coal (70%), wood pellets (25%) and olive cake (5%). Les Awirs power plant has been converted to use 100% wood pellets. The joint pellet consumption of both is about 700,000 tonnes a year [636]. In total, about 15% of the feedstock is expected to originate from Belgium; the rest is shipped to the harbour of Antwerp and from there it is transported on flat boats to the power plants [637].

In order to obtain green certificates for the electricity produced, each supplier of pellets to GDF-SUEZ/Electrabel is required to undergo an audit. Each supply chain is analysed by a local independent inspectorate, and approved by SGS Belgium, the latter being accepted as an independent body by Belgian authorities for the granting of green certificates. First, the energy balance and GHG emissions of the whole supply chain are investigated. SGS checks the sourcing of the wood (hardwood, softwood, sawdust, shavings, coppices) and the transport of feedstock to the pellet plant. If primary feedstocks are used, the entire energy consumption needed for planting, fertilising, harvesting etc. must be taken into consideration. Also, energy consumption during the pellet production process (e.g. electricity for densification and auxiliaries as well as fossil fuel or biomass for drying) and during the final transportation to the sea harbour (train, truck, ship) is taken into account. All energy used in the process is finally subtracted from the number of granted green electricity certificates [638].

Also Belgian authorities require the sustainable character of the forestry resources to be proven. Evidence of sustainability can be delivered according to a traceable chain management system at the supplier's end, and by forest certificates safeguarding sustainability of sources (FSC, PEFC systems or equivalent). In case a delivery is found not to meet the generic sustainability principle, the Belgian regulatory bodies have the right to cancel the granted green certificates.

Finally, it should be stated that in addition to this Belgian initiative, other initiatives have been developed in recent years to certify and guarantee a sustainable solid biomass trade, such as the Green Gold Label [639].

10.11.6.5 Technical requirements for industrial wood pellets

In addition to these sustainability requirements, the certification procedure also informs a potential supplier of wood pellets about all requirements of the utility concerning the technical specifications of the product for firing in a thermal power plant. As an example, specifications for clean biomass pellets (wood based) are shown in Table 10.10).

All this is concentrated in one single document called the "Pellet Supplier Declaration Form". This document is signed by a representative of the producer and is verified and stamped by a certified inspection body (local SGS representative) before being delivered to the Belgian authorities. This certification procedure has been used since 2006, and more than 30 suppliers have already been screened by SGS for the delivery of feedstock, providing a unique view on fossil energy inputs and GHG emissions related to pellet plants located everywhere in the world [638]. Table 10.11 shows CO_2 balances of pellet supply from different countries to a power plant in Belgium. All pellets are transported by large sea and river going vessels to the harbour of Antwerp. There the pellets are transferred to a river barge and then transported to the power plant. Total calculated emissions range from 18 to 32 kg CO_2/MWh_{NCV}, mainly depending on the country of origin (the lower value corresponding to pellets from Germany, the higher to pellets from Canada). Heat for drying is generated mainly from local biomass resources so that drying does not contribute to GHG emissions. Local transport of the wood residues to the pellet production plant is generally estimated to be always less than 2 kg CO_2/MWh_{NCV}.

Practical experience with this system has shown that the procedure is fast. Acceptance of a new supplier by the authorities is obtained within two weeks. The procedure is also relatively inexpensive, as certification costs are typically less than 0.1% of the biomass fuel cost.

Table 10.10: Example of a quality standard for pellets to be used in a large power plant

Explanations: data source [640]

Parameter	Unit	Value
Diameter	mm	4 - 10
Length	mm	10 - 40
Volatile matter	wt.% (d.b.)	> 65
Moisture content	wt.% (w.b.)	< 10
Bulk density	kg/m^3	> 600
NCV	GJ/t (w.b.)	> 16
Ash content	wt.% (d.b.)	< 2
Bark content	wt.% (d.b.)	< 5
Initial melting temperature (red cond)	°C	> 1,200
Cl	wt.% (d.b.)	< 0.03
S	wt.% (d.b.)	< 0.2
N	wt.% (d.b.)	< 0.5
F	ppm	< 70
P	ppm	< 300
Additives: paste (including only additives from vegetal origin), vegetable oils	qualitative	Vegetal origin only
Recycled wood	qualitative	Forbidden
Heavy metals As+Co+Cr+Cu+Mn+Ni+Pb+Sb+V	ppm	< 800
As	mg/kg (d.b.)	< 2
Cd + Ti	mg/kg (d.b.)	< 1
Cr	mg/kg (d.b.)	< 15
Cu	mg/kg (d.b.)	< 20
Hg	mg/kg (d.b.)	< 0.1
Pb	mg/kg (d.b.)	< 20
Zn	mg/kg (d.b.)	< 20
Halogenated organic compounds		
Benzo-a-pyrene	mg/kg (d.b.)	< 0.5
Pentachlorphenol	mg/kg (d.b.)	< 3
Durability	wt.%	94 - 98
Particle size distribution before pellets are milled		
< 4.0 mm	wt.%	100
< 3.0 mm	wt.%	> 99
< 2.0 mm	wt.%	> 95
< 1.5 mm	wt.%	> 75
< 1.0 mm	wt.%	> 50

Table 10.11: CO_2 balance of pellet supply from different countries and regions

Explanations: in kg CO_2/MWh (related to NCV); [1]...by large sea and river going vessels to the harbour of Antwerp; [2]...by river barge to a Belgian power plant; data source [638]

Phase	Germany	Baltic states	Sweden	Russia	Canada
Local transport	1	1	2	2	2
Pelletising	11	13	15	20	13
Sea / river transport[1]	4	6	5	7	15
River transport[2]	2	2	2	2	2
Total	18	22	24	31	32

To sum up, the procedure provides (a minimal level of) guarantees on the traceability and the sustainability of raw material sourcing and illustrates that certification of international wood pellet supply chains is feasible and can offer clear incentives to minimise energy inputs and GHG emissions in supply chains. The Belgian example illustrates that, while in the first instance sustainability criteria and certification requirements can be seen as an additional hassle and barrier to trade, in the long term it may be a necessary and positive measure to ensure and demonstrate the sustainability of wood pellet production, trade and use.

10.11.6.6 Logistics

Biomass often has a low energy density (especially compared to fossil fuels) and a high moisture content (up to 55 wt.% (w.b.)). Wood pellets offer an increased energy density, and improved physical properties compared to many other woody biomass types, which has undoubtedly contributed to their success. However, there remains the fact that wood pellets face significant logistical challenges, for example, they have to be kept dry during loading, unloading and storage. Other issues include fire precautions, absorbance of oxygen during storage of large volumes of pellets and dust control during loading and unloading. To solve some of these issues, the construction of special pellet terminals at major harbours was deemed to be an important step, as was the development of further advanced pre-treatment options such as torrefaction (cf. Figure 10.49).

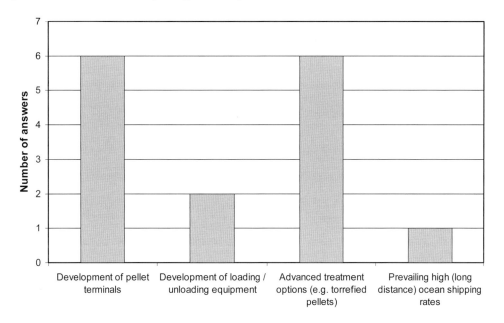

Figure 10.49: Anticipated logistical challenges to be tackled for more efficient wood pellet supply chains as estimated by wood pellet experts

Explanations: June 2008 [631]

10.11.7 Case study of a supply chain of western Canadian (British Columbia) wood pellets to power plants in Western Europe

This case study describes the logistical wood pellet chain and specifications for wood pellets produced in western Canada to be used at an industrial scale in Western Europe. Specific stages and steps were chosen along the supply chain, which relate to the practices and means most commonly used in today's wood pellet export from western Canada to Western Europe. A visualisation of the logistical chain of western Canadian wood pellets to Western Europe is shown in Figure 10.50. The single steps are discussed in the following sections.

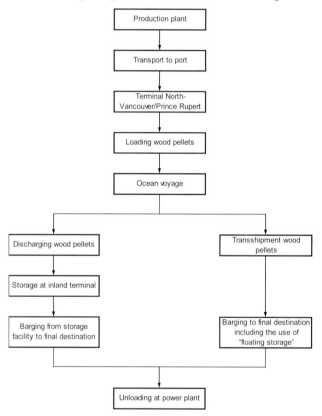

Figure 10.50: Logistical chain of western Canadian wood pellets to Western Europe

10.11.7.1 Production plant

The majority of the wood pellet production plants for this case study are located in the interior of British Columbia, Canada, a region rich in forestry (cf. Figure 10.51). These geographical settings create a vast resource of raw materials to be used in wood pellet production.

Direct access to raw materials in the form of sawmill residues (sawdust, chips and shavings) is a crucial part of the quality management within the pellet production sites today and serves to ensure ongoing stable quality and pellet characteristics, allowing optimum balance between resource utilisation, production capacity and meeting the specific quality parameters as required by customers. However, the dependency on the fibre supply from sawmills is obvious and should be considered as a potential threat.

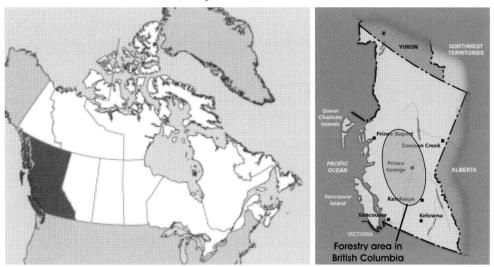

Figure 10.51: Forestry area in British Columbia, Canada

Explanations: data source [641; 642]

In the British Columbia interior, temperature changes substantially throughout the seasons from 40°C during the summer to as low as - 40°C in the winter. A typical British Columbian forest consists of a mixture of softwood species that are used by the lumber and wood pellet industry. Due to climatic conditions in this region, trees grow relatively slowly, which makes the wood fibres dense in composition. As a result, the energy density of pellets is relatively high compared to wood pellets produced in other parts of the world. In addition, the ambient temperature changes throughout the year have significant effects on the energy input needed to dry the raw material in the wood pellet plant and create challenges regarding consistent pellet quality.

Typical British Columbian wood pellet production plants for wood pellet export have annual production capacities of 150,000 to 200,000 t per year (corresponding to capacities of between about 20 and 30 t/h), which seems an appropriate size considering the sourcing of raw materials. Depending on the performance of the plants, actual produced tonnages range from approximately 60 to 95% of the capacity.

Depending on the storage availability at the seaport and frequency of loading operations for export, the pellet production plant has wood pellet storage silos on site. Currently, storage silos on the plant site have a capacity to store several days of production. The wood pellets are retained in the silos until railcars are available to load the cargo.

10.11.7.2 Transport to port

Due to the vastness of British Columbia, the only way to transport wood pellets from the interior to the coast in a feasible way is by using railway. Typically, distances from the pellet production plants to the main ports of British Columbia (Prince Rupert and Vancouver) range from 500 to 1,000 km by rail. Sawmills have been predominantly using the railway to ship their lumber products to seaports and other remote markets and were keen to locate their mills adjacent to the railway system. In the same way, pellet production plants in British Columbia are located directly next to the railway system, which allows them to load railcars directly from their production facility. Typically, railcars in British Columbia have a capacity of 85 to 100 tonnes (cf. Section 4.2.2.5) of wood pellets each and can be lined up in strings of 100 to 120 cars. Thus, up to 8,500 to 12,000 t of wood pellets can be transported per train.

In British Columbia, two major railway companies provide cargo transportation services. Wood pellet producers often have long-term lease contracts with the railway and railcar providers, which secures a certain volume of wood pellets to be transported each month from plant to port. Since the suppliers of both the railway and the railcars are few, a great dependency exists on these railway related companies. Moreover, the capacity of railways can sometimes be at its limits and waiting for railcars can occur, which sometimes results in late delivery of wood pellets to port.

10.11.7.3 Terminal north Vancouver/Prince Rupert

There are two terminals in north Vancouver that have the ability to store and load wood pellets in bulk onto ocean vessels. The total storage capacity in those terminals is currently approximately 42,000 t and silos alone are used. In Prince Rupert, another terminal is handling and storing wood pellets (approximately 10,000 t). In case vessels of large loading volumes come to load, the three terminals in British Columbia are used to drain all the cargo. If the storage silos do not contain enough cargo, loading proceeds directly from railcars.

Railcars are discharged upon delivery to the terminals and most cargo will be stored in the silos for several weeks before being loaded onto an ocean vessel (cf. Figure 10.52).

Figure 10.52: Discharging railcar

Explanations: hopper of railcar opened and sampled; data source [643]

Wood pellets are produced under high pressure and high temperatures. Cooling down the pellets to ambient air temperature is important to stabilise the product. Since the British Columbian climate shows a rather high variability in temperatures, the final product also shows big differences in temperatures throughout the year. Monitoring the temperature of the wood pellets is important to conserve the product's quality, whereby parameters such as ambient temperature, moisture content of pellets and humidity in the air are important factors that may influence the degradation of the pellets.

Especially during hot periods in the summer (30°C and more), the risk of self-heating increases for the pellets stored in the silos (cf. Section 5.2.3). Temperatures above 50°C are undesired or not accepted by the buyers of the product. Above this threshold, temperature dynamics of wood pellets are relatively unpredictable. Both terminals have temperature monitoring devices installed that set off an alarm when critical temperatures are reached (typically between 40 and 50°C).

The terminals apply different techniques to reduce the product's temperature. A ventilation system is used that increases circulation of ambient air into the silos. Another method is to utilise an extensive belt system that runs warm product from one silo through a series of conveyor belts into another silo. Temperature reductions of approximately 10 to 20°C can be achieved by these methods.

By storing the product in dedicated wood pellet silos, the risk of contamination and water damage is limited.

Wood pellets are composed of very small compressed wood particles. Circumstances such as friction, collision and drop impacts may cause breakage of the wood pellets, which creates free particles within the product, called fines. To avoid breakage and dust formation in the wood pellets, gently handling the product at the terminal is vital to maintain the good condition of the wood pellets. Different measures are in place at the three terminals in British Columbia to accomplish this. At transfer points of the conveyor belts, dust suction devices are installed. Also, big plastic tarps are wrapped around transfer points to prevent the wind from blowing the dust particles away. One of the terminals uses a cascade system in the silo and a ship loader spout, which slows down the falling speed of the pellets to reduce the impact of the drop and thus breakage.

10.11.7.4 Loading wood pellets

Before arrangements are made to load wood pellets on a vessel, the suppliers and buyers agree upon a certain quantity of wood pellets, a specified quality and the loading or delivery date. Often these specifications are set out in contracts between the sellers and the buyers. Based on the terms of loading and transport (for instance FOB or CIF) one of the involved parties arranges ocean transport (the charterer) with the carrier (ship owner) who is responsible for providing a suitable vessel. Specifications are written out in a contract between the charterer and the ship owner, also called the "chartering party". In conjunction with a carrier, a vessel is nominated that meets the requirements for loading wood pellets in bulk and a laycan (period to load the vessel according to predetermined schedule) is determined. The laycan sets out a time window in which the vessel is scheduled to arrive at the loadport.

Once ocean transport has been arranged, the supplier of the wood pellets makes sure that sufficient cargo is delivered to the loading terminal on time (prior to arrival of the vessel at loadport), which is often a logistical challenge. Production tonnages at the pellet production plants need to be streamlined with railway availability and storage capacities at the terminals. Furthermore, other competing vessels might be scheduled to load around the same time at the same terminal, which causes a line up of vessels (mostly multi-product terminals face this problem). Sometimes other vessels for other commodities have priority, while some terminals have the policy of first come first served.

Before loading of an ocean vessel commences, a dedicated surveyor checks the holds of the vessel to certify their cleanliness. A hose-test is performed on some occasions to make sure that the seals of the hatches are watertight before the vessel's voyage. This is done by means of spraying high pressured water on the seals with a hose from the outside and after that assessing whether water has leaked inside the hold. In addition, in case there is no accurate terminal belt scale (scale to determine the weight of the cargo on the conveyor belt) available, an initial draft survey is performed in order to determine the loaded quantity upon completion of loading on the basis of the buoyancy of the vessel (Archimedes Law).

Furthermore, the surveyor checks the entire supply line on the loading site and examines whether everything is suitable for loading wood pellets, from dumping railcars or silos to the spout of the ship loader. To avoid any contamination, conveyor belts should be clean and free of water and no other products should be present while running the conveyor belts.

Loading wood pellets onto vessels is a delicate process. Once the wood pellets are on the conveyor belt at full speed it is very hard to stop quickly when something is wrong with the cargo. Surveyors monitor most wood pellet shipments continuously and look for anomalies (mainly temperature increases, dust formation, deterioration of pellets, moisture in pellets, colour) in the wood pellet quality and make sure that wood pellets are loaded under the correct conditions. Samples are taken at predetermined tonnage intervals (e.g. one sample each 25 to 50 tonnes), in order to analyse the apparent quality on site (above stated parameters) and get more detailed specifications in an official laboratory. Currently, Panamax vessels are loaded partially with up to 35,000 t of wood pellets and full dedicated vessels are loaded with up to 47,000 t ("full dedicated vessels" are vessels where each hold is loaded with wood pellets, whereas partial load means that some holds are loaded with wood pellets and other holds with other products).

Nominal loading speeds at the Vancouver terminals range from an average of 600 to 1,500 t/h depending on the continuity of the loading operations. In Prince Rupert, loading speeds can reach up to 2,500 t/h. Typical causes of loading delay are weather (precipitation, wind) and technical problems at the terminal, such as belt problems, terminal scale problems (i.e. problems with reading the correct weight from the belt scale, which is used by terminals to determine the loaded quantity) and problems with the dust collector (at several points of the loading line dust loaded air is discharged and the dust is then separated from the air in a filter or cyclone; the separated dust is disposed of in a landfill). Loading during rain must be avoided in order to prevent the wood pellets from disintegrating during the voyage. In the Vancouver and Prince Rupert area, rain is very common in autumn and winter, which can sometimes cause idle times of several days.

Breakage of the pellets can be noticed predominantly when the pellets come out of the spout and drop into the vessel's hold. Both terminals have a periscope spout that in part which reduces the free fall of the wood pellets. One terminal has a cascading system integrated into the spout, which slows down the speed of the falling wood pellets into the vessel's hold and achieves the best result in reducing breakage during the loading process (cf. Figure 10.53).

Upon completion of loading, the hatches are sealed and a final draft survey is performed resulting in a certified quantity agreement, which is reflected in the Bill of Lading.

Figure 10.53: Cascading spout in vessel's hold

Explanations: data source [643]

10.11.7.5 Ocean voyage

The shipment time from British Columbian load ports to northwestern Europe takes approximately four to six weeks and the route leads through the Panama Canal over a distance of approximately 17,000 km. This shipment time depends on weather conditions and whether the vessel will call at multiple ports to load or unload parts of the pellets or in case of partial cargo, load or unload of other products on its route.

During the ocean transport, vessels can face some rough weather that can potentially cause seawater to seep into the holds. This salt seawater contains chlorine and may cause problems down the supply chain in the power plant since it enhances corrosion of metals. Furthermore, self-heating (cf. Section 5.2.3) can lead to dangerous situations when warm wood pellets have the opportunity to heat even more over four to six weeks. It is not uncommon to find 10 to 20°C increases of wood pellet temperatures after the ocean voyage from Vancouver to the Netherlands.

The shipping costs for wood pellets depend on the contract specifications with the ship owners and can therefore not be stated generally.

10.11.7.6 Discharging wood pellets

After the ocean voyage, the vessel arrives at one of the ports of the VARAGT (Vlissingen/Amsterdam/Rotterdam/Antwerp/Gent/Terneuzen) region. There are two main pathways the wood pellets go (cf. Figure 10.50). The first is from the ocean vessel into storage, while the second option is from the ocean vessel directly into a barge for further transport to the power plant, including the use of the barge as "floating storage".

The discharge of wood pellets is done by clam buckets (cf. Figure 4.34) that feed the pellets onto a conveyor belt system. The conveyor belts transport the wood pellets to the indoor storage facility.

Surveyors continuously check the discharging operations and make sure that things are as expected. Often a vessel draft survey is performed to determine the discharged tonnage.

Again, whilst discharging the wood pellets it is important to ensure the right conditions are met, especially the absence of precipitation and gentle handling to reduce the amount of breakage and dust formation. Wood pellet dust can cause equipment to fail and may cause complaints from neighbouring businesses or environmental groups.

10.11.7.7 Transhipment of wood pellets

The transhipment from an ocean vessel directly onto a barge is performed by stevedores either using floating cranes that are positioned in between the vessel and the barge or using terminal equipment if the vessel is berthed on a quay. In case of the utilisation of floating cranes, barges are gauged (reading draft) before and after loading in order to determine the loaded tonnage per barge. Typical tonnages per barge range from 1,500 to 2,000 t. A typical barge used for pellet transport is shown in Figure 10.54.

Figure 10.54: Typical barge used for pellet transport

Explanation: data source [643]

Discharging speeds from the vessel with clam buckets are in the range of 500 to 1,000 t/h and two cranes may be used at the same time to speed up this process. Risks are involved when heavy winds make the cranes move and therefore, beyond a certain wind speed, operations are ceased.

10.11.7.8 Storage at inland terminal (in VARAGT zone/Western Europe)

The current storage capacity for wood pellets in the VARAGT zone is well over 80,000 t, however, plans are being made to increase the capacity and in this way wood pellets are becoming a more flexible commodity to use and trade.

When pellets are stored, temperature monitoring is one of the main tasks. In the past, several fire incidents in storage facilities have made people very aware of the apparent risk of storing wood pellets. Temperature monitoring is performed in several storage facilities by a wireless system that allows clients to see the temperature status of their cargo in real time through the internet.

10.11.7.9 Barging to final destination

Once the wood pellets are loaded onto the barge, the voyage leads to the power plant (typical distances 75 to 125 km). There, the barges are usually kept next to the power plant as a floating storage until the wood pellets are needed.

Barging is common practice in Western Europe and the distances to the power plants are relatively small. However, during very dry periods, water levels of the rivers can drop substantially, which can cause problems with draft. In order to resolve this problem, the barges are not loaded fully.

10.11.7.10   Unloading at power plant

The barges are unloaded by crane with clam bucket and the wood pellets are put onto the conveyor belts for transport to the burners. Air borne dust during loading the wood pellets onto the conveyor belts and further into the power plant's processes can automatically shut down the system. Therefore, dust and breakage of the wood pellets should be prevented as much as possible throughout the supply chain.

10.12 Socio-economic aspects of pellet production and utilisation

Socio-economic impact studies are commonly used to evaluate the local, regional and/or national impacts of implementing particular development decisions. Typically, these impacts are measured in terms of economic variables, such as employment, revenue and taxes, but a complete analysis must also include social, cultural and environmental aspects. These last three elements are not always suitable for quantitative analysis and, therefore, were precluded from many impact assessments in the past, even though at the local level they may be very significant. In reality, local socio-economic impacts are diverse and will differ according to such factors as the nature of the technology, local economic structures, social profiles and production processes.

The use of wood pellets as fuel in all areas of energy generation from domestic stoves and boilers to co-firing in thermal power plants has been an amazing success story over the past 20 years. Pellets made of wood waste were actually first produced in the late 1970s in the USA and remained a small niche market for two decades before rapid market development started in Europe. Today it is the most advanced and a widely used biomass fuel. This development was caused by a number of socio-economic factors and has triggered a set of interesting socio-economic effects.

In many ways, the social implications arising from local pellet production or any activity concerned with bioenergy represent the least clear and least concrete output of impact studies. Nevertheless they can be divided into two categories, namely those relating to an increased standard of living and those that contribute to increased social cohesion and stability. In economic terms, "standard of living" refers to a household's consumption level, or its level of income. However, other factors contribute to a person's well-being, which may have no immediate economic value. These include such factors as education, the surrounding environment and healthcare, and, accordingly, they should be given consideration. Moreover, the introduction of an employment and income generating source, such as bioenergy production, could help to counteract adverse social and cohesion trends (e.g. high levels of unemployment, rural depopulation, etc.). Rural areas in some countries are suffering from significant levels of outward migration, which has a negative impact on population stability. Consequently, given bioenergy's propensity for rural locations, the erection of pellet production plants may have positive effects on rural labour markets by, first, introducing direct employment and, second, by supporting related industries and the employment therein (e.g. forestry).

Similarly, by securing a heat and power supply system based on domestic resources, exposure to international fuel price fluctuations is minimised, thus the risk of rising costs of production, transport, etc., is reduced.

The issue of security of energy supply has become very important in European countries within the last few years and moved into particular focus through the natural gas crisis and the Russia-Ukraine dispute in the winter of 2008/2009. In that regard, the increased use of pellets that exhibits a broad geographical distribution could secure long-term access to energy supplies at relatively constant costs.

Moreover, the use of domestic resources implies that much of the expenditure on energy provision is retained locally and recirculated within the local/regional economy. However, it is also important to take into consideration that the increased use of pellets for electricity production and the corresponding increase in demand for pellets could cause temporary shortages of supply during periods of high demand. Households are particularly vulnerable in this respect.

The nature and extent of any particular pellet production plant's socio-economic impact will depend on a number of factors including the level and nature of capital investment, the availability of local goods and services, the degree to which money can be kept in the region rather than being spent outside the region, the time scale of plant construction and many others. Also, pellets are a product of large-scale international trade making the whole picture even more complex. This adds issues such as international fair trade, macroeconomic aspects and trade balance to the overall socio-economic analysis.

A summary of the socio-economic aspects associated with local pellet production and utilisation is listed in Table 10.12 [644].

In order to more precisely define the importance of each factor noted in the table and possibly quantify its impact, case studies and analyses for each geographical situation (local, regional, national, international) must be carried out.

There are a variety of approaches and methodologies used to integrate socio-economic criteria in the overall assessment framework of bioenergy use. A commonly used methodology is multi criteria analysis (MCA), which has been widely applied in the bioenergy related fields over the past 15 years. Generally, MCA is concerned with the establishment of an adequate framework for the evaluation of a specific project by considering a number of different factors. These factors comprise technical, economic, social and environmental criteria and MCA is typically used to compare several different project options (for example using renewable or conventional energy sources to meet energy demand). However, the specific techniques and tools to apply the MCA methodology are quite varied and, based on the selected tool, different results can be obtained [645].

Table 10.12: General socio-economic aspects associated with local pellet production and utilisation

Dimension	Aspect

Social aspects	• Increased standard of living
	   ○ Environment
	   ○ Health
	   ○ Education
	• Social cohesion and stability
	   ○ Migration effects (mitigating rural depopulation)
	   ○ Regional development
	   ○ Rural diversification
	   ○ Poverty reduction

Macro level	• Security of supply / risk diversification • Regional growth • Reduced regional trade balance deficit • Export potential
Supply side	• Increased productivity • Enhanced competitiveness • Labour and population mobility (induced effects) • Improved infrastructure
Demand side	• Employment • Income and wealth • Induced investment • Support of related industries

[646] suggest a method to account for the social dimension of projects. They suggest a semi-quantitative approach based on stakeholder involvement to assess eight social criteria such as societal product benefit and social dialogue, along with others such as those noted above in Table 10.12.

According to von Geibler's [646] theme of societal product benefit, assessments can also include benefits beyond local boundaries such as contribution to GHG reduction or helping to meet a nation's international commitments. Events beyond one's boundary can have a profound effect on other local economies. For example, the rapid growth of the Canadian pellet market has been fuelled by European efforts to decrease GHGs. Tracking national and even international benefits associated with pellet production can help to obtain government assistance that can be crucial in starting a new project in an undeveloped or underdeveloped market.

Other impact assessment approaches include the development of individual tools and methodologies focused on the assessment of socio-economic interaction with specific aspects of bioenergy such as biodiversity [647], changes in rural land use [648] and others.

The assessment and quantification of the macroeconomic impacts in terms of GDP, trade balance and employment of large-scale pellet production can be carried out by means of various economic models. For example, a model based on the input–output or computable general equilibrium (CGE) methodology can be developed to evaluate the direct, indirect and induced macroeconomic impacts of pellet production [649].

While the cost structure of pellets appears stable at present, impact analyses should examine the potential effects of price changes in key input materials such as wood residues and in competitive fuels such as oil, natural gas and electricity. While residues are inexpensive at present, as the pellet market expands or as other uses are found for residues, the price could change dramatically. This has occurred with cooking oil, which went from being a waste product to a valuable commodity. One must also assess long-term supply. For example, beetle killed timber in British Columbia, Canada, is a raw material for pellet production for the European market. How long will it be available and what will be the effect on the pellet market when it is no longer available? Prices for oil and natural gas have fallen well below their record highs; what effects does this have on the competitive position of pellets?

The common feature of all methodologies that take socio-economic aspects into account in the overall framework of pellet production and utilisation is that obtaining extensive feedback from local stakeholders, usually by organising several workshops, round tables and similar meetings at every phase of the project, is of great relevance. It is often crucial because basic economic information is often not available from national statistics agencies. This information can then be structured into appropriate assessment criteria that can be used to analyse the potential impacts but also to estimate the bioenergy potential from a socio-economic point of view.

A variety of tools exists to measure these impacts ranging from simple cost–benefit analysis to CGE modelling and MCA, among others. Choosing the appropriate tool is critical to carry out relevant analysis.

From the development of several national pellet markets (cf. Sections 10.1 to 10.9), the following critical socio-economic factors in developing pellet markets can be identified:

- Financial incentives for investing in wood pellet heating rapidly increase uptake, even when pellets are already competitive with alternative fuels. This is because pellet technologies and markets are immature (new), so increased upfront investment is required to ensure their further development and ability to compete with other proven technologies.

- The existence of a strong sawmill industry is important to provide, at least initially, a low cost and readily available source of raw material. As the pellet industry continues to grow, competitive uses for sawmill residues could have a profound effect on the cost structure of the pellet industry.

- Stringent quality and sustainability requirements for pellet boilers with regard to emissions, efficiency and safe operation should be imposed as low quality boilers can permanently damage the market, cause environmental concerns to become an issue and cause major operational problems.

- Effective quality control mechanisms for wood pellets should be established. National or international tracking systems that allow identification of the origin of pellets should be introduced. This could include certification that the pellets are made of wood from sustainable forestry.

- Installers of pellet heating systems should be qualified and, if possible, certified. Installers have a decisive influence on consumer confidence and must be qualified in order to ensure trouble free operation

- Quality requirements for boilers and certification of installers should be linked to subsidies – this is very efficient in driving development in the right direction.

- Publicly supported promotion campaigns are recommended. This is particularly important in the early stages when industries are unlikely to have the required capital.
- Wood pellet heating systems should be installed in public buildings in order to demonstrate their applicability and to act as an example.
- Incentives for utilities to enter the biomass heating market should be developed. For example, a possible measure could be to allow green electricity obligations to be satisfied by certified green heat deliveries. In this way, utilities would build up significant interest in offering green heat services to reduce their green electricity obligations.

A complete impact analysis must investigate the effects on all market actors in the economy. Will consumers, be they households or firms, embrace a relatively new technology such as pellets? They will require sufficient information to make an informed decision. Can firms produce pellets profitably over the long term? Are the benefits to society sufficient to get governments involved in providing subsidies or other forms of assistance to consumers and producers in the short or long term? Knowing the level of commitment of the various levels of government to support the pellet industry is essential in conducting an accurate assessment of social and economic impacts.

These questions indicate that, in addition to geographically defined factors such as the location of existing wood industries, the availability of material and the price of alternative fuels, less quantitative factors such as political motivation and development of novel support mechanisms are equally important.

Strong growth can be expected, with political support at the EU level playing a major role in the extension of the pellet industry, especially in new member states. The ambitious EU target of achieving 20% of energy supply from renewable energy by the end of 2020 is impossible without appropriate policies. In addition, the on-going oil price fluctuations and carbon dioxide reduction targets also encourage the expansion of the pellet markets.

10.13 Summary/conclusions

The first steps to introduce pellets as a biological fuel were undertaken at the beginning of the 1980s. Since the second half of the 1990s, the pellet markets in a number of European countries as well as North America and even worldwide have exhibited rapid growth and there is no end to this development in sight. Several factors were crucial for this development. The cornerstone was the automation of furnaces, which created similar user comfort to that previously only possible with gas or oil heating systems. In addition, national funding schemes, price rises in the oil and gas sector and marketing and public information campaigns by both national and international pellet and biomass associations contributed to the success of pellets.

Continuous expansion of production capacities has gone in tandem with market development. In order to make use of synergies, pellet production plants are preferably located at existing production sites of the wood industry. Manufacturers of pellet furnaces (especially small-scale furnaces) have also expanded their capacities so as to meet increasing demand [650; 651].

What is interesting is the different development of different markets. Whereas pellet use is limited to small-scale applications in some countries (e.g. Austria, Germany and Italy), it is large-scale plants for the most part that are fired with pellets in other countries (e.g. Belgium and the Netherlands). Moreover, in countries such as Sweden or Denmark, pellet utilisation takes place in small-, medium- and large-scale applications whereas other countries solely produce large quantities of pellets but have no or negligible domestic markets (e.g. Canada and some Eastern European countries). Worldwide, around 11 to 12 million tonnes of pellets are used (basis 2008/2009), of which around 65% are applied in small-scale systems and 35% in power plants and other medium- and large-scale applications. Pellet requirements vary in the different countries. With regard to large-scale applications, pellets are produced for the sole purpose of reducing transport and storage costs. In part, the pellets are in most cases even ground again before firing. Quality plays a subordinate role in these applications. Large plants for instance can manage a higher fuel ash content, which renders the use of other raw materials in pelletisation, such as bark or straw, possible. So, raw material potential is broadened. Pellets that are used in small-scale applications must be of superior quality (especially concerning durability and purity) in order to safeguard high user comfort and operational reliability of systems.

The largest pellet producers worldwide are the USA, Canada, Sweden, Germany and Russia. They all have annual productions above 1 million tonnes per year and together produce about 9.2 million tonnes (2009). This represents about two thirds of worldwide production.

The largest pellet consumers are Sweden, the USA, Italy, Germany, Denmark and the Netherlands, all of them with consumption above 1 million tonnes per year. Together they consume 8.4 million tonnes of pellets per year (2009), which is about 75% of worldwide pellet consumption.

There are activities concerned with pellet production and/or use going on in many other countries worldwide. Regarding the use of pellets, China (with the ambitious goal of 50 million t (w.b.)$_p$/a in 2020), Japan, Korea and New Zealand, all with nascent pellet markets, should be mentioned. South Africa, Brazil, Argentina and Chile produce pellets with the main purpose of exporting to Europe. Activities are reported Turkey and Mongolia.

If all pellet production of the different countries is totalled (cf. Sections 10.1 to 10.9), pellet production of Europe amounts to around 8.1 million t (w.b.)$_p$/a and 14.2 million t (w.b.)$_p$/a are produced worldwide (2007 to 2009). Total consumption is around 8.9 million t (w.b.)$_p$/a in Europe (without Russia) and about 11.3 million t (w.b.)$_p$/a worldwide. The gap of 2.9 million t (w.b.)$_p$/a between production and consumption might be explained by the inaccurateness of some production and consumption data, as they are often based on rough estimations due to the lack of exact data. Moreover, inconsistent data might occur due to the fact that production and consumption data are sometimes based on different years. Prognoses for worldwide pellet production in 2020 are between 130 and 170 million tonnes per year.

Specific pellet consumption in tonnes per 1,000 capita is shown for the different countries in Figure 10.55. On the basis of this figure, Sweden and Denmark are the largest pellet consumers, whereby in these countries the pellets are used in small-, medium- and large-scale applications. They are followed by Belgium, Austria and the Netherlands. In Belgium and the Netherlands, pellets are fired almost exclusively in large-scale power plants. Austria proves to have the fourth largest pellet consumption per capita, whereby pellets are almost exclusively used for residential heating. Looking at the rest of the countries that have very low pellet consumption per capita, Finland has the highest consumption per capita with 28 t (w.b.)$_p$/1,000 inhabitants.

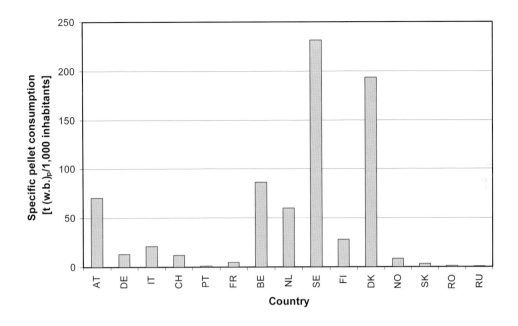

Figure 10.55: Specific pellet consumption in different countries

Explanations: base year depending on available data either 2007, 2008 or 2009; data according to Sections 10.1 to 10.9

The strong growth in pellet markets worldwide requires consideration concerning pellet production potential. Appropriate raw materials must first be available and second, production sites have to be erected. Evaluations at European and global scales showed that there are regions in Europe with an already high concentration of pellet production plants with limited potential for further plants based on sawdust and wood shavings as raw materials, such as in Austria, Bavaria and middle Sweden. Other areas still have a low density of pellet production plants, such as the UK, northern Germany, western Poland, France and Spain, which consequently have higher potential for further plants with wood shavings and sawdust as raw materials. Concerning raw material potential for pellet production in Europe, there are regions in Europe where a shortage of sawdust, still the most important raw material for pellet production, has already occurred, but there are regions where a further increase of pellet production based on sawdust is still possible, for instance in Germany, Finland and Poland. In addition to sawdust, alternative raw materials such as forest residues, log wood or even energy crops have great potential in many countries worldwide and are already used.

Over the past decade, the growth of wood pellet supply and demand has been closely linked to international trade and, as a rough estimate, between one third and one half of all wood pellets consumed are traded over an international border. This can vary from short distance trade (e.g. from Austria to Italy) by truck to long distance trade of more than 10,000 kilometres (from British Columbia to Japan or northwestern Europe) by vessel. Transport by train is almost unknown in Europe but common practice in North America (e.g. in British Columbia, Canada and the USA). The largest part of international wood pellet trade is carried out by means of ocean vessels. The first intercontinental shipment occurred in 1998 from Canada via the Panama Canal to Sweden and intercontinental trade reached 1.2 million tonnes in 2007. A case study describing the logistical route of wood pellets produced in western Canada to be used at industrial scale in Western Europe was shown as an example. Specific stages and steps are chosen throughout the supply chain, which refer to the practices and means most commonly used in today's wood pellet export from western Canada to Western Europe. Each step encompasses challenges that relate to the quality of the product as well as to the cooperation of involved parties. Many predictable and unpredictable factors play a role, which may cause potential risks for the owner of the cargo. Through planning and constant supervision, many significant problems can be avoided, which makes wood pellet trading from western Canada to Western Europe feasible.

Pellets are either imported or exported from almost each country in the EU and beyond. Trade routes from outside to the EU mainly come from Canada, the USA, South Africa and Russia. North America, north western Russia, Chile, South Africa and Australia are countries and regions that are expected to increase their pellet exports in the near future. Increasing demands are expected in new and emerging markets such as the UK, Japan, China or France as well as in traditional markets such as Austria, Germany or the USA.

A typical freight rate for a pellet shipment has been calculated to be 31.7 €/t (w.b.)$_p$, calculated for a 22,000 t bulk transport from the Far East to Europe. Taking a wood pellet price of approximately 130 €/t (w.b.)$_p$ into account (CIF ARA), shipment accounts for about 25% of the price (the remaining 75% contains pellet production including raw material and local transport from the pellet production plant to the port as well as margins). However, charter rates for ocean vessels vary considerably. Prices dropped dramatically by 92 to 99% between late 2007/early 2008 and the end of 2008, which makes shipment currently relatively cheap.

The environmental impact of pellet transport by ocean vessels expressed in CO_2 emissions per MWh pellets can increase by up to 78%, if pellets are imported from Canada to Europe (from 18 to 32 kg CO_2/MWh$_{NCV}$). However, it also has to be emphasised that even with long-distance transport, the avoided emissions when replacing, for example, coal for electricity are typically substantial and around or above 90% [652].

Several opportunities and barriers for international pellet trade have been identified such as fossil fuel prices, policy support measures, raw material availability and costs, sustainability and technical requirements. Finally, logistical challenges such as the development of pellet terminals, loading and unloading equipment, advanced treatment options such as torrefaction or shipping rates will also play a significant role.

With the increasing scarcity of residue feedstocks such as bark and sawdust in Europe, large-scale sourcing of raw material from special plantations is already in place (e.g. on the east coast of the USA). This implies that in order to guarantee a sustainable production of pellets, additional measures may have to be implemented. One possible solution could be robust certification schemes, such as the Belgian Pellet Supplier Declaration Form. How fast this shift to new and internationally sourced feedstocks will develop is highly uncertain. During the wood energy workshop in Utrecht, in the Netherlands in June 2008 within the framework of the Pellets@las project, the high oil price was seen as a major driver for wood pellet trade, while the intercontinental trade was suffering from high dry bulk shipping rates. By the end of 2008, however, oil prices had fallen to levels lower than 40 US$/barrel, but also the Baltic Exchange Dry Index had fallen by more than 90%. How this (and the general economic crisis) will affect global pellet markets in the years to come is yet to be seen. However, it seems certain that with increasing production of and demand for wood pellets, the international trade will continue to flourish.

In order to evaluate local, regional and national impacts of pellet utilisation, socio-economic impact studies can be used, where economic variables (e.g. employment, revenues and taxes) as well as social (e.g. increased standard of living, social cohesion and stability), cultural and environmental aspects are measured. A variety of approaches and methodologies exist to integrate socio-economic criteria in the overall assessment framework of bioenergy use (e.g. MCA). Choosing the appropriate tool is critical to carry out robust analysis. The common conclusion of all methodologies that take socio-economic aspects into account in the overall framework of pellet production and utilisation is that obtaining extensive feedback from local stakeholders is of great relevance. This information can then be structured into appropriate assessment criteria that can be used to analyse the potential impacts of pellet production and utilisation but also to estimate the bioenergy potential from a socio-economic point of view. Financial incentives, the existence of a strong sawmill industry, stringent quality and sustainability requirements both for pellets and for pellet heating systems, qualified and certified installers of pellet heating systems, promotion campaigns, installation of pellet heating systems in public buildings and incentives for utilities to enter the biomass heating market were identified as crucial socio-economic factors of pellet market development.

11 Case studies for the use of pellets for energy generation

In this chapter examples of existing plants where pellets are used for energy generation in all fields of applications are presented. The applications include simple pellet stoves for room heating with thermal outputs of a few kilowatts and annual pellet consumptions of some 100 kg up to large power plants with thermal and electric outputs in the range of several megawatts and annual consumptions of over 100,000 t of pellets. The aim of this chapter is to provide an overview of the different possibilities for thermal utilisation of pellets by means of interesting and representative case studies.

For all case studies, plant descriptions, technical data and economic information are provided. Moreover, all plant descriptions comprise name and location of the plant, detailed functional description, descriptions of important components such as furnace and boiler, storage facilities, fuel feeding system, de-ashing, flue gas cleaning and control system. As far as available, information concerning emission limits and actual emissions of particulate matter, CO, NO_x and hydrocarbons is also provided. The starting year of operation and operating hours already achieved are also indicated and pictures of the plants are presented.

The technical data are given not only for the pellet unit but also for medium and/or peak load units and the district or process heating network (as far as such units are applied).

The economic data comprise at least information on investment costs and annual fuel costs. If available, additional economic data concerning investment funding, costs of other fuels used for peak load coverage as well as consumption, operating, maintenance and other costs are presented.

11.1 Case study 1 – small-scale application: pellet stove (Germany)

11.1.1 Plant description

The pellet stove (cf. Figure 11.1) of a private homeowner is located in Straubing, Germany. It was put into operation in November 2007 and it is the sole heating system of a small, old single family house without central heating system in an urban area. The stove is placed in the hallway between kitchen and living room and thus provides most heat to these two rooms. Due to very low insulation levels, the rest of the house can only be kept free of frost by opening the door to the living room.

The pellets that are purchased in bags are stored in the garden shed. Before each heating period, fuel for the whole period is stored, which means 65 bags or 1,000 kg. The pellet reservoir of the stove is filled by hand every day or every two days. The pellet reservoir can hold 16 kg; fuel consumption is 1.8 kg/h at the most.

The pellet stove (without water jacket) is a product of the Austrian manufacturer RIKA, type Memo, with a nominal thermal capacity of 8 kW, which is solely released to the room where it stands. The power output is controlled by a thermostat in a continuous way. The pellets are taken from the pellet reservoir by a conveyor system with feeding screw and fed into the burner according to demand. Combustion is started automatically by a lighting-up cartridge, whereby the phase of ignition takes about five to eight minutes. Switching off the stove induces a burnout phase where the pellets that are still in the retort are combusted as fast as possible by raised air supply.

The supply of combustion air from the room to be heated to the burner is controlled by an air flow sensor (primary air). In addition, combustion is optimised by further air supply to the combustion chamber (secondary air). This air flow is led directly alongside the front window and thus safeguards a clean glass surface. The flue gases are led through a 3 m long flue gas tube into the chimney with the support of a fan.

During the heating period, the (cooled down) retort has to be cleaned every day with a normal hoover in order to free the air nozzles from ash and clinker and prevent back burning. The flue gas ducts, the flue gas collector and the suction fan are cleaned twice a year; the pellet reservoir is cleaned every now and then.

Figure 11.1: Pellet stove (8 kW) in the living room

11.1.2 Technical data

Table 11.1: Technical data of the pellet stove in Straubing

Explanations: [1]...according to type test, expected to be a bit lower in field operation

Parameter	Unit	Value
Fuel power input (nominal conditions)	kW_{NCV}	8.5
Pellet consumption at nominal load	kg/h	1.8
Annual fuel demand	kg/a	900
Nominal thermal capacity	kW_{th}	8.0
Minimum thermal output	kW_{th}	2.0
Thermal efficiency at nominal load[1]	%	94.5
Storage capacity	kg	16
Storage capacity at nominal load	h	8.7

The power output of the pellet stove is controllable in steps of 5%, whereby the heat is released solely to the room surrounding the stove. The annual pellet demand is around 900 kg, which is equivalent to 1.4 m³ when a bulk density of 650 kg/m³ is considered. The most important technical data of the stove are summarised in Table 11.1.

11.1.3 Economy

The pellet stove was subsidised according to the guidelines of the market incentive programme for renewable energy (MAP) by the federal office of economics and export control (BAFA) with 1,500 €, which is equivalent to nearly 58% subsidies at a system price of 2,590 €.

Since the described pellet stove is the only heating system of the house but does not heat the entire house to a full extent, the economy of the system cannot be determined in a detailed way but has to be estimated. Hereby, 200 heating days and 5 h operation per day at 50% part load were assumed, resulting in 500 annual full load operating hours.

With a utilisation period of 20 years for the technical equipment and an interest rate of 6% p.a., capital costs of 106.8 € p.a. result (with subsidies taken into account). Operating costs that consist of costs of maintenance and chimney sweep amount to about 61 € p.a. With an average fuel price of 214 €/t for wood pellets (incl. VAT) and a fuel input of 4,140 kWh/a, consumption costs are 192.6 €/a. So specific costs of 87.1 €/MWh$_{NCV}$ result. Due to the lack of annual efficiency data for the pellet stove, it is not possible to quantify the specific costs related to useful energy. Table 11.2 provides an overview of economic data of the pellet stove.

Table 11.2: Economic data of the pellet stove in Straubing

Explanations: all prices incl. VAT; [1)]...due to the lack of annual efficiency data for the pellet stove, the specific costs are related to the annual fuel power input

Parameter	Unit	Value
Interest rate	% p.a.	6
Utilisation period	a	20
Annual fuel power input	kWh/a	4,140
Net investment costs	€	1,225
Pellet stove	€	2,590
Chinmey connection	€	135
Funding	€	1,500
Operating costs	€/a	61.0
Consumption costs	€/a	192.6
Capital costs	€/a	106.8
Total annual costs	€/a	360.4
Specific costs [1)]	€/MWh$_{NCV}$	87.1

11.2 Case study 2 – small-scale application: pellet central heating (Austria)

11.2.1 Plant description

This pellet central heating plant is located in the village of St. Lorenzen/Mürztal in Styria, Austria. It is installed in a detached house built in 1949, extended in 1970 and renovated from 2005 to 2007. Previously, the house was heated by two stoves, one on each floor, fired with coal and firewood. In 1988, an oil central heating system was installed which was replaced by the pellet central heating system during the renovation from 2005 to 2007.

The renovated detached house is mainly equipped with wall and some floor heating surfaces. Radiators are not installed. Therefore, the feed temperature is very low.

The plant was put to operation in autumn 2006 when the house was still a construction site. Renovation was completed in 2007 and since then the house has been inhabited again.

A picture of the pellet furnace is shown in Figure 11.2. It is placed in the cellar of the house in a separate boiler room. The storage room is directly behind the wall at the back of the picture.

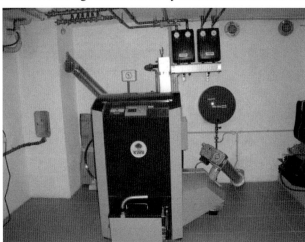

Figure 11.2: Pellet central heating system in St. Lorenzen/Mürztal, Austria

Pellets are fed from the storage room with a conventional screw conveyor to the furnace. The difference in height between storage room and furnace is overcome with a cardan joint in the feeding screw where the feeding direction is changed upwards.

The ash from pellet combustion in the retort falls into an ash collecting space directly below the burner from where the ash is transported with a screw conveyor into an external ash box, which can be seen in front of the boiler in Figure 11.2. This happens once a day at the same time when the automatic heat exchanger cleaning system is in operation. The automatic heat exchanger cleaning system is based on spiral scrapers in the boiler tubes, which remove deposits. The ash falls into the ash collecting space below and is discharged together with the ash from the retort to the external ash box. The external ash box is dimensioned in such a way that emptying is necessary only once a year. Its volume is 33 l. It is equipped with wheels so that it can easily be moved. The ash is used as a fertilising and liming agent in the garden of the home owner.

Due to the fact that the furnace is able to operate down to 26% part load, a heat buffer storage was not installed. For hot water supply, a hot water boiler with a volume of 300 l was installed.

CO, NO_x, OGC and particulate matter emissions according to the type test of the furnace as well as a comparison with Austrian emission limits are shown in Table 11.3. It can be seen that the actual emissions are far below the corresponding emission limits. However, it must be expected that the emission levels in field operation are considerably higher (cf. Section 9.5).

Table 11.3: Emissions and emission limits of the pellet central heating system

Explanations: values in mg/MJ_{NCV}; emissions according to type test (emission levels in field operation are expected to be considerably higher, cf. Section 9.5); data source: user manual

Parameter	Load	Emission	Limiting value
CO	Nominal	16	500
	Part	257	750
NO_x	Nominal	60	150
OGC	Nominal	1	40
	Part	2	40
Particulate matter	Nominal	8	60

The furnace is regulated by a micro processor based control system. The feed temperature is determined by the outdoor and room temperatures based on heating curves and is regulated by fuel and air supply. Three power levels are possible, namely nominal, medium and minimum load. For each power level a rotational speed of combustion air fan and induced draught fan is predefined.

To date (October 2009) about 4,000 full load operating hours have been achieved without major operational problems, though the automatic ignition has caused two failures. The problem has, however, been solved by the repair service of the furnace and boiler manufacturer. One drawback of the system is related to phases of very low heat demand in autumn and spring and even in mild winters. Start-ups and shutdowns are frequently necessary during these phases, as a low heat demand often leads to shutdowns because the maximum water temperature is reached, even though the furnace operates at part load. This fact also results in a comparatively low annual efficiency of the system. Taking the nominal thermal capacity, the average annual full load operating hours and the average annual pellet consumption of the first three years of operation into account (cf. Table 11.4), an average annual efficiency of 68.6% results (based on a NCV of the pellets used of 4.7 kWh/kg $(w.b.)_p$). It must be pointed out that the actual annual efficiency might be somewhat different, as the determination of the annual heat output is based on the full load operating hour counter of the control system and not on a heat meter, and the actual NCV of the pellets might be slightly different. However, the result must be regarded as realistic, as similar annual efficiencies were found in field measurements (cf. Section 9.7). This shows the importance of an appropriate process control system and an optimised hydronic system adapted to the pellet boiler.

The annual efficiency of the system will be improved as soon as the house is equipped with a solar heating system, including a heat buffer storage, which is planned in the near future.

11.2.2 Technical data

The technical data of the pellet central heating system are shown in Table 11.4. The nominal thermal capacity of the plant is 10 kW. The feed temperature of the heating circuit depends on the outside and the room temperatures and usually varies between 30 and 35°C. The maximum amounts to 45°C. In order to prevent condensation of flue gases in the boiler, a return flow temperature increase is installed, which raises the return temperature to at least 50°C before boiler inlet. The storage capacity of the storage space amounts to 6 t (w.b.)$_p$ or 9.2 m^3, which is equivalent to about 1.45 times the annual fuel demand (based on the average of the first three years of operation).

Table 11.4: Technical data of the pellet central heating system

Explanations: [1]...average of the first three years of operation based on the full load operating hour counter; [2]...based on average consumption of the first three years of operation; [3]...according to type test, expected to be a bit lower in field operation

Parameter	Unit	Value
Fuel power input (nominal conditions)	kW$_{NCV}$	10.9
Pellet consumption at nominal load	kg/h	2.3
Nominal thermal capacity	kW$_{th}$	10.0
Minimum thermal output	kW$_{th}$	2.6
Thermal efficiency at nominal / minimum load[3]	%	91.8 / 90.0
Annual full load operating hours	h p.a.	1,330[1]
Feed temperature heating circuit	°C	max. 45 average 30 to 35
Return temperature heating circuit	°C	25 to 30
Storage capacity	t	6
Storage capacity[2]	a	1.45

11.2.3 Economy

Table 11.5: Economic data of the pellet central heating system

Explanations: all prices incl. VAT; [1]...including feeding system from the storage room to the furnace, hot water boiler, all fixtures in the storage space, installation and start-up

Parameter	Unit	Value
Interest rate	% p.a.	6
Utilisation period	a	20
Annual fuel power input	kWh/a	19,400
Net investment costs	€	10,200
Pellet boiler[1]	€	11,000
Chimney renovation	€	1,000
Funding	€	1,800
Operating costs	€/a	311
Consumption costs	€/a	787
Capital costs	€/a	889
Total annual costs	€/a	1,987
Specific costs	€/MWh$_{UE}$	149.0

The investment costs of the pellet furnace including feeding system from the storage room to the furnace, hot water boiler (300 l) and all fixtures in the storage space as well as installation and start-up amounted to approximately 11,000 € (incl. VAT, valid for all prices in this section). In addition, the chimney had to be renovated, which caused costs of about 1,000 €. The installation of the plant was funded by both the federal state of Styria and the municipality of St. Lorenzen/Mürztal with 1,800 € in total.

To date, the annual fuel costs have been between 616 and 960 €, depending on pellet price and pellet demand; the average over the first three years of operation was 787 €.

Economic calculations for this pellet central heating system are shown in Table 11.5. The specific heat generation costs result in 149 €/MWh useful energy. However, taking a reasonable annual efficiency of 85% into account, the specific heat generation costs could be reduced to 137.6 €/MWh useful energy.

11.3 Case study 3 – small-scale application: retrofitting existing boiler with a pellet burner (Sweden)

11.3.1 Plant description

The owners of this house bought it in 2006 and they decided to change from oil to pellet heating. The house was built in 1969 and the standard of insulation is typical for the area and building time. The house is located in Ulricehamn, about one hour drive east of Gothenburg, Sweden. The residential space is 130 m² with another space in the basement of 80 m² that is kept between 5 to 10°C.

Figure 11.3: Old combination boiler with new pellet burner

The existing boiler was a TMW Alfa 1 installed in 1993. It was a combination boiler with one compartment for wood logs and one compartment for the oil burner. It is still possible to use wood logs. It is also equipped with an electrical coil at 6 kW that is used as a reserve. The boiler is still in good condition and therefore retrofitting with a pellet burner was the most economic choice and needed only a minor changes to the existing installation. The boiler is equipped with a 120 litre water tank that acts as a small heat buffer storage. Heat is supplied to the house by a central heating system based on radiators. The feed temperature depends on the outdoor temperature and is controlled by an outdoor sensor.

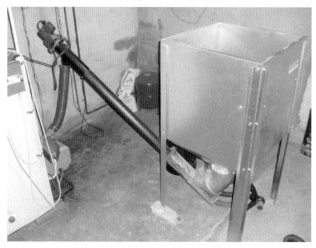

Figure 11.4: Week's storage connected to the pellet burner by a screw conveyor

In order to choose the pellet burner the house owner was helped both by the boiler and the burner manufacturer. Since the combustion compartment in the existing boiler is quite small, only a forward burning pellet burner could be used. A eurofire burner manufactured by Ekosystem i Gävle AB was selected. The burner was connected to the boiler by a new compartment door. The burner is P-marked, which is a voluntary marking system in Sweden to guarantee high quality, safety and efficiency of the equipment.

The pellets are purchased in bags and delivered to the house on pallets – 52 bags with 16 kg pellets each on one pallet. Each pallet is covered in plastic and can be stored outdoors until the plastic is removed. In the boiler room a MAFA week's storage for 180 kg (11 bags) was installed. In winter the storage is filled once a week and in summer once a month. The pellets are conveyed from the storage to the burner by a feeding screw according to demand.

The burner was installed by the house owner himself. The combustion was adjusted by a professional who had the equipment to measure carbon monoxide, carbon dioxide and oxygen in the flue gas. In winter, the bottom ash has to be removed once a week because of the small combustion compartment. The ash is used as a fertiliser in the garden of the home owners. Filling the storage and cleaning out the bottom ash takes about 30 minutes. The heat exchanger in the boiler is cleaned manually by the home owners twice every year. There have been no problems with the burner so far and no professional service was ever needed. Several enterprises in the region offer this service, however, should it be needed.

The organic gaseous carbon (OGC) emission level for a pellet burner in Sweden is 100 mg per Nm^3 at 10 vol.% O_2. The eurofire burner was tested in combination with nine different boilers by the organisation Swedish Consumers. The emissions varied depending on the boiler construction. For the three combination boilers included in the test, the OGC emissions at full load were 17, 91 and 93 mg. At lower load the emissions were 44, 88 and 155 mg.

11.3.2 Technical data

The main technical data of the retrofitted burner are summarised in Table 11.6. The power output from the burner is 20 kW. There is no modulation of the power but the operation is on–off. The burner can be adjusted by the owners to a lower power, namely 12 kW. Data for fuel power input, pellet consumption and thermal efficiency at nominal load are not available, as these data have not been measured for this specific burner/boiler combination. Heat is supplied to the central heating circuit according to a signal from a sensor for indoor temperature, and the feed temperature depends on the outdoor temperature. The hot tap water is heated to 65°C. Since the installation of the burner in 2006, about 5.8 pallets every year, i.e. 4.8 t (w.b.)$_p$/a, have been used. Pellets are used all year round, including in summer when only hot water is needed. The pellet burner has three safety systems against burn-back, i.e. a drop chute, a temperature sensor and a combustible hose where the pellets are fed into the burner.

Table 11.6: Technical data of the retrofitted burner in Sweden

Explanations: [1]…on–off operation (no modulation); [2]… adjustment by the owner necessary

Parameter	Unit	Value
Nominal thermal capacity[1]	kW$_{th}$	20.0
Minimum thermal output[2]	kW$_{th}$	12.0
Annual fuel demand	t/a	4.8
Storage capacity	kg	180

11.3.3 Economy

The costs for the pellet burner were 1,800 € (in 2006, incl. VAT, valid for all prices in this section). In 2006 it was possible for house owners to get a subsidy of up to 140 € in Sweden to change from oil to pellet heating.

About 4.8 tonnes of pellets per year (as an average) are used, which corresponds to fuel costs of about 1,400 €/a. Based on the assumption (no actual data are available) of an annual efficiency of the heating system of 70% for pellets and 80% for oil, the total costs for oil would be 3,051 €/a. This means that they save 1,659 € each year. The prices are based on November 2009.

11.4 Case study 4 – medium-scale application: 200 kW school heating plant Jämsänkoski (Finland)

11.4.1 Plant description

Koskenpää school is located in Koskenpää village, a part of Jämsänkoski town (cf. Figure 11.5). The school was founded in 1954 and was initially heated with firewood, with two cast iron Högfors A5 boilers in the boiler room located in the basement of the building. In 1967 these boilers were modified to oil fired boilers by mounting a light fuel oil burner on them. Annual oil consumption was 45,000 to 50,000 litres.

Figure 11.5: Koskenpää elementary school in Jämsänkoski town

In 2002, the ESCO (energy service company) Enespa Oy signed a contract with Jämsänkoski town for the replacement of oil with wood pellets for heating the school building. This was due to studies made by the Energy Agency of Central Finland and the Forestry Centre of Central Finland, which focused on the possibilities of increasing biofuel use in Central Finland.

Enespa Oy is the first ESCO in Finland. It is a professional business providing a broad range of comprehensive energy solutions including design and implementation of energy savings projects, energy conservation, energy infrastructure outsourcing, power generation and energy supply and risk management. The modification of Koskenpää School to pellet firing was the first investment of the company in the use of biofuels. Other activities are being planned.

The modified heating centre can be seen in Figure 11.6. In the background of the picture there is the new pellet boiler, which acts a base load boiler. The cast iron boiler, located in the foreground of the figure, is equipped with an oil burner. It is used as an auxiliary boiler and, if needed, as a peak load boiler.

When the heating centre was renovated, one of the cast iron boilers was dissembled and exported to Estonia. A Tulimax boiler, with a nominal output of 200 kW, replaced it. The boiler is designed for combustion of biomass fuels and is equipped with a large combustion chamber and vertical convection, which reduces the need for cleaning. The boiler was manufactured by HT Enerco Oy. The combustion equipment is a stoker burner made by Säätötuli Oy. The boiler is equipped with a burner that is charged with fuel by a horizontal screw feeder. Combustion air is led into the boiler by two separate fans. Primary air is led through the grate of the combustor head and the secondary air enters the combustion chamber above the fuel layer. The amount of combustion air can be controlled manually by using the flap valves mounted on the fans. Fuel feed is adjusted by controlling the ratio of the operating and stop times (on–off control) of the feeding screw. During off periods, a little fuel and air are added at times to keep the fire going. So there is no ignition system. When operating, the screw rotates at a constant speed. The length of the fuel feeding screw is about 3.5 m, and the one end is placed under the pellet silo. The risk of burn-back into the feeding system is prevented by temperature measurements in the feeding system in combination with a water injection system. The operating motor of the screw also drives the moving plates on the walls of the fuel silo. The purpose of these plates is to ensure the falling of the fuel onto the conveyor screws so that bridging is prevented. The boiler is equipped with a heat buffer storage with a volume of 5,000 litres. Ash removal is carried out manually. However, the furnace is constructed so that an automatic ash removal system can be retrofitted either with a screw conveyor or with a de-ashing fan (big vacuum cleaner). There are no flue gas cleaning devices installed. Particulate emissions are between 40 and 50 mg/Nm^3.

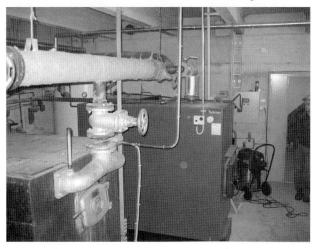

Figure 11.6: Modified heating centre of Koskenpää elementary school

A picture of the new pellet boiler is shown in Figure 11.7. Ash removal has to be carried out manually through the opening at the bottom of the boiler.

Figure 11.7: New pellet boiler of Koskenpää elementary school

The pellet silo was placed underground at the side of the school building (cf. Figure 11.8). The silo is filled pneumatically by a pellet lorry. The angle of inclination of the silo walls is 45°. Vapo Oy Energia delivers the pellets.

Figure 11.8: Underground pellet silo

11.4.2 Technical data

The nominal thermal capacity of the pellet boiler is 200 kW. Its annual operating hours are between 4,000 and 5,000 hours per year. The storage capacity of the underground pellet silo is 25 m^3. This is sufficient for two weeks when operated at the nominal output of 200 kW.

11.4.3 Economy

The total costs for renovation of the heating centre were 44,500 € (excl. VAT, valid for all prices in this section). Säätötuli Oy delivered and installed the fuel feeding system, the stoker burner and the boiler. The costs of these deliveries were 24,426 €. The rest of the expenses were caused by dissembling of the old boiler, construction of the fuel silo, assembling of a new hot water reservoir, the new fire doors of the heating centre, underdrainage costs of the silo and heating centre, and the surface drainage system of the schoolyard, constructed simultaneously.

The share of the ESCO investments covered by Enespa Oy was 28,200 €. The difference between the total costs for renovation and the ESCO investments, namely 16,300 €, was directly funded by the town. Jämsänkoski gets the investment back by savings in fuel costs. The payback time, defined in the ESCO contract, is 10 years (related to the total investment costs of 44,500 €). The typical pellet price is about 70% of oil (related to NCV).

The pellet heating system of Koskenpää school has been in operation since early 2004.

11.5 Case study 5 – medium-scale application: 500 kW heating plant in Straubing (Germany)

11.5.1 Plant description

The pellet heating plant described in this section is located at and operated by the Institute for Aurally Handicapped Persons (IFH) in Straubing, Germany. This modern furnace, which started operation in October 2008, provides heat to the school and adjacent institute complex, which comprises of a double size sport hall, an indoor swimming pool, a boarding school and special congress facilities. The two building complexes are connected via district heating lines. Due to the relatively high hot water demand, a year round heat supply is required. For peak heat demand an oil boiler was also installed. It is operated at very low temperatures in winter or during times of maintenance or pellet boiler failures.

The whole district heating station including the pellet storage is located in the basement of the boarding school. The filling status of the pellet storage can be monitored via inspection windows from the heating station. Heating oil is stored in an underground steel tank. In the primary heating circuit (boiler circuit) a heat buffer storage is installed in order to optimise furnace operation. The two secondary heating circuits for both the school and the housing are connected to the primary circuit via heat exchangers. All three circuits are independently equipped with pressure sustainment and degassing.

The flue gas of the pellet boiler passes through a cyclone for dust separation. Combustion control is performed using a lambda based system, which is adapted to the actual heat power. Primary air is supplied via the grate and secondary air is conducted above the grate into the fire clay combustion chamber.

The pellets are transported from the storage via a screw conveyor. The risk of burn-back into the feeding system is prevented by a valve in the dropshaft. The fuel is then transported into the combustion chamber via a stoker screw (on–off operation). The stoker screw is equipped with a sprinkler for extinguishing any fire for additional safety. In the furnace itself, the fuel is moved via a hydraulic moving grate, which is separated into several aeration areas. Fuel ignition is achieved via a hot air fan. Primary air is adjusted according to the actual heat demand. The moving grate ensures an even horizontal fuel distribution in the combustion area with homogeneous air supply. Secondary air is injected via several injection nozzles, which are distributed over the full length of the grate.

All components of the heating station in the basement can be removed or replaced via a large hopper. Ash is collected in ash bins and transported upwards by a lift for collection by a waste disposal truck.

According to the calculations made in the planning phase (data on a full year of operation are not yet available), the pellet furnace is supposed to run for about 3,200 full load operating hours annually. The share of biomass of the annual fuel consumption is 83%. A total of 3,500 h of full load operation was achieved by summer 2009. The emission limits of the German emission directive are met. For the nominal heat power class of a maximum of 500 kW, this means that the CO concentration has to stay below 1 g/Nm^3 and the maximum for particle emission is 150 mg/Nm^3, both based on 13 vol.% O$_2$ concentration in the flue gas.

The combustion plant is operated by the district administration of Lower Bavaria. In addition, a maintenance contract was signed and in the case of severe failures a message is automatically sent to the service contractor via the central building control system (e.g. by SMS to a mobile phone).

Figure 11.9: Pellet boiler at the IFH in Straubing

11.5.2 Technical data

Table 11.7: Technical data of the pellet heating plant at the IFH in Straubing

Explanations: [1]...according to type test, expected to be a bit lower in field operation

Parameter	Unit	Value
Pellet unit		
Fuel power input (nominal conditions)	kW_{NCV}	561
Pellet consumption at nominal load	kg/h	115
Annual fuel demand	t/a	407
Nominal thermal capacity	kW_{th}	500
Minimum thermal output	kW_{th}	150
Thermal efficiency at nominal load[1]	%	89.1
Annual heat supply (boiler output)	GWh/a	1.6
Annual full load operating hours	h p.a.	3,200
Feed temperature heating circuit	°C	75
Return temperature heating circuit	°C	55
Storage capacity	m^3	70
Storage capacity at nominal load	days	16.5
Peak load and reserve unit		
Fuel		heating oil
Nominal thermal capacity	kW_{th}	485
Minimum thermal output	kW_{th}	320

The main technical data are given in Table 11.7. The nominal thermal power output of the pellet boiler is 500 kW and the heat buffer storage tank has a volume of 25,000 litres, which results in a specific buffer volume of 50 l/kW. Additional heat can be provided by a 485 kW heating oil boiler, which also serves as a backup system in case of a wood boiler failure. The district heating line has a length of 94 m. With a nominal pipe diameter of 100 mm, it is dimensioned for a heat transport capacity of 900 kW. The storage capacity of the pellet storage (a concrete bin) is 70 m^3 and it can thus store around 47 t of pellets at a bulk density of 650 kg/m^3. The annual fuel demand is around 400 t. For heating oil, a subterraneous steel tank of 20,000 l is available.

11.5.3 Economy

The project was financially supported by the renewable raw materials programme of the Bavarian State with 79,150 €. The total investment costs (excl. VAT) were 583,000 €, including oil boiler and district heating network. Taking only the pellet unit (without oil boiler and district heating network) and the funding into account, the net investment was about 440,000 €.

The economy of the plant was calculated according to the annuity method (VDI guideline 2067), taking only the pellet unit into account (without oil boiler and district heating network). Assuming a useful life of 50 and 20 years for buildings and heating technology, respectively, and an interest rate of 6% p.a., the annual capital costs are 34,145 €. Operational costs (repair and maintenance, electricity, labour, disposal and chimney sweep costs) are in the order of 21,500 €/a. Assuming a pellet price of around 187 €/t (excl. VAT), the total annual fuel costs are about 76,000 €. With these data and the annual efficiency of 80.6% (which results from the technical data shown in Table 11.7), the costs of the heat produced (ex pellet boiler) amount to around 82.30 €/MWh (cf. Table 11.8).

Table 11.8: Economic data of the pellet heating plant at the IFH in Straubing

<u>Explanations</u>: all prices excl. VAT; calculation related to the pellet unit only (without consideration of peak load unit); specific costs related to useful energy ex pellet boiler

Parameter	Unit	Value
Interest rate	% p.a.	6
Utilisation period buildings	a	50
Utilisation period (technical installations and planning)	a	20
Annual fuel power input	kWh/a	1,985,000
Net investment costs	€	439,712
Biomass unit	€	153,400
Buildings	€	176,520
Hydraulics	€	115,120
Planning and other costs	€	73,822
Funding	€	79,150
Operating costs	€/a	21,500
Consumption costs	€/a	76,057
Capital costs buildings	€/a	11,199
Capital costs (technical installations and planning)	€/a	22,946
Capital costs total	€/a	34,145
Total annual costs	€/a	131,703
Specific costs	€/MWh	82.3

11.6 Case study 6 – medium-scale application: 600 kW district heating plant in Vinninga (Sweden)

11.6.1 Plant description

The district heating network in Vinninga, just outside the municipality of Lidköping in southwest Sweden, was built and put into use in 2008. Vinninga is a small community with about 1,000 inhabitants. The pellet fired heating boiler supplies heat to three schools, a home for older people, ten private houses and a few small companies. The investment in the local heating plant of Vinninga is part of the municipality's explicit goal of phasing out oil as a source of energy. Extending the downtown district heating network to the smaller communities outside the municipality's main town would have been too expensive, and instead, small-scale local heating plants and plants to heat single units were built. The pellet firing plant in Vinninga is one of six plants in the vicinity of Lidköping. Operation, maintenance and fuel at all six plants are taken care of by one company. In Vinninga, more than 150,000 l of oil have been replaced by pellets. The switch from oil to pellets has reduced carbon dioxide emissions by 370 tonnes a year.

The plant has a boiler from Osby Parca, which is equipped with a patented rotating ceramic pellet burner from Janfire AB and an automatic ash handling system. The flue gas of the boiler passes through a cyclone for dust separation. Combustion air is supplied as primary air inside the burner and as secondary air in a slit close to the rim of the burner drum. Combustion is controlled by a lambda sensor. Fuel is supplied to the burner by a screw. The burner drum is rotating and so the fuel bed is rotated too, which ensures the fuel bed is stirred and warrants a good burnout. Security against back firing is ensured by a water sprinkler system, a temperature sensor and by using a hose in the fuel feed system that burns off without an open flame to stop the fuel feed. The burner is ignited manually. The plant has an oil burner as a reserve.

Figure 11.10: District heating plant in Vinninga behind the pellet silo

The plant is located next to the school playground. It is a red building, looks like a farmyard barn, measures a few metres in width, and has a 10 m silo at one end (cf. Figure 11.10). This height is necessary in order to have room to store the pellets and provide the right pressure for the water heated in the boiler. There is no heat accumulator. Water is heated in the boiler and then transported through pipes. A full kilometre of pipes was laid. A heat exchanger was installed in each building, where the incoming heat is used to heat the building's own water. Since the plant went into operation, 510 tonnes of pellets have been used. The plant is used all year around. Positive aspects are high availability and little and simple maintenance.

11.6.2 Technical data

The pellet boiler and the reserve oil boiler have a nominal thermal capacity of 600 kW each. The storage capacity of the silo is 90 m^3 and the annual fuel demand is 350 tonnes of pellets. The feed temperature of the district heating network is 80°C; its return temperature is 60°C. Technical data of the district heating plant in Vinninga are summarised in Table 11.9.

Table 11.9: Technical data of the district heating plant in Vinninga

Explanations: [1]…according to type test, expected to be a bit lower in field operation

Parameter	Unit	Value
Pellet unit		
Fuel power input (nominal conditions)	kW$_{NCV}$	630
Pellet consumption at nominal load	kg/h	132
Annual fuel demand	t/a	350
Nominal thermal capacity	kW$_{th}$	600
Minimum thermal output	kW$_{th}$	60
Thermal efficiency at nominal load[1]	%	95.2
Feed temperature district heating network	°C	80
Return temperature district heating network	°C	60
Storage capacity	m^3	90
Storage capacity at nominal load	days	18.5
Peak load and reserve unit		
Fuel		heating oil
Nominal thermal capacity	kW$_{th}$	600
Total plant		
Annual heat supply (plant output)	GWh/a	1.8

11.6.3 Economy

The total investment costs for the plant, including the district heating network, were 0.9 million € (excl. VAT). No funding was gained. The annual fuel costs are estimated to be 42,000 €. Heating costs have fallen substantially because pellets are cheaper than oil. The estimated consumption is 350 tonnes of pellets per year and this means savings of 60,000 €/a compared to oil.

11.7 Case study 7 – large-scale application: 2.1 MW district heating plant Kåge (Sweden)

11.7.1 Plant description

The district heating network in Kåge supplies a school, several larger buildings and detached houses with heat and hot water. Kåge is a small municipality situated by the Baltic Sea in the north of Sweden. The district heating network and the plant are owned by Skellefteå Kraft, one of the largest energy companies and pellet producers in Sweden. The plant consists of two boilers and the fuel is wood pellets.

The larger boiler is a container boiler with reciprocating grate made by Hotab. A container boiler was chosen to make a possible exchange for a larger boiler easy if more customers connect to the system. It was built in 2000 and contains fuel feeding system, boiler, fans for air and flue gas, a multi-cyclone, ash handling systems, chimney and pumps for the district heating network. A reserve and peak load pellet boiler from Linka was installed in 2006 in an existing building. This boiler has a fixed grate.

Fuel is supplied from the storage to the boilers by screws. In order to avoid back firing, there are a cell feeder and a water sprinkler system that are controlled by the temperature in the fuel screw. The combustion air is supplied as primary air through the grates and as secondary air above the grates. The combustion is controlled by lambda sensors in the flue gas. There is no automatic system to start the boilers. If they are stopped, they have to be ignited manually.

Emission limits for the plant are 100 mg/Nm^3 at 13 vol.% O_2 for particulate matter and 110 mg/MJ for CO_2 and NO_x.

Figure 11.11: District heating plant in Kåge

The plant is situated close to the buildings it provides heat to (cf. Figure 11.11). The advantage of a location close to buildings is that the district heating system can be more cost effective.

Figure 11.11 shows the pellet silo (white building) and the container boiler behind the silo with the door to the boiler room to the right of the silo. This boiler room contains the larger boiler in Kåge. The building left of the silo contains the smaller boiler. The two buildings behind the container and to the right edge of the picture are connected to the district heating network (in addition to 60 to 70 other buildings).

11.7.2 Technical data

The technical data of the district heating plant in Kåge are shown in Table 11.10. The nominal thermal capacities are 1.5 MW for the larger pellet boiler and 600 kW for the smaller one. There is no heat buffer storage tank. The large boiler has a minimum load of 300 kW and the small one a minimum load of 100 kW. The plant is operated all year around. The storage capacity of the silo is 50 m³ and the annual fuel demand is 1,200 tonnes of pellets. Together the two boilers provide 5.0 GWh of heat p.a. to the district heating network.

Table 11.10: Technical data of the district heating plant in Kåge

Explanations: [1]...according to type test, expected to be a bit lower in field operation

Parameter	Unit	Value
Pellet unit		
Fuel power input (nominal conditions)	kW_{NCV}	1,650
Pellet consumption at nominal load	kg/h	345
Nominal thermal capacity	kW_{th}	1,500
Minimum thermal output	kW_{th}	300
Thermal efficiency at nominal load[1]	%	90.9
Feed temperature district heating network	°C	100
Return temperature district heating network	°C	50
Peak load and reserve unit		
Fuel		pellets
Nominal thermal capacity	kW_{th}	600
Minimum thermal output	kW_{th}	100
Total plant		
Annual fuel demand	t/a	1,200
Storage capacity	m³	50
Annual heat supply (plant output)	GWh/a	5.0

11.7.3 Economy

The total investment costs for the plant installed in 2000 were 3.2 million SEK (about 0.32 million €, excl. VAT, valid for all prices in this section), which includes the large pellet furnace and boiler itself, the fuel feeding system, fans for air and flue gas, a multi-cyclone, the ash handling system, the chimney and pumps for the district heating network (excluding the district heating network). The reserve and peak load pellet boiler was installed in 2006 in an existing building and the investment costs were 1.8 million SEK (about 0.18 million €).

11.8 Case study 8 – large-scale application: 4.5 MW district heating plant Hillerød (Denmark)

11.8.1 Plant description

In the city of Hillerød in Denmark, a new wood pellet fired plant was built in 2004. The plant is connected to the district heating network covering the city, and is located approximately 3 km west of the city centre, at Krakasvej in Ullerød-byen. The plant is owned and operated by Hillerød Varme A/S, a limited liability company, which is fully owned by the municipality of Hillerød.

The plant went into operation in 2005 with the purpose of supplying the Hillerød district heating network with heat – mainly intermediate and peak load. Hillerød Kraftvarmeværk (ownerd by Vattenfall) is a natural gas fired combined cycle plant that supplies the system with the base load of heat. In case of a stoppage at the base load plant, however, the pellet boiler can also supply base load to the district heating network.

The plant consists of one pellet fired boiler and a separate natural gas fired boiler for peak load, both located in one building. Today, the operational economics for the pellet boiler are more favourable than anticipated when the plant was designed. As a consequence, the pellet boiler is in operation most of the year while the gas boiler still acts as peak heat supplier.

Pellets are delivered by trucks and dumped into a reception bin with a nominal capacity of approximately 50 m^3 (33 tonnes). Pellets fall from the reception bin onto the conveyor. From the conveyor, pellets are transported in a cup elevator to one of the two main cylindrical storage silos outdoors in a cup elevator, each with a capacity of 100 m^3 (65 tonnes). The reception system has a capacity to receive a truckload of 35 m^3 of pellets every half hour.

A series of five screw conveyors transport the pellets from the bottom of the two silos to the boiler feeding system. Through a rotary valve, the pellets are fed into the furnace by a stoker screw. Combustion takes place on a fixed hearth, i.e. a horizontal, cylindrical combustion chamber made of steel.

Combustion air is supplied as primary air in the hearth bottom and as secondary air higher up in the furnace. Air volume for each of the three combustion air fans, one for primary air and two for secondary air, is controlled by frequency controllers.

Combustion control is based on oxygen in the flue gas. A lambda sensor is applied. Start-up ignition is manual.

Flue gas is cleaned in two steps. The first stage is a multi-cyclone to remove coarse ash particles and the second stage is a baghouse filter.

The emission limits are 40 mg/Nm^3 for particulate matter, 625 mg/Nm^3 for CO and 300 mg/Nm^3 for NO_x (related to dry flue gas and all at 10% oxygen reference).

From the boiler bottom, the ash is conveyed by a screw and transported into an outdoor ash container together with the ashes from the cyclone and baghouse filter. The mixed ashes from the plant are deposited in a nearby controlled landfill.

Produced heat is supplied into the Hillerød district heating network. No heat buffer storage tank is installed as the full capacity of the plant can at all times be utilised in the very large (compared to the pellet plant capacity) district heating network.

The pellet boiler is operated most years for more than half of the year, which is more than anticipated during the design of the plant. Actual annual heat production is also higher than anticipated due to the fact that the heat production from the base load plant is lower than had been expected.

The boiler building is a light steel construction. The pellet reception facility, the main pellet silos and the ash container are located outside the main building. Architects were involved in the building design in order to avoid a rough industrial look of the buildings (earlier nearby constructions had given rise to complaints from private households).

The furnace and boiler system was delivered by the Danish manufacturer Linka, with significant supplies of boiler parts from Danstoker (also Danish). FORCE Technology was responsible for engineering.

A few problems were observed during commissioning of the plant:

- Slagging was observed in the bottom ash screw conveyor system. The problem was solved by acquiring only high quality wood pellets with a high ash melting point. Motors in conveyors were replaced.
- Screw conveyor systems from silos to stoker should have been more robust; the equipment installed was adopted from farm scale feeding systems for grain and was not suitable for an industrial-scale boiler system.

A picture of the district heating plant in Hillerød is shown in Figure 11.12. The reception bin can be seen at the left, the main storage silos in the centre and the boiler building in the background.

Figure 11.12: District heating plant in Hillerød

Explanations: photo: Carsten Monrad, FORCE Technology

11.8.2 Technical data

The technical data of the district heating plant in Hillerød are shown in Table 11.11. The nominal thermal capacity of the pellet boiler is 4.5 MW and its thermal efficiency at nominal load is 90.9%.

Table 11.11: Technical data of the district heating plant in Hillerød

Explanations: [1]…according to type test, expected to be a bit lower in field operation

Parameter	Unit	Value
Pellet unit		
Fuel power input (nominal conditions)	kW_{NCV}	4,950
Pellet consumption at nominal load	kg/h	1,020
Annual fuel demand	t/a	10,800
Nominal thermal capacity	kW_{th}	4,500
Minimum thermal output	kW_{th}	1,750
Thermal efficiency at nominal load[1]	%	90.9
Feed temperature district heating network	°C	75
Return temperature district heating network	°C	45
Storage capacity	m³	200
Storage capacity	days	5.3
Peak load and reserve unit		
Fuel		natural gas
Nominal thermal capacity	kW_{th}	5,000
Minimum thermal capacity	kW_{th}	900

The boiler building height is 7.5 m and the boiler building area is 230 m². Wood pellet consumption in 2008 amounted to 10,800 tonnes.

11.8.3 Economy

The total investment in the plant was 13.5 million DKK (1.81 million €) in 2005 (excl. VAT, valid for all prices in this section). This includes storage facilities, building, surface work, stack, boilers for wood pellets and natural gas, engineering, property etc. Total investment divides approximately into:

- Buildings cost: 4.3 million DKK (0.58 million €) including reception bin, surface work and architects fees;
- Mechanicals contract: DKK 6.6 million DKK (0.89 million €);
- Natural gas boiler: 0.9 million DKK (0.12 million €);
- District heating connection pipelines: 1.0 million DKK (0.13 million €);
- Engineering: 0.7 million DKK (0.09 million €).

In May 2009, the wood pellet price in Denmark was approximately 1,300 DKK (175 €) per tonne delivered to an industrial scale plant by truck, the annual fuel costs for pellets were approximately 14 million DKK (1.9 million €). Assuming the same heat production for peak and reserve load produced in a natural gas fired boiler, fuel costs including natural gas tax and carbon dioxide tax would be around 22 million DKK (2.9 million €). If the heat was bought from the Vattenfall CHP plant, the costs would be around half way between the natural gas based costs and the pellets costs, i.e. somewhere in the order 16 to 20 million DKK.

Other operational costs (personnel, electricity, maintenance, service contracts etc.) were not taken into account. However, a rapid payback of the investment is anticipated because of annual fuel cost savings in the order of 4 to 7 million DKK (0.5 to 1.0 million €).

11.9 Case study 9 – CHP application: CHP plant Hässelby (Sweden)

11.9.1 Plant description

The CHP plant Hässelby [653] beside lake Mälaren near Stockholm supplies the three district heating networks of Stockholm with heat. It consists of three identical pulverised fuel boilers each with four burners. Three steam turbines generate electricity. A picture of the CHP plant is shown in Figure 11.13.

Figure 11.13: CHP plant Hässelby

Explanations: data source: Hässelby plant

The plant was put into operation in 1959 and run with oil until 1982. In 1982 there was a changeover to coal, which was used until 1994. From 1990 to 1993 test runs with pellets were performed and in 1994, the plant changed from coal to wood and bark pellets owing to the economic pressure that arose from the introduction of the CO_2 tax on coal. For start-up and shutdown oil is still used.

Pellet delivery is carried out solely by ship. Unloading of the ships and the filling of the storage facilities take place fully automatically with a capacity of 250 t (w.b.)$_p$/h. There are two different possibilities for unloading. If the ships have the appropriate equipment, discharge can take place directly from the ship. This kind of discharge is dust free. Otherwise there is a crane available, but the method involves considerable dust emissions and cannot be carried out in all weather conditions. The deliveries predominantly come from Sweden but pellets are also imported from the Baltic states, the Netherlands, Finland, the USA and Canada.

Before the pellets are combusted, they are ground in the six available coal mills (to a particle size of about 0.5 to 3 mm) and in two hammer mills with a capacity of 20 t/h each. The powder from the hammer mills is transported to the silos and then to the bends after the coal mills pneumatically.

Initially, Hässelby held shares in a pellet producer (BioNorr). However, the shares were then sold and since then pellets are exclusively purchased on the pellet market.

The fly ash produced is disposed of in a landfill. The operating experience includes high amounts of bottom ash with unburned parts (currently transported to another plant for combustion) and deposit formation in air pre-heaters.

11.9.2 Technical data

The technical data of the CHP plant Hässelby are shown in Table 11.12. The nominal fuel power input of each of the three boilers is 100 MW. Every boiler is equipped with four burners. The CHP plant produces around 868 GWh of heat per year (loco plant). The customer with the greatest distance to the plant is 40 km away (Sigtuna). Average losses of the district heating network are 6%. The feed temperature is 80°C in summer and 120°C in winter and the return temperature 40°C. For electricity generation three steam turbines are used, with a total nominal electric capacity of 75 MW$_{el}$. The steam parameters are 510°C and 80 bar. The CHP plant is operated in heat controlled mode and for combustion control a Siemens PSC7 system is used. Particles are removed in an ESP, and SNCR is used for NO$_x$ reduction. Typical emission data for "white pellets" are 55 to 70 mg/MJ for NO$_x$ depending on boiler and 3.5 mg/MJ particulate matter. The corresponding emission limits are 75 mg/MJ (annual average) NO$_x$, 180 mg/MJ (hourly average) CO and 13 mg/MJ (monthly average) particulate matter. To date (2009) about 50,000 operating hours for each boiler on wood pellets have been accumulated.

Table 11.12: Technical data of the CHP plant Hässelby

Explanations: [1]...summer/winter

Parameter	Unit	Value
Fuel power input (nominal conditions)	kW$_{NCV}$	300,000
Pellet consumption at nominal load	kg/h	61,900
Annual fuel demand	t/a	300,000
Nominal thermal capacity	kW$_{th}$	189,000
Nominal electric capacity	kW$_{el}$	75,000
Thermal efficiency at nominal load	%	63.0
Electric efficiency at nominal load	%	25.0
Annual heat supply (boiler output)	GWh$_{th}$/a	868
Annual electricity generation (gross)	GWh$_{el}$/a	300
Annual full load operating hours	h p.a.	4,600
Feed temperature district heating network	°C	80 / 120[1]
Return temperature district heating network	°C	40
Storage capacity	m^3	15,000
Storage capacity at nominal load	days	6.6

Two storage spaces with a capacity of 5,000 t (w.b.)$_p$ each are available. Approximately 62 tonnes of pellets are consumed per hour at full load, hence storage capacity is enough for almost one week of operation. Annual pellet consumption is around 300,000 t (w.b.)$_p$.

The loading capacity of the ships is usually between 1,500 and 2,000 t (w.b.)$_p$. Ships up to a capacity of 3,000 t (w.b.)$_p$ can be unloaded.

11.9.3 Economy

The costs for the change from coal to pellets of around 10 million € (excl. VAT) were mainly caused by storage space adjustments. All other parts of the plant could be used for pellet utilisation more or less unchanged.

One of the main reasons for the utilisation of pellets in this CHP plant is the relatively high CO_2 tax on coal and fossil energy carriers in general. Moreover, the purchase of such large quantities of pellets renders the acquisition of the fuel more economic than would be the case for smaller applications.

11.10 Case study 10 – large-scale power generation application: power plant Les Awirs (Belgium)

11.10.1 Plant description

Electrabel, the largest power company in the Benelux, undertook the full retrofitting of an existing pulverised coal power plant located near Liège [50; 654]. Unit 4 of Les Awirs power plant (cf. Figure 11.14) was commissioned in 1967 for generating 125 MW of electricity firing heavy oil and natural gas. A first retrofitting of the plant was carried out for firing pulverised coal from 1982 onwards.

Figure 11.14: Les Awirs power plant near Liège

A second retrofitting was realised in 2005 for firing 100% biomass. The former power level was lowered to 80 MW_{el} for technical reasons (lower heating value, high volatile content, higher residence time needed in the boiler for wood pellets). Electricity generation takes place in a steam turbine.

The plant fires pelletised woody biomass only. The capacity of the renewed plant is 350,000 t of wood pellets per year. The feedstock originates from around the world. About a third is shipped from overseas to Antwerp harbour. Another third is transported by boat from northeastern Europe to Antwerp. From here the pellets are loaded onto barges to deliver the pellets just-in-time up the river Maas to the plant. Apart from a buffer silo with a storage capacity of 7,000 t, no pellets are stored at the plant. The final third originates from nearby areas in southern Belgium and is transported by truck to the power plant.

The biomass consists of pelletised woody biomass that is then ground to wood dust. The pulverisation unit was completely redesigned for the change from coal to pellets. Two hammer mills with a capacity of 30 tonnes per hour each are now used instead of the former roller mills. The wood pellets are not only broken down coarsely but further milled in order to feed the boiler with particles all smaller than 3 mm, while 75% is under 1.5 mm.

All on-site logistics were modified. Unloading of pellets from the barges into the receiving hopper is carried out by means of a mobile crane. The existing coal bins for intermediate storage of the raw fuel before milling are also used and were equipped with more sophisticated anti-explosion systems.

The existing coal boiler has remained unchanged with the exception of new burners designed for the use of pulverised wood. The mixture of primary air with wood dust and secondary air are injected via separate concentric tubes. A mixture of dust with cold air is injected in the centre while hot secondary air is injected along the periphery of the dust burners. The injection systems use compressed air and primary air is kept at a temperature under 60°C in order to prevent fires and explosions. Steam temperature was reduced from 545°C to 510°C, steam pressure being 145 bar.

Belt and conveyor chains are fully covered and equipped with dust suction and filtering at any transition level. Some of them are new conveyors but a number of coal conveyors are still used.

Wood dust can spread to the surroundings along the handling system and is injected as such into the boiler after milling the pellets. Under certain mass ratios of mixtures with air, wood dust is known to be explosive and it might self-ignite. Thanks to compliance with European ATEX legislation as regards exposure to dust in the workplace, the production process affords the necessary protection. ATEX refers to two European directives governing explosive atmospheres. The first, 94/9/EC, relates to the acceptable equipments in these atmospheres. It aims at standardising the legislation of the Member States relating to the devices and the protection systems intended for zones of risks. The second, 99/92/EC, defines the minimum health protection measures for the safety of workers likely to be exposed to the risks.

Many prevention devices were included in the newly designed plant:

- Metal and spark detection;
- Earthing of the equipment;
- Micro-pulverisation of water in critical places with sprinklers along the conveyors;
- Anti-explosion bottles injecting sodium bicarbonate into the bins, etc.

In August 2005, the conversion of Unit 4 of Les Awirs coal power plant into a biomass power plant was achieved after seven months of work, including the necessary studies – considered to be a very speedy conversion. Since then, the power plant has been extensively tested to optimise every component of the new wood dust handling system from the unloading hopper up to the new pulverised fuel burners of the former coal boiler.

Today, the renewed plant is operated at nominal load and generates both electricity and green certificates as expected.

In order to deal with technical issues as well as all environmental limitations when firing wood pellets, Electrabel designed their own fuel specifications (cf. Section 10.11.6.4). These specifications are defined according to the most stringent requirements already existing in European standards, such as the Swedish SS 18 71 20, the German DIN 51731 and the Austrian ÖNORM M7135.

Intensive work was demanded by the Belgian authorities for certifying the origin of the imported wood fuel delivered to the power plant from all parts of the world, i.e. Canada, Asia, South Africa, Latin America and Eastern Europe. The granting of green certificates in Belgium is linked to very strict conditions related to the energy balance of the supply chain as well as the guarantee that forest resources are managed on a sustainable basis. Every supplier has to accept an extended audit carried out by an independent body.

Flue gas measurements have shown that acid gas (NO_x and SO_x) as well as particulate matter emissions were much reduced in comparison to coal firing and are at least five times lower that the limits imposed from 2008 onwards by the LCP-directive 2001/80/EC (cf. Table 11.13).

Table 11.13: Emissions of the Les Awirs power plant fired with wood pellets

Explanations: data related to mg/Nm^3 at 6 vol.% O_2

	Dust	SO_x	NO_x
Measured values	19	30	120
Limit before 1st January 2008	350	1,700	1,100
Limit after 1st January 2008	100	1,200	600

The retrofit from coal to pellets led to almost CO_2 free electricity generation. However, a drawback of the plant is that the heat produced is not utilised and consequently its overall efficiency is rather low.

The ashes produced are disposed of in a landfill.

11.10.2 Technical data

The technical data of the Les Awirs power plant near Liège in Belgium are shown in Table 11.14. The nominal electric capacity of the power plant is 80 MW and its electric efficiency amounts to 34%. The plant operates at 7,000 annual full load operating hours and produces around 560 GWh of electricity per year (electricity only operation).

Table 11.14: Technical data of the Les Awirs power plant

Parameter	Unit	Value
Fuel power input (nominal conditions)	kW_{NCV}	235,000
Pellet consumption at nominal load	kg/h	48,500
Annual fuel demand	t/a	350,000
Nominal electric capacity	kW_{el}	80,000
Electric efficiency at nominal load	%	34.0
Annual electricity generation (gross)	GWh_{el}/a	562
Annual full load operating hours	h p.a.	7,000
Storage capacity	t	7,000
Storage capacity at nominal load	days	6.0

11.10.3 Economy

Total investment costs to retrofit the coal fired power plant to use 100% wood pellets amounted to about 6.5 million € (excl. VAT).

11.11 Case study 11 – co-firing application: Amer power plant units no. 8 and no. 9 in Geertruidenberg (the Netherlands)

11.11.1 Plant description

Essent's Amer power station is situated in Geertruidenberg, the Netherlands (cf. Figure 11.15). Currently it consists of two units (unit no. 8, in service since 1980 and unit no. 9, in service since 1993). Both units were built as coal fired power plants and are now equipped with an extensive flue gas cleaning system consisting of a $DeNO_x$ (SCR), ESP filters and a (wet) desulphurisation unit.

Essent started co-firing 75,000 t/a of paper sludge in 2000. Since then, Essent has gained experience with co-firing of different fuels (e.g. olive kernels, wood pellets). Today, the Amer plant has a permit to co-fire a total of up to 1.2 million tonnes of biomass per year. Only wood based fuel has been used since 2006 due to reduced subsidies for agricultural residues.

Unit no. 8 is a 645 MW_{el}/250 MW_{th} coal/biomass fired power plant. The boiler is a subcritical boiler (live steam conditions: 178 bar/540°C, reheat conditions: 40 bar/540°C) with six burner levels fed with coal (four burners per level) and two burner levels fed with biomass (four burners per level). Steam is delivered to a high pressure, intermediate pressure and three condensing low pressure turbines. Heat is delivered to a district heating system (households and greenhouses).

In contrast to unit no. 9 where two existing coal mills were modified, two separate hammer mills were installed in unit no. 8 in 2003. Each mill has a capacity of 160,000 t/a. The biomass fuel supplies 37 MW_{el} output (5.7% of total plant output).

Amer unit no. 9 is a 600 MW_{el}/350 MW_{th} coal/biomass fired power plant. The boiler is a tangentially fired forced once through supercritical type (live steam conditions: 270 bar/540°C, reheat conditions: 55 bar/568°C). There are seven burner levels with four burners per level. Four burner levels are fed with coal, two with biomass and one with syngas from biomass gasification. Steam is delivered to a high pressure, intermediate pressure and three condensing low pressure turbines. Heat is delivered to a district heating system (households and greenhouses).

Unit no. 9 uses biomass both directly and indirectly. Part of the biomass is fed into the boiler by direct co-firing. Pellets are ground by two coal mills that were modified in 2003 and 2005. This results in a total power output of 136 MW_{el} from biomass (23% of total plant output). The capacity of each mill is 300,000 t/a. Furthermore, next to unit no. 9, a wood gasification unit is in operation. By wood gasification, syngas is produced that is combusted in the main boiler, resulting in an output of 34 MW_{el}.

Up to June 2009, over 3 million tonnes of biomass were co-fired at Amer power station. Wood pellets made of sawdust are purchased in large volumes, mainly from overseas suppliers and delivered via the port of Rotterdam. A special biomass unloading station was erected especially for pellet delivery via ship (cf. Figure 11.16).

Figure 11.15: Amer co-firing power plant in Geertruidenberg

Explanations: source: © Aerocamera BV

Figure 11.16: Biomass unloading station at Amer power plant in Geertruidenberg

Explanations: source: © Aerocamera BV

The biomass is pneumatically unloaded from the ships and transported to four storage bunkers. From there, the biomass is transported mechanically to the daily storage bunkers in units no. 8 and no. 9.

As already described, the co-firing processes in units no. 8 and no. 9 are different. For each process, a number of technical issues have been the subject of study. Below a summary of the different process steps and issues is given.

Both units 8 and 9 have a wet bottom ash discharge system and ESPs for fly ash removal. Fly ash is sold as a useful by-product to the cement industry.

Table 11.15: Technical issues related to different process steps of Amer power station units no. 8 and no. 9

Process	Unit no. 8	Unit no. 9
Fuel logistics	Availability is reduced due to dust problems	Availability is reduced due to dust problems
Storage in day bunker	Good experience, no cases of self-heating or ignition are known	Good experience, no cases of self-heating or ignition are known
Fuel supply	Screw feeder, good experience	Coal feeder, good experience
Milling/drying	No further reduction of the original particle size of the raw material by milling (pellets are just broken up to the original material). Availability is reduced due to wear of hammer mills. Flue gas recirculation is used for inertisation (to reduce explosion risk)	No further reduction of the original particle size of the raw material by milling (pellets are just broken up to the original material). Modified coal mills have a very high availability.
Pneumatic conveying	Conveying by primary air. Low transport velocity. Large pressure drop causes a limited conveying capacity as larger biomass particles require higher transport velocities.	Conveying by primary air. Low transport velocity. Pressure drop probably limits the milling capacity (conveying system was originally designed for transporting pulverised coal).
Combustion	Fair burnout and fly ash quality. Co-firing only takes place above 45% load.	Good burnout and fly ash quality. Co-firing only takes place above 45% load.
Flue gas cleaning	Good experience	Good experience

11.11.2 Technical data

In Table 11.16, the key co-firing data for Amer units no. 8 and no. 9 are shown.

Table 11.16: Technical data of the Amer units no. 8 and no. 9

Parameter	Unit	Amer no. 8	Amer no. 9
Fuel power input (pellets, nominal conditions)	kW_{NCV}	185.000	336.000
Fuel power input (total, nominal conditions)	kW_{NCV}	1.613.000	1.412.000
Pellet consumption at nominal load	kg/h	42.000	75.000
Nominal thermal capacity	kW_{th}	250.000	350.000
Nominal electric capacity	kW_{el}	645.000	600.000
Electric efficiency at nominal load	%	40,0	42,5
Annual electricity generation (gross)	GWh_{el}/a	4.900	4.980
Annual full load operating hours	h p.a.	7.600	8.300
Feed temperature district heating network	°C	120	120
Return temperature district heating network	°C	65	65
Storage capacity	m^3	400	970
Storage capacity at nominal load	days	0,26	0,35
Co-firing percentage	$\%_{NCV}$	11,5	23,0

11.11.3 Economy

Taking subsidies and CO_2 benefits (because of less use of coal) into account, large-scale co-firing of wood pellets has become economically feasible in the Netherlands.

11.12 Summary/conclusions

The case studies in this chapter illustrate the broad range of possible applications for the use of pellets. From the very small-scale pellet stoves located in rooms to be heated with pellet demands of some tonnes per year, to medium-scale applications for large buildings or district heating networks, to large-scale power and CHP plants with pellet consumption of some 100,000 tonnes per year, pellets offer attractive and economic fields of applications.

For all plants in all power ranges, high annual utilisation rates should be achieved, which can be reached by proper dimensioning of the plants, adequate control systems and proper integration in heating or district heating systems. For CHP plants, the utilisation of heat and electricity produced are important. Heat controlled operation is the optimum from an energetic point of view. Power only generation is not recommended due to the rather poor overall annual utilisation rate achievable.

In the field of large-scale applications, retrofitting existing coal or oil fired CHP plants to burn pellets is an attractive possibility to use large amounts of biomass for energy generation. Moreover, co-firing of pellets in coal fired power stations bears similar benefits without major costs for retrofitting. One of the biggest advantages of pellets in this field is their high energy density, which increases transport and storage efficiency substantially in comparison to other biomass fuels.

For small-scale applications, the importance and relevance of pellets is their comparatively high energy density, their homogeneity and their standardised chemical composition. These are very important parameter to ensure high availability and automatic operation of small-scale systems.

12 Research and development

Many research teams work on pellet related issues in the world and national R&D programmes that are focused on pellet production and utilisation are carried out in many countries. By 2005, more than 110 research teams were identified in Europe alone [655]). The purpose of this chapter is to present a thematic overview of the most important R&D topics. As R&D in the field of pellet production and energetic utilisation is very dynamic, this overview does not cite all ongoing R&D projects but provides relevant R&D trends, objectives and information on ongoing projects (basis spring 2010).

12.1 Pellet production

12.1.1 Use of raw materials with lower quality

12.1.1.1 Herbaceous biomass

Herbaceous biomass, such as straw, hay, grass cuttings, different sorts of crops etc., are available at high quantities in many countries and thus have great potential to enter the market of biomass fuels. However, many problems have yet to be overcome before opening up this market as both the pellet production and the combustion behaviour of herbaceous biomass is not comparable to that of wood.

Different research projects follow different approaches to counter the drawbacks of herbaceous biomass. Investigations into quality improvement of pellets made of straw and hay were carried out within the R&D work of [656]. Different binding agents and additives such as molasses, starch, dolomitic lime and sawdust were used. It was found that the qualities of pellets made of straw and hay could be improved by these additives. Molasses increased bulk density and the mechanical durability of the pellets. Lime actually increased the ash softening temperature. Sintering of ash during combustion of straw pellets could not be avoided, however. In addition, lime had a negative impact on the mechanical durability of the pellets. Promising results were achieved in pelletisation of defibrated hay and straw, produced by means of a special technique. The pellets made of defibrated straw achieved the highest durabilities, and emissions of particulate matter during combustion were notably reduced. However, preparation of the defibrated straw costs around 40 €/t according to manufacturers, so the method is not economically efficient under present framework conditions.

Another approach to render herbaceous biomass suitable for pellet production is to mix herbaceous and woody biomass in order to reduce the problems arising from the use of herbaceous biomass alone [115; 120; 657]. With regards to abrasion behaviour according to prEN 14961-2, being one of the most important quality criteria of pellets, promising results were achieved by use of a blend of willow and wheat straw or shredded wheat. This has yet to be verified at industrial scale as lab results cannot be assigned directly to industrial scale without further work. Combustion trials were not carried out with pellets produced in this way. Other pelletising trials were carried out with sawdust, straw, sunflower hulls, grains and nutshells as well as mixtures of these materials. The addition of aluminium hydroxide, kaolinite, calcium oxide and limestone should prevent slagging. These trials also led to good pellet qualities. In combustion trials, however, wood pellets that were used as a reference proved to be in a class on their own when looking at combustion behaviour, slag and deposit formation. All other pellets caused problems concerning slagging and deposit formation to some degree.

These trials all show that the combustion behaviour of pellets made of herbaceous biomass can be improved by a number of measures, but the characteristics of wood pellets cannot be achieved. In order to deal with herbaceous biomass fuels, appropriate adaption and optimisation of furnace technology will be indispensable. Producing a quality pellet out of herbaceous biomass that can be used without problems in the pellet furnaces currently available on the market is probably impossible.

In Southern European countries, the use of pellets made of herbaceous biomass in larger furnaces is an objective [658]. There are two reasons for that. First, woody biomass is not sufficiently available in these countries and second, the overabundance of herbaceous biomass creates a waste disposal problem. Nevertheless, deposit formation, increased emissions (especially particulate matter and NO_x), corrosion and amount of ash (about 10 to 15 times more) and thus the increase in cleaning and service efforts needed, still pose problems that are yet to be solved.

In Denmark, for example, pellet production from agricultural residues such as straw was studied and tested but no commercial breakthrough could initially be achieved (project "Quality Characteristics of Biofuel Pellets" at the Danish Technological Institute in co-operation with FORCE Technology, DONG Energy and Sprout Matador in 2002). Since 2004, however, experience with large-scale straw pellet utilisation has been gained in the Amager power station near Copenhagen (cf. Section 10.6) [659; 660; 661]. These activities are planned to be continued in the new unit number 2 of the Amager power station from 2010 onwards and straw pellet consumption is expected to increase. Accompanying R&D activities are ongoing.

It must be noted that the abovementioned disadvantages of herbaceous biomass fuels are in particular relevant for small-scale furnaces. Co-firing of certain amounts of herbaceous biomass fuels in large CHP or power plants is possible and already carried out (cf. Section 6.5.4.2).

12.1.1.2 *Short rotation crops*

The availability of raw materials is increasingly an issue due to growth in the pellet market (cf. Section 10.1.3). The use of energy crops could gain more importance when the freely available quantities of wood shavings, sawdust and other low-cost raw materials become scarce. Herbaceous as well as woody energy crops could become relevant.

With regard to herbaceous energy crops, the factors as stated in Section 12.1.1.1 apply. As concerns woody energy crops, the use of the fast growing tree species willow or poplar seems of interest. Endeavours in this direction are carried out in Germany where SRC (willow) have been grown for utilisation as a raw material in pelletisation [662]. Similar activities are ongoing in Austria with SRC (using willow and poplar), but these plantations are still in trials [663; 664]. Pellet production with SRC has not yet occurred (cf. Section 3.4.3). Owing to the higher ash content of pellets made of SRC, producing class A1 pellets according to prEN 14961-2 is not possible. Pellets made of SRC used as industrial pellets could relieve market pressure on standardised pellets, however.

Making use of energy crops requires a more holistic consideration than needed for pellet production from sawdust or wood shavings, since the entire raw material supply chain, including planting, fertilising and harvest as well as pre-treatment and logistics must be taken into account. Obviously, pellet production becomes more sophisticated and more expensive in this way – the entire raw material supply chain has to be optimised accordingly. Moreover, pellet production out of energy crops will most likely be rendered economical by framework conditions such as increasing oil and gas prices. Such developments are foreseeable. Due to the variety of ways in which energy crops can be used, they are not treated in depth here (cf. [59; 451; 665; 666].

12.1.1.3 Increasing the raw material basis

In order to secure the supply of pellets when market demand continues to increase, a larger raw materials base is necessary. At the same time it is necessary to define different pellet qualities for different end users. In the Swedish R&D project "Evaluation of combustion characteristics of different pellet qualities from new raw materials" [667] pellets from new raw materials from forest and agriculture are evaluated in combustion experiments on a domestic and commercial scale. The experimental results are generalised in mathematical models to provide input for the selection of raw materials and to control the quality of the pelletising process. The specific goals for the project are the following:

- Develop well defined, appropriate and firmly established criteria for definition of pellet quality from different raw materials and applicable to the whole range of combustion plants using pellets;

- Provide feedback to fuel producers regarding development of methods and choice as well as proper admixing of raw materials for production of specific pellet qualities;

- Develop principles and data for a cost efficient and reliable quality assurance system taking the entire product chain from the raw material to the burner into account;

- Develop theoretical models to describe critical parameters for the fuel conversion process concerning ash related operational problems as well as emissions, for example in order to control the pellet quality towards ideal characteristics.

The project is running from 2007 to 2010.

12.1.2 Pellet quality and production process optimisation

12.1.2.1 Influence of production process parameters

Different R&D activities aim to further reduce abrasion and the hygroscopic properties of pellets [90; 668; 669]. The influence the different process steps of pelletisation have on the quality of pellets is examined. In particular, the influence of storage, drying, thermal activation (conditioning) and cooling on pellet quality are being investigated. In addition, the effects of adding an activating substance to the raw material and the use of coating technology are being looked at. Adding the right activating substance can break up binding sites in the wood structure, thus creating new possibilities for binding. Coating primarily aims to improve the hygroscopic attributes of pellets [92]. Research activities are still in progress but first outcomes show that the drying temperature influences throughput, energy consumption of pelletising and abrasion. Adding hydrogen peroxide as an activator increased throughput in the pellet mill. Parameters such as abrasion, particle density and moisture content remained almost constant when hydrogen peroxide was added, but gross calorific value (GCV) went down slightly. Cooling was also found to have an influence on pellet quality. The influence of different drying technologies and drying parameters on pellet quality are also being investigated [670; 671]. A certain potential for quality improvements lies in appropriate optimisation measures of the pelletisation process [672].

Clear reduction of the hygroscopic property of pellets is achieved by torrefaction as the hygroscopic attributes of biomass change during the torrefaction process to hydrophobic (cf. Section 4.1.4.2).

Concerning the influence of the particle size distribution of sawdust (scotch pine) on several quality parameters of pellets (8 mm in diameter) investigations were carried out by [673]. It was found that the particle size distribution of raw materials does have a certain effect on the energy consumption of the pellet mill and on the pressure resistance of pellets, but it had no traceable effect on bulk density, particle density, moisture content, hygroscopic properties during storage or mechanical durability. From that it was concluded that raw material grinding could be omitted for particle sizes below 8 mm. Particles that are bigger should be screened out and then be ground on their own or used elsewhere (e.g. for briquetting or as a fuel in a biomass furnace). It must be noted, however, that these findings apply only to 8 mm pellets made of scotch pine sawdust. Pellets are made of a great number of different wood species and, at least in German speaking countries, pellets of 6 mm in diameter are normally used in small-scale furnaces. Moreover, the conclusions contradict the assertions of a number of pellet producers who claim that the particle size of raw materials should not exceed 4 mm (cf. Section 4.1.1.1).

Test trials with other raw materials are recommended and they should be carried out for each individual case because an optimisation potential with regards to energy consumption may even render the grinding step unnecessary. Looking at pellet production using raw materials that require much grinding effort, such as log wood or wood chips, the results seem relevant for the optimisation of grinding.

Ongoing investigations are also concerned with the influence of raw material composition on the combustion behaviour of pellets [674]. It was found for instance, that the burnout time of a single pellet is dependent not on particle density but on raw material composition. Although higher particle density prolongs the time required for complete charcoal burnout, the kind of woody biomass used as the raw material is more influential. In order to obtain detailed knowledge on the share of charcoal, charcoal burnout and single fuel components as well as their interaction, more research needs to be carried out.

In Germany and Austria, system failures in small-scale furnaces were often caused by slagging in the past. In order to find out the reasons for this, research projects were also carried out in this area.

Research that was carried out in Austria [675] showed that, regardless of the furnace being used (five different furnaces were used in the trials), pellets adhering to the ÖNORM M 7135 did not show a tendency for slagging. Pellets that did not adhere to the standard led to slagging in all furnaces. These results leave one with the conclusion that the tendency for slagging depends mainly on the fuel used and not on the type of furnace.

In Germany, however, system failures occurred in cases where standardised pellets were used too. It was conjectured that biological additives, as allowed by prEN 14961-2, might be responsible for this. Within a research project, pellets were produced with a series of different biological additives that are normally used in practice at an industrial scale and tried in two different pellet boilers (overfeed and underfeed furnace) [676]. The trials showed that the influence of biological additives on slagging behaviour is negligible.

Investigations by [677] showed that different places where pellet raw materials are grown and different pellet length could prove responsible for slagging. The origin of the raw materials has an impact on ash melting behaviour and thus slagging behaviour insofar as raw materials stemming from soils with a low pH value have lower ash softening temperatures than raw materials from soils with higher pH values. It was found that the CaO content of the ash correlates quite well with the pH value of soils, whereby high CaO contents of the ash lead to elevated ash softening temperatures. Pellet length probably influences the ash melting behaviour insofar as short pellets have higher bulk densities and hence the bed of embers is more dense leading to increased temperatures, which in turn promote softening and melting of the ash. Combustion trials with different pellet lengths and different kinds of burners could not establish a clear correlation, however, since the aforementioned effect occurred to varying extents and not in all burners. Increased temperatures in the bed of embers were noted in four out of five kinds of burners when short pellets were used, leading to the assumption that short pellets cause high temperatures in the bed of embers, thus bringing a danger of slagging.

12.1.2.2 Mitigation of self-heating and off-gassing

In order to ensure safe handling and storage of pellets, there is need for further research to better understand the phenomena of off-gassing and self-heating, which under certain circumstances can result in spontaneous ignition. Real incidents, some with fatal consequences, have shown the importance of these issues [224; 225]. Based on present knowledge, there are many factors influencing the off-gassing [238; 239; 241; 242; 243; 250; 252; 253] and self-heating [227; 228; 229; 230; 231; 235; 236] properties of a specific pellet quality. The type of raw material (e.g. content of fatty/resin acids), handling and storage of the raw material, the pellet production process, the moisture content of the pellets and water adsorption processes are some of these factors, but the interconnection between these and possible ways to mitigate these problems are not fully understood. It is also important to investigate oxidation reaction mechanisms and kinetics to describe how degradation of fatty/resin acids leads to the formation of non-condensable gases such as CH_4, CO and CO_2. For example, there is no direct evidence (published results) that shows which fatty/resin acids may be decomposed to CH_4, CO and CO_2. There is also a need for simple test methods that could be used by pellet producers to control pellet production on a regular basis to ensure that the pellets produced do not pose a great risk. The increased use of pellets also increases the size of pellet storages, both in heaps and silos. The larger the storage volume is, the larger is the effect of self-heating and the possibilities for spontaneous ignition as the overall heat loss from a large storage volume is less. Improved methods to predict the risk of spontaneous ignition for various storage volumes are therefore of great importance. Suitable fire detection and fire fighting methods also need to be developed further, especially for storage in large indoor heaps (A-frame flat storages). An increasing number of raw materials not previously used for pellet production are gradually being introduced to replace the dwindling supply of the traditional feedstock such as sawdust and wood shavings. Therefore, characterisation and handling of new raw materials (e.g. by-products from bio-refinery processes) for pellet production and optimisation of process parameters are important fields to be investigated in order to cope with the risk of self-heating, off-gassing and eventual fire caused by pellets made of the new feedstock.

12.1.2.3 Pellet production process optimisation

R&D activities in Denmark concentrate on optimisation of the supply chain, including the production of pellets. Focus is on friction, pressing energy and capacity in pellet mills in order to reduce production costs and energy consumption. Related R&D projects are the project "Basic understanding of the pelletising process" at the Institute for Mechanics, Energy and Construction (MEK-DTU) in cooperation with Energi E2, ReaTech and the Danish Technological Institute (2005–2006) and the ongoing project "Advanced understanding of the pelletising process" at the National Laboratory for Sustainable Energy of the Technical University of Denmark in co-operation with several industrial and scientific partners (until 2011) [104; 678].

12.1.3 Torrefaction

Several different groups and organisations are engaged in R&D activities concerning torrefaction as a pre-treatment step before pelletisation [180; 181; 186; 187; 188]. Different concepts and reactor types are under investigation and some groups claim to be close to market introduction. Therefore, it is likely that torrefaction technology will be demonstrated for the first time at an industrial scale in the near future. A detailed description of the basics of torrefaction as well as descriptions of the most relevant technological concepts and reactors can be found in Section 4.1.4.2.

Many R&D activities concerning production of torrefied biomass pellets are concentrated in the Netherlands. Process development being led by at least two organisations, and the first commercial production plant should be in operation in 2010 [188; 191; 193]. Another pilot plant for pelletisation of torrefied biomass is planned in Austria. It should be put into operation at the end of 2010 [194; 195; 627]. In addition to optimising the production process, remaining questions are related to the fate of chlorine and alkaline fuel components during the torrefaction process since these may have a negative impact on boiler operation due to corrosion and ash deposition. One possible option is the TORWASH process, under development at ECN. In this process, biomass is heated in pressurised hot water, so that it both torrefies and releases these components as these are usually soluble.

Torrefied pellets are also studied and tested in Denmark. An R&D project entitled "Upgrading fuel properties of biomass fuel and waste by torrefaction" is currently being carried out (until 2012) at the Danish Technological Institute in cooperation with the National Laboratory for Sustainable Energy of the Technical University of Denmark, DONG Energy Power, University of Copenhagen and Forest & Landscape.

In Sweden, the newly founded company BioEndev is planning a research and demonstration plant for torrefaction with a capacity of 21 to 24 MW torrefied product (based on NCV) in cooperation with Umeå University. The plant will be in operation at the end of 2010. In a first stage, forest residues will be used followed by more challenging raw materials. A pilot torrefaction plant (continuous flexible rotary kiln, 30 kg/h) has also been in operation since 2008 by ETPC at Umeå University.

12.1.4 Decentralised pellet production

In contrast to the ongoing trend to build increasingly large pellet production plants, frequently with production capacities of 100,000 t (w.b.)$_p$/a or more (cf. Chapter 4), a Finnish project [679] follows a decentralised approach of pellet production precisely because there is often a lack of knowledge and technological and economic limits in this area (cf. Section 7.3). Within the framework of this project, barriers to produce pellets at a small- or medium-scale are tackled by best practice examples, exchange of required knowledge and proper networking by all players.

A Swedish company [680] has specialised in small-scale pellet production plants and provides complete solutions with capacities from 250 up to 1,000 kg (w.b.)$_p$/h.

With the knowledge that potential raw materials for pelletisation are often available in small amounts at small- and medium-sized enterprises and have to be transported to central larger pellet production plants, a decentralised approach surely presents an effective means to complement large-scale pellet production.

12.2 Pellet utilisation

12.2.1 Emission reduction

With regard to emission reduction from pellet furnaces, positive developments have already been achieved through technical progress, as demonstrated by Section 9.5.2. Numerous activities are concerned with further emission reduction from pellet furnaces (cf. for instance [681; 682; 683; 684]). Further reduction of particulate matter in general and especially fine particulate matter and aerosols is a focus (cf. Section 9.9). In addition, measures to reduce other emissions such as CO, NO_x or C_xH_y are being examined.

Activities are ongoing to introduce pellets from herbaceous biomass fuels into the market (cf. Section 12.1.1.1). However, combustion of herbaceous biomass is associated with increased particulate matter emissions and especially the emission of fine particulate matter. The contents of a number of undesired elements such as nitrogen, sulphur and chlorine (cf. Section 3.5 in this respect) are higher than in wood, which results in increased NO_x, SO_x and HCl emissions. Increased chlorine contents also augment the corrosion risk. The ash content of herbaceous biomass is also higher and the ash melting point is lower, which causes the probability of slagging and deposit formation to rise.

Within the framework of the international workshops "Aerosols in Biomass Combustion" in March 2005 [428] in Graz and "Fine particulate emissions from small-scale biomass combustion systems" in January 2008 in Graz [685], the issue of fine particulate matter and aerosol emission reduction was discussed in great depth.

12.2.1.1 Fine particulate emissions

12.2.1.1.1 Fine particulate formation and characterisation

There are relatively high emissions of particles < 1 µm (PM1) arising from biomass combustion. When complete combustion takes place, these particles are mainly salts of ash forming elements (KCl, K_2SO_4, K_2CO_3) [434]. If softwood pellets are used, K_2SO_4 is the dominating compound of the inorganic ash fraction in all cases of investigated particulate matter from pellet combustion [405]. In modern, automatic wood furnaces, the share of organic components out of total particulate matter is usually below 5 wt.%, with ideal combustion conditions even below 1 wt.% (related to operation at nominal load).

Measurements of fine particulate matter emissions (PM10) in old and new firewood boilers as well as old and new pellet furnaces were carried out by [435]. Regardless of the type of furnace, the fuel quality and the operating conditions, the particulate emissions were dominated by the fractions of the smallest particles lying within the range of micrometers both with regard to mass and quantity. Similar results were achieved by [686], claiming that around 90% of total particulate matter emissions are of less than 1 µm in diameter. Mass concentrations were the highest in old firewood furnaces and lowest in modern pellet furnaces. The number of measured particles was scattered in a wide range, which led to quantities of particles from pellet combustion being in part more than those of the firewood combustion. On the whole, pellet furnaces exhibited lower concentrations here too. In addition, a definite correlation between mass concentrations of particulate matter and the sum of hydrocarbons (C_xH_y or OGC) was found. The rise of particulate emissions under poor combustion conditions had been found already by previous work [687; 688]. It occurs as a result of fine particle formation during incomplete combustion (soot and condensed hydrocarbon compounds). This supports the theory that the toxicity of particulate matter is increased by a rise in the unburned share of particulates because this raises the share of organic matter in total particulate matter.

Within the framework of the project "Fine particulate emissions from small-scale biomass furnaces" [689] at the Institute for Process and Particle Engineering (IPPT) at the Graz University of Technology in co-operation with the Austrian bioenergy competence centre BIOENERGY 2020+, previously unavailable data on the main characteristics of fine particulate emissions from modern small-scale furnaces (pellet boilers, wood chip boilers, firewood boilers) should be acquired and evaluated by field and test stand measurements. For comparative reasons, the same measurements were carried out in oil furnaces. The project confirmed the fact that modern small-scale biomass furnaces have notably lower fine particulate emissions compared to old systems and that the share of fine particulate emissions out of total particulate emissions was above 90 wt.% in all systems that were investigated. Emission peaks were found to arise during unsteady operation (start-up, load change), but were less accenttuated in pellet and wood chip furnaces than in firewood furnaces. Emission peaks are the result of incomplete combustion and are caused by the formation of organic aerosols and incomplete oxidation of soot particles. If operating conditions at nominal load are steady and if complete flue gas burnout is achieved, the formed aerosols contain less than 10 wt.% organic carbon and soot. The fine particles consist mostly of potassium, sulphur and chlorine and to some extent sodium and zinc. Thus, at steady operating conditions, it is mainly the composition of the fuel that determines the amount of fine particulate emissions. When operating conditions are unsteady and often also at partial load, fine particulate emissions increase due to non-ideal combustion conditions that lead to incomplete flue gas burnout. Aerosol emissions from furnaces fired with extra light fuel oil mainly consist of sulphur and organic compounds, soot and to a small extent heavy metals. Fine particulate emissions of such oil furnaces amount not even to 1 mg/Nm³ (related to 13 vol.% O_2, dry flue gas), so they are considerably lower than those of biomass combustion plants but show a very different chemical composition.

12.2.1.1.2 Primary measures for particulate emission reduction

In Switzerland, wood pellets are mainly used for residential heating in stoves and boilers < 70 kW. R&D for residential applications of wood pellets focuses on the reduction of pollutant emissions, especially on particulate matter. Besides type test measurements, current investigations focus on emissions under practical operating conditions including cold start and part load operation. In order to minimise emissions and increase annual efficiencies of pellet boilers in practice, the influence of the system integration of pellet boilers for residential heating is being evaluated and recommendations for improved control technologies are being developed. In addition to applications of wood pellet systems alone, combinations with solar heating and heat storage tanks are being evaluated. Furthermore, the influence of cold start and part load on emissions in practice are being evaluated.

In addition to applications in residential heating, wood pellets are also used in boilers > 70 kW and up to 1 MW heat output, for example in applications where the storage room is not sufficient for wood chips. For such applications, specific boilers for wood pellets have been developed. To guarantee the emission limit values of such applications, primary measures to reduce particulate matter emissions have been developed using staged combustion ("low-particle combustion") [690; 691].

The Austrian bioenergy competence centre BIOENERGY 2020+, Graz, in co-operation with partners from Finland, Germany, Sweden, Poland, Ireland and Denmark, is currently working on the ERA-NET Bioenergy R&D project "Future low emission biomass combustion systems" (FutureBioTec) [692], which focuses on the further development of wood stoves for significantly decreased PM emissions by air staging and optimised air distribution, grate design and implementation of automated process control systems. It also focuses on the improvement of automated furnaces in the small- to medium-scale capacity ranges to achieve lower PM emissions by extremely staged combustion, use of additives and fuel blending with new biomass fuels.

12.2.1.1.3 Fine particulate precipitation

Several fine particulate precipitation systems were evaluated with regard to their basic suitability for small-scale biomass furnaces within the framework of [689]. In principal, there are four different technologies available:

- Gravitational separation (cyclones or multi-cyclones);
- Scrubbers and flue gas condensation systems;
- Filters (baghouse filters and metal fibre filters);
- Electrostatic precipitators.

Analysis of these fine particulate precipitation systems showed that electrostatic precipitators are likely to be the best option. Developments of electrostatic precipitators that are designed especially for use in small-scale biomass furnaces are in progress. Cyclones are not suitable for aerosol precipitation due to their limitations with regard to separation size. Scrubbers and flue gas condensation systems exhibit moderate precipitation efficiencies according to the experience that is available. Baghouse filters are prone to blockages, danger of fire and they have high pressure losses.

Within IEA Bioenergy, Task 32 "Biomass Combustion and Co-firing", the study "Review of small-scale particle removal technologies" is ongoing (2010), where systems for furnaces up to 50 kW$_{th}$ are evaluated. It shows that current developments are all based on electrostatic precipitators (ESPs), mostly on dry ESPs but in part also on wet ESPs in combination with heat recovery (flue gas condensation). Approximately 15 manufacturers are active in such R&D projects.

Within the aforementioned ERA-NET project FutureBioTec at the Austrian bioenergy competence centre BIOENERGY 2020+ [692] (cf. Section 12.2.1.1.2), not only primary measures for particulate emission reduction are being investigated but also secondary measures for residential biomass combustion systems are being evaluated and tested.

A system for fine particulate precipitation in small-scale wood furnaces was developed on the basis of the electrostatic precipitator's working principle [693]. Precipitation efficiencies of beyond 80% were achieved at test stands. In the field, precipitation efficiencies of up to 60% were achieved, even though the technology had not been adapted or optimised with regard to the system as a whole. As a next step towards market introduction, a first small series will be installed. Investment costs are too high for wide ranging application of the system. Series production of large numbers could reduce these costs and rendering the technology an interesting possibility for reduction of fine particulate emissions in the area of small-scale furnaces.

Within the framework of studies by [435], a new kind of electrostatic precipitator was developed that exhibited precipitation efficiencies of 84 ± 4% in old wood furnaces. The particulate emissions of 4 mg/Nm³ were the lowest of all furnaces that were investigated. More information on fine particulate emission reduction can be found in [334; 694; 695; 696; 697; 698; 699; 700; 701; 702].

R&D activities are ongoing with the aim to combine the Schräder Hydrocube, a flue gas condensation system already developed (cf. Section 6.1.2.9.1.3.2.4), with a wet electrostatic precipitator. The Schräder Hydrocube already exhibits a certain particulate precipitation effect, which should be improved and extended to particles with aerodynamic diameters below about 1 μm. First measurements at a prototype have shown that particulate removal efficiency can be improved [703].

12.2.1.1.4 Health effects of fine particulate emissions

Tests with regard to the health effects of fine particulate emissions were carried out by [434; 704]. Inorganic particulate matter from wood combustion under ideal conditions (modern, automatic pellet furnace), particulate emissions from the incomplete combustion of wood (old, manual wood furnace) and diesel soot were examined. In addition, volatile organic components such as polycyclic aromatic hydrocarbons (PAHs) can be absorbed at the surface of all kinds of particulate matter.

The filters that were charged with flue gases from an almost complete combustion of wood and from diesel combustion (in combustion engines) showed completely different colours after sampling. The filters laden with particles from diesel combustion were black due to the soot particles in the diesel exhaust gas. The filters laden with particles from ideal wood combustion were white. The particulate matter of ideal wood combustion hence contains an irrelevant amount of black soot particles and mainly consists of inorganic salts.

The tests revealed that the inorganic fine particulate matter of an almost complete combustion of natural wood in an automatic wood furnace has 5 to 10 times less biological reactivity by cell toxicity than diesel. Fine particulate matter from incomplete combustion of natural wood in a poorly operated old wood furnace has a reactivity around 10 times as high and around 20 times as high a PAH content as diesel exhaust gas. Thus, in comparison with inorganic particulate matter from almost complete combustion of natural wood, fine particulate matter of incomplete combustion under very poor conditions has about 100 times the biological reactivity. The results are derived from first trials and should not be regarded as certainties as concerns toxicity of different particulate matters as more detailed investigations and replications are required.

The differences between particulate matters of diesel and wood combustion under ideal and poor conditions are probably caused by the differences in chemical composition of the particulate substances. Diesel particulate matters and particulate matters from incomplete combustion of wood consist mainly of unburned carbonic substances with low inorganic contents. Particulate matters in state-of-the-art automatic wood furnaces consist of inorganic compounds for the most part (mainly potassium salts). It is very probable that incomplete combustion of biomass with high concentrations of organic substances, as it is the case in old, poorly designed or poorly operated furnaces, leads to higher toxicity of the emitted particulate matters than is the case in optimised combustion processes.

Results presented at the Central European Biomass Conference 2008 in Graz also showed that health effects of particulate emissions seem to strongly depend on the concentrations of carbonic substances in the fine particulate matter. Finnish in-vitro tests, in which lung cells were exposed to fine particulate matter sampled from wood combustion, revealed that particle samples of incomplete combustion caused much stronger reactions as well as more dead cells than particulate emissions of an almost complete combustion, as takes place in modern biomass furnaces [425]. In-vivo studies (inhalation tests) with rats, which were carried out in Germany with particulate matter from complete combustion had almost no adverse effects [426]. To conclude, it is important to look at the chemical composition of fine particulate emissions in a toxicological evaluation.

It can be derived from these studies and investigations that the health risk posed by fine particulate emissions of modern small-scale biomass furnaces seems much less than that posed by old, poorly controlled small-scale biomass furnaces. One continuing research project aiming at the investigation of the correlation between the chemical composition of fine particulate emissions from small-scale biomass furnaces and their toxicity is currently being carried out by the Institute for Process and Particle Engineering at the Graz University of Technology and the bioenergy competence centre BIOENERGY 2020+ in Graz in co-operation with Finnish research institutes.

The ongoing ERA-NET project "Health effects of particulate emissions from small-scale biomass combustion" (BIOHEALTH) [705] aims to produce new scientific data for the assessment of potential health risks of different combustion technologies and biomass fuels in order to guide the development of clean small-scale combustion systems and to support authorities in the development of guidelines and legislation via an upcoming database from this project.

12.2.1.2 Gaseous emissions

Apart from reducing fine particulate emissions from pellet furnaces, measures for further reduction of gaseous emissions, such as CO, NO_x or OGC, are being examined within the framework of several R&D activities, showing some interactions between single measures. For instance, measures to reduce fine particulate emissions by thorough mixing of combustion air and flue gas as well as sufficient residence time of the flue gas at certain high temperatures in the combustion chamber are also able to reduce CO and organic carbon emissions since ideal combustion and complete flue gas burnout can be achieved (cf. Section 9.10.3). A staged air supply also improves burnout. This can be done by separating the combustion chamber into primary and secondary combustion zones. This spatial separation inhibits re-mixing of primary with secondary air and the primary combustion zone can be operated as a gasification zone with a substoichiometric air ratio, which is of great relevance for NO_x emissions since the formation of N_2 is favoured under substoichiometric conditions. Complete oxidation of the flue gas takes place in the secondary combustion zone, thorough mixing of flue gas and secondary combustion air being of major importance. This is achieved by appropriate combustion chamber geometries and nozzle design. Moreover, a long residence time of the hot flue gas and hence a sufficiently large-sized combustion chamber is necessary for full burnout of the flue gas.

A research team of the Austrian bioenergy competence centre BIOENERGY 2020+, Graz, in co-operation with the Institute for Process and Particle Engineering at the Graz University of Technology is engaged in the issue of NO_x formation and further NO_x reduction, mainly focusing on primary measures such as air staging [706; 707; 708; 709; 710]. Currently, primary and secondary gaseous emission reduction measures are aimed at within the ERA-NET project FutureBioTec at the Austrian bioenergy competence centre BIOENERGY 2020+ [692].

12.2.2 New pellet furnace developments

12.2.2.1 Pellet furnaces with very low nominal boiler capacities

Recent developments are exploring the lower capacities of pellet furnaces, reflecting the trend towards low energy houses. In Germany new building standards were recently introduced that drastically reduce the allowed heating load for houses. One possibility for such applications is pellet stoves that are equipped with a hot water heat exchanger coupled with a heat buffer storage system. The system can be combined with hot water supply by solar energy, creating an innovative heating concept for low energy houses. Systems such as these are already on the market.

Other developments move towards very small furnaces for pellet central heating systems. An ongoing Austrian development aims at a pellet boiler with a nominal capacity of 3 kW [711]. A German development aims to achieve operation with a very low nominal boiler capacity with a furnace based on a small rotary grate [287; 712].

An Austrian pellet furnace manufacturer recently presented a pellet boiler with a nominal thermal output of just 7 kW, which is designed to be wall mounted. Its minimum thermal output is only 2 kW and the pellet boiler is therefore well suited for modern low energy houses. Pellets are supplied via a pneumatic feeding system from the storage room to the furnace and so fully automatic operation is achieved [713; 714].

12.2.2.2 Multi fuel concepts

Multi fuel boilers are a relatively new concept. Multi fuel systems allow the use of pellets and firewood. If the furnace is operated with pellets, the user benefits from all advantages of a conventional, fully automatic pellet furnace. In addition, the user retains the possibility of using firewood, in most cases without any modifications of the furnace (e.g. by simply putting it in a grate). Innovative boiler systems automatically recognize the fuel used. So, when firewood is fed, the pellet supply is reduced accordingly. When the firewood has burned down, pellets are automatically fed again, if needed. Systems such as these are suitable for users who can easily obtain firewood and are not put off by the increased operational effort, but at the same time do not want to do without automatic operation [282].

12.2.3 Increase of annual efficiencies

As shown in Section 9.7, the annual efficiencies achieved by pellet boilers in operation are in some cases quite low. Values as low as 69.9% were measured and it must be expected that even poorer installations exist (cf. Section 11.2). Several activities are ongoing to improve existing pellet boiler installations with regard to their annual efficiencies. The load control strategy and system integration seem to play a major role concerning annual efficiencies. The load control has to better adapt the boiler output to the actual heat demand of the building. In this way, repeated start-ups and shutdowns could be avoided and losses reduced. Proper system integration should be worked at, for example, by appropriately dimensioned heat storage systems. The measures must be taken into account under consideration of the constraints of different residential heating systems. Low temperature heating systems such as wall or floor heating systems have different requirements than heating systems based on radiators. Whether or not a solar heating systems is installed for just hot water supply or to support room heating, again requires a different strategy. Thus, sound solutions to optimally adapt the hydronic circuit and the control of residential houses to the pellet boiler and vice versa are important. In general, there is great potential for optimisation with regard to the annual efficiency of pellet boilers.

Further field tests are being carried out to identify the reasons for low annual efficiencies that have been observed in practice and to find possible solutions to this problem [715].

Another issue is related to the development of test stand methods for the determination of annual efficiency. Currently, type tests for heating systems are performed on test stands where the nominal efficiency as well as the part load efficiency are determined, both at steady state conditions. These efficiencies, however, provide limited information about the annual utilisation rates achievable with any one system. Therefore, a method is under development where not only steady state conditions but also start-ups, shutdowns, load changes, stand-by periods and the characteristics of the load control system are taken into account [414]. The test stand results for annual efficiencies are in close agreement with results from field tests.

12.2.4 Micro- and small-scale CHP systems based on pellets

CHP technologies were examined in Section 6.4. In the field of micro-scale CHP generation, thermoelectric generators prove a good option. Thermoelectric generators exploit a thermoelectric effect in which a current flows in an electric circuit made of two different metals or semi-conductors, as long as the electric contacts have different temperatures (the principle is shown in Section 6.4.1). Application in pellet furnaces is currently tested by means of prototypes within the framework of an R&D project of the Austrian bioenergy competence centre BIOENERGY 2020+ at their site in Wieselburg [169; 345; 346; 352; 353; 354; 355]. The development aims at generating sufficient energy for the pellet furnace itself so that the system can operate autonomously, i.e. without the need for an external electrical grid, which, in contrast to classic naturally draught firewood furnaces, is not yet possible in automated pellet furnaces. The main challenge is to make such systems economical and to achieve a better lifespan.

Another option in this field is Stirling engines. An Austrian company worked on the development of a micro-scale CHP system based on a 1 kW_{el} Stirling engine [347; 349; 350; 351]. However, this development was recently discontinued [716]. A German company [717] is also developing a Stirling engine as a micro-scale CHP technology with an electric capacity of 3 kW. However, technical problems currently hamper the development [718]. The Stirling engine is also a promising option for medium-scale biomass CHP systems. Demonstration projects with Stirling engines with 35 and 70 kW_{el} are ongoing [356; 357; 358; 359] (cf. Section 6.4.2).

12.2.5 Utilisation of pellets with lower quality

R&D activities with regard to utilisation of pellets made of herbaceous biomass presently concentrate on measures to improve the quality of the fuel itself and on manipulating the fuel's combustion behaviour by using certain additives (cf. Section 12.1.1).

In addition, some Austrian producers are working on concepts of pellet furnaces that may be fed with pellets made of herbaceous biomass. In part, such systems are already available on the market [141]. R&D activities of the kind are also carried out at the Austrian bioenergy competence centre BIOENERGY 2020+ at their site in Wieselburg [719; 720].

Owing to strong pellet market growth, using raw materials of lower quality for pellet production and for the use in small-scale systems is repeatedly contemplated. Herbaceous biomass and SRC such as poplar and willow are often mentioned in this respect. These raw materials and hence the pellets made of these raw materials would have higher ash contents and lower ash softening and melting points than the wood pellets currently used.

A Swedish development is trying to accommodate the higher ash contents and lower ash softening and melting points of pellets made of herbaceous biomass with a new type of burner [721]. First trials with problematic fuels showed promising results when compared to conventional burners.

The basic problems with thermal utilisation of herbaceous biomass at residential scale (10 to 15 times higher ash content than woody biomass, low ash melting temperature and thus increased dangers of slagging and deposit formation, increased risk of corrosion and higher fine particulate emissions, elevated N contents resulting in higher NO_x emissions) have until now not been solved to a satisfactory extent for small-scale applications. Further R&D work is thus needed and whether these problems can be solved at all remains unclear. At present, the use of herbaceous biomass in small-scale furnaces cannot be recommended. Indeed, using herbaceous biomass in pellet boilers that are presently available on the market must be strongly advised against as these boilers are not made to operate with these fuels. Most likely it is more reasonable to use such fuels in medium- and large-scale systems that can be better adapted to the use of these comparatively difficult fuels in terms of process complexity.

Numerous projects are occupied with this subject, which examine the characteristics of pellets made of herbaceous biomass, looking for appropriate technological solutions in the area of small- and medium-scale systems. In Scandinavian countries in particular, but also in Austria, projects concerned with pellets made of herbaceous biomass have already been carried out or they are in progress [722; 723; 724; 725; 726; 727; 728; 729].

12.2.6 Furnace optimisation and development based on CFD simulations

Despite the ecologic benefits of using renewable energy resources, technical and economic optimisations of biomass furnaces are necessary to render these systems competitive against heat and electricity generation systems based on fossil fuels. Development of biomass furnaces, especially within the small-scale power range, is often based on information gained empirically, involving long development times and excessive test effort. CFD simulation allows for shorter development times and less test effort and it also increases the reliability of developments.

Even though combustion of solid biomass in a fixed bed and with a turbulent, burnable flow in a combustion chamber of complex geometry is complicated, BIOS BIOENERGIESYSTEME GmbH has successfully developed and optimised combustion chambers of several biomass furnaces using CFD simulations [710; 730; 731; 732; 733; 734]. Large- (10 to 30 MW_{th}), medium- (0.3 to 10 MW_{th}) and small-scale (< 300 kW_{th}) furnaces have been developed and optimised, among them several pellet furnaces.

The CFD model for design and optimisation of biomass furnaces and boilers was developed in co-operation with researchers from the Institute for Process and Particle Engineering at Graz University of Technology. It consists of an empiric combustion model developed in-house as well as validated sub-models of the CFD software Fluent for turbulent and reactive combustion air flow. Tests and verifications of the whole CFD model for biomass grate furnaces were carried out on furnaces at pilot and industrial scale.

CFD is the spatially and chronologically resolved calculation of flow processes (laminar, turbulent, chemically reactive and multiphase). A three dimensional visualisation of the turbulent, chemically reactive flow in the combustion chamber is thereby possible. Such a visualisation appears in Figure 6.20 and is described in Section 6.1.2.4. Figure 12.1 presents another successful optimisation of a biomass furnace on the basis of the iso-surfaces of the flue gas temperature in horizontal cross sections of a medium-scale pellet furnace. The reduction of temperature peaks that could be achieved is clearly visible. Figure 12.2 shows the iso-surfaces of the CO concentrations in the flue gas in horizontal cross sections of the same furnace both before and after optimisation.

The aims of CFD simulations of pellet furnaces can be summarised as follows:

- Achieving an efficient air staging as a basis for staged combustion and thus NO_x emission reduction.
- Ensuring ideal mixing of unburned flue gas with secondary air to achieve complete burnout of flue gas at nominal and partial load (low CO emissions).
- Improved utilisation of furnace and boiler volume by optimisation of furnace geometry.
- Reduction of local velocity and temperature peaks to minimise material erosion and deposit formation.

- Evaluation of the sensitivities of, for example, load condition, moisture content or change of air staging of the combustion, as a basis for optimisation of the control system.

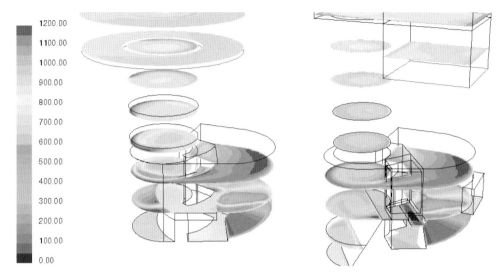

Figure 12.1: Iso-surfaces of flue gas temperature [°C] in horizontal cross sections of the furnace

Explanations: left: basic design, right: optimised design; level of cross sections: 0.0, 0.2, ..., 1.4 m above reference surface (reference surface = upper edge of the grate/lower edge of the primary combustion zone); data source [735]

By means of CFD simulations of pellet furnaces the following activities can be carried out:

- Design and optimisation of furnace and boiler geometry (incl. convective section).
- Design and optimisation of boiler cleaning systems.
- Design and optimisation of secondary air nozzles.
- Efficient reduction of CO and NO_x emission in nominal and partial load operation.
- Evaluation and optimisation of operating conditions for furnaces and boilers with regard to efficiency, plant availability, part load behaviour and multi fuel use.
- Reduction of local temperature peaks by furnace cooling or improved operating conditions.
- Simulation and reduction of deposition and material erosion tendencies caused by fly ash.
- Calculation of the heat exchange and the influence of deposits in the radiative section of steam and thermal oil boilers.
- Calculation of residence times and flue gas temperatures as a basis for the modelling of fine particulate and NO_x emissions.

Furnace development supported by CFD simulations shows clear advantages concerning reduced emissions, higher efficiencies, lower volumes, reduced material wear, increased plant availability, reduced development time and reduced test efforts as well as increased reliabilities of developments, which have already been verified by practical experience. The application of CFD simulation for the development and optimisation of small-scale pellet furnaces has successfully been established; it contributes to a better understanding of the combustion processes in the furnace and is gaining increasing importance as an innovative furnace development tool.

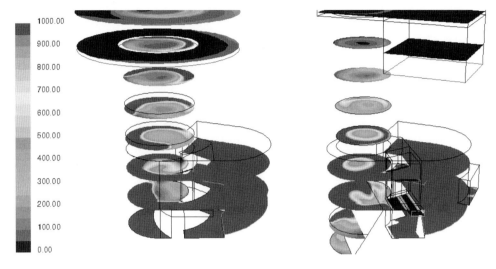

Figure 12.2: Iso-surfaces of CO concentration in the flue gas [ppmv] in cross sections of the furnace

Explanations: left: basic design, right: optimised design; CO emissions at the inlet to the heat exchanger: 63 mg/Nm³ dry flue gas, 13 vol.% O_2 in the basic design and 6 mg/Nm³ dry flue gas, 13 vol.% O_2 in the optimised design; level of cross sections: 0.0, 0.2, ..., 1.4 m above reference surface (reference surface = upper edge of the grate/lower edge of the primary combustion zone); data source [735]

Moreover, several R&D activities to further develop and improve the used CFD models are ongoing with a main focus on the modelling of the combustion process in the fixed bed as well as on NO_x and deposit formation modelling. The ERA-NET project "Scientific tools for fuel characterisation for clean and efficient biomass combustion" (SciToBiCom) [736], coordinated by the Technical University of Denmark with partners in Denmark, Finland, Norway and Austria, intends to develop advanced standard characterisation methods for biomass fuels in various combustion systems as well as advanced CFD based simulation routines that consider single particle conversion and solid biomass combustion for different applications.

12.2.7 Pellet utilisation in gasification

In Denmark gasification of pellets in small-scale gasifiers (20 to 300 kW$_{el}$) was tested from 2007 to 2008 within the project "Development and demonstration of combined heat and power on wood pellets in a staged open core gasifier" by the company BioSynergi Proces ApS. The test gasifier ran successfully on wood pellets for 700 hours, including 160 hours with the gas engine in operation. Stable gas production was achieved, however, stability during engine tests was less satisfactory. Wood pellet gasifier applications may be the right choice in some instances, however, because wood pellets are significantly more expensive than wood chips, wood chips are expected to be the preferred option.

12.3 Support of market developments

To date, the use of pellets has been limited to just a few countries. The technology for pellet utilisation is available to a satisfactory extent but knowledge of it and the acceptance of the technology remain limited. The project BIOHEAT [737], financed by the European Commission, tried to raise awareness and provide information on a European scale by involving chief market players such as national energy agencies as well as providers of renewable energy systems that were interested. Thanks to the experience that had been gained over many years in some countries, a knowledge transfer can be achieved by appropriate information exchanges. Thus, mistakes that were made early in some countries can be avoided in the development of new markets, and market development can proceed more rapidly. The EU-ALTENER project "Pellets for Europe" tried to support and foster market development by collecting and distributing important R&D activities and results [738].

At present, market actors from the pellet sector are linked through an internet platform and supplied with suitable information, such as on current market data or other reports, within the framework of the project Pellets@las [496].

Projects and initiatives such as this should support the distribution of pellets in all areas of application as well as in newly or slowly developing markets in the future.

However, it is also in well developed markets such as Germany or Austria that there are hindrances that slow down market development. A study [739] showed that environmental aspects were important in decision making for German consumers that purchased a pellet central heating system. Investment costs played a subordinate role. However, investment costs did play a role for consumers opting for another heating system. This leads to the conclusion that the comparatively high investment costs of a pellet heating system are still a hindrance to market development, despite possible subsidies, and faster market growth cold be achieved by cheaper pellet heating systems. A study on Austrian consumers led to similar conclusions [740], with more than 92% of interviewees claiming environmental aspects to be "important" or "very important" in decision making, while most of all the purchasers of pellet furnaces opted for "very important". It is interesting in this respect that purchasers of oil or gas heating systems also often answered with "important" or "very important". It can be assumed that the fine particulate emissions issue, in which pellets have often received bad press, contributed to the fact that purchasers of oil or gas heating systems estimated their heating systems to be very environmentally friendly. This perception may probably have changed because the public image of pellet heating systems with regard to fine particulate emissions has been improved through appropriate public awareness work. In addition, investment costs play a key role in decision making regarding heating systems, indeed, it is often just the investment costs alone that inform decisions and not the operating costs, consumption costs and other costs. Low investment costs were said to be "very important" or "important" by more than 50% of questioned individuals. Only 13% stated these costs to be "not important at all" or "not important". Purchasers of gas heating systems in particular named investment costs as "important". A relative majority of 23% of purchasers stated the investment costs to be the decisive factor in their decision for a heating system. The results of this study show that there is a need for action and raising awareness with regard to the total costs of heating systems since the decision in favour of a seemingly cheap heating system can result in much higher costs when looking at the system's lifespan, as has been shown on the basis of the full cost calculations in Chapter 8.

In an ongoing project studying Swedish consumers, it has been found that the most important parameters for choosing a (central) heating system are, in decreasing order, economy, reliability, easy maintenance, price stability (of the fuel) and clean indoor environment [741]. It seems to be important for the future to make the public aware of the fact that a full cost calculation and not a comparison of investment costs is an important and proper basis for a decision.

Another key research area is related to ensuring the sustainability of imported biomass. Sustainability criteria have been issued by the Dutch government, and in future all imported biomass has to comply with these. Practical experience is currently being gained to validate the feasibility of such a quality certification scheme.

12.4 Summary/conclusions

There remains a continuing need for R&D in many areas of pellet production and utilisation, which is being met by numerous activities and projects on national and international scales.

Concerning pellet production, alternative raw materials are one focus because strong growth of the worldwide pellet market using conventional raw materials, i.e. wood shavings and sawdust, has already been subjected to shortages of these raw materials. Market growth will only be possible on the basis of alternative raw materials in the medium term. In addition to log wood, which is already used for pelletisation to some extent, the focus lies on herbaceous raw materials and SRC. Looking at these raw materials, it is especially the higher ash content, the lower ash softening and melting temperatures and the higher nitrogen content that cause problems, the solutions to which are being worked on. Moreover, the further improvement of the quality of wood pellets is an R&D issue. Production technology is very advance and the general trend is to have large-scale production plants, often with capacities of 100,000 t (w.b)$_p$/a and more. Decentralised pellet production in small-scale units confront many technical and economic problems. Activities are in progress to overcome these constraints and they aim to supply relevant market players with appropriate information. As potential raw materials for pelletisation are often available in small amounts in small- and medium-sized enterprises and have to be transported to central, larger plants, decentralised pellet production must be regarded as a good means to complement large-scale pellet production.

Additional problems in pellet production are off-gassing, self-heating and self-ignition in raw material and pellet storages. Research projects are also in progress that aim to investigate the basics of this phenomena and derive suitable measures for prevention and, if the case should arise, for fast detection of possible off-gassing, self-heating or self-ignition processes. It is important to investigate chemical degradation reactions in pellets during storage to learn more about the mechanisms and kinetics and possibly be able to completely reduce or prevent off-gassing, self-heating and self-ignition during transport and storage. Emissions of aldehydes or ketones sometimes cause bad and strong smells in pellet storage rooms, which may negatively affect pellet marketing. Therefore, R&D in this area must be continued to secure future pellet production and consumption.

A rather new raw material pre-treatment step for pelletisation is torrefaction. Several R&D activities are known focusing on this technology, which is close to being demonstrated for the first time at an industrial scale. Pellets from torrefied biomass are characterised by higher net calorific values and bulk densities and consequently by higher energy densities. Moreover, they can be ground easily and are of a hydrophobic nature, which makes their storage and logistics easier. Their main market use will be co-firing in coal power plants, as torrefied pellets can be handled together with coal without any additional changes or investments in storage or feeding systems. The use of torrefied pellets in the residential heating sector is also a target.

Concerning pellet utilisation, emission reduction, new pellet furnace developments and the utilisation of pellets made of new biomass fuels (e.g. herbaceous biomass, SRC) are the chief issues. Activities concerning emission reduction are especially concerned with the reduction of fine particulate emissions, whereby it has to be noted that modern pellet furnaces already emit far less fine particulate matter than do conventional wood furnaces, and hence pellet furnaces are particularly suitable for reducing fine particulate emissions by replacing old, often poorly controlled, wood furnaces. Both primary and secondary measures are investigated. The newest developments within the field of pellet furnaces are especially furnaces with heat recovery (flue gas condensation), boilers for the combined use of firewood and pellets, pellet furnaces with very low nominal heat capacities for the use in low energy houses, the sensible combination of pellet and solar heating systems and the development of micro-scale CHP systems. Concerning the use of pellets made of herbaceous biomass, there are activities aiming to improve the combustion behaviour of herbaceous biomass by appropriate mixing or biological additives and activities aiming at the development of furnaces that can deal with the characteristics of herbaceous biomass. Both issues have yet to be resolved.

In order to improve the annual efficiencies of pellet boilers in the residential heating sector, several R&D activities are ongoing. The main focus in this field lies on the optimisation of the load control strategy and the proper integration of pellet boilers in the hydronic systems of residential houses. Also, test stand methods for the determination of the annual efficiency are under development. In general, there is great potential for optimisation with regard to the annual efficiency of pellet boilers.

Besides combustion, pellet utilisation in gasification is also under investigation.

Another key issue within the field of small-scale pellet furnaces is their CFD supported development and optimisation. Development times and test efforts can be clearly reduced and the reliability of developments can be improved by this innovation. Several R&D activities dealing with the improvement of CFD models and their application in furnace optimisation are under way.

In addition to all these R&D trends, there are several activities to develop strategies and take measures to boost the use and further distribution of pellets. On an international level there is a special need and there are possibilities to foster market development by knowledge and information transfers from established markets to newly and slowly developing markets. Finally, possibilities for ensuring the sustainability criteria in particular of imported pellets from outside Europe are being investigated in R&D projects.

Appendix A: Example of MSDS – pellets in bulk

In this Appendix, the Canadian Material Safety Data Sheet (MSDS) for bulk pellets is shown as an example. This example is from the Wood Pellet Association of Canada (WPAC) and should be regarded as a self-contained document. Units, abbreviations and syntax do not necessarily comply with the rest of this book.

LOGO   (header to be completed by the party issuing the document)   Issued DATE
Company name (full legal name)

MATERIAL SAFETY DATA SHEET

WOOD PELLETS IN BULK

> For Wood Pellets in Bags, see
> **MATERIAL SAFETY DATA SHEET for Wood Pellets in Bags**
> issued by the producer

1. Product Identification and Use

Product name/trade name:	Wood Pellets
Producer's Product Code:	*(to be filled in by the party issuing the document)*
Synonyms:	Wood Pellets, Fuel Pellets, Whitewood Pellets, Softwood Pellets, Hardwood Pellets, Bark Pellets
Product appearance:	*Light to dark blond or chocolate brown, glossy to semi-glossy, cylinder with ¼ inch diameter (6.35 mm referred to as 6 mm pellets) and 5 to 25 mm in length. (proposed text to be adapted accordingly by the party issuing the document)*
Product use:	Fuel for conversion to energy, animal bedding, absorbent
HS Product Code:	44013090
United Nations Number:	Not allocated
Hazchem:	Not allocated
IMO Safety Code:	Material Hazardous in Bulk (MHB) Group B (IMO-260E)

Manufacturer: *(to be filled in by the party issuing the document)*
Name of company (full legal name with no abbreviations)
Visiting address
Place and postal code
Country

Tel.:	number incl. country code
Fax:	number incl. country code
Website:	to be filled
Email:	to be filled

Emergency contact:

Tel (direct):	**number incl. country code**
Tel (mobile):	**number incl. country code**
Fax:	**number incl. country code**

Member of Wood Pellet Association of Canada (WPAC)

Appendix A: Example of MSDS – pellets in bulk

II. Composition and Physical Properties

Wood Pellets are manufactured from ligno-cellulosic saw dust, planer shavings or bark by means of one or any combination of the following operations; drying, size reduction, densification, cooling and dust removal. The chemical composition of Wood Pellets varies between species of raw material, components of the wood, soil conditions and age of the tree. Wood Pellets are typically manufactured from a blend of feedstock with the following composition;

Feedstock	Oxygenated compounds (indicative composition in % of weight)		
	Cellulose		30 - 40
	Hemi-cellulose		25 - 30
	Lignin		30 - 45
	Extractives (terpene, fatty acids, phenols)		3 - 5
Additives	None except as stated in Wood Pellets Product Specification		
Binders	None except as stated in Wood Pellets Product Specification		

Classification as per CEN/TC 14961 Standard; D06/M10/A0.7/S0.05/DU97.5/F1.0/N0.3

Many pellet products consist of a blend of white wood and bark feedstock which may affect the characteristics of the pellets. For more detailed information about the properties, see the latest version of Wood Pellets Product Specification issued by the manufacturer. This MSDS includes the major differences in the characteristics of the Dust from pure whitewood and pure bark pellets.

III. Health Hazard Data

Wood Pellets emit dust and gaseous invisible substances during handling and storage as part of the normal degradation of all biological materials. Ambient oxygen is typically depleted during such degradation. The sizes of the particulate matter range from crumbs to extremely fine airborne dust. The dust normally settles on surfaces over time. Emitted gases are immediately diluted by the air in the containment and escape with ventilation air. If the Wood Pellets are stored in a containment which is not ventilated (naturally or forced) the concentration of emitted gases, or the oxygen depletion, may pose a health threat for humans present in the containment and the containment should be ventilated and precautions should be taken as specified in this MSDS. Section IX includes a method of estimating the concentration of gases. The gases emitted at normal indoor temperature include carbon-monoxide (CO), carbon-dioxide (CO_2), methane (CH_4) and hydrocarbons with Permissible Exposure Levels (PEL) and symptoms as follows;

Member of Wood Pellet Association of Canada (WPAC)

Entry	Substance	Permissible Exposure Level and symptom		Remedial action
Swallow	Dust	Dry sensation, see Section IX.		Rinse mouth thoroughly with water. Do not induce vomiting.
Inhale	Dust	Coughing, dry throat. For toxicological data, see Section X.		Rinse mouth thoroughly with water. Do not induce vomiting.
	Carbon monoxide (CO)	Toxic invisible and odorless gas. Living space TLV-TWA 9 ppmv (ASHRAE). Work space TLV-TWA 25 ppmv (OSHA).		If hygiene level is exceeded, evacuate and ventilate thoroughly, see Section IX for estimation of ventilation requirement.
		50 ppmv	Max 15 minutes.	
		200	Mild headache.	Evacuate.
		400	Serious headache.	Evacuate and seek medical attention.
		800	Dizziness, convulsion, unconscious in 2 hours, death in 2-3 hours.	Evacuate and seek medical attention.
		1,600	Dizziness, convulsion, unconscious, death in 1 - 2 hours.	Evacuate and seek medical attention.
		3,200	Dizziness, convulsion, unconscious, death in 1 hour..	Evacuate and seek medical attention.
		6,400	Dizziness, convulsion, unconscious, death in 25 minutes.	Evacuate and seek medical attention.
		12,800	Dizziness, convulsion, unconscious, death in 1 – 3 minutes.	Evacuate and seek medical attention.
	Carbon dioxide (CO_2)	Asphyxiating invisible and odorless gas. Occupational TLV-TWA 5,000 ppmv (OSHA)		If hygiene level is exceeded, ventilate thoroughly, see Section IX for estimation of ventilation requirement.
	Methane (CH_4)	Asphyxiating invisible and odorless gas.		Ventilate
	Hydrocarbons	See Section IX. Odor.		Ventilate
	Oxygen depleted air	Oxygen level is normally 20.9 % at sea level in well ventilated space. Minimum hygiene level is 19.5 % in work space (NIOSH)		If oxygen level is less than hygiene level, evacuate and ventilate thoroughly.
Skin contact	Dust	Itching for some people. For toxicological data, see Section X.		Remove contaminated clothing. Rinse skin thoroughly with water.
Eye contact	Dust	Tearing, burning. For toxicological data, see Section X.		Flush with water and sweep out particles inward towards the nose

IV. First Aid Procedures

Wood Pellets are considered a benign product for most people. However, individuals with a propensity for allergic reactions may experience reactions and should contact their physician to establish the best remedial action to take if reaction occurs.

In case Wood Pellets are not handled or stored in accordance with recommendations in Section VII the risk of harmful exposure increases, particularly exposure to concentration of CO higher than stipulated PEL in Section III. In case of exposure it is important to quickly remove the victim from the contaminated area. Unconscious persons should immediately be given oxygen and artificial respiration. The administration of oxygen at an elevated pressure has shown to be beneficial, as has treatment in a hyperbaric chamber. The physician should be informed that the patient has inhaled toxic quantities of carbon monoxide. Rescue personnel should be equipped with self-contained breathing apparatus when entering enclosed spaces with gas.

Carbon monoxide is highly toxic by means of binding with the hemoglobin in the blood to form carboxyhemoglobin which can not take part in normal oxygen transport, greatly reducing the blood's ability to transport oxygen to vital organs such as the brain.

Asphyxiating gases like carbon dioxide and methane (sometimes called simple asphyxiant) are primarily hazardous by means of replacing the air and thereby depriving the space of oxygen. Person exposed to oxygen depleted conditions should be treated the same as a person exposed to carbon monoxide.

V. Fire and Explosion Measures

Wood Pellets is a fuel and by nature is prone to catch fire when exposed to heat or fire. During handling of Wood Pellets there are three phases with various levels of stability, reactivity (see section IX) and decomposition products:
- solid intact Wood Pellets
- crumbs or dust
- non-condensable (primarily CO, CO_2 and CH_4) and condensable gases (primarily aldehydes, acetone, methanol, formic acid)

Extinguishing a fire in Wood Pellets require special methods to be successful as follows;

State of Wood Pellets	Extinguishing measures	Additional information
General	Restrict oxygen from entering the space where the Wood Pellets are stored.	
	Cover exposed pellets with foam or sand to limit exposure to air.	
	Be prepared for an extended period of extinguishing work. An industrial size silo may take a week to fully bring under control.	
Storage in enclosed space	Seal openings, slots or cracks where Wood Pellets may be exposed to air.	
	Inject nitrogen (N_2) or carbon dioxide (CO_2) in gaseous form at the bottom or in the middle of the pile of Wood Pellets or as close as possible to the fire if exposed. N_2 is preferred. Dosage of gas depends on the severity of the fire (how early detection is made). Recommended injection speed is 5 – 10 kg/m^2/hour (m^2 refers to the cross section of the storage containment such as a silo) with a total injected volume throughout the extinguishing activity of 5 – 15 kg/m^3 for less severe fires and 30 – 40 kg/m^3 for more advanced fires.	Recommended values developed by SP Technical Research Institute of Sweden Specific volume for N_2 is 0.862 m^3/kg and for CO_2 0.547 m^3/kg (at NTP)
Storage in open flat storage	Cover the pile of Wood Pellets with foam or sand if available or spray water. Dig out the pile to reach the heart of the fire and remove effected material.	
During handling	Restrict oxygen from entering the space where the Wood Pellets are present	
	Cover the Wood Pellets with foam or sand if available or spray water. Dig out the material to reach the heart of the fire and remove effected material.	

VI. Accidental Release Measures

If Wood Pellets are released in a populated area, the material should be removed by sweeping or vacuuming as soon as possible. Wood Pellets are a fuel and should preferably be disposed of by means of burning. Deposition of Wood Pellets or related dust should be such that gas from the material does not accumulate. Wear a protective mask to prevent inhaling of dust during cleanup (see Section VIII).

VII. Safe Handling and Storage

Precautionary measures are recommended to avoid hazardous conditions by the reactivity as outlined in Section IX developing when handling Wood Pellets.

State of Wood Pellets	Precautionary measures	Additional information
General	Always store Wood Pellets in containment with a minimum of one (1) air exchange per 24 hours at + 20°C and a minimum of two (2) air exchanges per 24 hours at + 30°C and above.	One air exchange corresponds to the volume of the containment.
	For long period storage in large bulk containment shall be as air tight as possible. Fires tend to migrate towards air (oxygen) supply. For shorter period open storage, ventilate to eliminate gas and odor.	Early warning sensors for heat and gas detection enhances the safety of storing Wood Pellets
	Protect the Wood Pellets from contact with water and moisture to avoid swelling, increased off-gassing, increased microbial activity and subsequent self-heating.	For large enclosed storage, label the points of entry to storage containment or communicating spaces containing Wood Pellets with a sign such as "Low Oxygen Risk Area, Ventilate thoroughly before Entry".
	Always protect Wood Pellets and dust from exposure to heat radiators, halogen lamps and exposed electrical circuitry which may generate ignition energy and set off a fire or explosion.	See Section IX Explosibility and applicable ATEX directives.
	Always segregate the Wood Pellets from oxidizing agents (e.g. poly-oxides capable of transferring oxygen molecules such as permanganate, per-chlorate) or reducing agent (e.g. chemical compounds which includes atoms with low electro-negativity such as ferrous ions (rust), sodium ions (dissolved sea salt)).	Schedule for Wood Pellets, Code of Safe Practice for Solid Bulk Cargoes, 2004, IMO 260E.
	Do not expose Wood Pellets to rain.	
	Do not smoke or extinguish cigarettes in the vicinity of Wood Pellets or wood dust.	Install heat and gas detectors with visible and audible alarm.
Storage in enclosed space	For large enclosed storage entry should be prohibited by means of secured lock and a well established written approval process for entry, only AFTER ventilation has been concluded and measurement with gas meter has confirmed safe atmosphere in the space. Alternatively, use self-contained breathing apparatus when entering space. Always make sure backup personnel are in the immediate vicinity monitoring the entry.	Label points of entry to enclosed storage areas containing Wood Pellets with "Carbon monoxide Risk Area, Ventilate thoroughly before Entry".
	Install N_2 or CO_2 sprinklers as per applicable fire regulations.	A Shipper Cargo Information Sheet (SCIS) must be used when shipping Wood Pellets in ocean vessels as per international regulations issued by IMO, see SCIS issued by Producer.

Appendix A: Example of MSDS – pellets in bulk

Storage in open space	For large storage spaces install water sprinklers. For smaller storage spaces, contact your local fire department for recommendations.	Sand or foam has proven to be effective to limit access of oxygen in case of fire.
During handling	Avoid breakage caused by dropping the Wood Pellets. Be aware of potential dust generation during high pressure pneumatic handling of pellets.	Monitor temperature at bearings, pulleys, augers or other heat generating machinery.
	Avoid friction generated by rough surfaces such as worn out conveyor belts as much as possible.	
	Suppress dust generation and accumulation at transfer points and in areas close to mechanical moving parts which may dissipate heat.	
	Apparatus exposed to dust generated during the handling should be rated accorded to applicable safety standards, see ATEX directives. Warning signs should be posted in areas where dust tends to remain suspended in air or settle on hot surfaces, see Section IX Explosibility.	Example of labels and pictogram: HIGH DUST CONCENTRATION OR ACCUMULATION ON SURFACES MAY CAUSE EXPLOSIONS OR FIRES. VENTILATE AND KEEP SURFACES CLEAN.

VIII. Exposure Control and Personal Protection

The following precautionary measures shall be taken for personal protection:

Activity	Precautionary measure	Additional information
Entering space containing Wood Pellets	Ventilate thoroughly all communicating spaces before entering.	For estimation of ventilation requirement, see Section IX.
	In the event the space is enclosed, always measure both level of carbon monoxide and oxygen.	Oxygen level at sea level shall be 20.9 % in well ventilated space. Space with carbon monoxide level > 25 ppmv shall not be entered into without caution, see Section III.
	When door to space is labeled with warning sign, make sure to follow instructions and obtain permit in writing to enter.	Examples of labels and pictogram: LOW OXYGEN RISK AREA. VENTILATE BEFORE ENTRY. ALWAYS MEASURE CARBONMONOXIDE AND OXYGEN.
	Use self-contained breathing apparatus if entry is required before proper ventilation has been completed.	CARBONMONOXIDE RISK AREA. VENTILATE BEFORE ENTRY. ALWAYS MEASURE CARBONMONOXIDE AND OXYGEN.
Exposure to dust from Wood Pellets	Wear protective glasses and dust respirator. Wear gloves during continuous or repetitious penetration.	

Member of Wood Pellet Association of Canada (WPAC)

IX. Stability and Reactivity Data

The stability and reactivity properties of Wood Pellets are as follows:

Parameter	Measure	Value
Odor	°C	Above + 5 °C, fresh Wood Pellets in bulk smells like aldehydes in poorly ventilated space and more like fresh softwood in ventilated space.
Off-gassing	Emission Factor (g/tonne)	Emission of CO, CO_2 and CH_4 from Wood Pellets contained in a space is a function of temperature, ambient air pressure, bulk density, void in Wood Pellets, access to oxygen, relative humidity in air (if ventilated) as well as the age and composition of the raw material (unique for the product as specified in the Wood Pellet Product Specification). The emission rate in grams (g) of off-gassing per tonne of stored Wood Pellets given below are from measurements of gas generated within a sealed containment filled with Wood Pellets at approximately constant pressure without ventilation over a period of > 20 days. The emission factors values are only valid for sealed containment without sufficient oxygen available to support oxidation of the Wood Pellets (see Oxidation in this Section). The numbers should not at any time be substituted for actual measurements. The following examples illustrate how the emission factors can be used for estimating a rough order of magnitude of the gas concentration in a non-ventilated as well as a ventilated containment with Wood Pellets, assuming the ambient air pressure is constant.
		Non-ventilated (sealed) containment
		<table><tr><th>Gas species</th><th>Temperature °C</th><th>Emission factor (±10 %) g/tonne/>20 days</th></tr><tr><td>Carbon-monoxide (CO)</td><td>+ 20</td><td>12</td></tr><tr><td></td><td>+ 30</td><td>15</td></tr><tr><td></td><td>+ 40</td><td>16</td></tr><tr><td></td><td>+ 50</td><td>17</td></tr><tr><td></td><td>+ 55</td><td>17</td></tr><tr><td>Carbon-dioxide (CO_2)</td><td>+ 20</td><td>20</td></tr><tr><td></td><td>+ 30</td><td>54</td></tr><tr><td></td><td>+ 40</td><td>80</td></tr><tr><td></td><td>+ 50</td><td>84</td></tr><tr><td></td><td>+ 55</td><td>106</td></tr><tr><td>Methane (CH_4)</td><td>+ 20</td><td>0.2</td></tr><tr><td></td><td>+ 30</td><td>1.0</td></tr><tr><td></td><td>+ 40</td><td>1.3</td></tr><tr><td></td><td>+ 50</td><td>1.5</td></tr><tr><td></td><td>+ 55</td><td>1.9</td></tr></table>
		Example A. - Mass of Wood Pellets = 1000 tonne - Bulk density of Wood Pellets = 700 kg/m³ (0.7 tonne/m³) - Solids in bulk Wood Pellets including 0.5 % fines = 50 % - Size of containment = 2800 m³ - Temperature = +20 °C (constant) - Emission factor for CO (>20 days storage time) = 12 g/tonne (see table above) **Calculation of concentration of CO (g/m³) in containment;** 12 (g/tonne)*1000 (tonne)/[2800 (m³)-50%*1000 (tonne)/0.7 (tonne/m³)] = 5.8 g/m³ **Calculation of concentration of CO (ppmv) in containment** - Ambient pressure = 101.325 kPa (1 atm) - Molecular weight of CO (Mwt) = 28 (g/mol)

Appendix A: Example of MSDS – pellets in bulk

		$(g/m^3)*(20(°C)+273.1(C°))/Mwt(g/mol)/0.012 = 5.8*293.1/28/0.012 = $ <u>5060 ppmv</u> after > 20 days of storage in sealed containment. PEL (TLV-TWA = 15 minutes, See Section III) = 50 ppmv which means a person shall not be exposed to the atmosphere in the non-ventilated containment. **Ventilated containment** | Gas species	Temperature °C	Emission rate factor (±10 %) g/tonne/day
Carbon-monoxide (CO)	+ 20	0.9
	+ 30	2.2
	+ 40	8.0
	+ 50	18.0
	+ 55	25.0
Carbon-dioxide (CO$_2$)	+ 20	1.3
	+ 30	4.8
	+ 40	17.0
	+ 50	29.0
	+ 55	119.0
Methane (CH$_4$)	+ 20	0.01
	+ 30	0.04
	+ 40	0.18
	+ 50	0.38
	+ 55	1.10 | **Example B** - Volume of Wood Pellets = 1000 tonne - Size of containment = 2800 m^3 - Storage time = 5 days - Temperature = +20°C (constant) - Ambient pressure = 101.325 kPa (1 atm) - Emission of CO = 0.9 g/tonne/day (see Table above) - Ventilation rate = 1 air exchanges (2800 m^3) /day - Molecular weight of CO (Mwt) = 28 (g/mol) - Conversion factor (g/m^3 to ppmv) = 0.012 **Calculation of concentration of CO:** 0.9 (g/tonne/day)*1000 (tonne)/[2800 (m^3/day)]*[1-exp(-2800 (m^3/day)/2800 (m^3)*5 (days)] = 0.32 g/m^3 **Conversion to ppmv:** (g/tonne)*(T+273.1(C°))/Mwt(g/mol)/0.012 = 0.32*293.1/28/0.012 = <u>279 ppmv</u> To keep the concentration below PEL the containment needs to be ventilated with more than one air exchange per day. For more accurate estimation of gas concentrations in containment with variations in temperature and pressure, see "Report on Off-gassing from Wood Pellets" to be issued by Wood Pellet Association of Canada (www.pellet.org) when results from on-going research becomes available.
Oxidization	Rate	It is believed oxidation of fatty acids contained in the woody material is the primary cause for depletion of oxygen and emission of gas species as exemplified above during storage of Wood Pellets or related dust. The depletion ratio is a function of temperature, pressure, bulk density, void in Wood Pellets, relative humidity in air (if ventilated) as well as the age and composition of the raw material (unique for the product as specified in the Wood Pellet Product Specification). The numbers below are from

Member of Wood Pellet Association of Canada (WPAC)

		measurements of gas generated within the space of the Wood Pellets at approximately constant pressure. The numbers should not at any time be substituted for actual measurements.

Temperature °C	(±10 %) Depletion of oxygen in %/24h
+ 20	0.7 – 1.2
+ 30	
+ 40	1.5 – 2.5
+ 50	
+ 55	

		For more accurate estimation of oxygen concentrations in containment with variations in temperature and pressure, see "Report on Off-gassing from Wood Pellets" issued by Wood Pellet Association of Canada (www.pellet.org) when results from on-going research becomes available.
Melting temperature	-	Not applicable.
Vaporization	-	Emit hydrocarbons as vapors above + 5 °C.
Boiling temperature	-	Not applicable.
Flash point temperature	-	Not applicable.
Auto-ignition temperature	°C	Auto-ignite of Wood Pellets at temperatures > + 260 °C in the presence of oxygen. For dust, see Section Explosibility Dust deflagration below.
Pyrophorocity	Rate	Wood Pellets or dust are not classified as pyrophoric solids as defined by UN MTC Rev.3, 2000, Class 4.2 Test N.4.
Flammability	Rate	Wood Pellets or dust are not classified as flammable solids as defined by UN MTC Rev. 3, 2000, Class 4.1 Test N.1. (Burning rate < 200 mm/2 min.) Burning rate; Airborne Wood Pellet Dust = 20 mm/2 min. Airborne Bark Pellet Dust = 22 mm/2 min.
Self-heating	Rate	Propensity to start self-heating in presence of oxygen.
Bio-degradability	%	100.
Corrosivity		Not applicable.
pH		The **p**otential for **H**ydrogen ions (pH) varies depending on species of wood.
Solubility	%	If penetrated by water Wood Pellets will dissolve into its feedstock fractions.
Mechanical stability	-	If exposed to wear and shock Wood Pellets will disintegrate into smaller fractions and dust.
Incompatibility	-	Always segregate the Wood Pellets from oxidizing agents (e.g. poly-oxides capable of transferring oxygen molecules such as permanganate, per-chlorate) or reducing agent (e.g. chemical compounds which includes atoms with low electro-negativity such as ferrous ions (rust), sodium ions (dissolved sea salt)). (See Schedule for Wood Pellets, Code of Safe Practice for Solid Bulk Cargoes, 2004, IMO 260E), see Section VII.
Swelling	Rate	If penetrated by water Wood Pellets will swell about 3 to 4 times in volume.
Shock	Rate	The mechanically integrity of Wood Pellets will degrade if exposed to an external force as a result of for example a drop in height.
Mechanical ware	Rate	Wood Pellets are sensitive to friction between the Wood Pellets and a transportation causeway or conveyor belt and may generate dust.
Explosibility	Dust deflagration	Sieving of dust for testing purposes; 230 mesh < 63 μm. Moisture content for whitewood pellets dust = 5.6 % of weight. Moisture content for bark pellets dust = 7.9 % of weight. ASTM E11-04 Standard.

		The following data is not necessarily intrinsic material constants for Dust from Wood Pellets.
		Minimum Ignition Temperature for dust cloud (T_c) Whitewood dust = + 450 °C. Bark dust = + 450 °C. ASTM E 1491 Standard.
		Minimum Ignition Temperature for dust layer 5 mm (T_{L5}) Whitewood dust = + 300 °C. Bark dust = + 310 °C. ASTM E 2021 Standard.
		Minimum Ignition Temperature for dust layer 19 mm (T_{L19}) Whitewood dust = + 260 °C. Bark dust = 250 °C. ASTM E 2021 Standard.
		Auto - Ignition Temperature for dust layer (T_{AUTO}) Whitewood dust = +225 °C. Bark dust = +215 °C. US Bureau of Mines RI 5624 Standard.
		Minimum Ignition Energy for dust cloud (MIE_c) Whitewood dust = 17 mJ. Bark dust = 17 mJ. ASTM E 2019 Standard.
		Maximum Explosion Pressure of dust cloud (P_{max}) Whitewood dust = 8.1 bar (gauge). Bark dust = 8.4 bar (gauge). ASTM E 1226 Standard.
		Maximum Explosion Pressure Rate of dust cloud $(dP/dt)_{max}$ Whitewood dust = 537 bar/sec. Bark dust = 595 bar/sec. ASTM E 1226 Standard.
		Specific Dust Constant (KSt) Whitewood dust = 146 bar.m/sec. Bark dust = 162 ba.m/sec. ASTM E 1226 Standard.
		Explosion Class (St) Whitewood dust = St 1. (> 0 to 200 bar.m/sec). Bark dust = St 1. (> 0 to 200 bar.m/sec). ASTM E 1226 Standard.
		Minimum Explosible Concentration for dust cloud (MEC_{dc}) Whitewood dust = 70 g/m^3 Bark dust = 70 g/m^3 ASTM E 1515 Standard.
		Limiting Oxygen Concentration for dust cloud (LOC_c) Whitewood dust = 10.5 %. Bark dust = 10.5 %. ASTM E 1515 Standard (modified).
	Gas	Carbon monoxide (CO) is potentially explosive in concentration > 12 % by volume (120,000 ppmv) when mixed with air. Wood Pellets are not known to generate this level of concentration.
		Methane (CH_4) is flammable in concentration > 20 % (LFL 20) by volume (200,000 ppmv) when mixed with air. Solid Wood Pellets are not known to generate this level of concentration.

Member of Wood Pellet Association of Canada (WPAC)

X. Exposure and Toxicological Data

The feedstock is the basis of the toxicological characteristics of Wood Pellets. The available data does not make a clear distinction between whitewood and bark material. The toxicological data applies primarily to the material in form of dust.

Feedstock	PEL (OSHA)	REL (NIOSH)	TLV (ACGIH)	Health Effects
Softwood such as fir, pine, spruce and hemlock.	15 mg/m^3 Total Dust 5 mg/m^3 Respirable Dust	TWA = 1 mg/m^3 for 10 hours @ 40 hours week	TWA = 5 mg/m^3 for 8 hours @ 40 hours week STEL = 10 mg/m^3 for 15 minutes, max 4 times/day, each episode max 60 minutes	Acute or chronic dermatitis, asthma, erythema, blistering, scaling and itching (ACGIH).
Hardwood such as alder, aspen, cottonwood, hickory, maple and poplar.	15 mg/m^3 Total Dust 5 mg/m^3 Respirable Dust	TWA = 1 mg/m^3 for 10 hours @ 40 hours week	TWA = 5 mg/m^3 for 8 hours @ 40 hours week STEL = 10 mg/m^3 for 15 minutes, max 4 times/day, each episode max 60 minutes	Acute or chronic dermatitis, asthma, erythema, blistering, scaling and itching (ACGIH). Suspected tumorigenic at site of penetration (IARC).
Oak, walnut and beech.	15 mg/m^3 Total Dust 5 mg/m^3 Respirable Dust	TWA = 1 mg/m^3 for 10 hours @ 40 hours week	TWA = 1 mg/m^3 for 8 hours @ 40 hours week	Suspected tumorigenic at site of penetration (ACGIH).
Western Red Cedar.	15 mg/m^3 Total Dust 5 mg/m^3 Respirable Dust	TWA = 1 mg/m^3 for 10 hours @ 40 hours week TWA = 1 mg/m^3 for 10 hours @ 40 hours week	TWA = 5 mg/m^3 for 8 hours @ 40 hours week STEL = 10 mg/m^3 for 15 minutes, max 4 times/day, each episode max 60 minutes	Acute or chronic rhinitis, dermatitis, asthma (ACGHI).

Respirable Dust means particles with an AED<10 µm capable of deposition in nasal, thoracic and respiratory regions.

Dust from certain hardwoods has been identified by IARC as a positive human carcinogen. An excess risk of nasal adeno-carcinoma has been reported mainly in those workers in this industry exposed to wood dusts. Some studies suggest workers in the sawmilling, pulp and paper and secondary wood industries may have an increased incidence of nasal cancers and Hodgkin's disease. However, IARC concludes that the epidemiological data does not permit a definite assessment.

Dust from Western Red Cedar is considered a "Nuisance Dust" (= containing less than 1% silicates (OSHA)) with no documented respiratory cancinogenic health effects (ACGIH). Cedar oil is a skin and respiratory irritant.

XI. Notice to Reader

The information contained in this MSDS is based on consensus by occupational health and safety professionals, manufacturers of Wood Pellets and other sources believed to be accurate or otherwise technically correct. It is the Reader's responsibility to determine if this information is applicable. This MSDS is updated from time to time, and the reader has the responsibility to make sure the latest version is used. We do not have an obligation to immediately update the information in the MSDS.

Product data available from the manufacturer of the Wood Pellets includes;
- MSDS for Wood Pellets Packaged in Bag Smaller than 25 kg
- MSDS for Wood Pellets in Bulk
- Wood Pellet Product Specification
- Shipper Cargo Information Sheet (SCIS)

Member of Wood Pellet Association of Canada (WPAC)

Contact the manufacturer to order the latest version of these documents.

Notice that some of the information in this MSDS applies only to Wood Pellets manufactured by the Manufacturer identified on the first page of this MSDS and may not necessarily be applicable to products manufactured by other producers.

While we have attempted to ensure that the information contained in this MSDS is accurate, we are not responsible for any error or omissions, or for the results obtained from the use of this information.

We are not responsible for any direct, indirect, special, incidental, or consequential damage, or any other damages whatsoever and however caused, arising out of or in connection with the use of the information in this MSDS, or in reliance on that information, whether the action is in contract, tort (including negligence) or other tortious action. We disclaim any liability for unauthorized use or reproduction of any portion of this information in this MSDS.

XII. Abbreviations Used in This Document

ACGIH	American Conference of Governmental Industrial Hygienists
AED	Aerodynamic Equivalent Diameter
ASHRAE	American Society of Heating Refrigerating and Air-conditioning Engineers
ATEX	ATmosphere EXplosible
atm	atmosphere pressure
bar	10^5 Pascal (Pa) or 100 kPa or 0.9869 atm
CCOHS	Canadian Center for Occupational Health and Safety
CEN/TC	European Committee for Standardization/Technical Committee Comité Européén De Normalisation
g	gram = 0.001 kg
mg	milligram = 0.000001 kg
HS	Harmonized System Code
IARC	International Agency for Research on Cancer
IMO	International Maritime Organization (UN)
m^3	cubic meter
μm	micrometer = 0.000001 meter
MSDS	Material Safety Data Sheet
NTP	National Toxicology Program
LEL	Lower Explosible Limit (MEC=LFL=LEL)
LFL	Lean Flammability Limit (MEC=LFL=LEL)
MEC	Minimum Explosible Concentration (MEC=LFL=LEL)
NFPA	National Fire Protection Association (USA)
NIOSH	National Institute for Occupational Safety and Health (USA)
NTP	Normal Temperature and Pressure (+20°C, 101.325 kPa or 1 atm)
OSHA	Occupational Safety and Health Administration (USA)
PEL	Permissible Exposure Level
ppmv	parts per million on a volume basis. For example, 5,000 ppmv means 5,000 molecules per 1 million molecules of gas, which also corresponds to 0.5 %. A concentration of 10,000 ppmv corresponds to 1 % of volume
REL	Recommended Exposure Limit
SCIS	Shipper Cargo Information Sheet
sec	second
STEL	Short Term Exposure Limit
STP	Standard Temperature and Pressure (0°C, 101.325 kPa or 1 atm)

Member of Wood Pellet Association of Canada (WPAC)

TLV	Threshold Limit Value
tonne	1000 kg
TWA	Time weighted Average
WPAC	Wood Pellet Association of Canada

Appendix B: Example of MSDS – pellets in bags

In this Appendix, the Canadian Material Safety Data Sheet (MSDS) for bagged pellets is shown as an example. This example is from WPAC and should be regarded as a self-contained document. Units, abbreviations and syntax do not necessarily comply with the rest of this book.

LOGO	(header to be completed by the party issuing the document)	
Company name (full legal name)		Issued DATE

MATERIAL SAFETY DATA SHEET

WOOD PELLETS IN BAGS

> This MSDS is valid for Wood Pellets in bags up to 25 kg in size and stored in ventilated space with minimum one air exchange per 24 hours). If bag or multiple bags are stored in unventilated space smaller than 10 times the volume of the bag or bags, see <u>MATERIAL SAFETY DATA SHEET for Wood Pellets in Bulk</u> issued by the producer.

1. Product Identification and Use

Product name/trade name:	Wood Pellets.
Producer's Product Code:	*(to be filled in by the party issuing the document)*
Synonyms:	Wood Pellets, Fuel Pellets, Whitewood Pellets, Softwood Pellets, Hardwood Pellets, Bark Pellets.
Product appearance:	*Light to dark blond or chocolate brown, glossy to semi-glossy, cylinder with ¼ inch (6.35 mm referred to as 6 mm) diameter and 5 to 25 mm in length.* *(proposed text to be adapted accordingly by the party issuing the document)*
Product use:	Fuel for conversion to energy, animal bedding, absorbent.
HS Product Code:	44013090.
United Nations Number:	Not allocated.
Manufacturer:	*(to be filled in by the party issuing the document)* *Name of company (full legal name with no abbreviations)* *Visiting address* *Place and postal code* *Country*
	Tel.: *number incl. country code*
	Fax: *number incl. country code*
	Website: *to be filled*
	Email: *to be filled*
Emergency contact:	***Tel (direct):*** ***number incl. country code***
	Tel (mobile): ***number incl. country code***
	Fax: ***number incl. country code***

Member of Wood Pellet Association of Canada (WPAC)

Appendix B: Example of MSDS – pellets in bags

LOGO (header to be completed by the party issuing the document)
Company name (full legal name)   Issued DATE

II. Composition and Physical Properties

Wood Pellets are manufactured from ligno-cellulosic saw dust, planer shavings or bark by means of drying, size reduction, densification, cooling and dust removal. During the densification the feedstock material is compressed 3 to 4 times and heats up during compression resulting in a plasticized surface appearance. The chemical composition of Wood Pellets varies between species, components of the wood, soil conditions and age of the tree. Wood Pellets are typically manufactured from a blend of feedstock with the following composition;

Feedstock	Oxygenated compounds (indicative composition in % of weight)	
	Cellulose	30 - 40
	Hemi-cellulose	25 - 30
	Lignin	30 - 45
	Extractives (terpene, fatty acids, phenols)	3 - 5
Additives	None except as stated in Wood Pellets Product Specification	
Binders	None except as stated in Wood Pellets Product Specification	

Classification as per CEN/TC 14961 Standard; D06/M10/A0.7/S0.05/DU97.5/F1.0/N0.3

For more detailed information about the properties, see the latest version of Wood Pellets Product Specification issued by the manufacturer.

III. Health Hazard Data

Wood Pellets emit dust and gaseous invisible substances during handling and storage as part of the normal degradation occf all biological materials. Ambient oxygen is typically depleted during such degradation. Emitted gases are immediately diluted by the air in the containment and escape with ventilation air. If the Wood Pellets are stored in

    a) bulk, or
    b) unventilated space <u>smaller than 10 times</u> the packaged volume
       of the Wood Pellets, or
    c) containment with less than 1 air exchange per 24 hours,

the concentration of emitted gases, or the oxygen depletion, may pose a health threat for humans present in the containment and precautions should be taken as specified in **MATERIAL SAFETY DATA SHEET for Wood Pellets in Bulk** issued by the manufacturer. The gases emitted at normal indoor temperature include carbon-monoxide (CO), carbon-dioxide (CO_2), methane (CH_4) and hydrocarbons with Permissible Exposure Level (PEL) and symptoms as follows;

Member of Wood Pellet Association of Canada (WPAC)

Entry	Substance	Permissible Exposure Level and symptom	Remedial action
Swallow	Dust	Dry sensation	Rinse mouth thoroughly with water. Do not induce vomiting
Inhale	Dust	Coughing, dry throat (see Section X.)	Rinse mouth thoroughly with water. Do not induce vomiting.
	Carbon monoxide (CO)	Toxic invisible and odorless gas. Indoor living space TLV-TWA 9 ppmv (ASHREA) Occupational TLV-TWA 25 ppmv (OSHA)	If hygiene level is exceeded, evacuate and ventilate thoroughly.
	Carbon dioxide (CO_2)	Asphyxiating invisible and odorless gas. Occupational TLV-TWA 5,000 ppmv (OSHA)	If hygiene level is exceeded, evacuate and ventilate thoroughly.
	Methane (CH_4)	Asphyxiating invisible and odorless gas.	Ventilate
	Hydrocarbons	(see Section IX. Odor)	Ventilate
	Oxygen depleted air	Oxygen level is normally 20.9 % at sea level in well ventilated space. Minimum hygiene level is 19.5 % in work space (NIOSH)	If oxygen level is less than hygiene level, evacuate and ventilate thoroughly.
Skin contact	Dust	Itching for some people	Remove contaminated clothing. Rinse skin thoroughly with water.
Eye contact	Dust	Tearing, burning	Flush with water and sweep out particles inward towards the nose

IV. First Aid Procedures

Wood Pellets are considered a benign product if handled properly. However, individuals with a propensity for allergic reactions may experience reactions and should contact their physician to establish the best remedial action to take if reaction occurs.

In case Wood Pellets are not handled or stored in accordance with recommendations in Section VII the risk of harmful exposure increases, particularly exposure to concentration of CO higher than stipulated PEL in Section III. In case of exposure it is important to quickly remove the victim from the contaminated area. Unconscious persons should immediately be given oxygen and artificial respiration. The administration of oxygen at an elevated pressure has shown to be beneficial, as has treatment in a hyperbaric chamber. The physician should be informed that the patient has inhaled toxic quantities of carbon monoxide. Rescue personnel should be equipped with self-contained breathing apparatus when entering enclosed spaces with gas.

Carbon monoxide is highly toxic by means of binding with the hemoglobin in the blood to form carboxyhemoglobin which can not take part in normal oxygen transport, greatly reducing the blood's ability to transport oxygen to vital organs such as the brain.

V. Fire and Explosion Measures

Wood Pellets are a fuel and by nature is prone to catch fire when exposed to heat or fire. During handling of Wood Pellets there are three phases with various levels of stability, reactivity (see Section IX) and decomposition products:
- solid intact Wood Pellets
- crumbs or dust
- non-condensable (primarily CO, CO_2 and CH_4) and condensable gases (primarily aldehydes, acetone, methanol, formic acid)

 Member of Wood Pellet Association of Canada (WPAC)

Appendix B: Example of MSDS – pellets in bags

Extinguishing a fire in Wood Pellets require special methods to be successful as follows;

State of Wood Pellets	Extinguishing measures
General	Restrict oxygen from entering containment where the Wood Pellets are stored
	Be prepared for an extended period of extinguishing work.
Storage in enclosed space	Seal openings, slots or cracks where Wood Pellets may be exposed to air. Inject carbon dioxide (CO_2), nitrogen or foam.
Storage in open space	Cover the pile of Wood Pellets with foam or sand if available or spray water. Dig out the pile to reach the heart of the fire and remove effected material.
During handling	Restrict oxygen from entering the space where the Wood Pellets are present
	Cover the Wood Pellets with foam or sand if available or spray water. Dig out the material to reach the heart of the fire and remove affected material.

VI. Accidental Release Measures

If Wood Pellets are released in a populated area, the material should be removed by sweeping or vacuuming as soon as possible. Wood Pellets is a fuel and should preferably be disposed of by means of burning. Deposition of Wood Pellets or related dust should be such that gas from the material does not accumulate. Wear a protective mask to prevent inhaling of dust during cleanup (see Section VIII).

VII. Safe Handling and Storage

Precautionary measures are recommended to avoid hazardous conditions by the reactivity as outlined in Section IX developing when handling Wood Pellets.

State of Wood Pellets	Precautionary measures
General	Always store Wood Pellets in space with a minimum of 1 air exchange per 24 hours at + 20°C and a minimum of 2 air exchanges per 24 hours at + 30°C and above.
	Protect the Wood Pellets from moisture penetration to avoid swelling, increased off-gassing, self-heating and increased microbial activity.
	Always protect Wood Pellets from direct penetration by heat sources, sparks, halogen lamps and exposed electrical circuitry which could set off a fire or explosion.
	Always segregate the Wood Pellets from oxidizing agents or in-compatible materials.
	Do not expose Wood Pellets to rain.
	Do not smoke in the vicinity of Wood Pellets or wood dust.
	Install heat and gas detectors with alarm.
During handling	Avoid breakage caused by dropping the Wood Pellets.

VIII. Exposure Control and Personal Protection

The following precautionary measures shall be taken for personal protection:

Activity	Precautionary measure
Entering space with Wood Pellets	Thoroughly ventilate all communicating spaces before entering.
Exposure to dust from Wood Pellets	Wear protective glasses, dust respirator and gloves as deemed necessary.

Member of Wood Pellet Association of Canada (WPAC)

IX. Stability and Reactivity Data

The stability and reactivity properties of Wood Pellets are as follows:

Parameter	Measure	Value
Odor	-	Above + 5 °C, fresh Wood Pellets in bulk smells like aldehydes in poorly ventilated space and more like fresh softwood in ventilated space.
Off-gassing	-	The amount of gas emitted is dependt on raw material used during the production of Wood Pellets, storage temperature and access to oxygen (air). The concentration of gas in a containment depends on ventilation. If Wood Pellets are stored under conditions outlined in Section III, the **MSDS for Wood Pellets in Bulk** applies which specifies the emission factors and method of estimating gas concentrations as well as precautionary measures.
Oxidization	-	Decomposition of Wood Pellets consumes oxygen from the surrounding air and may cause health threat to humans entering a containment. If Wood Pellets are stored under conditions outlined in Section III, the **MSDS for Wood Pellets in Bulk** applies which specifies the emission factors and method of estimating gas concentrations as well as precautionary measures.
Vaporization	-	Emit hydrocarbons in vapors above + 5 °C.
Auto-ignition	°C	Auto-ignite in presence of oxygen at temperatures > + 260 °C.
Flammability	Rate	Not flammable (Class 4.1, UN MTC Rev. 3, 2000).
Explosivity	-	Wood dust may explode in concentration of 70 g/m^3 for particles < 0.63 μm depending on moisture content, atmospheric conditions and ignition energy
Self-heating	Rate	Propensity to start self-heating in presence of oxygen.
Bio-degradability	%	100.
Solubility	%	If penetrated by water Wood Pellets will dissolve into its feedstock fractions.
Mechanical stability	-	If exposed to wear and shock Wood Pellets will disintegrate into smaller fractions and dust.
Swelling	Rate	If penetrated by water Wood Pellets will swell about 3 to 4 times in volume.

X. Exposure and Toxicological Data

The feedstock is the basis of the toxicological characteristics of Wood Pellets.

Feedstock material (wood dust)	Permissible Exposure Level (PEL)	Toxicological information
Alder, aspen, cottonwood, hickory, maple, poplar,	TLV-TWA 5 mg/m^3	Non-allergenic (OSHA)
Oak, beech	TLV-TWA (8 hours) 10 mg/m^3	Tumorigenic, tumors at site of application (ACGIH).
Fir, pine, gum hemlock, spruce	TLV-TWA (8 hours) 5 mg/m^3 STEL (15 min) 1 mg/m^3	Non-allergenic (ACGIH).
Western red cedar	TLV-TWA (8 hours) 2.5 mg/m^3	Allergenic

XI. Notice to Reader

The information contained in this MSDS is based on consensus by occupational health and safety professionals, manufacturers of Wood Pellets and other sources believed to be accurate or otherwise technically correct. It is the Reader's responsibility to determine if this information is applicable. This MSDS is updated from time to time, and the reader has the responsibility to make sure the latest version is used. We do not have an obligation to immediately update the information in the MSDS.

Product data available from the manufacturer of the Wood Pellets includes;
- MSDS for Wood Pellets Packaged in Bag Smaller than 25 kg
- MSDS for Wood Pellets in Bulk

- Wood Pellet Product Specification
- Shipper Cargo Information Sheet (SCIS)

Contact the manufacturer to order the latest version of these documents.

Notice that some of the information in this MSDS applies only to Wood Pellets manufactured by the Manufacturer identified on the first page of this MSDS and may not necessarily be applicable to products manufactured by other producers.

While we have attempted to ensure that the information contained in this MSDS is accurate, we are not responsible for any error or omissions, or for the results obtained from the use of this information.

We are not responsible for any direct, indirect, special, incidental, or consequential damage, or any other damages whatsoever and however caused, arising out of or in connection with the use of the information in this MSDS, or in reliance on that information, whether the action is in contract, tort (including negligence) or other tortious action. We disclaim any liability for unauthorized use or reproduction of any portion of this information in this MSDS.

XII. Abbreviations Used in This Document

ACGHI	American Conference of Governmental Industrial Hygienists
ASHREA	American Association of Heating Refrigerating and Air-conditioning Engineers
CCOHS	Canadian Center for Occupational Health and Safety
CEN/TC	European Committee for Standardization/Technical Committee Comité Européén De Normalisation
HS	Harmonized System Code
IARC	International Agency for Research on Cancer
IMO	International Maritime Organization (UN)
NTP	National Toxicology Program
LEL	Lower Explosive Limit
LFL	Lower Flammable Limit
NFPA	National Fire Protection Association (USA)
NIOSH	National Institute for Occupational Safety and Health (USA)
OSHA	Occupational Safety and Health Administration (USA)
PEL	Permissible Exposure Level
ppmv	parts per million (volume/volume measure)
SCIS	Shipper Cargo Information Sheet
STEL	Short Term Exposure Limit
TLV	Threshold Limit Value
TWA	Time weighted Average
WPAC	Wood Pellet Association of Canada

Member of Wood Pellet Association of Canada (WPAC)

500 — Appendix B: Example of MSDS – pellets in bags

LOGO   (header to be completed by the party issuing the document)
Company name (full legal name)   Issued DATE

Member of Wood Pellet Association of Canada (WPAC)

References

1 Oxford Mini Dictionary, Thesaurus and Word Guide, Sara Hawker (ed.), Oxford University Press, Oxford, 2002

2 ÖNORM M 7135, 2000: *Compressed wood or compressed bark in natural state – pellets and briquettes – requirements and test specifications*, Austrian Standards Institute, Vienna, Austria.

3 ÖNORM M 7136, 2000: *Compressed wood or compressed bark in natural state – pellets – quality assurance in the field of logistics of transport and storage*, Austrian Standards Institute, Vienna, Austria.

4 ÖNORM M 7137, 2003: *Compressed wood in natural state – woodpellets – requirements for storage of pellets at the ultimate consumer*, Austrian Standards Institute, Vienna, Austria.

5 BioTech's life science dictionary, available at http://biotech.icmb.utexas.edu/search/dict-search.html

6 FAO, 2004: Unified Bioenergy Energy Terminology – UBET, Food and Agriculture Organization of the United Nations, Forestry Department, available at ftp://ftp.fao.org/docrep/fao/007/j4504e/j4504e00.pdf, retrieved [25.3.2010]

7 ALAKANGAS Eija, 2010: Written notice, VTT, Technical Research Centre of Finland, Jyväskylä, Finland.

8 IMSBC Code, International Maritime Solid Bulk Cargoes and Supplements, 2009 Edition, IMO Publications # IE260E.

9 SS 18 71 20, 1998: *Biofuels and peat – Fuel pellets – Classification*, Swedish Standards Institute, Stockholm, Sweden.

10 DIN CERTCO, 2009: Homepage, http://www.dincertco.de, retrieved [22.9.2009], DIN CERTCO Gesellschaft für Konformitätsbewertung mbH, Berlin, Germany

11 SS 18 71 80, 1999: *Solid biofuels and peat – Determination of mechanical strength for pellets and briquettes*, Swedish Standards Institution, Stockholm, Sweden.

12 CILES Jeremy Hugues Dit, 2002: French Pellet Club – Die französische Pelletbranche organisiert sich. In *Holzenergie*, no. 5 (2002), pp22-23, ITEBE, Lons Le Saunier Cedex, France.

13 GERARD Marie-Maud, 2003: Die Qualitätsnormen des ITEBE. In *Holzenergie*, no. 1 (2003), pp42-43, Mazzanti Editori srl, Venezia Mestre, Italy.

14 KOJIMA Ken'ichiro, 2006: Wood pellet fuel standardisation in Japan, in *Proceedings of the 2nd World Conference on Pellets* in Jönköping, Sweden, ISBN 91-631-8961-5, pp113-116, Swedish Bioenergy Association, Stockholm, Sweden.

15 SN 166000, 2001: *Testing of solid fuels – Compressed untreated wood – Requirements and testing*, Swiss Association for Standardisation, Winterthur, Switzerland.

16 NBN EN 303-5, 1999: *Heating boilers – Part 5 : Heating boilers for solid fuels, hand and automatically stocked, nominal heat output of up to 300 kW – Terminology, requirements testing and marking*, Belgian Institute for Standardisation, Brussels, Belgium.

17 PICHLER Wilfried, 2007: Neue europäische Normen für Holzpellets und deren Auswirkungen auf die Qualitätsstandards DINplus und ÖNORM M 7135. In *Proceedings of the 7th Pellets Industry Forum* in Stuttgart, Germany, pp118-122, Solar Promotion GmbH Pforzheim, Germany.

18 MÜLLER Norbert, 2008: Interview in *Fachmagazin der Pelletsbranche*, no. 1 (2008), Solar Promotion GmbH, Pforzheim, Germany.

19   TEMMERMAN M., RABIER F., DAUGBJERG JENSEN P., HARTMANN H., BÖHM T., 2006: Comparative study of durability test methods for pellets and briquettes. In *Biomass and Bioenergy*, ISSN 0961-9534, vol. 30 (2006), pp964-972, Elsevier Ltd., Oxford, UK.

20   HARTMANN Hans, 2010: Written notice, Technologie- und Förderzentrum (TFZ), Straubing, Germany.

21   RATHBAUER Josef, 2010: Written notice, BLT - Biomass · Logistics · Technology, Francisco Josephinum, Wieselburg, Austria.

22   GOLSER Michael, 2001: *Standardisierung von Holzpellets – Aktuelle nationale und internationale Entwicklungen*, conference contribution at the 2nd European Round Table "wood pellets", UMBERA Umweltorientierte Betriebsberatungs-, Forschungs- und Entsorgungs GmbH, Salzburg, Austria.

23   ÖKL-Merkblatt Nr. 66, 2008: *Planung von Pelletsheizanlagen im Wohngebäude*. Österreichisches Kuratorium für Landtechnik und Landentwicklung, Vienna, Austria.

24   BEHR Hans Martin, WITT Janet, BOSCH Jakob, 2009: Implementing the Europen biomass standards in Germany. In *Proceedings of the 9th Pellets Industry Forum* in Stuttgart, Germany, pp32-37, Solar Promotion GmbH, Pforzheim, Germany.

25   DEPI, 2010: Homepage, http://www.depi.de/, retrieved [8.3.2010], Deutsches Pelletinstitut GmbH, Berlin, Germany.

26   ISO, 2010: Homepage, http://www.iso.org, retrieved [8.3.2010], International Organization for Standardization, Geneva, Switzerland.

27   SJÖBERG Lars, 2010: Personal notice, SIS – Swedish Standards Institute, Stockholm, Sweden.

28   EN 303-5, 1999: Heating boilers – Part 5: Heating boilers for solid fuels, hand and automatically stocked, nominal heat output of up to 300 kw - Terminology, requirements, testing and marking, European Committee for Standardization (CEN), Brussels, Belgium.

29   ÖNORM EN 303-5, 1999: *Heating boilers – Part 5: Heating boilers for solid fuels, hand and automatically stocked, nominal heat ouput of up to 300 kW – terminology, requirements, testing and marking*, Austrian Standards Institute, Vienna, Austria.

30   BMU, 2001: *Erste Verordnung zur Durchführung des Bundes-Immissionsschutzgesetzes, Verordnung über kleine und mittlere Feuerungsanlagen – 1. BImSchV*, Bundesministerium für Umwelt, Naturschutz und Reaktorsicherheit, Berlin, Germany.

31   DEUTSCHER BUNDESTAG, 2009: *Erste Verordnung zur Durchführung des Bundes-Immissionsschutzgesetzes, Verordnung über kleine und mittlere Feuerungsanlagen – 1. BImSchV*. Drucksache 16/13100 vom 22.05.2009, Berlin, Germany.

32   EVALD Anders, 2009: written notice, emission limits for wood pellet fired energy systems in Denmark compiled from Ministerial orders by FORCE Technology in February 2009, FORCE Technology, Brøndby, Denmark.

33   SWAN LABELLING, 2007: Swan labelling of solid biofuel boilers, Version 2.0, 14 March 2007 – 30 June 2011, issued by Nordic Ecolabelling.

34   SCHWEIZERISCHER BUNDESRAT, 1985: Luftreinhalte-Verordnung (LRV), Fassung vom 1. September 2007, Bern, Switzerland.

35   DIRECTIVE 2005/32/EC OF THE EUROPEAN PARLIAMENT AND OF THE COUNCIL of 6 July 2005 establishing a framework for the setting of ecodesign requirements for energy-using products and amending Council Directive 92/42/EEC and Directives 96/57/EC and 2000/55/EC of the European Parliament and of the Council, Official Journal of the European Union, available at http://ec.europa.eu/enterprise/policies/sustainable-business/documents/eco-design/, retrieved [8.3.2010].

36   BEHNKE Anja, 2009: *The Eco-design Directive: Implementation for solid fuel small combustion installations*. In *Proceedings of the 9th Pellets Industry Forum* in Stuttgart, Germany, pp48-54, Solar Promotion GmbH, Pforzheim, Germany.

37   BRUSCHKE-REIMER Almut, 2010: Eine unendliche Geschichte geht weiter. In *Fachmagazin der Pelletsbranche*, no. 1 (2010), Solar Promotion GmbH, Pforzheim, Germany.

38   OBERNBERGER Ingwald, THEK Gerold, 2001: *An Integrated European Market for Densified Biomass Fuels – Country Report Austria*; Report within the framework of the EU ALTENER PROJECT AL/98/520, Institute of Chemical Engineering Fundamentals and Plant Engineering, Graz University of Technology, Austria.

39   OBERNBERGER Ingwald, THEK Gerold, 2002: Physical characterisation and chemical composition of densified biomass fuels with regard to their combustion behaviour. In *Proceedings of the 1st World Conference on Pellets* in Stockholm, Sweden, ISBN 91-631-2833-0, pp115-122, Swedish Bioenergy Association, Stockholm, Sweden.

40   OBERNBERGER Ingwald, THEK Gerold, 2004: Physical characterisation and chemical composition of densified biomass fuels with regard to their combustion behaviour. In *Biomass and Bioenergy*, ISSN 0961-9534, vol. 27(2004), pp653-669, Elsevier Ltd., Oxford, UK.

41   WOPIENKA E., GRIESMAYR S., FRIEDL G., HASLINGER W., 2009: Quality check for European wood pellets. In *Proceedings of the 17th European Biomass Conference & Exhibition* in Hamburg, Germany, ISBN 978-88-89407-57-3, pp1821-1823, ETA-Renewable Energies, Florence, Italy.

42   BE2020, 2010: Internal database, BIOENERGY 2020+ GmbH, Wieselburg location, Austria.

43   HAAS Johannes, HACKSTOCK Roger, 1998: Brennstoffversorgung mit Biomassepellets. In *Berichte aus Energie- und Umweltforschung*, no.6 (98), Bundesministerium für Wissenschaft und Verkehr, Vienna, Austria.

44   OTTLINGER Bernd, 1998: Herstellung von Biomassepellets - Erfahrungen mit der Flachmatrizenpresse, Anforderungen an den Rohstoff. In *Proceedings of the Workshop "Holzpellets - Brennstoff mit Zukunft"* in Wieselburg, Austria, Bundesanstalt für Landtechnik, Wieselburg, Austria.

45   HUBER Rudolf, 2001: Der Ausgleich zwischen kontinuierlicher Produktion und diskontinuierlicher Abnahme – Verfahren und Kosten. In *Proceedings of the 2nd European Round Table Woodpellets*, UMBERA Umweltorientierte Betriebsberatungs-, Forschungs- und Entsorgungs Gesellschaft mbH, Salzburg, Austria.

46   ZIEHER Franz, 1998: *Brandschutzaspekte bei Pelletsfeuerungen*. In *Proceedings of the workshop "Holzpellets - Brennstoff mit Zukunft"* in Wieselburg; Bundesanstalt für Landtechnik, Wieselburg, Austria.

47   HARTMANN H., BÖHM T., DAUGBJERG JENSEN P., TEMMERMAN M., RABIER F., JIRJIS R. HERSENER J.-L., RATHBAUER J., 2004: *Methods for Bulk Density Determination of Solid Biofuels*. In Van Swaaij, W. P. M.; Fjällström, T.; Helm, P.; Grassi, A. (Eds.): 2nd World Conference and Technology Exhibition on Biomass for Energy, Industry and Climate Protection, pp662-665.

48   HARTMANN H., BÖHM T., 2004: Physikalisch-mechanische Brennstoffeigenschaften. In: Härtlein, M.; Eltrop, L.; Thrän, D (Hrsg.): *Voraussetzungen zur Standardisierung biogener Festbrennstoffe, Teil 2: Mess- und Analyseverfahren*. Schriftenreihe Nachwachsende Rohstoffe (23), Fachagentur Nachwachsende Rohstoffe (Hrsg.), Landwirtschaftsverlag, Münster, Germany pp558-632.

49   RABIER F., TEMMERMAN M., BÖHM T., HARTMANN H., DAUGBJERG JENSEN P., RATHBAUEUR J., CARRASCO J., FERNANDEZ M., 2006: Particle density determination of pellets and briquettes. In *Biomass and Bioenergy*, vol. 30 (2006), pp954-963, Elsevier Ltd., Oxford, UK.

50   RYCKMANS Yves, ALLARD Patrick, LIEGEOIS Benoit, MEWISSEN Dieudonné, 2006: Conversion of a pulverised coal power plant to 100% wood pellets in Belgium. In *Proceedings of the 2nd World Conference on Pellets* in Jönköping, Sweden, ISBN 91-631-8961-5, pp51-53, Swedish Bioenergy Association, Stockholm, Sweden.

51 DAUGBJERG Jensen P., TEMMERMAN M., WESTBORG S., 2009: *Particle size distribution of raw material in biofuels pellets*, Fuel JFUE-D-09-00838, under review.

52 TEMMERMAN M., DDAUGBJERG Jensen P., 2009: *Wood particle size distribution from pellets raw material at production to particles used in power plants*, Fuel JFUE-S-09-00486, under review.

53 VAN LOO Sjaak, KOPPEJAN Jaap (Ed.), 2008: *The Handbook of Biomass Combustion and Co-firing*, ISBN 978-1-84407-249-1, Earthscan, London, UK.

54 OBERNBERGER Ingwald, 2004: *Thermische Biomassenutzung*, lecture at the Institute of Ressource Efficient and Sustainable Technology, Graz University of Technology, Graz, Austria.

55 FRIEDL A., PADOUVAS E., ROTTER H., VARMUZA K., 2005: Prediction of heating values of biomass fuel from elemental composition. In *Analytica Chimica Acta*, no. 544 (2005), pp191-198, Elsevier Ltd., Oxford, UK.

56 CORDERO T., MARQUEZ F., RODRIGUEZ-MIRASOL J., RODRIGUEZ J.J., 2001: Predicting heating values of lignocellulosics and carbonaceous materials from proximate analysis. In *Fuel*, no. 80 (2001), pp1567-1571, Elsevier Ltd., Oxford, UK.

57 THIPKHUNTHOD Puchong, MEEYOO Vissanu, RANGSUNVIGIT Pramoch, KITIYANAN Boonyarach, SIEMANOND Kitipat, RIRKSOMBOON Thirasak, 2005: Predicting the heating value of sewage sludges in Thailand from proximate and ultimate analyses. In *Fuel*, no. 84 (2005), pp849-857, Elsevier Ltd., Oxford, UK.

58 SHENG Changdong, AZEVEDO J.L.T., 2005: Estimating the higher heating value of biomass fuels from basic analysis data. In *Biomass and Bioenergy*, no. 28 (2005), pp499-507, Elsevier Ltd., Oxford, UK.

59 STOCKINGER Hermann, OBERNBERGER Ingwald, 1998: *Systemanalyse der Nahwärmeversorgung mit Biomasse*. Book series Thermische Biomassenutzung, vol.2, ISBN 3-7041-0253-9, dbv – publisher of Graz University of Technology, Graz, Austria.

60 OBERNBERGER Ingwald, THEK Gerold, 2009: Herstellung und energetische Nutzung von Pellets – Produktionsprozess, Eigenschaften, Feuerungstechnik, Ökologie und Wirtschaftlichkeit; book series Thermal Biomass Utilization, Volume 5, ISBN 978-3-9501980-5-8, published from BIOS BIOENERGIESYSTEME GmbH, Graz, Austria

61 OBERNBERGER Ingwald, DAHL Jonas, ARICH Anton, 1998: *Biomass fuel and ash analysis, report of the European Commission*, ISBN 92-828-3257-0, European Commission DG XII, Brussels, Belgium.

62 OBERNBERGER Ingwald, 2007: *Technical utilisation of sustainable resources*, lecture at the Institute of Ressource Efficient and Sustainable Technology, Graz University of Technology, Graz, Austria.

63 OBERNBERGER Ingwald, 1997: *Nutzung fester Biomasse in Verbrennungsanlagen unter besonderer Berücksichtigung des Verhaltens aschebildender Elemente*. Book series Thermische Biomassenutzung, vol. 1, ISBN 3-7041-0241-5, dbv – publisher of Graz University of Technology, Graz, Austria.

64 BIOBIB - A Database for Biofuels: http://www.vt.tuwien.ac.at, Institute of Chemical Engineering, Fuel and Environmental Technology, Vienna University of Technology, retrieved [24.7.2003], Vienna, Austria.

65 OBERNBERGER Ingwald, BRUNNER Thomas, BÄRNTHALER GEORG, 2005: Aktuelle Erkenntnisse im Bereich der Feinstaubemissionen bei Pelletsfeuerungen. In *Proceedings of the 5th Wood Industy Forum* in Stuttgart, Germany, pp54-64, Deutscher Energie-Pellet-Verband e.v. and Solar Promotion GmbH, Germany.

66 OBERNBERGER Ingwald, BIEDERMANN Friedrich, DAHL Jonas, 2000: database *BIOBANK, information about ashes from biomass combustion (physical and chemical characteristics)*, BIOS Consulting, Graz and the Institute of Chemical Engineering Fundamentals and Plant Engineering, Graz University of Technology, Graz, Austria.

67 WELTE Michael, 1980: *Untersuchungen über den Einfluß der Holzbeschaffenheit auf die Eigenschaften von thermomechanischen Holzstoffen (TMP)*, PHD Thesis, Institute of Biology, Hamburg University, Hamburg, Germany.

68 LEHTIKANGAS Päivi, 1999: Quality properties of pelletised sawdust, logging residues and bark. In *Biomass and Bioenergy,* no. 20 (2001), pp351-360, Elsevier Ltd., Oxford, UK.

69 HOLZNER H., OBERNBERGER I., 1998: *Der sachgerechte Einsatz von Pflanzenaschen im Acker- und Grünland*, Fachbeirat für Bodenfruchtbarkeit und Bodenschutz beim Bundesministerium für Land- und Forstwirtschaft, Vienna, Austria.

70 OBERNBERGER I., SCHIMA J., HOLZNER H., UNTEREGGER E., 1997: Der sachgerechte Einsatz von Pflanzenaschen im Wald, Fachbeirat für Bodenfruchtbarkeit und Bodenschutz beim Bundesministerium für Land- und Forstwirtschaft, Vienna, Austria.

71 BGBl. I S. 2524, 2008: Verordnung über das Inverkehrbringen von Düngemitteln, Bodenhilfsstoffen, Kultursubstraten und Pflanzenhilfsmitteln (Düngemittelverordnung – DüMV), Bundesministerium für Ernährung, Landwirtschaft und Verbraucherschutz, Bonn and Berlin, Germany.

72 PAJUKALLIO Anna-Maija, 2004: *The reuse of residues and recycled materials in Finland – general and legislative aspects*, Ministry of the Environment, Finland, presented at a workshop in Stockholm on 15th October 2004.

73 ISU, 2009: Radioactivity in Nature, Radiation Information Network, Homepage, www.phyics.isu.edu/radinf/natural.htm, Idaho State University, USA

74 SSM, 2009: Ionizing Radiation, Homepage, www.ssm.se, Swedish Radiation Safety Authority, Stockholm, Sweden

75 PICHL Elke, RABITSCH Herbert, 2006: Aktivitätskonzentrationen von Radiocäsium und Kalium 40 in ausgewählten Lebensmitteln der Steiermark, http://alumni.tugraz.at/tug2/alumnitalks006_pichl_rabitsch.pdf, retrieved [8.10.2009]

76 STEINHÄUSLER Friedrich, LETTNER Herbert, HUBMER Alexander Karl, ERLINGER Christian, ACHLEITNER Alois, MACK Ulrich, 2006: *Mess- und Tätigkeitsbereich des radiologischen Messlabors des Landes Salzburg*, Berichtszeitraum 01. Oktober 2005 – 31. März 2006, Salzburg, Österreich

77 KIENZL Karl, KAITNA Manuela, SCHUH Robert, STREBL Friederike, GERZABEK Martin H., 1998: *Wechselwirkung zwischen Radiocäsium-Bodenkontamination und Hydrosphäre*, Report to the Bundesministerium für Umwelt, Jugend und Familie, Umweltbundesamt GmbH, ISBN 3-85457-423-1, Vienna, Austria.

78 RAKOS Christian, 2009: Written notice of proPellets Austria - network to reinforce the distribution of pellet heating systems, Wolfsgraben, Austria.

79 HEDVALL Robert, ERLANDSSON Bengt, MATTSSON Sören, 1995: Cs-137 in Fuels and Ash Products from Biofuel Power Plants in Sweden. In *J. Environ. Radioactivity*, Vol. 31 No. 1, pp103-117, Elsevier Science Limited, Ireland.

80 MELIN S., 2005: *Radioactivity in Wood Pellets from British Columbia*. Wood Pellet Association of Canada, Prince George, British Columbia, Canada.

81 PAUL, AYCIK M., SMITH G., SEFERINOGLU M., SANDSTRÖM M., PAUL J., 2009: Leaching of Radio Cesium and Radium in Turkish and Swedish Bioash. In *Proceedings of the World of Coal Ash Conference 2009*, Lexington, Kentucky, USA.

82 GABBARD Alex, 2008: *Coal Combustion: Nuclear Resource or Danger*. Oakridge National Lab. www.ornl.gov/info/ornlreview/rev26034/text/colmain.html, retrieved [15.10.2009]

83 TORRENUEVA A., 1995: *Radioactivity in Coal Ash and Ash Products, A review of the published literature*. Ontario Hydro.

84 LETTNER Herbert, 2009: *Measurement report of the Division of Physics and Biophysics*, Department for Materials Research and Physics of the University of Salzburg, Austria.

85 SJÖBLOM R., 2009: *Cesium 137 in ash from combustion of biofuels.* Application of regulations from the Swedish Radiation Safety Authority. Värmeforsk Report 1080, Projektnummer Q6-614, ISSN 1653-1248.

86 SRPI, 1999: Policy för biobränsle, PM Dnr 822/504/99, Swedish Radiation Protection Institute, Stockholm, Sweden

87 BUNDESMINISTERIUM FÜR LAND- UND FORSTWIRTSCHAFT, UMWELT UND WASSERWIRTSCHAFT; BUNDESMINISTERIUM FÜR WIRTSCHAFT UND ARBEIT; BUNDESMINISTERIUM FÜR VERKEHR, INNOVATION UND TECHNOLOGIE; BUNDESMINISTERIUM FÜR BILDUNG, WISSENSCHAFT UND KULTUR; BUNDESMINISTERIUM FÜR GESUNDHEIT UND FRAUEN, 2006: Verordnung über allgemeine Maßnahmen zum Schutz von Personen vor Schäden durch ionisierende Strahlung (Allgemeine Strahlenschutzverordnung – AllgStrSchV), Bundesgesetzblatt 191/2006, Vienna, Austria.

88 ESTEBAN Luis, MEDIAVILLA Irene, FERNANDEZ Miguel, CARRASCO Juan, 2006: Influence of the size reduction of pine logging residues on the pelleting process and on the physical properties of pellets obtained. In *Proceedings of the 2nd World Conference on Pellets* in Jönköping, Sweden, ISBN 91-631-8961-5, pp19-23, Swedish Bioenergy Association, Stockholm, Sweden.

89 EU Project NNE5-2001-00158, Pre-normative work on sampling and testing of solid biofuels for the development of quality assurance systems - BIONORM

90 PICHLER Wilfried, GREINÖCKER Christa, GOLSER Michael, 2006: Pellet quality optimisation. In *Proceedings of the 2nd World Conference on Pellets* in Jönköping, Sweden, ISBN 91-631-8961-5, pp161-165, Swedish Bioenergy Association, Stockholm, Sweden.

91 HARTMANN H., 2004: Physical-Mechanical Fuel Properties – Significance and impacts. In Hein, M; Kaltschmitt, M (eds) *Standardisation of Solid Biofuels*, Int. Conf., Oct. 6-7, 2004, Institute for Energy and Environment (IE), pp106-115, Leipzig, Germany.

92 GREINÖCKER Christa, PICHLER Wilfried, GOLSER Michael, 2006: Hygroscopicity of wood pellets. In *Proceedings of the 2nd World Conference on Pellets* in Jönköping, Sweden, ISBN 91-631-8961-5, pp157-160, Swedish Bioenergy Association, Stockholm, Sweden.

93 SAMUELSSON Robert et al, 2006: *Effect of sawdust characteristics on pelletizing properties and pellet quality.* In *Proceedings of the 2nd World Conference on Pellets* in Jönköping, Sweden, ISBN 91-631-8961-5, pp25, Swedish Bioenergy Association, Stockholm, Sweden.

94 SAMUELSSON Robert, 2007: Written notice, Swedish University of Agricultural Sciences (SLU), Umeå, Sweden.

95 ARSHADI Mehrdad, GREF Rolf, GELADI Paul, DAHLQVIST Sten-Axel, LESTANDER Torbjörn, 2008: The influence of raw material characteristics on the industrial pelletizing process and pellet quality. In *Fuel Processing Technology*, no. 89 (2008), pp1442-1447, Elsevier Ltd., Oxford, UK.

96 MANN Markus, 2008: Pelletsproduktion: Effizienzsteigerung & Erfahrungsbericht. In *Proceedings of the World Sustainable Energy Days 2008*, O.Ö. Energiesparverband, Linz, Austria.

97 OBERNBERGER Ingwald et al, 2007: *Ash and aerosol related problems in biomass combustion and co-firing – BIOASH*, Final report of the EU FP6 Project SES6-CT-2003-502679, The Institute of Ressource Efficient and Sustainable Technology, Graz University of Technology, Graz, Austria.

98 BIEDERMANN Friedrich, 2000: *Fraktionierte Schwermetallabscheidung in Biomasseheizwerken*, PHD Thesis, Faculty of Mechanical Engineering, Graz University of Technology, Austria.

99 MICHELSEN Hanne Philbert, FRANDSEN Flemming, DAM-JOHANSEN Kim, LARSEN Ole Hede, 1998: Deposition and high temperature corrosion in a 10 MW straw fired boiler. In *Fuel Processing Technology* 54, pp95-108, Elsevier Ltd., Oxford, UK.

100 NIELSEN Per Halkjær, FRANDSEN Flemming, DAM-JOHANSEN Kim, 1999: Lab-Scale Investigations of High-Temperature Corrosion Phenomena in Straw-Fired Boilers. In *Energy & Fuels*, no. 13 (1999), pp1114-1121, American Chemical Society, Washington, DC, USA.

101 KRINGELUM Jon V. et al, 2005: *Large scale production and use of pellets – One year of operation experience*. In *Proceedings of the World Sustainable Energy Days 2005* in Wels, Austria, O.Ö. Energiesparverband, Linz, Austria.

102 FOSTER N. A., DRÄGER R., DAUBLEBSKY VON EICHHAIN C., WARNEKE R., 2007: *Wärmetechnische Auslegung von Kesseln für Verbrennung von Reststoffen – Grundlagen und Korrosionsdiagramm*. In *Unterlagen zum Seminar des VDI-Wissensforums "Beläge und Korrosion, Verfahrenstechnik und Konstruktion in Großfeuerungsanlagen (mit belasteten Brennstoffen)"*, Frankfurt am Main, Germany.

103 GAUR Siddharta, REED Thomas B., 1995: *An Atlas of Thermal Data for Biomass and Other Fuels*, NREL/TB-433-7965, UC Category:1310, DE95009212, National Renewable Energy Laboratory, Golden, Colorado, USA.

104 HOLM Jens K., HENRIKSEN Ulrik B., HUSTAD Johan E., SØRENSEN Lasse H., 2006: Toward an understanding of controlling parameters in softwood and hardwood pellets production. In *Energy & Fuels 2006*, no.20, pp2686-2694, American Chemical Society, Washington, DC, USA.

105 ÖNORM M 7133, 1998: *Chipped wood for energetic purposes – requirements and test specifications*, Austrian Standards Institute, Vienna, Austria.

106 PHYLLIS - A Database for Biomass and Waste: http://www.ecn.nl/phyllis, Netherlands Energy Research Foundation ECN, retrieved [21.9.2000], Petten, Netherlands.

107 ANONYMOUS, 2006: *Vom Brandenburger Wald in den Pelletkessel*. In *Holz-Zentralblatt*, no. 38(2006), pp1094-1095, DRW Weinbrenner GmbH & Co. KG, Leinfelden-Echterdingern, Germany.

108 MANN Markus, *2006: Einsatz von Hackgut für die Produktion von Pellets – Potenziale, Techniken, Kosten, Probleme*. In *Proceedings of the 6th Pellets Industry Forum* in Stuttgart, Germany, pp60-65, Solar Promotion GmbH, Pforzheim, Germany.

109 KISKER Jobst, 2006: Einsatz von Rundholz zur Herstellung von Pellets – das Beispiel Schwedt. In *Proceedings of the 6th Pellets Industry Forum* in Stuttgart, Germany, pp66-68, Solar Promotion GmbH, Pforzheim, Germany.

110 KISKER Jobst, 2006: First log pelletizing plant. In *The Bioenergy international*, no. 4(2006), Bioenergi Förlag AB, Stockholm, Sweden.

111 RINKE Greogor, 2005: Pelletsfertigung aus Waldfrischholz – eine technische Herausforderung. In *Proceedings of the World Sustainable Energy Days 2005* in Wels, Austria, O.Ö. Energiesparverband, Linz, Austria.

112 RINKE G., 2005: Waldholz als Alternative bei der Pelletsproduktion. Angepasste Maschinenausstattung ermöglicht Pelletsproduktion aus Sägewerksnebenprodukten und Rundholz. In *Holz-Zentralblatt*, no. 131(83), p1106, DRW Weinbrenner GmbH & Co. KG, Leinfelden-Echterdingern, Germany.

113 STOLARSKI Mariusz, 2005: Pellets production from short rotation forestry. In *Proceedings of the World Sustainable Energy Days 2005* in Wels, Austria, O.Ö. Energiesparverband, Linz, Austria.

114 KOOP Dittmar, 2006: Entwicklungsland Kurzumtrieb. In *Fachmagazin der Pelletsbranche*, no. 4 (2006), Solar Promotion GmbH, Pforzheim, Germany.

115 EDER Gottfried, 2003: *Wirtschaftliche und technische Möglichkeiten für die Herstellung und Nutzung einer neuen Generation von Biomassepellets*, Thesis, University of Applied Sciences Wiener Neustadt, Wieselburg, Austria.

116 SCHELLINGER Helmut, 2006: Neue Rohstoffpotenziale – langfristige Chancen für den Heizkesselmarkt. In *Proceedings of the 6th Pellets Industry Forum* in Stuttgart, Germany, pp53-59, Solar Promotion GmbH, Pforzheim, Germany.

117 HERING Thomas, 2006: Emissionsvergleich verschiedener Biomassebrennstoffe. In *Proceedings of the World Sustainable Energy Days 2006* in Wels, Austria, O.Ö. Energiesparverband, Linz, Austria.

118 SCHELLINGER Helmut, 2007: Pelletsrohstoff der Zukunft – Flächennutzungskonkurrenz der aktuellen Biomassenutzungspfade. Kosequenzen für den Pelletsmarkt. In *Proceedings of the 7th Pellets Industry Forum* in Stuttgart, Germany, pp52-59, Solar Promotion GmbH, Pforzheim, Germany.

119 NEUMEISTER Carsten, 2007: Pellets aus schnellwachsenden Bäumen – erste Erfahrungen aus Schweden. In *Proceedings of the 7th Pellets Industry Forum* in Stuttgart, Germany, Solar Promotion GmbH, Pforzheim, Germany.

120 NIKOLAISEN Lars et al, 2002: *Quality Characteristics of Biofuel Pellets*, ISBN 87-7756-676-9, Danish Technological Institute, Aarhus, Denmark.

121 HJULER K., 2002: Use of additives to prevent ash sintering and slag formation. In *Proceedings of the 12th European Biomass Conference* in Amsterdam, Netherlands, vol. 1, ISBN 88-900442-5-X, pp730-732, Amsterdam, Netherlands.

122 WITT Janet, LENZ Volker, 2007: Holzmischpellets: Eine Chance für den Kleinverbrauchermarkt?. In *Proceedings of the 7th Pellets Industry Forum* in Stuttgart, Germany, pp60-67, Solar Promotion GmbH, Pforzheim, Germany.

123 HILGERS Claudia, 2007: „Mischen" impossible?. In *Fachmagazin der Pelletsbranche*, no. 4 (2007), Solar Promotion GmbH, Pforzheim, Germany.

124 NIELSEN Niels Peter K., GARDNER Douglas, HOLM Jens Kai, TOMANI Per, FELBY Claus, 2008: The effect of LignoBoost kraft lignin addition on the pelleting properaties of pine sawdust. In *Proceedings Oral Sessions of the World Bioenergy 2008 Conference & Exhibition on Biomass for Energy* in Jönköping, Sweden, pp98-102, Swedish Bioenergy Association, Stockholm, Sweden.

125 ÖHMAN M., BOMAN C., HEDMAN H., NORDIN A., BOSTRÖM D., 2004: Slagging tendencies of wood pellet ash during combustion in residential pellet burners. In *Biomass & Bioenergy*, 27, pp585-596.

126 IVARSSON E., NILSSON C., 1998: *Smälttemperaturer hos halmaskor med respektive utan tillsatsmedel*, Special Report 153, Swedish University of Agricultural Sciences, Department of Farm buildings, Uppsala, Sweden.

127 STEENARI B. M., LINDQVIST O., 1998: High-temperature reactions of straw ash and the anti-sintering additives kaolin and dolomite. In *Biomass and Bioenergy*, 14, pp67-76.

128 WILEN C., STAAHLBERG P., SIPILÄ K., AHOKAS J., 1987: Pelletization and combustion of straw. In *Energy from Biomass and Wastes* 10 (1987), pp469-484.

129 TURN S. Q., 1998: A review of sorbent materials for fixed bed alkali getter systems in biomass gasifier combined cycle power generation applications. In *Journal of the Institute of Energy*, 71, pp163-177.

130 NORDIN A., LEVEN P., 1997: *Askrelaterade driftsproblem i biobränsleeldade anläggningar – Sammanställning av svenska driftserfarenheter och internationellt forskningsarbete*, Värmeforskrapportnr 607, Stockholm, Sweden

131 ÖHMAN M., GILBE R., LINDSTRÖM E., BOSTRÖM D., 2006: Slagging characteristics during residential combustion of biomass pellets. In *Proceedings of the 2nd World Conference on Pellets* in Jönköping, Sweden, ISBN 91-631-8961-5, pp93-100, Swedish Bioenergy Association, Stockholm, Sweden.

132 LINDSTRÖM E., LARSSON S., BOSTRÖM D., ÖHMAN M., 2009: *Slagging tendencies of woody biomass pellets made from a range of different Swedish forestry assortments*. Submitted to *Energy & Fuels*.

133 LINDSTRÖM E., ÖHMAN M., BOMAN C., BOSTRÖM D., DANIELSSON B., PALM L., DEGERMAN B., 2006: *Inverkan av additivinblandning i skogsbränslepelletskvaliteer för motverkande av slaggning i*

eldnings-utrustning, Slutrapport P 21464-1 inom STEM-programmet "Småskalig Bioenergianvändning". ISSN 15653-0551.

134 BÄFVER Linda, 2009: Slutrapport för delprojekt Partiklar från askrika bränslen i Energimyndighetsprojekt nr 30824-1, SP Arbetsrapport 2009, Borås, Sweden, ongoing work.

135 HEDMAN H., NYSTRÖM I-L, ÖHMAN M., BISTRÖM D., BOMAN C., SAMUELSSON R., 2008: *Småskalig eldning av torv-effekter av torvinblandning i träpellets på förbränningsresultatet i pelletsbrännare*. Rapport nr 9 i Torvforsks rapportserie, ISSN 1653-7955, Stockholm, Sweden.

136 BJÖRNBOM E., ZANZI R., GUSTAVSSON S-E, RUUSKA R., GUSTAVSSON P., BOSTRÖM D., BOMAN C., GRIMM A., LINDSTRÖM E., ÖÄHMAN M., BJÖRKMAN B., 2008: *Åtgärder mot korrosion och beläggningsbildning vid spannmålseldning*. SLF Projekt V064000, Rapport, Stockholm, Sweden.

137 BOSTRÖM D., LINDSTRÖM E., GRIMM A., ÖÄHMAN M., BOMAN C., BJÖRNBOM E., 2008: *Abatement of corrosion and deposits formation in combustion of oats*. In *Proceedings of the 16th European Biomass Conference & Exhibition* in Valencia, Spain, ISBN 978-88-89407-58-1, pp1528-1534, ETA-Renewable Energies, Florence, Italy.

138 LINDSTRÖM Erica, ÖHMAN Marcus, BOSTRÖM Dan, SANDSTRÖM Malin, 2007: Slagging characteristics during combustion of cereal grains rich in phosphorous. In *Energy & Fuels*, 21, pp710-717.

139 KOTRBA Ron, 2007: Closing the Energy Circle. In *Biomass Magazine*, no. 11(2007), pp24-29, BBI International, Grand Forks, North Dacota, USA.

140 MEIER Daniel, 2008: Grüne Energie für Europa. In *Fachmagazin der Pelletsbranche*, no. 4 (2008), Solar Promotion GmbH, Pforzheim, Germany.

141 PELLETS, 2006: *Pellets – Markt und Technik*. In *Fachmagazin der Pelletsbranche*, no. 3 (2006), Solar Promotion GmbH, Pforzheim, Germany.

142 FRUWIRTH Robert, 2007: Nassvermahlung von Hackschnitzeln. In *Proceedings of the 7th Pellets Industry Forum* in Stuttgart, Germany, pp126-128, Solar Promotion GmbH, Pforzheim, Germany.

143 RINKE Gregor, 2007: *Vom Holz zum Pellet*. In *Proceedings of the World Sustainable Energy Days 2007* in Wels, Austria, O.Ö. Energiesparverband, Linz, Austria.

144 RINKE Gregor, 2008: 240.000 t Pelletsproduktion/Jahr – eine Herausforderung für jeden Standort. In *Proceedings of the World Sustainable Energy Days 2008*, O.Ö. Energiesparverband, Linz, Austria.

145 NYSTRÖM I., DEDMAN H., BOSTRÖM D., BOMAN C., SAMUELSSON R., ÖHMAN M., 2008: Effect of peat addition on combustion characteristics in residential appliances. In *Proceedings Poster Session of the World Bioenergy 2008 Conference & Exhibition on Biomass for Energy* in Jönköping, Sweden, pp274-279, Swedish Bioenergy Association, Stockholm, Sweden.

146 SPROUT MATADOR, 2006: http://www.andritz.com, retrieved [11.9.2006], Sprout-Matador A/S, Esbjerg, Denmark.

147 REPKE Volker, 2006: Written notice, Dipl.-Ing. (FH) V. Repke Holzindustrieberatung, Olang/BZ, Italy.

148 HIRSMARK Jakob, 2002: *Densified Biomass Fuels in Sweden, Examensarbeten Nr 38*, Department of Forest Management and Products, Swedish University of Agricultural Sciences, Uppsala, Sweden.

149 SWISS COMBI, 2006: Oral notice and http://www.swisscombi.ch, retrieved [7.9.2006], SWISS COMBI W. Kunz dryTec AG, Dintikon, Switzerland.

150 LOHMANN Ulf, 1998: *Holzhandbuch*, DRW-Verlag Weinbrenner GmbH & Co. KG, Rosenheim, Germany.

151 PONNDORF, 2007: Written notice, Ponndorf Maschinenfabrik GmbH, Kassel, Germany.

152 ANHYDRO, 2007: http://www.anhydro.com, Anhydro GmbH, retrieved [28.9.2007], Kassel, Germany.

153 GRANSTRÖM K. M., 2006: Emissions of sesquiterpenes from spruce sawdust during drying. In *Proceedings of the 2nd World Conference on Pellets* in Jönköping, Sweden, ISBN 91-631-8961-5, pp121-125, Swedish Bioenergy Association, Stockholm, Sweden.

154 ÖHMANN Marcus, NORDIN Anders, HEDMAN Henry, JIRJIS Raida, 2002: Reasons for slagging during stemwood pellet combustion and some measures for prevention. In *Proceedings of the 1st World Conference on Pellets* in Stockholm, Sweden, ISBN 91-631-2833-0, pp93-97, Swedish Bioenergy Association, Stockholm, Sweden.

155 BÜTTNER, 2001: Company brochure, Büttner Gesellschaft für Trocknungs- und Umwelttechnik mbH, Krefeld, Germany.

156 ANDRITZ, 2007: Written notice, Andritz AG, Graz, Austria.

157 STELA, 2007: http://stela.de, retrieved [9.10.2007], STELA Laxhuber GmbH, Massing, Germany.

158 GRANSTRAND Lennart, 2006: Increased production capacity with new drying system. In *Proceedings of the 2nd World Conference on Pellets* in Jönköping, Sweden, ISBN 91-631-8961-5, pp13-15, Swedish Bioenergy Association, Stockholm, Sweden.

159 GRANSTRAND Lennart, 2009: Energy efficient drying of sawdust – examples of realised plants. In *Proceedings of the World Sustainable Energy Days 2009* in Wels, Austria, O.Ö. Energiesparverband, Linz, Austria.

160 LJUNGBLOM Lennart, 2009: New low temp drying system from SRE. In *The Bioenergy International*, no. 37 (2007), Bioenergi Förlag AB, Stockholm, Sweden.

161 STAHEL Roger, 2009: Low-temperature drying of raw materials for pellets production. In *Proceedings of the 9th Pellets Industry Forum* in Stuttgart, Germany, pp70-75, Solar Promotion GmbH, Pforzheim, Germany.

162 STORK, 1995: Company brochure, Stork Engineering, Göteborg, Sweden.

163 GRUBER Timon, KUNZ Werner, 2001: Spänetrocknung mit Dampf. In *Holz-Zentralblatt*, no. 147 (2001), p1886, DRW-Verlag Weinbrenner GmbH & Co. KG, Leinfelden-Echterdingern, Germany.

164 MÜNTER Claes, 2003: Exergy steam drying and its energy integration. In *Proceedings of the International Nordic Bioenergy Conference* in Jyväskylä, Finland, ISBN 952-5135-26-8, ISSN 1239-4874, pp271-273, Finnish Bioenergy Association, Jyväskylä, Finland.

165 MÜNTER Claes, 2004: Exergy steam drying for biofuel production. In *Proceedings of the World Sustainable Energy Days 2005* in Wels, Austria, O.Ö. Energiesparverband, Linz, Austria.

166 VERMA Prem, MÜNTER Claes, 2008: Exergy steam drying and energy integration. In *Proceedings Poster Session of the World Bioenergy 2008 Conference & Exhibition on Biomass for Energy* in Jönköping, Sweden, pp286-289, Swedish Bioenergy Association, Stockholm, Sweden.

167 NIRO, 2006: http://www.niro.de, retrieved [6.9.2006], Niro A/S, Soeborg, Denmark.

168 MBZ, 2000: Company brochure, MBZ Mühlen- und Pelletiertechnik, Hilden, Germany.

169 HASLINGER Walter, 2005: Pellets-Technologien – ein Überblick. In *Proceedings of the World Sustainable Energy Days 2005* in Wels, Austria, O.Ö. Energiesparverband, Linz, Austria.

170 GEISSLHOFER Alois et al., 2000: *Holzpellets in Europa, Berichte aus Energie- und Umweltforschung"*, no. 9 (2000), Bundesministerium für Verkehr, Innovation und Technologie Vienna, Austria.

171 CPM, 2001: Weit entwickelte Technologie zur Herstellung von Holzpellets steht bereit. In *Holz-Zentralblatt*, no. 147 (2001), p1885, DRW-Verlag Weinbrenner GmbH & Co. KG, Leinfelden-Echterdingern, Germany.

172 CPM, 2001: http://www.cpmroskamp.com, retrieved [10.1.2001], CPM/Europe B.V., Amsterdam, Netherlands.

173 BLISS, 2008: http://www.bliss-industries.com, retrieved [19.12.2008], Company brochure, Bliss Industries Inc., Ponca City, Oklahoma, USA.

174 BRUSLETTO Rune, 2006: Written notice, Arbaflame AS, Matrand, Norway.

175 BERGMANN Patrik C.A., BOERSMA Arjen R., KIEL Jacob H.A., 2007: Torrefaction for biomass conversion into solid fuel. In *Proceedings of the 15th European Biomass Conference & Exhibition* in Berlin, Germany, ISBN 978-88-89407-59-X, ISBN 3-936338-21-3, pp78-82, ETA-Renewable Energies, Florence, Italy

176 LENSSELINK Jasper, GERHAUSER Heiko, KIEL Jacob H.A., 2008: BO_2-technology for combined torrefaction and densification. In *Proceedings of the World Sustainable Energy Days 2008*, O.Ö. Energiesparverband, Linz, Austria.

177 ROLLAND Matthieu, REPELLIN Vincent, GOVIN Alexandre, GUYONNET René, 2008: Effect of torrefaction on grinding energy requirement: first results on Spruce. In *Proceedings Oral Sessions of the World Bioenergy 2008 Conference & Exhibition on Biomass for Energy* in Jönköping, Sweden, pp108-111, Swedish Bioenergy Association, Stockholm, Sweden.

178 LIPINSKY Edward S., ARCATE James R., REED Thomas B., 2002: Enhanced wood fuels via torrefaction. In *Fuel Chemistry Division Preprints 2002*, no. 47 (1), http://www.techtp.com/recent%20papers/acs_paper.pdf, retrieved [14.4.2008].

179 ZANZI R., TITO FERRO D., TORRES A., BEATON SOLER P., BJÖRNBOM E., 2004: Biomass torrefaction. In *Proceedings of the 2nd World Conference and Exhibition on Biomass for Energy, Industry and Climate Protection* in Rome, Italy, vol. 1, ISBN 88-89407-04-2, pp859-862, ETA-Florence, Italy.

180 PRINS Mark J., PTASINSKI Krzysztof J., LANSSEN Frans J.J.G., 2006: More efficient biomass gasification via torrefaction. In *Energy*, no. 31(2006), pp3458-3470, Elsevier Ltd, Oxford, UK.

181 SRIDHAR G., SUBBUKRISHNA D.N., SRIDHAR H.V., DASAPPA S., PAUL P.J., MUKUNDA H.S., 2007: Torrefaction of Bamboo. In *Proceedings of the 15th European Biomass Conference & Exhibition* in Berlin, Germany, ISBN 978-88-89407-59-X, ISBN 3-936338-21-3, pp532-535, ETA-Renewable Energies, Florence, Italy.

182 BRIDGEMAN T.G., ROSS A.B., JONES J.M., WILLIAMS P.T., 2007: Torrefaction: changes in solid fuel properties of biomass and the implications for thermochemical processing. In *Proceedings of the 15th European Biomass Conference & Exhibition* in Berlin, Germany, ISBN 978-88-89407-59-X, ISBN 3-936338-21-3, pp1320-1325, ETA-Renewable Energies, Florence, Italy.

183 ARIAS B., PEVIDA C., FERMOSO J., PLAZA M.G., RUBIERA F., PIS J.J., 2008: Influence of torrefaction on the grindability and reactivity of woody biomass. In *Fuel Processing Technology*, no. 89(2008), pp169-175, Elsevier Ltd, Oxford, UK.

184 HÅKANSSON K., OLOFSSON I., PERSSON K., NORDIN A., 2008: Torrefaction and gasification of hydrolysis residue. In *Proceedings of the 16th European Biomass Conference & Exhibition* in Valencia, Spain, ISBN 978-88-89407-58-1, pp923-927, ETA-Renewable Energies, Florence, Italy.

185 ROMEO Javier Celaya, BARNO Javier Gil, 2008: Evaluation of torrefaction + pelletization process to transform biomass in a biofuel suitable for co-combustion. In *Proceedings of the 16th European Biomass Conference & Exhibition* in Valencia, Spain, ISBN 978-88-89407-58-1, pp1937-1941, ETA-Renewable Energies, Florence, Italy.

186 ZANZI Rolando, MAJARI Mehdi, BJÖRNBOM Emilia, 2008: Biomass pre-treatment by torrefaction. In *Proceedings of the 16th European Biomass Conference & Exhibition* in Valencia, Spain, ISBN 978-88-89407-58-1, pp37-41, ETA-Renewable Energies, Florence, Italy.

187 KIEL J.H.A., VERHOEFF F., GERHAUSER H., MEULEMANN B., 2008: BO_2-technology for biomass upgrading into solid fuel – pilot-scale testing and market implementation. In *Proceedings of the 16th European Biomass Conference & Exhibition* in Valencia, Spain, ISBN 978-88-89407-58-1, pp48-53, ETA-Renewable Energies, Florence, Italy.

188 VAN DAALEN Wim, 2008: Torrifikationsbasierte BO_2-Technologie zur Veredelung von Biomasse in leicht handelbare feste Biomasse. In *Proceedings of the 8th Pellets Industry Forum* in Stuttgart, Germany, pp123-132, Solar Promotion GmbH, Pforzheim, Germany.

189 *Upgrading fuel properties of biomass fuel and waste by torrefaction*, R&D project of Teknologisk Institut, Vedvarende Energi og Transport, in cooperation with Danmarks Tekniske Universitet. Risø Nationallaboratoriet for Bæredygtig Energi (Risø DTU), Afdelingen for Biosystemer; DONG Energy Power, Kemi og Optimering and Københavns Universitet, Skov og Landskab

190 BERGMAN Patrick C.A., KIEL Jacob H.A., 2005: Torrefaction for biomass upgrading. In *Proceedings of the 14th European Biomass Conference & Exhibition* in Paris, France, ISBN 88-89407-07-7, pp206-209, ETA-Renewable Energies, Italy.

191 POST VAN DER BURG Robin, 2010: *Torrefied pellets: advantages and challenges.* Presentation at the World Sustainable Energy Days 2010 in Wels, Austria.

192 POST VAN DER BURG Robin, 2009: written notice, Topell Energy BV, The Hague, The Netherlands.

193 MAASKANT Ewout, 2009: Torrefaction. Presentation at the *Workshop on High cofiring percentages in new coal fired power plants*, June 30, 2009, Hamburg, Germany, available at http://www.ieabcc.nl/meetings/task32_Hamburg2009/cofiring/03%20Topell%20revised.pdf, retrieved [10.2.2010]

194 DEML Max, 2008: Invest in Pellets – Anlagemöglichkeiten in Pelletsproduktionen. In *Fachmagazin der Pelletsbranche*, no. 3 (2008), Solar Promotion GmbH, Pforzheim, Germany.

195 WILD Michael, 2008: Entwicklung der europäischen Märkte für Heiz- und Verstromungspellets – Rohstoffe und Verbraucherperspektiven. In *Proceedings of the 8th Pellets Industry Forum* in Stuttgart, Germany, pp34-39, Solar Promotion GmbH, Pforzheim, Germany.

196 BERNER Joachim, 2005: Leichte Fracht – der Transport von Holzpellets auf deutschen Flüssen nimmt zu. In *Fachmagazin der Pelletsbranche*, no. 4(2005), Solar Promotion GmbH, Pforzheim, Germany.

197 LJUNGBLOM Lennart, 2004: Pellets: booming business in Rotterdam. In *The Bioenergy International*, no. 9 (2004), Bioenergi Förlag AB, Stockholm, Sweden.

198 SWAAN John, 2003: Pellets from Canada to Europe. In *The Bioenergy International*, No 7, December 2003, Bioenergi Förlag AB (Ed.), Stockholm, Sweden

199 LJUNGBLOM Lennart, 2003: Pellets from Halifax. In *The Bioenergy international*, no. 7 (2003), Bioenergi Förlag AB, Stockholm, Sweden.

200 KOOP Dittmar, 2006: R-Bescheid in der Kombüse. In *Fachmagazin der Pelletsbranche*, no. 4 (2006), Solar Promotion GmbH, Pforzheim, Germany

201 KOOP Dittmar, 2007: Keimzelle Rotterdam. In *Fachmagazin der Pelletsbranche*, no. 4(2007), Solar Promotion GmbH, Pforzheim, Germany.

202 TCHARNETSKY Marina, 2008: Die Strukturen reifer Energiemärkte – Vorbild für die Pelletsbranche? In *Proceedings of the 8th Pellets Industry Forum* in Stuttgart, Germany, pp138-140, Solar Promotion GmbH, Pforzheim, Germany.

203 LJUNGBLOM Lennart, 2005: A fast grower BBG – Baltic Bioenergy Group. In *The Bioenergy international*, no. 17 (2005), Bioenergi Förlag AB, Stockholm, Sweden.

204 PFEIFER, 2009: Homepage, http://www.holz-pfeifer.com, retrieved [8.10.2009], Holzindustrie Pfeifer GesmbH, Kundl, Austria.

205 RAKOS Christian, SCHLAGITWEIT Christian, 2008: Written notice of proPellets Austria - network to reinforce the distribution of pellet heating systems, Wolfsgraben, Austria.

206 WILD Michael, 2009: written notice, European Bioenergy Services - EBES AG, Vienna, Austria

207 NAU, 2003: Company brochure, Stefan Nau GmbH & Co. KG, Moosburg, Germany.

208 MALL, 2003: http://www.mall.info, retrieved [31.7.2003], Mall GmbH, Donaueschingen, Germany.

209 KWB, 2005: Company brochure, Kraft & Wärme aus Biomasse GmbH, St. Margarethen/Raab, Austria.

210 BERNER Joachim, 2006: Flexible Pelletstanks. In *Fachmagazin der Pelletsbranche*, no. 1 (2006), Solar Promotion GmbH, Pforzheim, Germany.

211 JÄCKEL Günther, 2006: Holzpellets - eine Chance für den Brennstoffhandel in Deutschland. In *Proceedings of the European Pellets Forum 2006*, O.Ö. Energiesparverband, Linz, Austria.

212 RAKOS Christian, 2008: Versorgungssicherheit im Pelletsmarkt. In *Proceedings of the 8th Pellets Industry Forum* in Stuttgart, Germany, pp47-52, Solar Promotion GmbH, Pforzheim, Germany.

213 SCHONEWILLE Wijnand, 2008: Rotterdam: Entwicklung zum Drehkreuz der Holzpelletsverschiffung. In *Proceedings of the 8th Pellets Industry Forum* in Stuttgart, Germany, pp141-143, Solar Promotion GmbH, Pforzheim, Germany.

214 WINDHAGER, 2008: http://www.windhager.com, retrieved [9.12.2008], Windhager Zentralheizung AG, Seekirchen am Wallersee, Austria.

215 DAUGHERTY Jack, 1998: Assessment of Chemical Exposures, Calculation Methods for Environmental Professionals, ISBN 1-56670-216-X, CRC Press, Danvers, MA, USA

216 MELIN Staffan, 2008: *Testing of Explosibility and Flammability of Airborne Dust from Wood Pellets*. Wood Pellet Association of Canada, Prince George, Canada.

217 BARON Paul: Generation and behaviour of airborne particles (aerosols), National Institute of Occupational Health and Safety, Atlanta, GA, USA, available at http://www.cdc.gov/niosh/topics/aerosols/pdfs/aerosol_101.pdf, retrieved [26.1.2010]

218 BS 5958, 1991: *Code of practice for control of undesirable static electricity*, British Standards Institution, London, United Kingdom.

219 BARTON John, 2002: *Dust explosion prevention and protection: a practical guide*, ISBN 0-7506-7519-5, Gulf Publishing Company, Houston, Texas, USA.

220 ATEX-137, 1999: DIRECTIVE 1999/92/EC OF THE EUROPEAN PARLIAMENT AND OF THE COUNCIL of 16 December 1999 on minimum requirements for improving the safety and health protection of workers potentially at risk from explosive atmospheres (15th individual Directive within the meaning of Article 16(1) of Directive 89/391/EEC).

221 UNITED NATIONS, 1990: *Recommendation on the Transport of Dangerous Goods, Manuak of Tests and Criteria*, third revision, ISBN 92-1-139-068-0.

222 CFR 49, 2009: US Code of Federal Regulations, http://www.gpoaccess.gov/CFR.

223 MANI S., SOKHANSANJ S., HOQUE M., PETERSON J., 2007: *Moisture Sorption Isotherm for Wood Pellets*, ASABE 2007 Annual International Meeting, Minnesota, USA.

224 LÖNNERMARK A., PERSSON H., BLOMQVIST P., HOGLAND W., 2008: *Biobränslen och avfall - Brandsäkerhet i samband med lagring*, SP Sveriges Tekniska Forskningsinstitut, 2008:51, Borås, Sweden.

225 PERSSON H., BLOMQVIST P., 2004: *Släckning av silobränder*, SP Swedish National Testing and Research Institute, 2004:16, Borås, Sweden.

226 RUPAR-GADD K., 2006: *Biomass Pre-treatment for the Production of Sustainable Energy – Emissions and Self-heating*, Acta Wexionensia, No 88/2006, Bioenergy Technology, ISBN 91-7636-501-8, Växjö, Sweden.

227 BLOMQVIST P., PERSSON B., 2003: *Spontaneous Ignition of Biofuels - A Literature Survey of Theoretical and Experimental Methods*, SP Swedish National Testing and Research Institute, SP-AR 2003:18, Borås, Sweden.

228 BLOMQVIST P., PERSSON H., 2008: Self-heating in storages of wood pellets. In *Proceedings Oral Sessions of the World Bioenergy 2008 Conference & Exhibition on Biomass for Energy* in Jönköping, Sweden, pp138-142, Swedish Bioenergy Association, Stockholm, Sweden.

229 BLOMQVIST P., PERSSON H., HEES P. V., HOLMSTEDT G., GÖRANSSON U., WADSÖ L., SANATI M., RUPAR-GADD K., 2007: *An experimental study of sponaneousignition in storages of wood pellets*, Fire and Materials Conference, San Francisco, USA.

230 PERSSON H., BLOMQVIST P., 2007: *Fire and fire extinguishment in silos*, Interflam '07, pp365-376, London, England.

231 PERSSON H., BLOMQVIST P., 2009: Silo Fires and Silo Fire Fighting. In: *Proceedings of Bioenergy 2009*, pp693-702, Jyväskylä, Finland

232 NT ENVIR 010, 2008: *Guidelines for storing and handling of solid biofuels*, Nordic Innovation Centre, Oslo, Norway.

233 SAMUELSSON Robert, THYREL Michael, SJÖSTRÖM Michael, LESTANDER Torbjörn, 2009: Effect of biomaterial characteristics on pelletizing properties and biofuel pellet quality. In: *Fuel processing Technology*, pp1129-1134, Elsevier Ltd., Oxford, UK.

234 KUBLER H., 1987: Heat Generating Processes as Cause of Spontaneous Ignition in Forest Products. In *Forest Products Abstracts*, 10, 11, pp298-327, Oxford, UK.

235 ARSHADI Mehrdad, GELADI Paul, GREF Rolf, FJÄLLSTRÖM Pär, 2009: Emission of volatile aldehydes and ketones from wood pellets under controlled conditions. In *Ann. Occup. Hyg*, Vol. 53, No. 8, pp797-805, Elsevier Ltd., Oxford, UK.

236 LESTANDER T. A., 2008: *Water absorption thermodynamics in single wood pellets modelled by multivariate near-infrared spectroscopy*, Holzforschung, 62, pp429-434, Berlin, Germany.

237 BACK Ernst L., 1982: Auto-ignition in hygroscopic, organic materials - especially forest products - as initiated by moisture absorption from the ambient atmosphere. In *Fire Safety Journal*, 4, 3, pp185-196, Elsevier Ltd., Oxford, UK.

238 MELIN S., 2008: *Emissions from Woodpellets During Ocean Transportation (EWDOT)*, Research Report, Wood Pellet Association of Canada, January 16, Prince George, British Columbia, Canada.

239 SVEDBERG U., SAMUELSSON S., MELIN S., 2008: Hazardous Off-gassing of Carbon Monoxide and Oxygen Depletion during Ocean Transportation of Wood Pellets. In *Annual Occupational Hygiene*, Vol. 52, No. 4, pp259-266.

240 ARSHADI M., GREF R., 2005: Emissions of volatile organic compounds from softwood pellets during storage. In *Forest Products Journal*, 55, 12, pp132-135, Madison, WI, USA.

241 KUANG X., SHANKAR T. J., BI X. T., SOKHANSANJ S., LIM C. J., MELIN S., 2008: Characterization and Kinetics Study of Off-Gas Emissions from Stored Wood Pellets. In *Annals of Occupational Hygiene*, 52, 8, pp675-683, Oxford, UK.

242 KUANG X., SHANKAR T., BI T., SOKHANSANJ S., LIM CJ., MELIN S., 2008: *Rate and Peak Concentrations of Emissions in Stored Wood Pellets – Sensitivities to temperature, Relative Humidity and headspace volume.* Annals of Occupational Hygiene, Oxford, UK.

243 KUANG X., SHANKAR T., BI SOKHANSANJ S., T. LIM CJ., MELIN S., 2009: *Effects of Headspace Volume Ratio and Oxygen level on Off-gas Emissions from Wood Pellets in Storage*, Annals of Occupational Hygiene, Oxford, UK.

244 BC Code, 2004: *Code of Safe Practice for Solid Bulk Cargoes,* 2005 Edition, IMO document # ID260E. ISBN 13-978-92-801-4201-3, International Maritime Organization (IMO), London, UK.

245 PRATT Thomas H., 2000: *Electrostatic Ignitions of Fires and Explosions,* Center for Chemical Process Safety, ISBN 0-8169-9948-1, American Institute of Chemical Engineers, New York, USA.

246 NFPA 70, 2008: National Fire Protection Association 70, National Electrical Code, Quincy, Massachusetts, USA

247 SVEDBERG U., HÖGBERG H., CALLE B., 2004: Emissions of hexanal and Carbon Monoxide from Storage of Wood Pellets, a Potential Occupational and Domestic Health Hazard. In *The Annals of Occupational Hygiene,* Vol 48, Oxford, UK.

248 SVEDBERG U., CALLE B.: *Användning av FTIR teknik för bestämning av gasformiga emissioner vid träpelletstillverkning, Evaluation of the Time Correlated Tracer (TCT) Method for Assessment of Diffuse Terpene Emissions from Wood Pellets Production.* Värmeforsk, April 2001, ISSN 0282-3772, Stockholm, Sweden.

249 HAGSTRÖM K., 2008: *Occupational Exposure During Production of Wood Pellets in Sweden.* Örebro University, Doctoral Dissertation, Örebro, Sweden.

250 ARSHADI M., NILSSON D., GELADI P., 2007: Monitoring chemical changes for stored sawdust from pine and spruce using gas chromatography-mass spectrometry and visible-near infrared spectroscopy. In *Near Infrared Spectroscopy,* 15, pp379-386, West Sussex, UK.

251 FINELL Michael, ARSHADI Mehrdad, GREF Rolf, SCHERZER Tom, KNOLLE Wolfgang, LESTANDER Torbjörn, 2009: Laboratory-scale production of biofuel pellets from electron beam treated scots pine (Pinus silvestris L.) sawdust. In *Radiation Physics and Chemistry,* pp281-287, Elsevier Ltd., Oxford, UK.

252 EMHOFER Waltraud, POINTNER Christian, 2009: *Report Lagertechnik und Sicherheit bei der Pelletslagerung,* Bioenergy2020+ GmbH, Wieselburg, Austria, unpublished

253 EMHOFER, Waltraud, 2009: *CO und VOC Freisetzung in Pelletslagern,* Lecture given at Highlights der Bioenergieforschung, 12. November 2009, Vienna, Austria

254 Investigation Report, 2006: *Combustible Dust Hazard Study,* US Chemical Safety and Hazards Investigation Board, Investigation Report, Report No 2006-H-1, November 2006, Washington, D.C., USA.

255 DEPV, 2009: *Empfehlung zur Lagerung von Holzpellets,* Informationsblatt 01-2005-A, Stand 2008/2009, Deutscher Energie-Pellet Verband (DEPV), Berlin, Germany

256 GUO W., LIM CJ., SOKHANSANJ S., MELIN S., 2009: *Thermal Conductivity of Wood Pellets.* UBC February 2009, Vancouver, British Columbia, Canada.

257 YAZDANPANAH F., SOKHANSANJ S., LAU A., LIM C.J., BI X., MELIN S., 2009: Air flow versus pressure drop for bulk wood pellets. Submitted to *Biomass and Bioenergy,* Elsevier Ltd., Oxford, UK.

258 YAZDANPANAH F., SOKHANSANJ S., LAU A., BI T., LIM CJ., MELIN S., 2009: *Permeability of Bulk Wood Pellets in Storage,* Paper No CSBE08-105, March 2009, The Canadian Society for Bioengineering, Winnipeg, Manitoba, Canada.

259 PERSSON H., BLOMQVIST P., TUOVINEN H., 2009: *Inertering av siloanläggningar med kvävgas - gasfyllnadsförsök och simuleringar.* Bransdforsk projekt 602-071, SP Rapport 2009:10, SP Technical Research Institute of Sweden, Borås, Sweden.

260 PERSSON H., BLOM J., MODIN P., 2008: *Research experience decisive in extinguishing silo fires,* BrandPosten, Number 37/2008, SP Technical Research Institute of Sweden, Fire Technology, Borås, Sweden, available at http://www.sp.se/sv/units/fire/Documents/BrandPosten/BrandPosten_37_eng.pdf, retrieved [20.01.2010].

261 NORDSTRÖM T., SAMUELSSON A., 2009: *Sammanställning av händelseförloppet vid brand i cistern med stenkol på Stora Enso*, Hylte 2009-02-13, Räddningstjänsten i Halmstad, Sweden (in Swedish).

262 PERSSON Henry, BLOM Joel, 2008: *Research helps the fighting of a silo fire again,* BrandPosten #38/2008, SP Technical Research Institute of Sweden, Borås, Sweden.

263 ANONYMUS, 2004: *Rapport Silobrand Härnösand 8-13 september 2004. En beskrivning av olycksförloppet, olycksorsaken och våra erfarenheter från insatsen.* Räddningstjänsten Höga Kusten-Ådalen, Sweden (in Swedish).

264 IARC, 1995: Summaries and Evaluations, International Agency for Research on Cancer, *Wood Dust*, 62, 1995, www.iarc.fr, Lyon, France.

265 ACGIH, 2008: *Guide to Occupational Exposure Values*. American Conference of Governmental Industrial Hygienists (ACGIH), ISBN: 978-1-882417-80-3, Cincinnati, Ohio, USA.

266 NIOSH, 2005: NIOSH Pocket Guide to Chemical Hazards, National Institute for Occupational Safety and Health (NIOSH), Atlanta, Georgia, USA.

267 CCOHS, 2009: On-line publications for Occupational Cancer and Cancer Sites associated with Occupational Exposures, www.ccohs.com, Canadian Centre for Occupational Health and Safety, Hamilton, Ontario, Canada.

268 GUDMUNDSSON A., 2007: *Partiklar – Hälsa*, University of Lund, Lund. Sweden.

269 JOHANNSSON G., 2007: *Informationsmöte om uppkomst av farliga gaser i lastrum på fartyg*, Karolinska Institutet, Institute for Environmental Medicine, Stockholm, Sweden.

270 BREYSSE P. LEES S., 2006: *Particulate Matter*, John Hopkins University, School of Public health, Baltimore, Maryland, USA.

271 KUANG Xingya, SHANKAR Tumuluru Jaya, MELIN Staffan, BI Xiaotao, SOKHANSANJ Shahab, LIM Jim, 2008: *Characterization and Kinetics of Off-gas Emissions from Stored Wood Pellets*. Annals of Occupational Hygiene, Oxford, UK.

272 KUANG Xingya, SHANKAR Tumuluru Jaya, MELIN Staffan, BI Xiaotao, SOKHANSANJ Shahab, LIM Jim, 2008: Rate and Peak Concentrations of Emissions in Stored Wood Pellets – Sensitivity to temperature, relative humidity and Headspace Volume. In *Annual Occupational Hygiene*, 53, 8, pp789-796.

273 ARCHADI M., 2005: *Emission of Volatile Organic Compounds from Softwood Pellets During Storage*, Swedish University of Agricultural Sciences, Uppsala, Sweden.

274 American Society of Heating, Refrigerating and Air-conditioning Engineers, Inc. ANSI/ASHRAE Standard 62.1-2007, Table B-2. ISSN 1041-2336.

275 JOHANSSON G.: *Effects of Simultaneous Exposure to Carbon-monoxide and Oxygen Deficiency*. Karolinska Institutet, Faculty of Environmental Medicine, Stockholm, Sweden.

276 KEMI, 2009: FACTS – Safety Data Sheets, Swedish Chemicals Agency, Sundbyberg, Sweden, available at http://www.incopa.org/relatedDocs/FbSDSMarch09.pdf, retrieved [26.3.2010]

277 ASTM E1628 – 94, 2008: Standard Practice for Preparing Material Safety Data Sheets to Include Transportation and Disposal Data for the General Services Administration, ASTM International, West Conshohocken, Pennsylvania, USA.

278 WHMIS, 2006: Supplier's Guide to WHMIS, Preparing Compliant material Safety Data Sheets and Labels, Workplace Hazardous Materials Information System (WHMIS), Health Canada, Vancouver, British Columbia, Canada.

279 Regulation (EC) No 1907/2006 of the European Parliament and of the Council, dated December 18, 2006, concerning the Registration, Evaluation, Authorisation and Restriction of Chemicals (REACH), establishing a European Chemicals Agency, amending Directive 1999/45/EC and repealing Council

Regulation (EEC) No 793/93 and Commission Regulation (EC) No 1488/94 as well as Council Directive 76/769/EEC and Commission Directives 91/155/EEC, 93/67/EEC, 93/105/EC and 2000/21/EC.

280 OBERNBERGER Ingwald, THEK Gerold, 2002: The Current State of Austrian Pellet Boiler Technology. In *Proceedings of the 1st World Conference on Pellets* in Stockholm, Sweden, ISBN 91-631-2833-0, pp45-48, Swedish Bioenergy Association, Stockholm, Sweden.

281 OBERNBERGER Ingwald, 2004: Pellets-Technologien – ein Überblick. In *Proceedings of the World Sustainable Energy Days 2004* in Wels, Austria, O.Ö. Energiesparverband, Linz, Austria.

282 OBERNBERGER Ingwald, THEK Gerold, 2006: Recent developments concerning pellet combustion technologies – a review of Austrian developments. In *Proceedings of the 2nd World Conference on Pellets* in Jönköping, Sweden, ISBN 91-631-8961-5, pp31-40, Swedish Bioenergy Association, Stockholm, Sweden.

283 RIKA, 2005: http://www.rika.at, retrieved [1.4.2005], RIKA Metallwarenges.m.b.H. & Co KG, Micheldorf, Austria.

284 HERZ, 2003: http://www.herz-feuerung.com, retrieved [12.2.2003], HERZ Feuerungstechnik Ges.m.b.H., Sebersdorf, Austria.

285 HAGER, 2002: Company brochure, HAGER ENERGIETECHNIK GmbH, Poysdorf, Austria.

286 GUNTAMATIC, 2004: http://www.guntamatic.com/docs/main.htm, retrieved [7.4.2004], GUNTAMTIC Heiztechnik GmbH, Peuerbach, Austria.

287 STRAUSS Rolf-Peter, 2006: Neues von der Karussellfeuerung: Pelletstransport – Zündung – Betriebserfahrung. In *Proceedings of the 6th Pellets Industry Forum* in Stuttgart, Germany, pp151-155, Solar Promotion GmbH, Pforzheim, Germany.

288 KWB, 2008: http://www.kwb.at, retrieved [14.4.2008], Kraft & Wärme aus Biomasse GmbH, St. Margarethen/Raab, Austria.

289 SOMMERAUER & LINDNER, 2000: Company brochure, Sommerauer & Lindner Heizanlagenbau SL-Technik GmbH, St. Pantaleon, Austria.

290 AGRARNET AUSTRIA, 2008: http://www.agrarnet.info, retrieved [4.12.2008], Agrarnet Austria - Verein zur Förderung neuer Kommunikationstechnologien für den ländlichen Raum, Vienna, Austria.

291 WINDHAGER, 2002: Company brochure, Windhager Zentralheizung AG, Seekirchen am Wallersee, Austria.

292 KUNDE Robert, 2008: Emissions- und Leistungsmessungen an Pelletskesseln im Bestand. In *Proceedings of the 8th Pellets Industry Forum* in Stuttgart, Germany, pp59-65, Solar Promotion GmbH, Pforzheim, Germany.

293 SHT, 2006: Written notice and technical documentation: http://www.sht.at, retrieved [13.9.2006], Thermocomfort PN, sht – Heiztechnik aus Salzburg GmbH, Salzburg-Bergheim, Austria.

294 COMPACT, 2002: Company brochure, COMPACT Heiz- und Energiesysteme GesmbH, Gmunden, Austria.

295 EDER, 2001: Company brochure, Anton Eder GmbH, Bramberg, Austria.

296 BIOS, 2008: Company brochure, BIOS BIOENERGIESYSTEME GmbH, Graz, Östereich.

297 BIOS, 2001: Picture was taken at a visit of the Austrian biomass furnace manufacturer KWB Kraft und Wärme aus Biomasse GmbH, BIOS BIOENERGIESYSTEME GmbH, Graz, Östereich.

298 BRUNNER Thomas, BÄRNTHALER Georg, OBERNBERGER Ingwald, 2006: Fine particulate emissions from state-of-the-art small-scale Austrian pellet furnaces – characterisation, formation and possibilities of reduction. In *Proceedings of the 2nd World Conference on Pellets* in Jönköping, Sweden, ISBN 91-631-8961-5, pp87-91, Swedish Bioenergy Association, Stockholm, Sweden.

299 ORTNER Herbert, 2005: Brennwerttechnik – Die Anwendung der Brennwerttechnik bei der Pelletsheizung. In *Proceedings of the World Sustainable Energy Days 2005* in Wels, Austria, O.Ö. Energiesparverband, Linz, Austria.

300 BUNDESMINISTERIUM FÜR LAND- UND FORSTWIRTSCHAFT, UMWELT UND WASSERWIRTSCHAFT, 2000: Verordnung über die Begrenzung von Abwasseremissionen aus der Reinigung von Abluft und wässrigen Kondensaten (AEV Abluftreinigung), Bundesgesetzblatt 218/2000, Vienna, Austria.

301 BUNDESMINISTERIUM FÜR LAND- UND FORSTWIRTSCHAFT, UMWELT UND WASSERWIRTSCHAFT, 2005: Änderung der AEV Abluftreinigung, Bundesgesetzblatt 62/2005, Vienna, Austria.

302 NATIONALRAT, 1995: Vereinbarung zwischen dem Bund und den Ländern gemäß Art. 15a B-VG über die Einsparung von Energie, Bundesgesetzblatt 388/1995, Vienna, Austria.

303 DWA, 2003: *Kondensate aus Brennwertkesseln.* Deutsche Vereinigung für Wasserwirtschaft, Abwasser und Abfall e. V., Arbeitsblatt ATV-DVWK-A 251. Hennef: DWA, 20 Seiten, ISBN 978-3-924063-74-0.

304 BAYERISCHES LANDESAMT FÜR WASSERWIRTSCHAFT, 2000: *Einleiten von Kondensaten von Feuerungsanlagen in Entwässerungsanlagen,* Merkblatt Nr. 4.5/3, Stand: 30.08.2000, 6 Seiten, Munich, Germany.

305 OBERNBERGER Ingwald, 2004: Pelletfeuerungstechnologien in Österreich – Stand der Technik und zukünftige Entwicklungen. In *Proceedings of the World Sustainable Energy Days 2004* in Wels, Austria, O.Ö. Energiesparverband, Linz, Austria.

306 FRÖLING, 2010: Company brochure, Fröling Heizkessel- und Behälterbau GesmbH, Grieskirchen, Austria.

307 POWERcondens, 2006: http://www.powercondens.com, retrieved [20.9.2006], POWERcondens AG, Maienfeld, Switzerland.

308 ENERCONT, 2009: Oral notice and http://www.enercont.at, retrieved [12.2.2009], ENERCONT GmbH Energie- und Umwelttechnik Entsorgungstechnik, Golling, Austria.

309 BSCHOR, 2006: http://www.carbonizer.de, retrieved [27.4.2006], Bschor GmbH, Höchstädt/Donau, Germany.

310 BOMAT, 2006: Product information BOMAT Profitherm, BOMAT HEIZTECHNIK GMBH, Überlingen, Germany.

311 HARTMANN Hans, ROSSMANN Paul, LINK Heiner, MARKS Alexander, 2004: *Erprobung der Brennwerttechnik bei häuslichen Holzhackschnitzelfeuerungen mit Sekundärwärmetauscher,* ISSN 1614-1008, Technologie- und Förderzentrum (TFZ) in the Kompetenzzentrum für Nachwachsende Rohstoffe, Straubing, Germany.

312 SCHRÄDER, 2006: http://www.schraeder.com, retrieved [27.4.2006], Karl Schräder Nachf. Inh. Karl Heinz Schräder, Kamen, Germany.

313 RAWE Rudolf, 2006: Secondary heat exchanger and mass exchanger for condensing operation of biomass boilers – dust separation and energy recovery. In *Proceedings of the 2nd World Conference on Pellets* in Jönköping, Sweden, ISBN 91-631-8961-5, pp219-223, Swedish Bioenergy Association, Stockholm, Sweden.

314 RAWE Rudolf, KUHRMANN Hermann, NIEHAVES Jens, 2006: Die Schräder – HydroBox, Report, Fachhochschule Gelsenkirchen, Gelsenkirchen, Germany.

315 RAWE Rudolf, 2006: Schräder-HydroCube – Abgaswäscher/-wärmetauscher für Brennwertnutzung und Entstaubung bei Biomassekesseln. In *Proceedings of the 6th Pellets Industry Forum* in Stuttgart, Germany, pp113-121, Solar Promotion GmbH, Pforzheim, Germany.

316 RAWE Rudolf, KUHRMANN H., NIEHAVES J., STEINKE J., 2006: Brennwertnutzung und Staubabscheidung in Biomasse-Feuerungen. In *IKZ-FACHPLANER*, no. 10 (2006), STROBEL VERLAG GmbH & Co. KG, Arnsberg, Germany.

317 RAWE Rudolf, KUHRMANN Hermann, NIEHAVES Jens, 2006: Heat and mass exchanger for condensing biomass boilers energy recovery and flue gas cleaning. In *Proceedings of the Second International Green Energy Conference* in Oshawa, Canada, pp1216-1224, University of Ontario Institute of Technology, Ontario, Canada.

318 HARTMANN H., TUROWSKI P., ROSSMANN P., ELLNER-SCHUBERTH F., HOPF N., 2007: Grain and straw combustion in domestic furnaces – influences of fuel types and fuel pre-treatments. In *Proceedings of the 15th European Biomass Conference & Exhibition* in Berlin, Germany, ISBN 978-88-89407-59-X, ISBN 3-936338-21-3, pp1564-1569, ETA-Renewable Energies, Florence, Italy

319 BRUNNER Thomas, JÖLLER Markus, OBERNBERGER Ingwald, 2006: Aerosol formation in fixed-bed biomass furnaces - results from measurements and modelling. In *Proceedings of the Internat. Conf. Science in Thermal and Chemical Biomass Conversion* in Victoria, Canada, ISBN 1-872691-97-8, pp1-20, CPL Press, Berks, UK.

320 ÖKOENERGIE, 2007: *Zeitung für erneuerbare Energien*, no. 67 (2007), Österreichischer Biomasse-Verband & Ökosoziales Forum Österreich, Vienna, Austria.

321 MUSIL Birgit, HOFBAUER Hermann, SCHIFFERT Thomas, 2005: Development and analyses of pellets fired tiled stoves. In *Proceedings of the 14th European Biomass Conference & Exhibition* in Paris, France, ISBN 88-89407-07-7, pp1117-1118, ETA-Renewable Energies, Florence, Italy.

322 BEMMANN Ulrich, BYLUND Göran, 2006: Demonstration and optimisation of solar-pellet-combinations - SOLLET. In *Proceedings of the 2nd World Conference on Pellets* in Jönköping, Sweden, ISBN 91-631-8961-5, pp65-69, Swedish Bioenergy Association, Stockholm, Sweden.

323 PERSSON Tomas, FIEDLER Frank, RÖNNELID Mats, BALES Chris, 2006: Increasing efficiency and decreasing CO-emissions for a combined solar and wood pellet heating system for single family houses. In *Proceedings of the 2nd World Conference on Pellets* in Jönköping, Sweden, ISBN 91-631-8961-5, pp79-83, Swedish Bioenergy Association, Stockholm, Sweden.

324 NORDLANDER Svante, PERSSON Tomas, FIEDLER Frank, RÖNNELID Mats, BALES Chris, 2006: Computer modelling of wood pellet stoves and boilers connected to solar heating systems. In *Proceedings of the 2nd World Conference on Pellets* in Jönköping, Sweden, ISBN 91-631-8961-5, pp195-199, Swedish Bioenergy Association, Stockholm, Sweden.

325 BEMMANN Ulrich, GROSS B., BENDIECK A., 2006: EU-Projekt "SOLLET" - "Pellets + Solar" Demonstration und Optimierung. In *Proceedings of the European Pellets Forum 2006*, O.Ö. Energiesparverband, Linz, Austria.

326 KONERSMAN Lars, HALLER Michel, VOGELSANGER Peter, 2007: *Pelletsolar – Leistungsanalyse und Optimierung eines pellet-solarkombinierten Systems für Heizung und Warmwasser*, Final report, http://www.energieforschung.ch, retrieved [13.11.2008], SPF Institut für Solartechnik, Rapperswil, Switzerland.

327 FRANK Elimar, KONERSMANN Lars, 2008: PelletSolar – Optimierung des Jahresnutzungsgrades von Systemen mit Pelletkessel und Solaranlage. In *Proceedings of the 8th Pellets Industry Forum* in Stuttgart, Germany, pp66-72, Solar Promotion GmbH, Pforzheim, Germany.

328 BERNER Joachim, 2008: Solarenergie zuerst – wie Regelungen den gemeinsamen Betrieb von Pellets- und Solaranlagen steuern. In *Fachmagazin der Pelletsbranche*, no. 3 (2008), Solar Promotion GmbH, Pforzheim, Germany.

329 FIEDLER F., 2006: *Combined Solar and Pellet Heating Systems -Studies of Energy Use and CO-emissions*. Dissertation No 36, Mälardalen University, (PhD thesis) ISBN: 91-85485-30-6, Västerås, Sweden.

330 PERSSON T., 2006: *Combined solar and pellet heating systems for single-family houses – how to achieve decreased electricity usage, increased system efficiency and increased solar gains*. Doctoral thesis in Energy Thchnology, Trita REFR Report No. 06/56, ISBN: 91-7178-538-8, KTH – Royal Institute of Technology, Stockholm, Sweden.

331 PERSSON T., 2008: *Solar and Pellet heating Systems, Reduced Electricity Usage in Single-family Houses*. VDM Verlag Dr. Müller. ISBN: 978-3-639-12206-0, Saarbrücken, Germany.

332 DAHM Jochen, 1995, Chalmers University of Technology, Gothenburg, Sweden.

333 ENGLISCH Martin, 2005: *Konzepte und Erfahrungen mit Pelletsfeuerungen im mittleren Leistungsbereich*, VDI seminar, Salzburg, Austria.

334 KÖB Siegfried, 2006: Reduktion der Partikelkonzentration im Abgas von Pelletsfeuerungen > 150 kW. In *Proceedings of the 6th Pellets Industry Forum* in Stuttgart, Germany, pp122-127, Solar Promotion GmbH, Pforzheim, Germany.

335 KÖB, 2008: http://www.koeb-holzfeuerungen.com, retrieved [15.4.2008], KÖB Holzfeuerungen GmbH, Wolfurt, Austria.

336 SCHÖNMAIER Heinrich, 2005: Neue Anwendungsformen für Großanlagen: mobile Biomasse-Energiecontainer und Einsatz im Unterglasbau. In *Proceedings of the zum 5. Industrieforum Holzenergie* in Stuttgart, Germany, pp156-159, Deutscher Energie-Pellet-Verband e.v. und Deutsche Gesellschaft für Sonnenenergie e.V., Germany.

337 LUNDBERG Henrik, 2002: Combustion of Crushed Pellets in a Burner. In *Proceedings of the 1st World Conference on Pellets* in Stockholm, Sweden, ISBN 91-631-2833-0, pp43-44, Swedish Bioenergy Association, Stockholm, Sweden.

338 LJUNGDAHL Boo, 2003: Bioswirl® a wood pellets burner for oil retrofit. In *Proceedings of the International Nordic Bioenergy Conference* in Jyväskylä, Finland, ISBN 952-5135-26-8, ISSN 1239-4874, pp481-488, Finnish Bioenergy Association, Jyväskylä, Finland.

339 TPS, 2008: http://www.tps.se, retrieved [16.4.2008], TPS Termiska Processer AB, Nyköping, Sweden.

340 OBERNBERGER Ingwald, THEK Gerold, 2004: *Basic information regarding decentralised CHP plants based on biomass combustion in selected IEA partner countries*, Final report of the related IEA Task32 project, BIOS BIOENERGIESYSTEME GmbH, Graz, Austria.

341 RES LEGAL, 2009: *Rechtsquellen für die Stromerzeugung aus Erneuerbaren Energien*, http://res-legal.eu, retrieved [11.2.2009], Bundesministerium für Umwelt, Naturschutz und Reaktorsicherheit, Berlin, Germany.

342 OBERNBERGER Ingwald, THEK Gerold, 2004: *Techno-economic evaluation of selected decentralised CHP applications based on biomass combustion in IEA partner countries*, Final report of the related IEA Task 32 project, BIOS BIOENERGIESYSTEME GmbH, Graz, Austria.

343 BERNER Joachim, 2006: Mit dem Wärmekessel eigenen Strom produzieren. In *Fachmagazin der Pelletsbranche*, no. 3 (2006), Solar Promotion GmbH, Pforzheim, Germany.

344 BERNER Joachim, 2008: Heißer Kopf – Stirlingmotor nutzt die Wärme eines Pelletskessels zur Stromproduktion. In *Fachmagazin der Pelletsbranche*, no. 5 (2008), Solar Promotion GmbH, Pforzheim, Germany.

345 FRIEDL Günther, HECKMANN Matthias, MOSER Wilhelm, 2006: Small-scale pellet boiler with thermoelectric generator. In *Proceedings of the European Pellets Forum 2006*, O.Ö. Energiesparverband, Linz, Austria.

346 MOSER Wilhelm, FRIEDL Günther, HASLINGER Walter, 2006: Small-scale pellet boiler with thermoelectric generator. In *Proceedings of the 2nd World Conference on Pellets* in Jönköping, Sweden, ISBN 91-631-8961-5, pp85-86, Swedish Bioenergy Association, Stockholm, Sweden.

347 STANZEL K. Wolfgang et al., 2003: Business plan for a Stirling engine (1 kW$_{el}$) integrated into a pellet stove. In *Proceedings of the 11th International Stirling Engine Conference (ISEC)* in Rome, pp388-389, Department of Mechanical and Aeronautical Engineering, University of Rome "La Sapienza", Rome, Italy.

348 OBERNBERGER Ingwald, HAMMERSCHMID Alfred, 1999: Dezentrale Biomasse-Kraft-Wärme-Kopplungstechnologien – Potential, technische und wirtschaftliche Bewertung, Einsatzgebiete. In *Thermische Biomassenutzung*, vol.4, ISBN 3-7041-0261-X, dbv – publisher of Graz University of Technology, Graz, Austria.

349 STANZEL Karl Wolfgang, 2006: Strom und Wärme aus Pellets für Haushalte. In *Proceedings of the European Pellets Forum 2006*, O.Ö. Energiesparverband, Linz, Austria.

350 STANZEL Karl Wolfgang, 2006: Jedem sein Kraftwerk – Strom und Wärme aus Holzpellets. In *Proceedings of the 6th Pellets Industry Forum* in Stuttgart, Germany, pp136-138, Solar Promotion GmbH, Pforzheim, Germany.

351 SPM, 2008: http://www.stirlingpowermodule.com, retrieved [23.12.2008], Stirling Power Module Energie-umwandlungs GmbH, Graz, Austria.

352 FRIEDL Günther, MOSER Wilhelm, HOFBAUER Hermann, 2006: Micro-Scale Biomass-CHP – Intelligent Heat Transfer with Thermoelectric Generator. In *Proceedings of the 6th Pellets Industry Forum*, Oktober 2006, Stuttgart, Germany, pp139-144, Solar Promotion GmbH, Pforzheim, Germany.

353 MOSER Wilhelm, FRIEDL Günther, AIGENBAUER Stefan, HECKMANN Matthias, HOFBAUER Hermann, 2008: A biomass-fuel based micro-scale CHP system with thermoelectric generators. In *Proceedings of the Central European Biomass Conference 2008* in Graz, Austrian Biomass Association, Vienna, Austria.

354 HASLINGER Walter, GRIESMAYR Susanne, POINTNER Christian, FRIEDL Günther, 2008: Biomassekleinfeuerungen – Überblick und Darstellung innovativer Entwicklungen. In *Proceedings of the 8th Pellets Industry Forum* in Stuttgart, Germany, pp53-58, Solar Promotion GmbH, Pforzheim, Germany.

355 FRIEDL Günther, MOSER Willhelm, GRIESMAYR Susanne, 2008: *Pelletfeuerung mit thermo-elektrischer Stromerzeugung*, Presentation at the 10th Holzenergie-Symposium in Zurich, Switzerland, http://www.holzenergie-symposium.ch, retrieved [22.12.2008]

356 OBERNBERGER Ingwald, CARLSEN Henrik, BIEDERMANN Friedrich, 2003: State-of-the-Art and Future Developments Regarding Small-scale Biomass CHP Systems with a Special Focus on ORC and Stirling Engine Technologies. In *Proceedings of the International Nordic Bioenergy Conference* in Jyväskylä, Finland, ISBN 952-5135-26-8, ISSN 1239-4874, pp331-339, Finnish Bioenergy Association, Jyväskylä, Finland.

357 BIEDERMANN Friedrich, CARLSEN Henrik, SCHÖCH Martin, OBERNBERGER Ingwald, 2003: Operating Experiences with a Small-scale CHP Pilot Plant based on a 35 kW$_{el}$ Hermetic Four Cylinder Stirling Engine for Biomass Fuels. In *Proceedings of the 11th International Stirling Engine Conference (ISEC)* in Rome, Italy, pp248-254, Department of Mechanical and Aeronautical Engineering, University of Rome "La Sapienza", Rome, Italy.

358 BIEDERMANN Friedrich, CARLSEN Henrik, OBERNBERGER Ingwald, SCHÖCH Martin, 2004: Small-scale CHP Plant based on a 75 kW$_{el}$ Hermetic Eight Cylinder Stirling Engine for Biomass Fuels – Development, Technology and Operating Experiences. In *Proceedings of the 2nd World Conference and Exhibition on Biomass for Energy, Industry and Climate Protection* in Rome, Italy, vol. 2, ISBN 88-89407-04-2, pp1722-1725, ETA-Florence, Florence, Italy.

359 OBERNBERGER Ingwald, THEK Gerold, 2008: Combustion and gasification of solid biomass for heat and power production in Europe – state-of-the-art and relevant future developments (keynote lecture). In *Proceedings of the 8th European Conference on Industrial Furnaces and Boilers* in Vilamoura, Portugal, INFUB, Rio Tinto, Portugal.

360 TILT Yumi Koyama, 2001: *Micro-Cogeneration for Single-Family Dwellings*, Case study, Department of Intercultural Communication & Management, Copenhagen Business School, Copenhagen, Denmark.

361 OBERNBERGER Ingwald, BINI Roberto, NEUNER Helmut, PREVEDEN Zvonimir, 2001: *Biomass fired CHP plant based on an ORC cycle - Project ORC-STIA-Admont*, Final report of the EU-THERMIE project no. BM/120/98, European Commission, DG TREN, Brussels, Belgium.

362 THONHOFER Peter, REISENHOFER Erwin, OBERNBERGER Ingwald, GAIA Mario, 2004: Demonstration of an innovative biomass CHP plant based on a 1,000 kWel Organic Rankine Cycle - EU demonstration project Lienz (A). In *Proceedings of the 2nd World Conference and Exhibition on Biomass for Energy, Industry and Climate Protection* in Rome, Italy, vol. 2, ISBN 88-89407-04-2, pp1839-1842, ETA-Florence, Florence, Italy.

363 OBERNBERGER Ingwald, BINI Roberto, REISINGER Heinz, BORN Manfred, 2003: *Fuzzy Logic controlled CHP plant for biomass fuels based on a highly efficient ORC process*, Final publishable report of the EU-project no. NNE5/2000/475, European Commission, Brussels, Belgium.

364 KNOEF Harie A. M., 2003: *Gasification of biomass for electricity and heat production – a review*, BTG biomass technology group, Enschede, Netherlands.

365 BRIDGWATER Anthony V.: *Fast pyrolysis of biomass for liquid fuels and chemicals – a review*; Bio-Energy Research Group, Aston University, Birmingham, UK.

366 OBERNBERGER Ingwald, THEK Gerold, REITER Daniel, 2008: Economic evaluation of decentralised CHP applications based on biomass combustion and biomass gasification. In *Proceedings of the Central European Biomass Conference 2008* in Graz, Austria, Austrian Biomass Association, Vienna, Austria.

367 OVERGAARD Peter, SANDER Bo, JUNKER Helle, FRIBORG Klaus, LARSEN Ole Hede, 2004: Two years' operational experience and further development of full-scale co-firing of straw. In *Proceedings of the 2nd World Conference and Exhibition on Biomass for Energy, Industry and Climate Protection* in Rome, Italy, vol. 3, ISBN 88-89407-04-2, pp1261-1264, ETA-Florence, Florence, Italy.

368 LIVINGSTON W. R., MORRIS K. W., 2009: Experience with Co-firing Biomass in PC Boilers to Reduce CO_2 Emissions. In *Proc. Power-Gen International 2009*, Las Vegas, NV, USA

369 OTTLINGER Bernd, 2000: Oral notice, Amandus Kahl GmbH, Reinbek, Germany.

370 KREUTZER, 2000: Oral notice, Holzindustrie Preding GmbH, Preding, Austria.

371 PROHOLZ, 2009: *Holzpreise in der Steiermark*, http://www.proholz-stmk.at, retrieved [1.10.2009], proHolz Steiermark, Graz, Austria.

372 IWOOD, 2008: *Aus Sägemehl wird eine iwood Platte*, http://www.iwood.ch, retrieved [5.12.2008], Zug, Switzerland.

373 WIENER BÖRSE, 2008: Kursblatt der Wiener Warenbörse Holz, 3. Dezember 2008, http://www.wienerboerse.at, retrieved [4.12.2008], Wiener Börse AG, Vienna, Austria.

374 PABST, 2008: http://www.hackgut.at, retrieved [4.12.2008], Franz PAPST HACKK EXPRESS, Obdach, Austria.

375 HACKGUTBÖRSE, 2008: http://www.hackgutboerse.at, retrieved [4.12.2008], Agrarmanagement NÖ-Süd, Warth, Austria.

376 KETTNER Claudia, KUFLEITNER Angelika, LOIBNEGGER Thomas, PACK Alexandra, STEININGER Karl W., TÖGLHOFER Christian, TRINK Thomas, 2008: *Regionalwirtschaftliche Auswirkungen verstärkter Biomasse-Energie-Nutzung*, http://www.regionalmanagement.at, retrieved [5.12.2008], Wegener Center for Climate and Global Change at Graz University, Graz, Austria.

377 BERTAINA Fabiano, VIDALE Sergio, 2008: *Die Anpflanzung ausgewählter Pappelsorten für die Herstellung von Biomasse im Energiebereich im Norden Italiens*, Presentation within the framework of the

event Tag der Bioenergie, http://www.hafendorf.at, retrieved [7.12.2008], Land- und forstwirtschaftliche Fachschule Hafendorf, Kapfenberg, Austria.

378 DORNER Egon, 2008: *Energieholznutzung in hocheffizienten KWK-Anlagen*, Presentation within the framework of the event Tag der Bioenergie, http://www.hafendorf.at, retrieved [7.12.2008], Land- und forstwirtschaftliche Fachschule Hafendorf, Kapfenberg, Austria.

379 TOPPER, 2002: http://www.tropper.at, retrieved [30.1.2002], Tropper Maschinen und Anlagen GmbH, Schwanenstadt, Austria.

380 THEK Gerold, OBERNBERGER Ingwald, 2001: Produktionskosten von Holzpellets gegliedert nach Prozessschritten und unter Berücksichtigung österreichischer Randbedingungen. In *Proceedings of the 2nd European Round Table Woodpellets* in Salzburg, Austria, pp33-40, Umbera GmbH, St. Pölten, Austria.

381 THEK Gerold, OBERNBERGER Ingwald, 2002: Wood pellet production costs under Austrian and in comparison to Swedish framework conditions. In *Proceedings of the 1st World Conference on Pellets* in Stockholm, Sweden, ISBN 91-631-2833-0, pp123-128, Swedish Bioenergy Association, Stockholm, Sweden.

382 THEK Gerold, OBERNBERGER Ingwald, 2004: Wood pellet production costs under Austrian and in comparison to Swedish framework conditions. In *Biomass and Bioenergy*, ISSN 0961-9534, vol. 27 (2004), pp671-693, Elsevier Ltd., Oxford, UK.

383 WALLIN Mikael, 2002: Mikro Scale Pellet Production Technology. In *Proceedings of the 1st World Conference on Pellets* in Stockholm, Sweden, ISBN 91-631-2833-0, p73, Swedish Bioenergy Association, Stockholm, Sweden.

384 FENZ Bernhard, STAMPFER Karl, 2005: *Optimierung des Holztransports durch Einsatz von faltbaren Containern (LogRac)*, Final report of a study by the Bundesministeriums für Land- und Forstwirtschaft, Umwelt- und Wasserwirtschaft, the Styrian government and the Institute of Forest Engineering, Department of Forest and Soil Sciences, University of Natural Resources and Applied Life Sciences, Vienna, Austria.

385 DEPV, 2006: http://www.depv.de, retrieved [12.12.2006], Deutscher Energie-Pellet-Verband e.V. (DEPV), Mannheim, Germany.

386 PROPELLETS, 2006: *Gas-Konflikt Ukraine: Österreichs Pelletsproduktion steht Gewehr bei Fuß*, received [19.10.2006], Austrian Press Agency Originaltext Service GmbH, Vienna, Austria.

387 GILBERT Jeremy, 2006: World oil reserves – can supply meet demand? In *Proceedings of the 6th Pellets Industry Forum* in Stuttgart, Germany, pp8-9, Solar Promotion GmbH, Pforzheim, Germany.

388 SELTMANN Thomas, 2008: Vom Überfluss zur Knappheit: Die fossile Energiewirtschaft vor dem Scheitelpunkt. In *Proceedings of the 8th Pellets Industry Forum* in Stuttgart, Germany, pp9-14, Solar Promotion GmbH, Pforzheim, Germany.

389 E-CONTROL, 2008: http://www.e-control.at, retrieved [8.12.2008], Energie-Control GmbH, Vienna, Austria.

390 IWO, 2008: http://www.iwo-austria.at, retrieved [11.12.2008], IWO-Österreich, Institut für wirtschaftliche Ölheizung, Vienna, Austria.

391 BLT WIESELBURG, 2008: http://www.blt.bmlf.gv.at, retrieved [12.1.2009], BLT - Biomass · Logistics · Technology Francisco Josephinum, Wieselburg, Austria.

392 KPC, 2008: http://public-consulting.at, retrieved [9.12.2008], Kommunalkredit Public Consulting GmbH, Vienna, Austria.

393 STANZEL Wolfgang, JUNGMEIER Gerfried, SPITZER Josef, 1995: *Emissionsfaktoren und energietechnische Parameter für die Erstellung von Energie- und Emissionsbilanzen im Bereich der Raumwärmeversorgung*, Final report, Institute of Energy Research, Joanneum Research, Graz, Austria.

394 EBS, 2008: http://energieberatungsstelle.stmk.gv.at, retrieved [7.12.2008], Energieberatungsstelle Land Steiermark, Graz, Austria.

395 JILEK Wolfgang, KARNER Karin, RASS Andrea, 1999: *Externe Kosten im Energiebereich*, LandesEnergieVerein Steiermark, Graz, Austria.

396 POINT CARBON, 2009: http://www.pointcarbon.com, retrieved [11.2.2009], Point Carbon, Oslo, Norway.

397 HAAS Reinhard, KRANZL Lukas, 2002: *Bioenergie und Gesamtwirtschaft*, Bundesministerium für Verkehr, Innovation und Technologie, Vienna, Austria.

398 NUSSBAUMER Thomas, OSER Michael, 2004: *Evaluation of Biomass Combustion based Energy Systems by Cumulative Energy Demand and Energy Yield Coefficient*, Version 1.0, ISBN 3-908705-07-X, International Energy Agency IEA Bioenergy Task 32 and Swiss Federal Office of Energy, Verenum press, Zurich, Switzerland.

399 RAKOS Christian, TRETTER Herbert, 2002: *Vergleich der Umweltauswirkungen einer Pelletheizung mit denen konventioneller Energiebereitstellungssysteme am Beispiel einer 400 kW Heizanlage*, Energieverwertungsagentur – the Austrian Energy Agency (E.V.A.), Vienna, Austria.

400 SPITZER Josef et al., 1998: *Emissionsfaktoren für feste Brennstoffe*, Final report, Institute of Energy Research, Joanneum Research, Graz, Austria.

401 SPITZER Josef et al., 1998: *Emissionsfaktoren für feste Brennstoffe*, Data, Institute of Energy Research, Joanneum Research, Graz, Austria.

402 FNR, 2001: *Leitfaden Bioenergie, Datensammlung*, http://www.fnr.de, retrieved [20.12.2001], Fachagentur Nachwachsende Rohstoffe e.V., Gülzow, Germany.

403 BÖHMER Siegmund, FRÖHLICH Marina, KÖTHER Traute, KRUTZLER Thomas, NAGL Christian, PÖLZ Werner, POUPA Stefan, RIGLER Elisabeth, STORCH Alexander, THANNER Gerhard, 2007: *Aktualisierung von Emissionsfaktoren als Grundlage für den Anhang des Energieberichtes*, Report to the Bundesministerium für Wirtschaft und Arbeit and the Bundesministerium für Land- und Forstwirtschaft, Umwelt und Wasserwirtschaft, Umweltbundesamt GmbH, ISBN 3-85457-872-5, Vienna, Austria.

404 OBERNBERGER Ingwald, THOMAS Brunner, BÄRNTHALER Thomas, 2006: *Feinstaubemissionen aus Biomasse-Kleinfeuerungsanlagen*, Interim report of the research project with the same name of the Zukunftsfonds of the Styrian government (Projekt Nr. 2088), BIOS BIOENERGIESYSTEME GmbH, Graz, Austria.

405 OBERNBERGER Ingwald, 2008: State-of-the-art small-scale biomass combustion with respect to fine particulate emissions – Country report from Austria. In *Proceedings of the Central European Biomass Conference 2008* in Graz, Austrian Biomass Association, Vienna, Austria.

406 AUSTRIAN GOVERNMENT, 2003: *Energiebericht der Österreichischen Bundesregierung*, http://www.bmwa.gv.at, retrieved [12.1.2009], Bundesministerium für Wirtschaft und Arbeit, Vienna, Austria.

407 ANONYMUS, 2001: *10 Jahre Emissionsmessungen an automatisch beschickten Holzheizungen*, http://www.vorarlberg.at, retrieved [12.1.2009]

408 NAGL Christian, STERRER Roland, SZEDNYJ Ilona, WIESER Manuela, 2004: *Emissionen aus Verbrennungsvorgängen zur Raumwärmeerzeugung – Literaturarbeit der Umweltbundesamt GmbH*, http://www.iwo-austria.at/fileadmin/user_upload/MitgliedernImgs/EmissionenRaumw_rmeEndfassung060 904.pdf, retrieved [12.1.2009], IWO Österreich (Institut für wirtschaftliche Ölheizung), Vienna, Austria.

409 PRIEWASSER Reinhold, 2005: Feinstaubproblematik und Holzheizungen. In *Ländlicher Raum*, no. 10 (2005), www.energycabin.de/uploads/media/Priewasser_Publ_Feinstaub_ Holzheiz.pdf retrieved [12.1.2009], Bundesministeriums für Land- und Forstwirtschaft, Umwelt- und Wasserwirtschaft, Vienna, Austria.

410 NUSSBAUMER Thomas, CZASCH Claudia, KLIPPEL Norbert, JOHANSSON Linda, TULLIN Claes, 2008: *Particulate Emissions from Biomass Combustion in IEA Countries - Survey on Measurements and Emission Factors*. Report on behalf of International Energy Agency (IEA) Bioenergy Task 32 and Swiss Federal Office of Energy (SFOE), ISBN 3-908705-18-5, Zurich, Switzerland.

411 RATHBAUER Josef, LASSELSBERGER Leopold, WÖRGETTER Manfred, 1998: *Holzpellets – Brennstoff mit Zukunft*, Bundesanstalt für Landtechnik Wieselburg, Wieselburg, Austria.

412 JUNGMEIER Gerfried, GOLJA Ferdinand, SPITZER Josef, 1999: *Der technologische Fortschritt bei Holzfeuerungen – Ergebnisse einer statistischen Analyse der Prüfstandsmessungen der BLT Wieselburg von 1980 bis 1998*, vol.11 (1999), ISBN 3-901 271-98-8, BMUJF, Graz, Austria.

413 LASSELSBERGER Leopold, 2005: Holz- und Pelletsfeuerungen – Qualität mit Zukunft. In *Proceedings of the World Sustainable Energy Days 2005* in Wels, Austria, O.Ö. Energiesparverband, Linz, Austria.

414 FRIEDL Günther, HECKMANN Matthias, ROSSMANN Paul, 2009: Advancements in energy efficiency of small-scale pellets boilers. In *Proceedings of the 9th Pellets Industry Forum* in Stuttgart, Germany, pp84-92, Solar Promotion GmbH, Pforzheim, Germany.

415 KUNDE Robert, VOLZ Florian, GADERER Matthias, SPLIETHOFF Hartmut, 2009: Felduntersuchungen an Holzpellet-Zentralheizkesseln. In: *Brennstoff Kraft Wärme* (BWK) Bd. 61 (2009) Nr. 1/2, pp58-66, Springer-VDI-Verlag GmbH & Co. KG, Düsseldorf, Germany

416 SAROFIN A.F., HELBLE J.J., 1993: *The impact of ash deposition on coal fired plants*, Williamson J. and Wigley F. (ed.), Taylor & Francis, Washington, USA, pp567-582

417 UMWELTBUNDESAMT, 2009: http://www.umweltbundesamt.at, retrieved [13.1.2009], Umweltbundesamt GmbH, Vienna, Austria.

418 SPANGL Wolfgang, NAGL Christian, MOOSMANN Lorenz, 2008: Annual report of ambient air quality measurements in Austria 2007, Umweltbundesamt GmbH, ISBN 3-85457-950-0, Vienna, Austria.

419 UMWELTBUNDESAMT, 2006: *Schwebestaub in Österreich – Fachgrundlagen für eine kohärente österreichische Strategie zur Verminderung der Schwebestaubbelastung*, Report BE-277, ISBN 3-85457-787-7, Umweltbundesamt GmbH, Vienna, Austria.

420 RAPP Regula, 2008: *Eigenschaften und Gesundheitswirkungen von Feinstaub*. In *Proceedings of the 10th Holzenergie-Symposium* in Zurich, Switzerland, ISBN 3-908705-19-9, pp115-127, ETH Zurich, Thomas Nussbaumer (Ed.), Zurich, Switzerland.

421 SCHNEIDER J., 2004: *Gesundheitseffekte durch Schwebestaub*, Workshop *PMx- Quellenidentifizierung - Ergebnisse als Grundlage für Maßnahmenpläne*, Duisburg, Germany.

422 WHO, 2003: *Health Aspects of Air Pollution with Particulate Matter, Ozone and Nitrogen Dioxide*, WHO Regional Office for Europe, Copenhagen, Denmark.

423 WHO, 2004: *Meta-analysis of time-series studies and panel studies of Particulate Matter (PM) and Ozone (O_3)*, WHO Regional Office for Europe, Copenhagen, Denmark.

424 WHO, 2004: *Health Aspects of Air Pollution – answers to follow-up questions from CAFE*, Report on a WHO working group meeting in Bonn, Germany, WHO Regional Office for Europe, Copenhagen, Denmark.

425 HIRVONEN Maija-Riitta, JALAVA Pasi, HAPPO Mikko, PENNANEN Arto, TISSARI Jarkko, JOKINIEMI Jorma, SALONEN Raimo O., 2008: In-vitro Inflammatory and Cytotoxic Effects of Size-Segregated Particulate Samples Collected from Flue Gas of Normal and Poor Wood Combustion in Masonry Heater. In *Proceedings of the Central European Biomass Conference 2008* in Graz, Austria, Austrian Biomass Association, Vienna, Austria.

426 BELLMANN Bernd, CREUTZENBERG Otto, KNEBEL Jan, RITTER Detlef, POHLMANN Gerhard, 2008: Health effects of aerosols from biomass combustion plants. In *Proceedings of the Central European Biomass Conference 2008* in Graz, Austria, Austrian Biomass Association, Vienna, Austria.

427 OBERNBERGER Ingwald, BRUNNER Thomas, BÄRNTHALER Georg, 2007: Fine particulate emissions from modern Austrian small-scale biomass combustion plants. In *Proceedings of the 15th European Biomass Conference & Exhibitio* in Berlin, Germany, ISBN 978-88-89407-59-X, ISBN 3-936338-21-3, pp1546-1557, ETA-Renewable Energies, Florence, Italy.

428 OBERNBERGER Ingwald, BRUNNER Thomas (Eds.), 2005: Aerosols in Biomass Combustion. In *Proceedings of the international workshop* in Graz, Austria, ISBN 3-9501980-2-4, Institute of Resource Efficient and Sustainable Systems, Graz University of Technology, Graz, Austria.

429 OBERNBERGER Ingwald, BRUNNER Thomas, FRANDSEN Flemming, SKIFVARS Bengt-Johan, BACKMAN Rainar, BROUWERS J.J.H., VAN KEMENADE Erik, MÜLLER Martin, STEURER Claus, BECHER Udo, 2003: *Aerosols in fixed-bed biomass combustion – formation, growth, chemical composition, deposition, precipitation and separation from flue gas*, Final report, EU project No. NNE5-1999-00114, European Commission DG Research, Brussels, Belgium.

430 CHRISTENSEN K. A., 1995: *The Formation of Submicron Particles from the Combustion of Straw*, PHD Thesis, ISBN 87-90142-04-7, Department of Chemical Engineering, Technical University of Denmark, Lyngby, Denmark.

431 ANDERL Michael, GANGL Marion, KAMPEL Elisabeth, KÖTHER Traute, MUIK Barbara, PAZDEMIK Katja, POUPA Stephan, RIGLER Elisabeth, SCHODL Barbara, SPORER Melanie, STORCH Alexander, WAPPEL Daniela, WIESER Manuela, 2008: *Emissionstrends 1990–2006 – Ein Überblick über die österreichischen Verursacher von Luftschadstoffen (Datenstand 2008)*, ISBN 3-85457-959-4, Umweltbundesamt GmbH, Vienna, Austria.

432 BUNDESMINISTERIUM FÜR LAND- UND FORSTWIRTSCHAFT, UMWELT UND WASSER-WIRTSCHAFT, 2008: Änderung der Abfallverzeichnisverordnung, Bundesgesetzblatt 498/2008, Vienna, Austria.

433 Neurauter Rudolf, Mölgg Martin, Reinalter Matthias, 2004: *Aschen aus Biomassefeuerungsanlagen, Leitfaden der Tiroler Landesregierung*, www.tirol.gv.at, retrieved [14.1.2009], Abteilung Umweltschutz/Referat Abfallwirtschaft, Innsbruck, Austria.

434 NUSSBAUMER Thomas, KLIPPEL Norbert, OSER Michael, 2005: Health relevance of aerosols from biomass combustion by cytotoxicity tests. In *Proceedings of the work-shop "Aerosols in Biomass Combustion"* in Graz, Austria, Obernberger Ingwald, Brunner Thomas (eds), ISBN 3-9501980-2-4, pp45-54, Institute of Resource Efficient and Sustainable Systems, Graz University of Technology, Graz, Austria.

435 JOHANSSON Linda et al., 2005: Particle emissions from residential biofuel boilers and stoves – old and modern techniques. In *Proceedings of the work-shop "Aerosols in Biomass Combustion"* in Graz, Austria, ISBN 3-9501980-2-4, pp145-150, Obernberger Ingwald, Brunner Thomas (eds), Institute of Resource Efficient and Sustainable Systems, Graz University of Technology, Graz, Austria.

436 PVA, 2009: http://pva.studiothek.com, retrieved [19.1.2009], Pelletsverband Austria, Vienna, Austria.

437 PROPELLETS, 2009: http://propellets.at, retrieved [2.10.2009], Verein proPellets Austria - Netzwerk zur Förderung der Verbreitung von Pelletsheizungen, Wolfsgraben, Austria.

438 ANONYMOUS, 2005: In Europa produzieren bereits über 200 Pelletierwerke. In *Holz-Zentralblatt*, no. 70 (2005), p923, DRW-Verlag Weinbrenner GmbH & Co. KG, Leinfelden-Echterdingern, Germany.

439 RAKOS Christian, 2006: Die Entwicklung internationaler Pelletsmärkte im Vergleich. In *Proceedings of the 6th Pellets Industry Forum* in Stuttgart, Germany, pp29-35, Solar Promotion GmbH, Pforzheim, Germany.

440 RAKOS Christian, 2008: *Development of the Austrian pellet market 2007*, Presentation at the World Sustainable Energy Days 2008 in Wels, Austria.

441 EUROPEAN PELLET CENTRE, 2006: http://www.pelletcentre.info, retrieved [14.12.2006], Force Technology, Lyngby, Denmark.

442 HOLZKURIER, 2004: Biomasse woher und wohin. In *Holzkurier*, no. 48 (2004), pp18-19, Österreichischer Agrarverlag, Leopoldsdorf, Austria.

443 RAKOS Christian, 2010: Current developments in the Austrian pellet market. In *Proceedings of the World Sustainable Energy Days 2010* in Wels, Austria, O.Ö. Energiesparverband, Linz, Austria.

444 ANONYMUS, 2005: In Europa produzieren bereits über 200 Pelletierwerke. In *Holz-Zentralblatt*, no. 70(2005), p923, DRW-Verlag Weinbrenner GmbH & Co. KG, Leinfelden-Echterdingern, Germany.

445 SCHLAGITWEIT Christian, 2010: Written information, proPellets Austria - network to reinforce the distribution of pellet heating systems, Wolfsgraben, Austria.

446 HOLZKURIER, 2008: Pelletsproduktion. In *Holzkurier*, special edition 09.08, *Wärme und Kraft aus Biomasse*, 28.2.2008, pp4-5, Österreichischer Agrarverlag, Vienna, Austria.

447 MARGL Hermann D., 2000: Chancen und Grenzen des Pelletmarktes. In *Holz-Zentralblatt*, no. 144 (2000), pp1998-1999, DRW-Verlag Weinbrenner GmbH & Co. KG, Leinfelden-Echterdingern, Germany.

448 HAHN Brigitte, 2002: *Empirische Untersuchung zum Rohstoffpotenzial für die Herstellung von (Holz)Pellets unter besonderer Berücksichtigung der strategischen Bedeutung innerhalb der FTE-Aktivitäten auf nationaler und EU-Ebene*, Study for the Bundesministerium für Wirtschaft und Arbeit, UMBERA Umweltorientierte Betriebsberatungs-, Forschungs- und Entsorgungs-Gesellschaft m.b.H., St. Pölten, Austria.

449 HUBER Rudolf, 2004: Erfolgreiches Marketing für Holz-Pellets. In *Proceedings of the World Sustainable Energy Days 2004* in Wels, Austria, O.Ö. Energiesparverband, Linz, Austria.

450 PVA, 2003: http://www.pelletsverband.at, retrieved [19.3.2003], Pelletverband Austria Vertriebs- und Beratungsgesellschaft mbH, Weißkirchen, Austria.

451 HEIN Michaela, KALTSCHMITT Martin (Ed.), 2004: Standardisation of Solid Biofuels – Status of the ongoing standardisation process and results of the supporting research activities (BioNorm). In *Proceedings of the International Conference Standardisation of Solid Biofuels*, Leipzig, Germany.

452 JANSEN Hans, 2004: Timber from Russia – a sustainable resource? In *Proceedings of the World Sustainable Energy Days 2004* in Wels, Austria, O.Ö. Energiesparverband, Linz, Austria.

453 AKIM Eduard L., 2004: Wood pellets production in Russia. In *Proceedings of the World Sustainable Energy Days 2004* in Wels, Austria, O.Ö. Energiesparverband, Linz, Austria.

454 JONAS Anton, HANEDER Herbert, 2004: *Zahlenmäßige Entwicklung der modernen Holz- und Rindenfeuerungen in Österreich, Gesamtbilanz 1989 bis 2003*, Forstabteilung der Niederösterreichischen Landes-Landwirtschaftskammer, St. Pölten, Austria.

455 FURTNER Karl, HANEDER Herbert, 2008: *Biomasse – Heizungserhebung 2007*, NÖ Landes-Landwirtschaftskammer, Abteilung Betriebswirtschaft und Technik, St. Pölten, Austria.

456 FURTNER Karl, HANEDER Herbert, 2009: *Biomasse – Heizungserhebung 2008*, NÖ Landes-Landwirtschaftskammer, Abteilung Betriebswirtschaft und Technik, St. Pölten, Austria.

457 REGIONALENERGIE STEIERMARK, 2009: http://www.regionalenergie.at, retrieved [21.1.2009] and personal information, Regionalenergie Steiermark Beratungsgesellschaft für Holzenergiesysteme, Weiz, Austria.

458 KOOP Dittmar, 2008: Erstaunliche Ausmaße. In *Fachmagazin der Pelletsbranche*, no. 6 (2008), Solar Promotion GmbH, Pforzheim, Germany.

459 NEMESTOTHY Kasimir, 2006: *Abschätzung des Holzpelletsbedarfes in Österreich*, Austrian Energy Agency, Bundesministerium für Verkehr, Innovation und Technologie, Vienna, Austria.

460 HANEDER Herbert, FURTNER Karl, 2006: *Biomasse – Heizungserhebung 2005*, NÖ Landes-Landwirtschaftskammer, Abteilung Betriebswirtschaft und Technik, St. Pölten, Austria.

461 JAUSCHNEGG Horst, 2002: *Die Entwicklung der Holzpellets in Europa*. In *Holzenergie*, no. 5 (2002), pp24-26, ITEBE, Lons Le Saunier Cedex, France.

462 UMDASCH, 2001: Data per E-Mail from Umdasch AG, Vertrieb Bio-Brennstoffe, Amstetten, Austria.

463 RAKOS Christian, 2006: Pellets in Österreich – wohin geht die Reise. In *Holz-Zentralblatt*, no. 33 (2006), p939, DRW-Verlag Weinbrenner GmbH & Co. KG, Leinfelden-Echterdingern, Germany.

464 BASISDATEN BIOENERGIE ÖSTERREICH, 2009: Brochure, Österreichischer Biomasse-Verband, Vienna, Austria.

465 MAHAPATRA Krushna, GUSTAVSSON Leif, 2006: Small-scale pellet heating systems from consumer perspective. In *Proceedings of the 2nd World Conference on Pellets* in Jönköping, Sweden, ISBN 91-631-8961-5, p239, Swedish Bioenergy Association, Stockholm, Sweden.

466 SEDMIDUBSKY Alice, 2004: *Daten zu erneuerbarer Energie in Österreich 2004*. Energieverwertungs-agentur – the Austrian Energy Agency (E.V.A.), Vienna, Austria.

467 STATISTIK AUSTRIA, 2009: Ergebnisse des Mikrozensus 2004 und 2006 nach Bundesländern, verwendetem Energieträger und Art der Heizung, retrieved [13.1.2009], Statistik Austria, Vienna, Austria.

468 BASISDATEN BIOENERGIE ÖSTERREICH, 2006: Brochure, Österreichischer Biomasse-Verband, Ökosoziales Forum, Vienna, Austria.

469 HANEDER Herbert, 2010: personal information, NÖ Landes-Landwirtschaftskammer, Technik und Energie, St. Pölten, Austria.

470 FISCHER Joachim, 2006: *The Geman Pellet Market*, Presentation at the World Sustainable Energy Days 2006 in Wels, Austria.

471 PILZ Barbara, 2006: Überblick über die Pelletsproduktion in Germany. In *Proceedings of the 6th Pellets Industry Forum* in Stuttgart, Germany, pp39-43, Solar Promotion GmbH, Pforzheim, Germany.

472 FISCHER Joachim, 2002: Holzpelletmarkt Deutschland wächst kontinuierlich. In *Holz-Zentralblatt*, no. 146(2002), pp1754-1755, DRW-Verlag Weinbrenner GmbH & Co. KG, Leinfelden-Echterdingern, Germany.

473 SCHMIDT Beate, 2006: Aktuelle Entwicklung und Perspektiven des Pelletsmarktes in Germany. In *Proceedings of the 6th Pellets Industry Forum* in Stuttgart, Germany, pp10-28, Solar Promotion GmbH, Pforzheim, Germany.

474 PELLETS, 2008: *Fachmagazin der Pelletsbranche*, no. 6 (2008), p6, Solar Promotion GmbH, Pforzheim, Germany.

475 SCHMIDT Beate, 2009: Development of the German pellets market. In *Proceedings of the 9th Pellets Industry Forum* in Stuttgart, Germany, pp9-27, Solar Promotion GmbH, Pforzheim, Germany.

476 FISCHER Joachim, PILZ Barbara, 2004: Deutschland – Aktuelle Entwicklung des deutschen Pellets-marktes. In *Proceedings of the World Sustainable Energy Days 2004* in Wels, Austria, O.Ö. Energiesparverband, Linz, Austria.

477 SCHMIDT Beate, 2008: Entwicklung des deutschen Pelletsmarktes. In *Proceedings of the 8th Pellets Industry Forum* in Stuttgart, Germany, pp15-33, Solar Promotion GmbH, Pforzheim, Germany.

478 SCHMIDT Beate, 2010: Current developments in the German wood pellets market. In *Proceedings of the World Sustainable Energy Days 2010* in Wels, Austria, O.Ö. Energiesparverband, Linz, Austria.

479 BURGER Frank, 2006: Kurzumtriebswälder: Potenziale, Techniken, Kosten. In *Proceedings of the 6th Pellets Industry Forum* in Stuttgart, Germany, pp76-82, Solar Promotion GmbH, Pforzheim, Germany.

480 HIEGL Wolfgang, JANSSEN Rainer, 2009: *Development and promotion of a transparent European pellets market – creation of a European real-time pellets atlas – pellet market country report Germany*. WIP Renewable Energies, Munich, Germany, available at www.pelletsatlas.info, retrieved [17.11.2009].

481 BIZ, 2003: *Holzpellets in Deutschland – Marktstrukturen, Marktentwicklung*, Newsletter Biomasse Info-Zentrum, Stuttgart, Germany.

482 STEPHANI Gregor, GÖNNER Tanja, 2009: *Vorfahrt für erneuerbare Wärme – was politische Rahmenbedingungen bewirken können*. Presentation at the 9th Pellets Industry Forum in Stuttgart, Germany.

483 BAFA, 2009: http://www.bafa.de, retrieved [18.11.2009], Bundesamt für Wirtschaft und Ausfuhrkontrolle, Eschborn, Germany

484 DEPV, 2009: http://www.depv.de, retrieved [17.11.2009], Deutscher Energieholz- und Pellet-Verband e.V. (DEPV), Berlin, Germany.

485 SPILOK Kathleen, 2009: *Wir werden wachsen – Pelletsbranche und Pelletsmarkt in Deutschland*. In *Fachmagazin der Pelletsbranche*, spezial issue 2009/2010, Solar Promotion GmbH, Pforzheim, Germany.

486 AMT FÜR ENERGIEEINSPARUNG, 2009: http://www.provinz.bz.it, retrieved [22.1.2009], Bolzano, Italy.

487 VIVARELLI Filippo, GHEZZI Lorenzo, 2009: *Development and promotion of a transparent European pellets market – creation of a European real-time pellets atlas – pellet market country report Italy*. ETA Renewable Energies, Florence, Italy, available at www.pelletsatlas.info, retrieved [17.11.2009]

488 KOOP Dittmar, 2008: Geschätztes Land – der Pelletsmarkt in Italien. In *Fachmagazin der Pelletsbranche*, no. 3 (2008), Solar Promotion GmbH, Pforzheim, Germany.

489 BRASSOUD Julie, 2004: The ITEBE quality charter for wood pellets. In *Proceedings of the World Sustainable Energy Days 2004* in Wels, Austria, O.Ö. Energiesparverband, Linz, Austria.

490 ZAETTA Corrado, PASSALACQUA Fulvio, TONDI Gianluca, 2004: The pellet market in Italy: Main barriers and perspectives. In *Proceedings of the 2nd World Conference and Exhibition on Biomass for Energy, Industry and Climate Protection* in Rome, Italy, vol. 3, ISBN 88-89407-04-2, pp1843-1847, ETA-Florence, Florence, Italy.

491 PANIZ Annalisa, 2006: Vortrag am Internationalen Workshop der Pelletsverbände, 11. Oktober 2006, Stuttgart, Deutschland

492 WITT Janet, DAHL Jonas, HAHN Brigitte, 2005: Holzpellets – Ein Wachstumsmarkt in Europa, Ergebnisse und Untersuchungen des EU-Altener Projektes (2003-2005): Pellets for Europe 4.1030/C/02-160. In *Proceedings of the 5th Industrieforum Holzenergie* in Stuttgart, Germany, pp167-175, Deutscher Energie-Pellet-Verband e.v. and Solar Promotion GmbH e.V., Germany.

493 BERTON Marino, 2008: *Current developments on the Italian pellet market*, Presentation at the World Sustainable Energy Days 2008 in Wels, Austria.

494 RAKOS Christian, 2007: Entwicklungen der internationalen Pelletmärkte. In *Proceedings of the 7th. Pellets Industry Forum* in Stuttgart, Germany, pp36-43, Solar Promotion GmbH, Pforzheim, Germany.

495 PANIZ Annalisa, 2009: *Current trends in the Italian pellet market*. Presentation at the 9th Pellets Industry Forum in Stuttgart, Germany, www.pelletsforum.de, retrieved [18.11.2009].

496 PELLETS@TLAS, 2009: http://www.pelletsatlas.info, retrieved [24.11.2009], The PELLETS@LAS project, coordinated by WIP - Renewable Energies, Munich, Germany.

497 BERNER Joachim, 2007: Unterschiedliche Pelletswelten – Europäische Pelletskonferenz diskutiert über globalen Handel von Holzpellets und deren Einsatz zur Stromproduktion. In *Fachmagazin der Pelletsbranche*, no. 2 (2007), Solar Promotion GmbH, Pforzheim, Germany.

498 RAKOS Christian, 2007: Pellet market development in Austria & Italy. In *Proceedings of the World Sustainable Energy Days 2007* in Wels, Austria, O.Ö. Energiesparverband, Linz, Austria.

499 EGGER Christiane, 2007: Challenges & opportunities in a rapidly growing market. In *Proceedings of the World Sustainable Energy Days 2007* in Wels, Austria, O.Ö. Energiesparverband, Linz, Austria.

500 ORTNER Herbert, 2007: Market development for small scale pellet boilers in Europe. In *Proceedings of the World Sustainable Energy Days 2007* in Wels, Austria, O.Ö. Energiesparverband, Linz, Austria.

501 FRANCESCATO Valter, PANIZ Annalisa, ANTONINI Eliseo, 2006: *Italian wood pellet market: an overview*, Presentation at the World Sustainable Energy Days in Wels, Austria.

502 BFE, 2005: http://www.energie-schweiz.ch, retrieved [23.3.2005], Bundesamt für Energie, Bern, Switzerland.

503 KEEL Andreas, 2004: Entwicklung des Pelletmarktes in der Schweiz: Bisherige Erfahrungen, zukünftige Strategien. In *Proceedings of the World Sustainable Energy Days 2004* in Wels, Austria, O.Ö. Energiesparverband, Linz, Austria.

504 BFE, 2004: *Subventionsprogramm Lothar*, Bundesamt für Energie, Bern, Switzerland.

505 KEEL Andreas, 2008: *Pelletmarkt Schweiz – Erfahrungen und Preisgestaltung*, Presentation at the World Sustainable Energy Days 2008 in Wels, Austria.

506 NIEDERHÄUSERN Anita, 2009: The Swiss pellets market – facing strong competitors. In *Proceedings of the 9th Pellets Industry Forum* in Stuttgart, Germany, pp160-164, Solar Promotion GmbH, Pforzheim, Germany.

507 JANZING Bernward, 2005: Boom trotz billigen Heizöls. In *Fachmagazin der Pelletsbranche*, no. 3 (2005), Solar Promotion GmbH, Pforzheim, Germany.

508 ELBER U., 2006: *Konzept zur Nutzung des Waldes als Energiequelle*. Presentation at the *6. Schweizer Pelletforum* in Zurich, Switzerland

509 KEEL Andreas, 2008: *Nationale und Internationale Märkte*. Presentation at the *8. Schweizer Pelletforum* in Bern, Switzerland

510 HIEGL Wolfgang, JANSSEN Rainer, 2009: *Development and promotion of a transparent European pellets market – creation of a European real-time pellets atlas – pellet market country report Switzerland*. WIP Renewable Energies, Munich, Germany, available at www.pelletsatlas.info, retrieved [17.11.2009]

511 BIOENERGI, 2009: Tidningen Bioenergi no 1 2009, Stockholm, Sweden.

512 STATISTICS SWEDEN, 2008: *Yearbook of Housing and Building Statistics 2008*, Statistics Sweden, Stockholm, Sweden.

513 SRSA, 2006: *Yearly Report 2006 on the Civil Protection Act*, The Swedish Rescue Services Agency, Karlstad, Sweden.

514 SBBA, 2009: personal information; Swedish Heating Boilers and Burners Association, Stockholm, Sweden.

515 ANONYMOUS, 2004: Pellets burner 2004. In *The Bioenergy international*, no. 9 (2004), Bioenergi Förlag AB, Stockholm, Sweden.

516 BERNER Joachim, 2008: Einmal um den Pelletsglobus. In *Fachmagazin der Pelletsbranche*, no. 2 (2008), Solar Promotion GmbH, Pforzheim, Germany.

517 VINTERBÄCK Johan, 2007: Pellet market development in Sweden. In *Proceedings of the World Sustainable Energy Days 2007* in Wels, Austria, O.Ö. Energiesparverband, Linz, Austria.

518 ARKELÖV Olof, 2004: Sweden. In *Proceedings of the World Sustainable Energy Days 2004* in Wels, Austria, O.Ö. Energiesparverband, Linz, Austria.

519 HÖGLUND J., 2008: *The Swedish fuel pellets industry production; market and standardization*. SLU, Department for forest products. Master thesis nr. 14, 2008, Uppsala, Sweden.

520 STATISTICS SWEDEN, 2007: *Energy Statistics for one- and two dwelling buildings in 2006*. EN16 SM0701, Statistics Sweden, Stockholm, Sweden.

521 PIR, 2010: http://www.pelletsindustrin.org, retrieved [7.4.2010], Swedish Association of Pellet Producers-PiR, Stockholm, Sweden.

522 VINTERBÄCK Johan, 2000: *Wood Pellet Use in Sweden - A systems approach to the residential sector*, PHD Thesis, Department of Forest Management and Products, Swedish University of Agricultural Sciences, ISBN 91-576-5886-2, Uppsala, Sweden.

523 BJERG Jeppe, EVALD Anders, 2004: Denmark: Market and technology trends in the Danish pellet market. In *Proceedings of the World Sustainable Energy Days 2004* in Wels, Austria, O.Ö. Energiesparverband, Linz, Austria.

524 EVALD Anders, 2009: *Det danske træpillemarked 2008* (The Danish Wood Pellet Market 2008), FORCE Technology and Danish Energy Agency, Kgs. Lyngby, Denmark.

525 EVALD Anders, 2009: written notice, surveys for the Danish Energy Agency performed by FORCE Technology, FORCE Technology, Brøndby, Denmark

526 AMAGERVÆRKET, 2009: Amagerværket – grønt regnskab 2008, brochure of Vattenfall A/S, Copenhagen, Denmark, available at http://www.vattenfall.dk/da/file/amagervarket-gront-regnskab-2_7841612.pdf, retrieved [12.4.2010].

527 BJERG Jeppe, 2002: *Træpillehåndbogen*, dk-TEKNIK ENERGI & MILØ, Søborg, Denmark.

528 BERGGREN Anders, 2003: *Production and Use of Pellets & Briquettes for Energy in Denmark*, Country report for the INDEBIF- project, Swedish University of Agricultural Sciences, Uppsala, Sweden.

529 MARCHAL D., CREHAY R., VAN STAPPEN F., WARNANT G., SCHENKEL Y., 2006: Wood pellet use in Wallonia (Belgium): Evaluation and environmental impact. In *Proceedings of the 2nd World Conference on Pellets* in Jönköping, Sweden, ISBN 91-631-8961-5, pp225-227, Swedish Bioenergy Association, Stockholm, Sweden.

530 MARCHAL Didier, 2008: *Current developments on Belgian pellet market*, Presenation at the World Sustainable Energy Days 2008 in Wels, Austria.

531 GUISSON R., MARCHAL M., 2008: Belgium country report, IEA Bioenergy – Task 40, Sustainable International Bioenergy Trade Securing Supply and demand.

532 BAREL Christophe, 2009: *Development and promotion of a transparent European pellets market – creation of a European real-time pellets atlas – pellet market country report Belgium*. ADEME Agence de l'Environnement, Metz, France, available at www.pelletsatlas.info, retrieved [17.11.2009].

533 PIERET Nora, 2009: *Rapport d'activité facilitateur bois énergie – secteur particulier*, Valbiom asbl, Gembloux, Belgium.

534 CORNELIS Erwin, 2007: *Solid biofuels in the Flemish Region, Production, Consumption & Standardisation interest*, VITO, Oral Presentation at the Conference *Biofuel Standardisation in Belgium*, 6 September 2007 at Centre Wallon de Recherches Agronomiques, CRA-W, Gembloux, Belgium

535 JUNGINGER Martin, SIKKEMA Richard, 2009: *Development and promotion of a transparent European pellets market – creation of a European real-time pellets atlas – pellet market country report Netherlands*. Utrecht University, Utrecht, the Netherlands, available at www.pelletsatlas.info, retrieved [17.11.2009].

536 KOPPEJAN J., ELBERSEN W., MEEUSEN M., 2009: *Inventarisatie beschikbare biomassa voor energietoepassingen in 2020*, SenterNovem, The Hague, The Netherlands.

537 MUISTE Marek, HABICHT Maria, 2009: *Development and promotion of a transparent European pellets market – creation of a European real-time pellets atlas – pellet market country report Finland*. LETEK – South Estonian Centre of Renewable Energy, Märja, Tartu County, Estonia, available at www.pelletsatlas.info, retrieved [17.11.2009].

538 TEISKONEN Johanna, NALKKI Janne, 2006: *The Finnish Pellet Market*, Presentation at the World Sustainable Energy Days 2006 inWels, Austria.

539 RAUTANEN Juha, 2004: Present & future pellet markets in Finland. In *Proceedings of the World Sustainable Energy Days 2004* in Wels, Austria, O.Ö. Energiesparverband, Linz, Austria.

540 KOOP Dittmar, 2007: Staat auf dem sicheren Holzweg. In *Fachmagazin der Pelletsbranche*, no. 1 (2007), Solar Promotion GmbH, Pforzheim, Germany.

541 MARKKU Kallio, HEIKKI Oravainen, 2003: Pellet research and development at VTT. In *Proceedings of the International Nordic Bioenergy Conference* in , Finland, ISBN 952-5135-26-8, pp489-493, FINBIO – The Bioenergy Association of Finland, Jyväskylä, Finland.

542 LEHTINEN Toni, 2003: *Use of Wood Briquettes & Pellets in Finland*, http://www.pellets2002.st/index.htm, retrieved [18.12.2006], Seminar paper, Swedish University of Agricultural Sciences, Uppsala, Sweden.

543 HANSEN Morten Tony, 2009: *Development and promotion of a transparent European pellets market – creation of a European real-time pellets atlas – pellet market country report Norway.* FORCE Technology, Kongens Lyngby, Denmark, available at www.pelletsatlas.info, retrieved [17.11.2009]

544 PEDERSEN Fredrik Dahl-Paulsen, 2006: *The Norwegian Pellets Market.* Presenation at the World Sustainable Energy Days 2006 in Wels, Austria.

545 PEKSA-BLANCHARD Malgorzata, DOLZAN Paulo, GRASSI Angela, HEINIMÖ Jussi, JUNGINGER Martin, RANTA Tapio, WALTER Arnaldo, 2007: *Global Wood Pellets Markets and Industry: Policy Drivers, Market Status and Raw Material Potential, IEA Bioenergy Task 40 – Report*, http://bioenergytrade.org/downloads/ieatask40pelletandrawmaterialstudynov2007final.pdf, retrieved [14.11.2008].

546 NOBIO, 2007: Bioenergi i Norge - Markedsrapport 2007 (Bioenergy in Norway - Market report 2007; in Norwegian), www.nobio.no, NoBio, Norway.

547 NOBIO, 2005: Bioenergi i Norge - Markedsrapport 2005 (Bioenergy in Norway - Market report 2005; in Norwegian), www.nobio.no, NoBio, Norway.

548 JANZING Bernward, 2006: *Boom dank Förderprogramm und steigender Ölpreise.* In *Fachmagazin der Pelletsbranche*, no. 1(2006), Solar Promotion GmbH, Pforzheim, Germany.

549 DOUARD Frédéric, 2008: *Update on the French pellet market.* Presentation at the 8th Pellets Industry Forum in Stuttgart, Germany, www.pelletsforum.de, retrieved [1.12.2008].

550 DOUARD Frédéric, 2007: Challenges in the expanding French market. In *Proceedings of the World Sustainable Energy Days 2007* in Wels, Austria, O.Ö. Energiesparverband, Linz, Austria.

551 KOOP Dittmar, 2008: Der Zwei-Phasen-Markt. In *Fachmagazin der Pelletsbranche*, no. 1 (2008), Solar Promotion GmbH, Pforzheim, Germany.

552 DE CHERISEY Hugues, 2010: Last evolutions of the French pellets market. In *Proceedings of the World Sustainable Energy Days 2010* in Wels, Austria, O.Ö. Energiesparverband, Linz, Austria.

553 KOOP Dittmar, 2008: Markt mit Potenzialen – Pellets für eine überfällige Energiewende in Grossbritannien. In *Fachmagazin der Pelletsbranche*, no. 6 (2008), Solar Promotion GmbH, Pforzheim, Germany.

554 HAYES Sandra, 2009: *Development and promotion of a transparent European pellets market – creation of a European real-time pellets atlas – pellet market country report UK.* The National Energy Foundation, Milton Keynes, UK, available at www.pelletsatlas.info, retrieved [17.11.2009]

555 HAYES Sandra, 2009: *Development and promotion of a transparent European pellets market – creation of a European real-time pellets atlas – pellet market country report Ireland.* The National Energy Foundation, Milton Keynes, UK, available at www.pelletsatlas.info, retrieved [17.11.2009]

556 PUENTE-SALVE Francisco, 2008: *Pellets market in Spain*, Presentation at the World Sustainable Energy Days 2008 in Wels, Austria.

557 KOOP Dittmar, 2008: Noch ist Siesta – Pelletsmärkte auf der Iberischen Halbinsel: Spanien und Portugal. In *Fachmagazin der Pelletsbranche*, no. 2(2008), Solar Promotion GmbH, Pforzheim, Germany.

558 VIVARELLI Filippo, GHEZZI Lorenzo, 2009: *Development and promotion of a transparent European pellets market – creation of a European real-time pellets atlas – pellet market country report Spain.* ETA Renewable Energies, Florence, Italy, available at www.pelletsatlas.info, retrieved [17.11.2009]

559 PUENTE Francisco, 2010: Propellets association. In *Proceedings of the World Sustainable Energy Days 2010* in Wels, Austria, O.Ö. Energiesparverband, Linz, Austria.

560 PASSALACQUA Fulvio, 2004: Agri-Pellets – a new fuel for Southern Europe? In *Proceedings of the World Sustainable Energy Days 2004* in Wels, Austria, O.Ö. Energiesparverband, Linz, Austria.

561 ANTOLIN Gregorio et al., 2004: Spanish biomass pellets market. In *Proceedings of the World Sustainable Energy Days 2004* in Wels, Austria, O.Ö. Energiesparverband, Linz, Austria.

562 VIVARELLI Filippo, 2009: *Development and promotion of a transparent European pellets market – creation of a European real-time pellets atlas – pellet market country report Portugal.* ETA Renewable Energies, Florence, Italy, available at www.pelletsatlas.info, retrieved [17.11.2009]

563 VOULGARAKI Stamatia, BALAFOUTIS Athanasios, PAPADAKIS George, 2009: *Development and promotion of a transparent European pellets market – creation of a European real-time pellets atlas – pellet market country report Greece.* Agricultural University of Athens, Department of Natural Resources, Athens, Greece, available at www.pelletsatlas.info, retrieved [17.11.2009]

564 LIIB Aili, 2005: The Estonian pellet market in development. In *Proceedings of the World Sustainable Energy Days 2005* in Wels, Austria, O.Ö. Energiesparverband, Linz, Austria.

565 KOOP Dittmar, 2007: Pellets made in Baltikum – Lettland und Litauen setzen auf Export. In *Fachmagazin der Pelletsbranche*, no. 3 (2007), Solar Promotion GmbH, Pforzheim, Germany.

566 STEINER Monika, PICHLER Wilfried, GOLSER Michael, 2009: *Development and promotion of a transparent European pellets market – creation of a European real-time pellets atlas – pellet market country report Bulgaria.* Holzforschungs Austria, Vienna, Austria, available at www.pelletsatlas.info, retrieved [17.11.2009]

567 BASTIAN Malgorzata, WACH Edmund, 2009: *Development and promotion of a transparent European pellets market – creation of a European real-time pellets atlas – pellet market country report Czech Republic.* Baltic Energy Conservation Agency, Gdańsk, Poland, available at www.pelletsatlas.info, retrieved [17.11.2009].

568 GYURIS Peter, CSEKÖ Adrienn, 2009: *Development and promotion of a transparent European pellets market – creation of a European real-time pellets atlas – pellet market country report Hungary.* Geonardo Ltd, Budapest, Hungary, available at www.pelletsatlas.info, retrieved [17.11.2009]

569 MUISTE Marek, HABICHT Maria, 2009: *Development and promotion of a transparent European pellets market – creation of a European real-time pellets atlas – pellet market country report Baltic countries – Estonia/Latvia/Lithuania.* LETEK – South Estonian Centre of Renewable Energy, Märja, Tartu County, Estonia, available at www.pelletsatlas.info, retrieved [17.11.2009].

570 BASTIAN Malgorzata, WACH Edmund, 2009: *Development and promotion of a transparent European pellets market – creation of a European real-time pellets atlas – pellet market country report Poland.* Baltic Energy Conservation Agency, Gdańsk, Poland, available at www.pelletsatlas.info, retrieved [17.11.2009].

571 STEINER Monika, PICHLER Wilfried, GOLSER Michael, 2009: *Development and promotion of a transparent European pellets market – creation of a European real-time pellets atlas – pellet market*

country report Romania. Holzforschungs Austria, Vienna, Austria, available at www.pelletsatlas.info, retrieved [17.11.2009].

572 BASTIAN Malgorzata, WACH Edmund, 2009: *Development and promotion of a transparent European pellets market – creation of a European real-time pellets atlas – pellet market country report Slovakia*. Baltic Energy Conservation Agency, Gdańsk, Poland, available at www.pelletsatlas.info, retrieved [17.11.2009].

573 GYURIS Peter, CSEKÖ Adrienn, 2009: *Development and promotion of a transparent European pellets market – creation of a European real-time pellets atlas – pellet market country report Slovenia*. Geonardo Ltd, Budapest, Hungary, available at www.pelletsatlas.info, retrieved [17.11.2009].

574 ELLIOT Stan, 2009: The growing pellets industry in the U.S. and its impact on the international pellets market. In *Proceedings of the 9th Pellets Industry Forum* in Stuttgart, Germany, pp134-137, Solar Promotion GmbH, Pforzheim, Germany.

575 STRIMLING Jon, 2008: *Der US-amerikanische Pelletsmarkt*, www.pelletsforum.de, retrieved [1.12.2008], Presentation at the 8th Pellets Industry Forum in Stuttgart, Germany.

576 ELLIOT Stan, 2007: Pellet "pains", a US perspective on international cooperation. In *Proceedings of the World Sustainable Energy Days 2007* in Wels, Austria, O.Ö. Energiesparverband, Linz, Austria.

577 SWAAN John, 2008: Intercontinental trade with pellets. In *Proceedings of the World Sustainable Energy Days 2008*, O.Ö. Energiesparverband, Linz, Austria.

578 TUCKER Kenneth R., 2002: Bagged Pellets to One Million Users. In *Proceedings of the 1st World Conference on Pellets* in Stockholm, Sweden, ISBN 91-631-2833-0, p57, Swedish Bioenergy Association, Stockholm, Sweden.

579 SWAAN John, 2004: North America. In *Proceedings of the World Sustainable Energy Days 2004* in Wels, Austria, O.Ö. Energiesparverband, Linz, Austria.

580 SWAAN John, 2007: The future of pellet markets & technologies. In *Proceedings of the World Sustainable Energy Days 2007* in Wels, Austria, O.Ö. Energiesparverband, Linz, Austria.

581 NATUCKA Dorota, 2006: Wood pellet industry update from North America. In *The Bioenergy International*, no.17 (2005), Bioenergi Förlag AB, Stockholm, Sweden.

582 KOOP Dittmar, 2009: Lernen von Europa. In *Fachmagazin der Pelletsbranche*, no. 6 (2009), Solar Promotion GmbH, Pforzheim, Germany.

583 LI Dingkai, CHE Zhanbin, HU Peihua, NI Weidou, 2006: Development and characteristics of a cold biomass pelletising technology in China. In *Proceedings of the 2nd World Conference on Pellets* in Jönköping, Sweden, ISBN 91-631-8961-5, p17, Swedish Bioenergy Association, Stockholm, Sweden.

584 HONG Hao, 2009: Current development and prospects of the biomass briquette fuels in China. In *Proceedings of the 9th Pellets Industry Forum* in Stuttgart, Germany, pp138-148, Solar Promotion GmbH, Pforzheim, Germany.

585 PELLETS, 2006: Die Exportmärkte *im Blick*. In *Fachmagazin der Pelletsbranche*, no. 3 (2006), Solar Promotion GmbH, Pforzheim, Germany.

586 CELIKTAS M.S., KOCAR G., 2006: A perspective on pellet manufacturing in Turkey with a SWOT analysis. In *Proceedings of the 2nd World Conference on Pellets* in Jönköping, Sweden, ISBN 91-631-8961-5, pp133-136, Swedish Bioenergy Association, Stockholm, Sweden.

587 GANZORIG Nadmidtseden, 2006: Developing pellets in Mongolia. In *Proceedings of the 2nd World Conference on Pellets* in Jönköping, Sweden, ISBN 91-631-8961-5, pp137-140, Swedish Bioenergy Association, Stockholm, Sweden.

588 DEL PINO VIVANCO Ramón, 2008: The pellet market of Chile. In *Proceedings of the World Sustainable Energy Days 2008*, O.Ö. Energiesparverband, Linz, Austria.

589 JANZING Bernward, 2009: Südamerikas Anfänge. In *Fachmagazin der Pelletsbranche*, no. 1 (2009), Solar Promotion GmbH, Pforzheim, Germany.

590 RAKITOVA Olga, 2010: The development of the pellet production in Russia. In *Proceedings of the World Sustainable Energy Days 2010* in Wels, Austria, O.Ö. Energiesparverband, Linz, Austria.

591 KOOP Dittmar, 2008: Inlandsmarkt lässt auf sich warten. In *Fachmagazin der Pelletsbranche*, no. 5 (2008), Solar Promotion GmbH, Pforzheim, Germany.

592 LORENZ Emilia, 2009: Fact-based SWOT analysis of production and sourcing of wood pellets in Ukraine. In *Proceedings of the 9th Pellets Industry Forum* in Stuttgart, Germany, pp154-159, Solar Promotion GmbH, Pforzheim, Germany.

593 DAE-KYUNG Kim, 2009: Das Thema Holzpellets ist neu in Südkorea. In *Fachmagazin der Pelletsbranche*, no. 2 (2009), Solar Promotion GmbH, Pforzheim, Germany.

594 KOOP Dittmar, 2009: Pellets vom Ende der Welt. In *Fachmagazin der Pelletsbranche*, no. 5 (2009), Solar Promotion GmbH, Pforzheim, Germany.

595 PEKSA-BLANCHARD M., DOLZAN P., GRASSI A., HEINIMÖ J., JUNGINGER M., RANTA T., WALTER A., 2007: *Global Wood Pellets Markets and Industry: Policy Drivers, Market Status and Raw Material Potential*. IEA Bionergy Task 40, Copernicus Institute, Utrecht, Netherlands.

596 SIKANEN L., MUTANEN A., RÖSER D., SELKIMÄKI M., 2008: *Pellet markets in Finland and Europe-An overview*. Study report, Pelletime Project, Joensuu, Finland.

597 ALAKANGAS E., HEIKKINEN A., LENSU T., VESTERINEN P., 2007: *Biomass fuel trade in Europe*. Summary Report VTT-R-03508-07. EUBIONET II-EIE/04/065/S07.38628. Technical Research Centre of Finland, Jyväskylä Finland.

598 WORTON B. J., 1989: Kernel methods for estimating the utilization distribution in home-range studies. In: *Ecology* 70, pp164-168, Ithaca NY, USA.

599 GRILICHES Z., 1957: Hybrid corn: an exploration in the economics of technological change. In: *Econometrica* 25 (4), pp501-522, Princeton NJ, USA.

600 HÖGLUND J., 2008: *The Swedish fuel pellets industry: Production, market and standardization*. Swedish University of Agricultural Sciences, Examarbeten Nr 14, 2008. ISSN 1654-1367.

601 EGGER Christiane, ÖHLINGER Christine, 2002: Strategy and Methods to Create a New Market. In *Proceedings of the 1st World Conference on Pellets* in Stockholm, Sweden, ISBN 91-631-2833-0, pp35-36, Swedish Bioenergy Association, Stockholm, Sweden.

602 YLITALO E., 2008: *Puun energiakäyttö 2007*. Metsätilastotiedote 15/2008. Metsäntutkimuslaitos, Metsätilastollinen tietopalvelu, Vantaa, Finland.

603 LAPPALAINEN I., ALAKANGAS E., ERKKILÄ A., FLYKTMAN M., HELYNEN S., HILLERBRAND K., KALLIO M., MARJANIEMI M., NYSTEDT Å., ORAVAINEN H., PUHAKKA A., VIRKKUNEN M., 2007: *Puupolttoaineiden pienkäyttö*. TEKES, Helsinki, Finland.

604 ASIKAINEN A., LIIRI H., PELTOLA S., KARHALAINEN T., LAITILA J., 2008: *Forest energy potential in Europe (EU27)*. Working papers of the Finnish forest research institute, Nr 69. ISSN 1795-150 X, Joensuu, Finland.

605 KALLIO M., KALLIO E., 2004: *Pelletization of woody raw material*. Project report. PRO2/P6012/04. VTT Prosessit, Jyväskylä, Finland.

606 HETEMÄKI L., HARSTELA P., HYNYNEN J., ILVESNIEMI H. & UUSIVUORI J., 2006: *Suomen metsiin perustuva hyvinvointi 2015 – Katsaus Suomen metsäalan kehitykseen ja tulevaisuuden vaihtoehtoihin*. Metlan työraportteja 26. Retrieved at: http://www.metla.fi/julkaisut/workingpapers/2006/mwp026.htm [11.10.2008]

607 NÄSLUND M., 2007: *Pellet production and market in Sweden.* Conference presentation: Enertic Valorisation of Forest Biomass in the South Europe, Pamplona, Spain.

608 DAM J., FAAIJ A.P.C., LEWANDOWSKIA I., VAN ZEEBROECK B, 2009: Options of biofuel trade from Central and Eastern to Western European countries. In *Biomass and Bioenergy*, no. 33 (2009), pp728-744, Elsevier Ltd., Oxford, UK.

609 MOLA-YUDEGO B., 2007: Trends of the yields from commercial willow plantations in Sweden (1986-2000). In *Proceedings from the IEA Task 31 International Workshop: Sustainable Production Systems for Bioenergy:* Forest Energy in Practice, Joensuu, Finland.

610 MOLA-YUDEGO B., ARONSSON P., 2008: Yield models from commercial willow plantations in Sweden. In *Biomass and Bioenergy*, no. 32 (9), pp829-837, Elsevier Ltd., Oxford, UK.

611 FAO, 2003: *State of the World's Forests - 2003.* Food and Agriculture Organization of the United Nations, Rome, Italy.

612 IEA, 2004: *World Energy Outlook 2004*, International Energy Agency, Paris, France.

613 FAOSTAT, 2006: *Forestry data,* Food and agriculture organization of the United Nations, from http://faostat.fao.org, retrieved November 2006.

614 FAO, 1990: *Energy conservation in the mechanical forest industries*, Forestry paper 93, Food and Agriculture Organization of the United Nations, Rome, Italy.

615 STATISTICS FINLAND, 2005: *Energy Statistics 2004.* Official statistics of Finland. Energy 2005:2. Helsinki, Finland.

616 HEINIMÖ Jussi, ALAKANGAS Eija, 2009: *Market of biomass fuels in Finland*, Lappeenranta University of Technology, Research Report 3, p38, available at www.eubionet.net or www.bioenergytrade.org, retrieved [15.1.2010].

617 HEINIMÖ J., JINGINGER M., 2009: Production and trading of biomass for energy – an overview of the global status. In *Biomass and Bioenergy*, no. 33 (9), pp1310-1320, Elsevier Ltd., Oxford, UK.

618 ALA KIHNIÄ J., 2009: *Foreign Trade Statistics on Bioenergy, Methodology and Classification* (Combined Nomenclature). Presentation at the EUBIONETIII workshop, 12.3.2009 in Brussels, available at www.eubionet.net, retrieved [16.10.2009].

619 LENSU T., ALAKANGAS E., 2004: *Small-scale electricity generation from renewable energy sources - A glance at selected technologies, their market potential and future prospects.* OPET Report 13, VTT, May 2004, p144, Jyväskylä, Finland.

620 SIKKEMA Richard, STEINER Monika, JUNGINGER Martin, HIEGL Wolfgang, 2009: *Final report on producers, traders and consumers of wood pellets.* Deliverables 4.1/4.2/4.3 for the Pellets@las project, December 2009, available at www.pelletsatlas.info, retrieved [12.1.2010].

621 VERKERK B., 2008: *Current and future trade opportunities for woody biomass end-products from British Columbia, Canada.* Master thesis, Copernicus Institute, Utrecht University, March 2008, p137, Utrecht, The Netherlands.

622 SIHVONEN Matti, GURNER Darren, 2009: Pellet price discovery and the various commercial uses of price benchmarks. In *Proceedings of the 9th Pellets Industry Forum* in Stuttgart, Germany, pp170-172, Solar Promotion GmbH, Pforzheim, Germany.

623 VEER Sipke, 2009: Increased transparency in the industrial wood pellet market. In *Proceedings of the 9th Pellets Industry Forum* in Stuttgart, Germany, pp173-177, Solar Promotion GmbH, Pforzheim, Germany.

624 ARGUS, 2010: Homepage, http://www.argusmedia.com, retrieved [15.1.2010], Argus Media Ltd, London, UK.

625 WALLIS Keith, 2009: *Pacific panamax rates slump as cargoes dry up.* Loyds List, March 23 2009; available at: http://www.lloydslist.com/ll/news/pacific-panamax-rates-slump-as-cargoes-dry-up/20017631415.htm, retrieved [4.11.2009].

626 SEASURE, 2008: Seasure Weekly Report, 19th December 2008, available at http://www.seasure.co.uk/docs/Seasure-Shipping-19th-December-2008.pdf, retrieved [4.11.2009], SEASURE SHIPPING LIMITED, London, UK.

627 WILD Michael, 2010: Personal notice, European Bioenergy Services - EBES AG, Vienna, Austria.

628 PROINNOVA, 2010: Homepage, http://www.sofortsparen.info, retrieved [12.1.2010], Proinnova - Energie und Geld sparen, Winden, Germany

629 FE[m], 2010: Forest Energy monitor, Volume 1, Issue 5, January 2010, Hawkins Wright Limited, Richmond, UK.

630 BRADLEY D., DIESENREITER F., WILD M., TRØMBORG E., 2009: *World Biofuel Maritime Shipping Study*, Climate Change Solutions, Vienna University of Technology, EBES AG & Norwegian University of Life Sciences, July 2009, p38. Report for IEA Bioenergy Task 40, available at http://www.bioenergytrade.org/downloads/worldbiofuelmaritimeshippingstudyjuly120092df.pdf, retrieved [16.10.2009]

631 JUNGINGER M. SIKKEMA R., SENECHAL S., 2008: *The global wood pellet trade – markets, barriers and opportunities*, workshop report, the Netherlands, Pellets@las project, available at: http://www.pelletsatlas.info/pelletsatlas_docs/showdoc.asp?id=090420111608&type=doc&pdf=true, retrieved [16.10.2009]

632 WILD M., 2008: *Entwicklung der europäischen Märkte für Heiz- und Verstromungspellets Rohstoffe und VerbrauchsperspektiveVerbrauchsperspektiven.* Presentation at the 8th Pellets Industry Forum, Stuttgart, Germany, October 28-29. Stuttgart, Germany.

633 VAN DAM J., JUNDINGER M., FAAIJ A., JÜRGENS I., BEST G., FRITSCHE U., 2008: Overview of recent developments in sustainable biomass certification. In *Biomass and Bioenergy*, no. 32 (8), pp749-780, Elsevier Ltd., Oxford, UK.

634 MARCHAL D., VAN STAPPEN F., SCHENKEL Y., 2009: Critères et indicateurs de production "durable" des biocombustibles solides: état des lieux et recommandations. In : *Biotechnol. Agron. Soc. Environ.* 13 (1), pp165-176, Gembloux, Belgium.

635 MARCHAL D., RYCKMANS, Y., JOSSART J.-M., 2004: Fossil CO_2 emissions and strategies to develop pellet's chain in Belgium. In *Proceedings of the World Sustainable Energy Days 2004* in Wels, Austria, O.Ö. Energiesparverband, Linz, Austria.

636 MARCHAL D., 2008: Current developments on Belgian pellet market. In *Proceedings of the World Sustainable Energy Days 2008* in Wels, Austria, O.Ö. Energiesparverband, Linz, Austria.

637 RYCKMANS Y., MARCHAL D., ANDRÈ N, 2006: Energy balance and greenhouse gas emissions of the whole supply chain for the import of wood pellets to power plants in Belgium. In *Proceedings of the 2nd World Conference on pellets*, pp127-130, Jönköping, Sweden.

638 RYCKMANS Y., ANDRÈ N., 2007: Novel certification procedure for the sustainable import of wood pellets to power plants in Belgium. In *Proceedings of 15th European Biomass Conference and Exhibition*, pp2243-2246, Berlin, Germany.

639 GGL, 2009: *Description and documentation of the Green Gold Label, a track-and trace system for sustainable solid biomass.* Available at: http://certification.controlunion.com/certification/program/Program.aspx?Program_ID=19 , retrieved [16.10.2009].

640 ELECTRABEL, 2009: Documents for the biomass verification procedure, http://www.laborelec.com/content/EN/Renewables-and-biomass_p83, retrieved [10.8.2009].

641 WIKIPEDIA, 2010: http://upload.wikimedia.org/wikipedia/commons/b/b8/British_Columbia-map.png, retrieved [26.2.2010].

642 EDUCATIONWORLD, 2010: http://library.educationworld.net/canadafacts/maps/bc_map_eng.gif, retrieved [26.2.2010].

643 VERKERK Bas, 2010: Personal information, Control Union Canada Inc., Vancouver, Canada.

644 DOMAC J., RICHARDS K., et al., 2005: Socio-economic drivers in implementing bioenergy projects. In *Biomass and Bioenergy*, no. 28 (2), pp97-106, Elsevier Ltd., Oxford, UK.

645 BUCHHOLZ Thomas, RAMETSTEINER Ewald, VOLK Timothy A., LUZADIS Valerie A., 2009: Multi Criteria Analysis for bioenergy systems assessment. In: *Energy Policy*, 37, pp484-495, Elsevier Ltd., Oxford, UK.

646 VON GEIBLER Justus, LIEDTKE Christa, WALLBAUM Holger, SCHALLER Stephan, 2006: Accounting for the Social Dimension of Sustainability: Experiences from the Biotechnology Industry. In: *Business Strategy and the Environment*, 15, pp334-346, John Wiley & Sons, Ltd., West Sussex, UK.

647 HABERL Helmut, GAUBE Veronika, DIAZ-DELGADO Ricardo, KRAUZE Kinga, NEUNER Angelika, PETERSEIL Johannes, PLUTZAR Christoph, SINGH Simron J., VADINEANU Angheluta, 2009: Towards an integrated model of socioeconomic biodiversity drivers, pressures and impacts. A feasibility study based on three European long-term socio-ecological research platforms. In: *Ecological Economics*, 68, pp1797-1812, Elsevier Ltd., Oxford, UK.

648 HAUGHTON Alison J., BOND Alan J., LOVETT Andrew A., DOCKERTY Trudie, SÜNNENBERG Gilla, CLARK Suzanne J., BOHAN David A., SAGE Rufus B., MALLOTT Mark D., MALLOTT Victoria E., CUNNINGHAM Mark D., RICHE Andrew B., SHIELD Ian F., FINCH Jon W., TURNER Martin M., KARP Angela, 2009: A novel, integrated approach to assessing social, economic and environmental implications of changing rural land-use: a case study of perennial biomass crops. In: *Journal of Applied Ecology*, 46, pp315-322, British Ecological Society, London, UK.

649 WICKE B., et al., 2009: *Macroeconomic impacts of bioenergy production on surplus agricultural land - A case study of Argentina*. Renewable and Sustainable Energy Reviews, July 2009, Elsevier Ltd., Oxford, UK.

650 HUEMER Günther, 2005: High-Tech bei Holzpelletsanlagen. In *Proceedings of the World Sustainable Energy Days 2005* in Wels, Austria, O.Ö. Energiesparverband, Linz, Austria.

651 KOLLBAUER Stefan, 2005: Vom Rohblech zum Pelletskessel – innovative Kesselfertigung in Österreich. In *Proceedings of the World Sustainable Energy Days 2005* in Wels, Austria, O.Ö. Energiesparverband, Linz, Austria.

652 SIKKEMA R., JUNGINGER H.M., PICHLER W., HAYES S., FAAIJ A.P.C., 2009: *The international logistics of wood pellets for heating and power production in Europe; Costs, energy-input and greenhouse gas (GHG) balances of pellet consumption in Italy, Sweden and the Netherlands*. Submitted to BioFPR, May 2009.

653 EDSTEDT Mathias, 2002: *The Hässelby Operation: Large Scale Conversion From Coal to Pellets by Vertical Integration*; 1st World Conference on Pellets in Stockholm, Sweden.

654 WBCSD, 2006: From coal to biomass, available at http://www.wbcsd.org/DocRoot/ADXtsPNj4E1CAHMhp5KZ/suez_awirs_biomass_full_case_web.pdf, retrieved [11.11.2009], leaflet of the World Business Council for Sustainable Development, Conches-Geneva, Switzerland

655 PELLETS FOR EUROPE, 2005: Publishable extended summary of the EU ALTENER Project No. 4.1030/C/02-160 Pellets.

656 KIESEWALTER Sophia, RÖHRICHT Christian, 2004: Pelletierung von Stroh und Heu. In *Proceedings of the World Sustainable Energy Days 2004* in Wels, Austria, O.Ö. Energiesparverband, Linz, Austria.

657 NIKOLAISEN Lars, JENSEN Torben Nørgaard, 2004: Pellet recipes for high quality and competitive prices. In *Proceedings of the World Sustainable Energy Days 2004* in Wels, Austria, O.Ö. Energiesparverband, Linz, Austria.

658 VASEN Norbert N., 2005: Agri-pellets. In *Proceedings of the World Sustainable Energy Days 2005* in Wels, Austria, O.Ö. Energiesparverband, Linz, Austria.

659 OTTOSEN Per, 2007: The future of pellets markets & technologies. In *Proceedings of the World Sustainable Energy Days 2007* in Wels, Austria, O.Ö. Energiesparverband, Linz, Austria.

660 OTTOSEN Per, GULLEV Lars, 2005: Avedøre unit 2 – the world´s largest biomass-fuelled CHP plant. Danish Board of District Heating (DBDH), Frederiksberg, Denmark, available at http://www.cader.org/documents/avedore-unit-2.pdf, retrieved [12.4.2010]

661 PEDERSEN Niels Ravn, 2010: *Production and use of Bio Pellets*, available at http://www.northernwoodheat.net/htm/news/Scotland/Elginbiomass/biomassprespdf/ProductionandUseofBioPellets.pdf, retrieved [12.4.2010]

662 SCHELLINGER Helmut, 2008: Interview in the *Fachmagazin der Pelletsbranche*, special edition 2008/2009, Solar Promotion GmbH, Pforzheim, Germany.

663 MAYER Karl, 2007: *Energiehölzer – Kurzumtriebswälder Anbaumethode, Pflege, Ernte und Wirtschaftlichkeit*, http://bfw.ac.at/rz/bfwcms.web?dok=6133, retrieved [28.1.2009], Presentation within the framework of the event „Energieholz in Kurzumtrieb" at the Forstliche Ausbildungsstätte Ort (FAST Ort), Gmunden, Austria.

664 SCHUSTER Karl, 2007: *Energieholzproduktion auf landwirtschaftlichen Flächen (Kurzumtrieb, Short-Rotation-Farming) – Erfahrungen in Niederösterreich*, http://bfw.ac.at/rz/bfwcms.web?dok=6133, retrieved [28.1.2009], Presentation within the framework of the event „Energieholz in Kurzumtrieb" at the Forstliche Ausbildungsstätte Ort (FAST Ort), Gmunden, Austria.

665 NUSSBAUMER Thomas, 1993: *Verbrennung und Vergasung von Energiegras und Feldholz*, Annual report 1992, Bundesamt für Energiewirtschaft, Bern, Switzerland.

666 HUTLA Petr, KÁRA Jaroslav, JEVIĆ Petr, 2004: Pellets from energy crops. In *Proceedings of the World Sustainable Energy Days 2004* in Wels, Austria, O.Ö. Energiesparverband, Linz, Austria.

667 GUSTAVSSON M. and RÖNNBÄCK M., 2009: Pellets from a wide base of raw materials elaborating a well to wheel quality assurance system. In *Proceedings of the 9th Pellets Industry Forum* in Stuttgart, Germany, pp124-127, Solar Promotion GmbH, Pforzheim, Germany.

668 HERZOG Paul, GOLSER Michael, 2004: Forschung zur Verbesserung der Pelletsqualität. In *Proceedings of the World Sustainable Energy Days 2004* in Wels, Austria, O.Ö. Energiesparverband, Linz, Austria.

669 HARTLEY Ian D., WOOD Lisa J., 2008: Hygroscopic properties of densified softwood pellets. In *Biomass and Bioenergy*, no. 32 (2008), pp90-93, Elsevier Ltd., Oxford, UK.

670 STAHL M., GRANSTRÖM K., BERGHEL J., RENSTRÖM R., 2004: Industrial processes for biomass drying and their effects on the quality properties of wood pellets. In *Biomass and Bioenergy*, no. 27 (2004), pp557-561, Elsevier Ltd., Oxford, UK.

671 STAHL M., 2006: Drying parameter variations and wood fuel pellets quality – pilot study with a new pelleting equipment set up. In *Proceedings of the 2nd World Conference on Pellets* in Jönköping, Sweden, ISBN 91-631-8961-5, pp113-116, Swedish Bioenergy Association, Stockholm, Sweden.

672 HOLM Jens Kai, 2006: Pelletising different materials – an overview. In *Proceedings of the European Pellets Forum 2006*, O.Ö. Energiesparverband, Linz, Austria.

673 BERGSTRÖM Dan, ISRAELSSON Samuel, ÖHMAN Marcus, DAHLQVIST Sten-Axel, GREF Rolf, BOMAN Christoffer, WÄSTERLUND Iwan, 2008: Effects of raw material particle size distribution on the characteristics of Scots pine sawdust fuel pellets. In *Fuel Processing Technology*, no. 89 (2008), pp1324-1329, Elsevier Ltd., Oxford, UK.

674 RHÉN Christopher, ÖHMAN Marcus, GREF Rolf, WÄSTERLUND Iwan, 2007: Effect of raw material composition in woody biomass pellets on combustion characteristics. In *Biomass and Bioenergy*, no. 31 (2007), pp66-72, Elsevier Ltd., Oxford, UK.

675 FRIEDL Günther, WOPIENKA Elisabeth, HASLINGER Walter, 2007: Schlackebildung in Pelletsfeuerungen. In *Proceedings of the 7th Pellets Industry Forum* in Stuttgart, Germany, pp147-152, Solar Promotion GmbH, Pforzheim, Germany.

676 WITT Janet, SCHLAUG Wolfgang, AECKERSBERG Roland, 2008: Optimierung der Holzpelletproduktion für Kleinfeuerungsanlagen. In *Proceedings of the 8th Pellets Industry Forum* in Stuttgart, Germany, pp114-119, Solar Promotion GmbH, Pforzheim, Germany.

677 BEHR Hans Martin, 2007: Einflussfaktoren auf das Ascheschmelzverhalten bei der Verbrennung von Holzpellets. In *Proceedings of the 7th Pellets Industry Forum* in Stuttgart, Germany, pp153-160, Solar Promotion GmbH, Pforzheim, Germany.

678 HOLM Jens K., HENRIKSEN Ulrik B., WAND Kim, HUSTAD Johan E., POSSELT Dorthe, 2007: Experimental verification of novel pellet model using a single pelleter unit. In *Energy & Fuels*, vol. 21, pp2446-2449.

679 OKKONEN Lasse, KOKKONEN Anssi, PAUKKUNEN Simo, 2008: PELLETime – solutions for competitive pellet production in medium-size enterprises. In *Proceedings Poster Session of the World Bioenergy 2008 Conference & Exhibition on Biomass for Energy* in Jönköping, Sweden, pp184-187, Swedish Bioenergy Association, Stockholm, Sweden.

680 SPC, 2010: Homepage, http://www.pelletpress.com/, retrieved [22.1.2010], Sweden Power Chippers AB, Borås, Sweden.

681 OLSSON Maria, KJÄLLSTRAND Jennica, 2004: Emissions from burning softwood pellets. In *Biomass and Bioenergy*, no.27 (2004), pp607-611, Elsevier Ltd., Oxford, UK.

682 ESKILSSON David et al, 2004: Optimisation of efficiency and emissions in pellet burners. In *Biomass and Bioenergy*, no.27 (2004), pp541-546, Elsevier Ltd., Oxford, UK.

683 KJÄLLSTRAND Jennica, OLSSON Maria, 2004: Chimney emissions from small-scale burning of pellets and fuelwood – examples referring to different combustion appliances. In *Biomass and Bioenergy*, no. 27 (2004), pp557-561, Elsevier Ltd., Oxford, UK.

684 WIINIKKA Henrik, GEBART Rikard, 2004: Experimental investigations of the influence from different operating conditions on the particle emissions from a small-scale pellets combustor. In *Biomass and Bioenergy*, no. 27 (2004), pp645-652, Elsevier Ltd., Oxford, UK.

685 CEBC, 2008: *Proceedings of the Central European Biomass Conference 2008*, Austrian Biomass Association, Vienna, Austria.

686 BOMAN Christoffer, NORDIN Anders, BOSTRÖM Dan, ÖHMAN Marcus, 2003: Characterization of Inorganic Particulate Matter from Residential Combustion of Pelletized Biomass Fuels. In *Energy & Fuels*, no. 18 (2004), pp338-348, American Chemical Society, Washington, DC, USA.

687 MUHLBALER DASCH J., 1982: Particulate and Gaseous Emissions from Wood-Burning Fireplaces. In *Environmental Science and Technology*, no.16, pp639-645, University of Iowa, Iowa City, USA.

688 RAU J.A., 1989: Composition and Size Distribution of Residential Wood Smoke Particles. In *Aerosol Science and Technology*, 10, pp181-192, Taylor & Francis, Philadelphia, USA.

689 OBERNBERGER Ingwald, BRUNNER Thomas, BÄRNTHALER Georg, JÖLLER Markus, KANZIAN Werner, BRENNER Markus, 2008: *Feinstaubemissionen aus Biomasse-Kleinfeuerungsanlagen*, Final

report of the Zukunftsfonds Projekt Nr. 2088, Institute for Process Enigneering, Graz University of Technology, Graz, Austria.

690 Nussbaumer, Th.: Low-Particle-Konzept für Holzfeuerungen, Holz-Zentralblatt, 131. Jg., Nr. 1 (2005), 13–14

691 OSER, M., NUSSBAUMER Th., 2006: *Low particle furnace for wood pellets based on advanced staged combustion*, Science in Thermal and Chemical Biomass Conversion, Volume 1, CPL Press, 2006, ISBN 1-872691-97-8, pp215–227, Newbury Berks, United Kingdom

692 ERA-NET Bioenergy R&D project "Future low emission biomass combustion systems" (FutureBioTec), coordinated by the Austrian bioenergy competence centre BIOENERGY 2020+, Graz, in cooperation with partners from Finland, Germany, Sweden, Poland, Ireland and Denmark, duration October 2009 until September 2012.

693 SCHMATLOCH Volker, 2005: Exhaust gas aftertreatment for small wood fired appliances – recent progress and field test results. In *Proceedings of the work-shop "Aerosols in Biomass Combustion"* in Graz, Austria, ISBN 3-9501980-2-4, pp159-166, Obernberger Ingwald, Brunner Thomas (Eds.), Institute of Resource Efficient and Sustainable Systems, Graz University of Technology, Graz, Austria.

694 HEIDENREICH Ralf, 2006: Filteranlagen für Pelletsfeuerungen – Anforderungen, Techniken, Emissionsminderungspotenziale. In *Proceedings of the 6th Pellets Industry Forum* in Stuttgart, Germany, pp97-107, Solar Promotion GmbH, Pforzheim, Germany.

695 RÜEGG Peter, 2006: Partikelabscheider für Feinstaub bei Holz- und Pelletsfeuerungen. In *Proceedings of the 6th Pellets Industry Forum* in Stuttgart, Germany, pp108-112, Solar Promotion GmbH, Pforzheim, Germany.

696 RÜEGG Peter, 2006: Klein-Elektroabscheider für Holzfeuerungen: Stand der Entwicklung und Praxiserfahrung. In *Proceedings of the 9th Holzenergie-Symposium* in Zurich, Schweiz, Thomas Nussbaumer (Ed.), ISBN 3-908705-14-2, pp79-94, ETH Zürich, Verenum Zürich and Bundesamt für Energie, Bern, Switzerland.

697 BERNTSEN Morten, 2006: Elektroabscheider für häusliche Holzfeuerungen. In *Proceedings of the 9th Holzenergie-Symposium* in Zurich, Switzerland, Thomas Nussbaumer (Ed.), ISBN 3-908705-14-2, pp95-103, ETH Zürich, Verenum Zürich und Bundesamt für Energie, Bern, Switzerland.

698 SCHMATLOCH Volker, 2008: Integrierte und nachgeschaltete Elektroabscheider für Holzöfen. In *Proceedings of the 11th Holzenergie-Symposium* in Zurich, Switzerland, Thomas Nussbaumer (Ed.), ISBN 3-908705-19-9, pp157-170, ETH Zürich, Verenum, Zurich, Switzerland.

699 BRZOVIC Trpimir, 2008: *OekoTube: Elektroabscheider als Kaminaufsatz für kleine Holzheizungen*. In *Proceedings of the 10th Holzenergie-Symposium* in Zurich, Switzerland, Thomas Nussbaumer (Ed.), ISBN 3-908705-19-9, pp171-180, ETH Zürich, Verenum, Zurich, Switzerland.

700 BLEUEL Thomas, 2008: Elektroabscheider für Biomasse-Heizanlagen von 0 bis 150 kW. In *Proceedings of the 10th Holzenergie-Symposium* in Zurich, Switzerland, Thomas Nussbaumer (Ed.), ISBN 3-908705-19-9, pp181-184, ETH Zürich, Verenum, Zurich, Switzerland.

701 BOLLINGER Ruedi, 2008: *Elektroabscheider „Spider" für Holzfeuerungen bis 70 kW*. In *Proceedings of the 10th Holzenergie-Symposium* in Zurich, Switzerland, Thomas Nussbaumer (Ed.), ISBN 3-908705-19-9, pp185-189, ETH Zürich, Verenum, Zürich Switzerland.

702 SCHEIBLER Mátyás, OBERFORCHER Philipp, 2008: Metallgewebefilter für automatische Anlagen von 100 kW bis 540 kW. In *Proceedings of the 10th Holzenergie-Symposium* in Zurich, Switzerland, Thomas Nussbaumer (Ed.), ISBN 3-908705-19-9, pp185-189, ETH Zürich, Verenum, Zürich Switzerland.

703 RAWE Rudolf, 2009: Dust separation with conventional and different electrically charged spray scrubbers. In *Proceedings of the 9th Pellets Industry Forum* in Stuttgart, Germany, pp103-110, Solar Promotion GmbH, Pforzheim, Germany.

704 KLIPPEL Norbert, NUSSBAUMER Thomas, 2006: Feinstaubbildung in Holzfeuerungen und Gesundheitsrelevanz von Holzstaub im Vergleich zu Dieselruß. In *Proceedings of the 9th Holzenergie-Symposium* in Zurich, Switzerland, Thomas Nussbaumer (Ed.), ISBN 3-908705-14-2, pp21-40, ETH Zürich, Verenum, Zürich and Bundesamt für Energie, Bern, Switzerland.

705 ERA-NET Bioenergy R&D project "Health effects of particulate emissions from small-scale biomass combustion" (BIOHEALTH), coordinated by the University of Eastern Finland, Kuopio, in co-operation with partners from Finland, Austria, Sweden and France, duration November 2009 until October 2012.

706 WIDMANN Emil, SCHARLER Robert, STUBENBERGER Gerhard, OBERNBERGER Ingwald, 2004: Release of NO_x precursors from biomass fuel beds and application for CFD-based NOx postprocessing with detailed chemistry. In *Proceedings of the 2nd World Conference and Exhibition on Biomass for Energy, Industry and Climate Protection* in Rome, Italy, vol. 2, ISBN 88-89407-04-2, pp1384-1387, ETA-Florence, Florence, Italy.

707 OBERNBERGER Ingwald, WIDMANN Emil, SCHARLER Robert, 2003: Entwicklung eines Abbrandmodells und eines NO_x-Postprozessors zur Verbesserung der CFD-Simulation von Biomasse-Festbettfeuerungen. In *Berichte aus Energie- und Umweltforschung*, no. 31 (2003), Bundesministerium für Verkehr, Innovation und Technologie, Vienna, Austria.

708 WEISSINGER Alexander, 2002: *Experimentelle Untersuchungen und theoretische Simulationen zur NO_x-Reduktion durch Primärmaßnahmen bei Rostfeuerungen*, Thesis, Graz University of Technology, Graz, Austria.

709 SCHARLER Robert, WIDMANN Emil, OBERNBERGER Ingwald, 2006: CFD modelling of NO_x formation in biomass grate furnaces with detailed chemistry. In *Proceedings of the Internat. Conf. Science in Thermal and Chemical Biomass Conversion* in Victoria, Canada, ISBN 1-872691-97-8, pp284-300, CPL Press, Berks, UK.

710 SCHARLER Robert, OBERNBERGER, Ingwald 2002: Deriving guidelines for the design of biomass grate furnaces with CFD analysis – a new Multifuel-Low-NOx furnace as example. In *Proceedings of the 6th European Conference on Industrial Furnaces and Boilers* in Estoril, Portugal, ISBN 972-8034-05-9, INFUB, Rio Tinto, Portugal.

711 PADINGER Reinhard, 2005: Written notice, AUSTRIAN BIOENERGY CENTRE GmbH, Graz, Austria.

712 STRAUSS Rolf-Peter, 2005: Die Karussellfeuerung – Vergleich mit den bekannten Feuerungsarten und aktuelle Forschungsergebnisse. In *Proceedings of the zum 5. Industrieforum Holzenergie* in Stuttgart, Germany, pp10-19, Deutscher Energie-Pellet-Verband e.v. and Deutsche Gesellschaft für Sonnenenergie e.V., Germany.

713 ÖKOENERGIE, 2010: *Guntamatic bringt Weltsensation auf den Heizkessel-Markt*. In: *Ökoenergie, Zeitung für erneuerbare Energien*, no. 78 (2010), Österreichischer Biomasse-Verband, Vienna, Austria.

714 GUNTAMATIC, 2010: http://www.guntamatic.com, retrieved [9.4.2010], GUNTAMTIC Heiztechnik GmbH, Peuerbach, Austria.

715 KUNDE Robert, GADERER Matthias, 2010: *Concept improvement of system technology at small scale biomass heating systems*, R&D project at ZAE BAYERN, Würzburg, Germany

716 SPM, 2010: http://www.stirlingpowermodule.com, retrieved [12.11.2009], Stirling Power Module Energieumwandlungs GmbH, Graz, Austria.

717 SUNMACHINE, 2010: Homepage, http://www.sunmachine.com, retrieved [9.4.2010], Sunmachine GmbH, Kempten, Germany.

718 SONNE WIND & WÄRME, 2010: Branchenmagazin für alle erneuerbaren Energien, Issue 4/2010, Bielefelder Verlag GmbH & Co. KG Richard Kaselowsky, Bielefeld, Germany.

719 HASLINGER Walter, EDER Gottfried, WÖRGETTER Manfred, 2005: Straw pellets for small-scale boilers. In *Proceedings of the World Sustainable Energy Days 2005* in Wels, Austria, O.Ö. Energiesparverband, Linz, Austria.

720 WOPIENKA Elisabeth, 2006: Stand der Technik und Entwicklungen hin zu höherer Brennstoffflexibilität bei Pelletskesseln. In *Proceedings of the 6th Pellets Industry Forum* in Stuttgart, Germany, pp128-135, Solar Promotion GmbH, Pforzheim, Germany.

721 ÖRBERG Håkan, KALÉN Gunnar, 2008: Burner cup technology for ash rich and sintering pellets fuels. In *Proceedings Oral Sessions of the World Bioenergy 2008 Conference & Exhibition on Biomass for Energy* in Jönköping, Sweden, pp222-223, Swedish Bioenergy Association, Stockholm, Sweden.

722 EDER Gottfried, 2007: Energiepflanzen-Monitoring: Feldtest Verbrennung "neuer" Biomasse. In *Proceedings of the World Sustainable Energy Days 2007* in Wels, Austria, O.Ö. Energiesparverband, Linz, Austria.

723 ÖHMAN Marcus, GILBE Ram, BOSTRÖM Dan, BACKMAN Rainer, LINDSTRÖM Erica, SAMUELSSON Robert, BURVALL Jan, 2006: Slagging characteristics during residential combustion of biomass pellets. In *Proceedings of the 2nd World Conference on Pellets*, Mai/Juni 2006, Jönköping, Sweden, ISBN 91-631-8961-5, pp93-100, Swedish Bioenergy Association (Ed.), Stockholm, Sweden

724 OLSSON Maria, 2006: Residential biomass combustion – emissions from wood pellets and other new alternatives. In *Proceedings of the 2nd World Conference on Pellets* in Jönköping, Sweden, ISBN 91-631-8961-5, pp181-185, Swedish Bioenergy Association, Stockholm, Sweden.

725 ÖHMAN Marcus, LINDSTRÖM Erica, GILBE Ram, BACKMAN Rainer, SAMUELSSON Robert, BURVALL Jan, 2006: Predicting slagging tendencies for biomass pellets fired in residential appliances: a comparison of different prediction methods. In *Proceedings of the 2nd World Conference on Pellets* in Jönköping, Sweden, ISBN 91-631-8961-5, pp213-218, Swedish Bioenergy Association, Stockholm, Sweden.

726 WOPIENKA E., SCHWABL M., EMHOFER W., FRIEDL G., HASLINGER W., WÖRGETTER M., MERKL R., WEISSINGER A., 2006: Straw pellets combustion in small-scale boilers – part 1: emissions and emission reduction with a novel heat exchanger technology. In *Proceedings of the 16th European Biomass Conference & Exhibition* in Valencia, Spain, ISBN 978-88-89407-58-1, pp1386-1392, ETA-Renewable Energies Florence, Italy.

727 EMHOFER W., WOPIENKA E., SCHWABL M., FRIEDL G., HASLINGER W., WÖRGETTER M., KÖLSCH T., WEISSINGER A., 2006: Straw pellets combustion in small-scale boilers – part 2: corrosion and material optimization. In *Proceedings of the 16th European Biomass Conference & Exhibition* in Valencia, Spain, ISBN 978-88-89407-58-1, pp1500-1503, ETA-Renewable Energies, Florence, Italy.

728 REZEAU A., DIAZ M., SEBASTIAN F., ROYO J., 2006: Operation and efficiencies of a new biomass burner when using pellets from herbaceous energy crops. In *Proceedings of the 16th European Biomass Conference & Exhibition* in Valencia, Spain, ISBN 978-88-89407-58-1, pp1458-1463, ETA-Renewable Energies, Florence, Italy.

729 HARTMANN Hans, ROSSMANN Paul, TUROWSKI Peter, ELLNER-SCHUBERTH Frank, HOPF Norbert, BIMÜLLER Armin, 2007: *Getreidekörner als Brennstoff für Kleinfeuerungen*, ISSN 1614-1008, Technologie- und Förderzentrum (TFZ), Straubing, Germany.

730 SCHARLER Robert, OBERNBERGER Ingwald, 2000: Numerical optimisation of biomass grate furnaces. In *Proceedings of the 5th European Conference on Industrial Furnaces and Boilers* in Porto, Portugal, ISBN 972-8034-04-0, INFUB, Rio Tinto, Portugal.

731 SCHARLER Robert, OBERNBERGER Ingwald, 2000: CFD analysis of air staging and flue gas recirculation in biomass grate furnaces. In *Proceedings of the 1st World Conference on Biomass for Energy and Industry* in Sevilla, Spain, ISBN 1-902916-15-8, vol. 2, pp1935-1939, James&James Ltd., London, UK.

732 SCHARLER Robert, 2001: *Entwicklung und Optimierung von Biomasse-Rostfeuerungen mittels CFD-Analyse*, Thesis. Graz University of Technology, Graz, Austria.

733 SCHARLER Robert, FORSTNER Martin, BRAUN Markus, BRUNNER Thomas, OBERNBERGER Ingwald, 2004: Advanced CFD analysis of large fixed bed biomass boilers with special focus on the convective section. In *Proceedings of the 2nd World Conference and Exhibition on Biomass for Energy, Industry and Climate Protection* in Rome, Italy, vol. 2, ISBN 88-89407-04-2, pp1357-1360, ETA-Florence, Florence, Italy.

734 SCHARLER Robert, 2005: CFD-gestützte Entwicklung und Optimierung von Pellet- und Hackgutfeuerungen für den kleinen und mittleren Leistungsbereich. In *Proceedings of the World Sustainable Energy Days 2005* in Wels, Austria, O.Ö. Energiesparverband, Linz, Austria.

735 SCHARLER Robert, WEISSINGER Alexander, SCHMIDT Wilhelm, OBERNBERGER Ingwald, 2005: CFD-gestützte Entwicklung und Optimierung einer neuen Feuerungstechnologie für feste Biomasse für den kleinen und mittleren Leistungsbereich. In *Proceedings of the World Sustainable Energy Days 2005* in Wels, Austria, O.Ö. Energiesparverband, Linz, Austria.

736 R&D project "Scientific tools for fuel characterization for clean and efficient biomass combustion" (SciToBiCom), coordinated by Technical University of Denmark, Lyngby, in co-operation with partners from Denmark, Finland, Norway and Austria, duration December 2009 until November 2012.

737 RAKOS Christian, 2004: The BIOHEAT project: developing the market for heating large buildings with biomass. In *Proceedings of the World Sustainable Energy Days 2004* in Wels, Austria, O.Ö. Energiesparverband, Linz, Austria.

738 OLSSON Maria, VINTERBÄCK Johan, 2005: Pellets R&D in Europe – An Overview. In *Proceedings of the World Sustainable Energy Days 2005* in Wels, Austria, O.Ö. Energiesparverband, Linz, Austria.

739 DECKER Thomas, 2007: Motive für den kauf einer Heizung – Ergebnisse einer Verbraucherbefragung mit dem Schwerpunkt „Holzpelletheizungen". In *Proceedings of the 7th Pellets Industry Forum* in Stuttgart, Germany, pp30-35, Solar Promotion GmbH, Pforzheim, Germany.

740 KLUG Siegrun, PERNKOPF Teresa, DÖRFELMAYER Daniela, HOFBAUER Verena, 2008: *Motivstudie „Heizsysteme"*, Study, http://www.energyagency.at, retrieved [30.1.2009], University of Applied Sciences Wiener Neustadt/Campus Wieselburg for the Austrian Energy Agency, Vienna, Austria.

741 PAULRUD Susanne, 2010: Personal communication, SP Swedish National Testing and Research Institute, Borås, Sweden.

Index

abrasion   66–68
additives   9
aerosols   327
A-frame flat storage   124
airborne dust   136
   flammability of   142
ash content   58
ash deformation temperature (DT)   9
ash flow temperature (FT)   9
ash formation   64, 321–322
ash fractions   63–64, 84, 321
ash hemisphere temperature (HT)   9
ash shrinkage starting temperature (SST)   9
auto-ignition temperature for dust cloud (T_C)   133
auto-ignition temperature for dust layer (T_L)   134
auxiliary energy   308

Baltic Dry Index (BDI)   397
Baltic Freight Index (BFI)   397
barging   410, 417
bark pellets   7, 21, 75, 136, 138, 142–143, 449
BC Code *see* Code of Safe Practice for Solid Bulk Cargoes
belt dryer   94
biofuel   10
BIOHEAT   476
biological additives   7, 9, 47, 60, 65, 69, 78–79, 82, 84, 99, 246–247, 267, 463, 478
biomass   1–3, 7, 10, 409, 452
   combustion systems   321, 330
     district heating   290
     fractions,   339
     furnaces   316, 326
blue angel ("Blauer Engel") label   43
boiler efficiency, 35, 37, 43, 198, 202–203, 213, 282, 296, 318
boiler performance

biomass firing, impact of   236
   co-firing, impact of   232, 236
BOMAT Profitherm   209
burn-back protection   192

calorific value   10
case study   427
   supply chain   410
   wood pellets   410
CEN fuel specifications and classes   13
CEN solid biofuels terminology   7
CEN/TC 335   7, 9
central heating system   292, 341
cereal straw co-firing system   233
certification system ENplus   34
CFD simulations   473
chain management   406
chemical treatment   10
CHP
   plants,   449, 457
   system   223
   technology   222
coal and co-milling, premixing   231
coal mills, conversion of   229
Code of Safe Practice for Solid Bulk Cargoes (the BC code)   21
co-firing
   application   454
   of biomass pellets   227
combi system   213
combustion chamber materials   195
combustion technology   216, 427, 446
comparative study, heating systems   291
compress wood   7
condensable gases   150
conditioning   99
consumption potential   354
control strategies   196
conversion efficiencies   318
conveyor systems   188, 238
cooling   102, 247

corrosion potential   69
cost calculation methodology   241

de-ashing   199
dedicated biomass burners   232
deflagration   133
demolition wood   10
densified biofuel   6, 10
detonation   133
disc chippers   88
discharging wood pellets   415
drum chippers   88
drum dryer   92
dry solid biomass fuels   146
drying   89, 243
dust inhalation   168

Ecodesign directive   44
ecological evaluation   242, 305, 317
electric energy consumption   245
emission factors   305, 320
   final energy supply   313
   test stand measurements   311
   field measurements   309
emission limits   37, 40–41, 431, 444
emission reduction   465
EN 14961-1   25
EN 15210-1   25
energy crops   76
energy
   density   10, 55
   generation   427
   production, by products in   387
   utilisation of   309
enronmental evaluation   305
ERA-NET project   468
EU-ALTENER project   47, 476
European Committee for Standardization (CEN)   7
European measurement standard   25
explosion Severity (ES)   134
external ignition sources   153
extinguishing fire   158

feed-in system   183
feeding screw   191
feedstock availability and costs   405

fine particulate emissions   323, 326, 328, 465
fine particulate precipitation   467
fine particulates   324
fire risks   152
fixed bed gasification   226
flat storage   125
flue gas condensation   201, 204, 282, 299, 446
   furnaces, types of   205
forest and plantation wood   10
forest biomass resources   383
forest industry
   liquid by-products (black liquor)   386
   solid by-products   386
forest logging residues   381
fossil fuel prices   404
fruit biomass   10
fuel classification   11
fuel conveyor systems   216
fuel specification   11, 15
fuel/heat supply   306
furnace geometry   193
furnace type   179
   integrated burners   183
   inserted burners   183
   external burners   180, 208
   solar heating combination   213
future trade routes   400

gas boilers   355
gas detection   154
gaseous emissions   470, 475
general investments   243
German Pellet Institute (DEPI)   34
global raw material   388
greenhouse gas (GHG)   24
   emissions   313
   reductions   406
grinding   245
gross calorific value (q_{gr})   11, 54–55, 71

harmonized system (HS)   20
health concerns   165
health effects   325
   on humans   169
heat and power applications   220

Index

heat buffer storage   213
heat costs   279
heat exchanger cleaning systems   199, 430
heating systems   315
heavy metals   61
herbaceous biomass   11, 14, 212, 459, 465
herbaceous raw materials   77
horizontally fed burner   185
hot surface ignition temperature for dust layer (T_S)   134
HS code   20
hydrocarbon emission   310, 315–316, 331
hygroscopic property   462

ignition   192
impurities   11
industrial pellets,   5
industrial wood chips   271, 340, 407
innovative concepts   216
inorganic additives   79
integrated condenser   206
intercontinental wood pellet trade   394
International convention on the harmonized commodity description and coding system (HS convention)   20
International Maritime Organization (IMO) code   20
International pellet trade   391
ISO solid biofuels standardisation   35

KWB TDS Powerfire   150, 217

lambda control   196
large-scale power generation
    application   218, 226, 238, 451
    2.1 MW district heating plant   444
    4.5 MW district heating plant   446
ligno-cellulosic raw materials   72
Ligno-Tester   25
Limited Oxygen Concentration for Dust Cloud (LOC)   134
loading wood pellets   413
logistics   409
low temperature dryer   95
lower heating value (LHV)   11

maintenance cost   260
market developments   476
material safety data sheet (MSDS)   20
maximum explosion pressure   133
mechanical durability   11
medium- and large-scale pellet storage   123
medium-scale systems   216, 222
    500 kW heating plant   439
    600 kW district heating plant   442
    school heating   435
micro- and small-scale CHP   471
mineral contamination   61
minimum explosible concentration for dust cloud (MEC)   134
minimum ignition energy for dust cloud (MIE)   133
mitigation measures   140
moisture content   57, 67
moisture sorption   143
MSDS   172
Multi fuel concepts   211, 471

natural binding agents, content of   60
natural gas heating system   286, 299
    flue gas condensation   286
net calorific value (q_{net})   11, 54
non-condensable gases   148

ocean transport   413
ocean voyage   415
off-gassing emissions   170
off-gassing   148, 151, 463
oil central heating system   284–285
Öko-Carbonizer   208
Organic Rankine Cycle process   224
Organic additives   78
Overfeed burner   186
oxygen depletion   151, 171

Panamax and Handymax Indices   397
particle density   66
particle size distribution   12
particulate emission reduction   467
peat   81
pelletisation   7, 57, 100, 246
pellet

analysis standards in Europe   24
angle of repose   50
angle of drain   50
associations, Austria   335
associations, Germany   350
boilers   319
bulk density   48
burner design   187
central heating system   280, 430, 432
contents of   52–54
consumer   363
consumption potential   346
consumption   315, 345, 361
dimensions   48
distribution costs   255
fired tiled stoves   212
furnace
   developments   470
    installation   347
    flue gas condensation   201
heating systems   23, 352
internal particle size distribution   52
market   371, 378, 423
mechanical durability   51
particle density   50, 52
physio-chemical character   47
production   30, 85, 336, 350, 359, 370, 372, 417, 459
   standards, in Europe   21
   costs   253, 241, 264
   plants, in Europe   374
   potentials   374
   process optimisation   464
quality   461
   assurance standards, in Europe   28
reservoir   427
specifications   31
storage   118, 181
stove   180, 427, 429
supplier declaration form   407
transport and storage, standards for   31
use of   427
utilisation   341, 351, 359, 417, 465
Pelletsverbrand Austria (PVA)   335
permissible exposure limits (PEL)   135
pneumatic conveying   239
pneumatic feeding system   190

Policy support measures   405
pollution   305
pressing aid   12
pressurised steam   104
prices and logistic requirements   398
production plant   410
   economic comparison   268
production potential   351
production process   461
pulverised coal pipework   233
pulverised fuel burners   219
PYROT   217

quality assurance   12, 29

Racoon   208
radioactive materials   62
radionuclides   62–64, 84
raw material basis   461
raw material   85, 251, 266, 272, 380, 462
   contamination of   61, 66
   handling and storage   108
   physio-chemical character   47
   pre-treatment   87, 121
   size distribution of   47
recommended exposure limits (REL)   135
renewable energy sources   345
research and development   459
residential heating sector   31, 275, 278
   pellet furnaces, standards   36
   fuel retail prices   275
retort furnaces   184
retrofitted burner   435
retrofitting   181, 238, 451, 433, 457

safety and health aspects   133
safety classification   134
safety measures   152
sample preparation   12
sawdust   339
   excess   389
   potential   383
Schräder Hydrocube   210
screening   103
screw chippers   88
self-heating   144, 147, 463

sensitivity analysis   258
short rotation crops   244, 460
short term exposure limit (STEL) TLV   134
silo fires, anatomy of   163
skin contact   168
small-scale pellet storage   118
small-scale systems   179, 220, 238, 427
    retrofitting,   433
    pellet central heating, 430
softwood and hardwood   72
solid biofuels   12, 16
Solid residues (ash)   330
starch content   68
steam explosion reactor   104
stemwood   12
Stirling engine process   222
Stirling engine   220, 472
storage and peripheral
    equipment   248
storage at inland terminal   416
stowage factor   49, 82
straw pellets   38, 70, 84
superheated steam dryers   97, 245
supply chain   12, 410
supply security   127, 135
Swan-labelling   23
Swedish standard   28
Swiss pellet market   357

temperature and moisture control   154
thermal energy consumption   245

threshold limit value (TLV)   134
time weighted average (TWA)   134
Torbed reactor   108
torrefaction   104, 464
total dust   324
total suspended particulate matter
    (TSP)   324
transhipment of wood pellets   416
transport   411
transportation and distribution   109
tube bundle dryers   91, 245

underfeed burners   184
underfeed stoker   184
unloading   417

vertical silo
    with flat bottom   124
    with tapered (hopper) bottom   123

waste heat   267
wet basis   13
wet solid biomass fuels   144
wheel shipper   88
wood chips central heating system   288
wood fuels   13
wood pellets   7, 16, 144, 409–410, 417
    combustion technologies   179
    shipping prices   395
wood shavings   14, 339
World Customs Organization (WCO)   20